MAMMALS OF THE SOUTH-WEST PACIFIC

For Marita, Mila and Santiago

MAMMALS OF THE SOUTH-WEST PACIFIC

Tyrone Lavery   Tim Flannery

CSIRO PUBLISHING

CRC Press
Taylor & Francis Group
Boca Raton   London   New York

CRC Press is an imprint of the
Taylor & Francis Group, an **informa** business

A catalogue record for this book is available from the National Library of Australia.

ISBN: 9781486312627 (hbk)
ISBN: 9781486312634 (epdf)
ISBN: 9781486312641 (epub)

Published in print in Australia and New Zealand, and in all other formats throughout the world, by CSIRO Publishing.

CSIRO Publishing
Private Bag 10
Clayton South VIC 3169
Australia

Telephone: +61 3 9545 8400
Email: publishing.sales@csiro.au
Website: www.publish.csiro.au
Sign up to our email alerts: publish.csiro.au/earlyalert

Published in print only, throughout the world (except in Australia and New Zealand), by CRC Press, with ISBN 9781032254401.

CRC Press
6000 Broken Sound Parkway NW, Suite 300, Boca Raton, FL 33487-2742
and
4 Park Square, Milton Park, Abingdon, Oxon, OX14 4RN
Website: www.routledge.com

CRC Press is an imprint of Taylor & Francis Group, LLC

Front cover: (top) *Pteropus capistratus* (illustration by Gustav Mützel), (bottom) *Pseudochirulus canescens* (illustration by Gustav Mützel)
Title page: *Spilocuscus maculatus* (illustration by Madison Erin Mayfield)
Back cover: (top) *Distoechurus pennatus* (illustration by Gustav Mützel), (bottom) *Melomys rufescens* (illustration by Joseph Smit)
Other cover illustrations: compiled from Wikimedia Commons and Etsy (AntiqueStock and SheShedPrintables)

Edited by Joy Window (Living Language)
Cover design by Cath Pirret
Typeset by Envisage Information Technology
Printed in Malaysia by Papercraft

Feb23_01

Foreword

It is a pleasure and an honour to write this foreword for *Mammals of the South-west Pacific*. Since the first publication that focused on mammals of this region, there have been numerous new species added to the list. Over 180 described mammals occupy these islands, and more than half of them are endemic, a statistic that is astounding given the small land areas that make up the region. For millennia, parts of the South-west Pacific have acted as factories of biodiversity. They are source pools that not only spur colonisation of the remotest parts of the Pacific, but also feed new forms of life upstream towards New Guinea and the continents. Hence, these insular ecosystems of the South-west Pacific are not only regionally important, but are also critical sites in the context of global mammal diversity.

Almost two decades ago, I was conducting mammal-focused field research on my mother's island of Choiseul and my father's island of Vangunu in Solomon Islands. I was seeking to understand how mammal diversity varied between different forest types and habitats, recording the deep associated ethnobiology data along the way. On two occasions on Choiseul, we recorded and photographed the cryptic and threatened rodent called *kumaka* in Choiseul's Sengga language (Bougainville giant rat *Solomys salebrosus*). On Vangunu island we noted a then undescribed native species of rat the locals referred to as *vika* in both the Marovo and Vangunu languages (spoken on the north and south of the island respectively). Locals knew it intimately as a species that inhabited strangler figs, the canopy of emergent forest trees, burrows in the forest floor, *Metroxylon* palms, and *Canarium* trees in primary lowland and hill forests. The existence of this mythical rat also had been noted by anthropologists and Australian Museum researchers decades earlier. However, it was not until 2017, when an animal was caught by conservation rangers from Zaira Village, that the world came to know of this rare endemic rat as Vangunu giant rat (*Uromys vika*).

What's heartbreaking is that the last spaces of forested landscapes, habitats for some of the rarest and most threatened mammals on Earth, are under threat from logging or mining. Indeed, the lone scientific record of *vika* came from a commercial logging coupe on southern Vangunu. As more and more habitats are altered or thoroughly replaced by anthropogenic landscapes, this will have (or in some cases already has had) serious negative impacts on the mammals of our islands. Further reducing populations of mammals is cause for alarm. While we have been provided with a checklist of mammals here, what's unnerving is that we hardly know much about the ecology and life history of many species in our region. Further, the recent discovery of *vika* shows we face a genuine risk of losing species to extinction before we even know they are there.

The advances of new tools such as molecular sequencing are adding to our knowledge of the life history, taxonomy and cryptic diversity of many of our region's mammals. But a high proportion of Indigenous peoples in the South-west Pacific still live traditional lifestyles, are rural subsistence farmers or are professionals with deep connections to place, and there remains a wealth of traditional ecological knowledge. Together with Western scientific census methods and new technologies, this traditional knowledge can equally contribute to the discovery of new mammal species. Further, there is a paucity of publications that harness this ethnobiological knowledge of mammal ecology and life history. These would be extremely beneficial to our understanding of human, environmental and wildlife linkages in the South-west Pacific and would be useful for achieving specific conservation outcomes. Here is an impetus for further research, and for conservation and management efforts.

I have had the opportunity of working closely with Tyrone Lavery in the field and laboratory. Our fieldwork has been both purely for scientific research and for conservation commissions. These adventures have taken us to some of the most remote insular ecosystems in the South-west Pacific. In some cases, we have been trekking in the footsteps of Tim Flannery, as we sampled some of his research sites from the 1990s. Dr Lavery's passion for scientific research and his publication record in the region are noteworthy, and complemented by his efforts in leveraging the challenges and opportunities of mammal conservation and management in the region. As a local scientist, I've come to appreciate his involvement of local collaborators. This publication is a testament to both Lavery's and Flannery's impeccable scientific quality and those close partnerships with Indigenous communities.

This book provides an overview of the taxonomy and natural history of South-west Pacific mammals. But, more importantly, it is a reference book that may assist scientists, conservation organisations, landowners, land managers, Indigenous communities, developers, universities and governments to make better informed choices and decisions. May the conservation and better management of our mammal diversity be the ultimate contribution of this important publication.

Dr Patrick Pikacha
Pacific Adventist University
Port Moresby, Papua New Guinea

Contents

Foreword	*v*
About the authors	*ix*
Preface	*x*
Acknowledgements	*xi*
Introduction	1
Preamble to species accounts	4
Geology	8
Flora and vegetation	19
Human presence, translocation and extinction	22
Palaeontology	25
Zoogeography	30
Conservation of South-west Pacific mammals	34
Species accounts	
Monotremata	40
Tachyglossidae	40
Dasyuromorphia	42
Dasyuridae	42
Peramelemorphia	49
Peroryctidae	49
Diprotodontia	56
Acrobatidae	56
Macropodidae	57
Phalangeridae	64
Petauridae	78
Pseudocheiridae	83
Rodentia	86
Muridae	86
Chiroptera	119
Pteropodidae	119
Emballonuridae	205
Hipposideridae	216
Miniopteridae	228
Molossidae	236

Mystacinidae 241
Rhinolophidae 244
Vespertilionidae 246
Introduced species 265
Artiodactyla 266
Bovidae 266
Cervidae 268
Suidae 274
Carnivora 276
Viveriidae 276
Felidae 279
Canidae 280
Diprotodontia 282
Macropodidae 282
Phalangeridae 286
Erinaceomorpha 288
Erinaceidae 288
Lagomorpha 289
Leporidae 289
Primates 291
Cercopithecidae 291
Rodentia 292
Muridae 292
Soricomorpha 297
Soricidae 297

Glossary *298*
Appendix A: Annotated faunal list for the nations and territories of the region *301*
References *308*
Index *340*

About the authors

Tyrone H. Lavery
Tyrone Lavery is a mammalogist first drawn to the South-west Pacific by a desire to search for undescribed species and others not documented since they were first described in the 1800s. He has studied many of the rare and unique mammals of the Pacific, filling gaps in our knowledge of how they are related, where they occur and the threats they are facing. Through this research, Tyrone endeavours to support Pacific Island communities working to conserve their forests and wildlife.

(Photograph by Kate Robertson)

Tim F. Flannery
Tim Flannery is a palaeontologist, explorer and conservationist. His early research was in palaeontology on the evolution of kangaroos. From the late 1980s, Tim's focus shifted towards the living mammals of Melanesia and the Pacific Islands. In 1995 he published comprehensive works on the biologically rich regions of New Guinea and the Pacific. Tim maintains a role in Pacific Island conservation efforts today via relationships with organisations and communities in Melanesia.

(Photograph by Alex Bloom)

Preface

The preface to *Mammals of the South-west Pacific and Moluccan Islands*, published in 1995, ended with a caveat that the book was a first attempt, as well as a hope that readers would be tolerant of any inadvertent omissions that may have marred it. *Mammals of the South-west Pacific and Moluccan Islands* was in fact a bold endeavour. No checklist of mammals nor any kind of publications encompassing this broad, geographically and biologically diverse region were previously available. Assembling natural history and morphological data for over 200 indigenous taxa, along with distribution maps and photographs (many of species that had been never photographed before), was an unenviable task.

Mammals of the South-west Pacific and Moluccan Islands proved an invaluable reference for what was known of the taxonomy, distribution and ecology of a diverse mix of species spanning from just east of Wallace's Line into remote Oceania. But, perhaps just as importantly as presenting what *was* known, the work also laid bare the scientific knowledge that was severely lacking. Species accounts enveloped anecdotes of searching for animals not documented since they were first described over 100 years earlier – possibly encountered and then frustratingly gone in an instant. Others provided intimate details of the natural history of striking species, still yet to be scientifically described. As a reader, it was difficult to avoid being enticed into Pacific Island mammalogy.

This book aims to provide a much-needed update to South-west Pacific mammalogy. The focus here is primarily on the living native species, with introduced and prehistorically extinct species discussed only briefly. It was challenging to define a western cut-off point, as the Melanesian influence on fauna extends well into Asia. On this occasion we have chosen to omit the Moluccan islands of Wallacea, to focus solely on the islands commonly grouped together under the banners of Melanesia, Micronesia and Polynesia.

There have been exciting recent advances in South-west Pacific mammalogy. Many of the taxa that lacked scientific names in 1995 have now been described. Others that were completely unknown to the scientific community have been encountered, described and actions set in motion to conserve them. Yet several species, missing since first encountered by naturalists, remain 'lost' and in desperate need of dedicated rediscovery efforts. Moreover, numerous undescribed species (cryptic or otherwise) still await scientific attention.

Once again, we hope the reader will be tolerant of any inadvertent omissions that may mar this work. This book must be considered an overdue second attempt, but one that we anticipate will again provide an important reference point and spur continued fascination towards the mammals of the South-west Pacific.

Tyrone H. Lavery
The Australian National University
February 2022

Acknowledgements

The ongoing collection of data for this work, its analysis, and compilation has been a mammoth task. First, and most importantly, many residents of the South-west Pacific shared their valuable time, skills and knowledge to guide field expeditions and make data collection possible. This embraces a period commencing in the late 1980s and continuing today. To experience their hospitality, knowledge and good humour has been transformative on professional and personal levels. Their names and those of their villages, in addition to those given below, can be found dotted throughout the pages of this book. We would like to thank Masaafi Alabai, Tommy Esau, Folofou, Reverend Green Jino, Nixon Jino, Hikuna Judge, Benjamin Kaniotolu, Esau Kekeubata, Pandakai Onga, John Masiu, Justine Masiu, and the Vatohi extended family, especially Logino, late Ana, Lillian, Lynes, Jae, Brian and Nelson.

Numerous researchers, professional biologists and government officials of Pacific Island nations have been instrumental to this work. Key people included Ian Aujare, Rose Babaua, Ernest Bani, David Boseto, M. Boulet, M. Brochot, David Esrom, John Fasi, John Genolani, Nicho Gowep, Piokera Holland, Josef Huruturau, Iamo Ila, Henry Isa, Felix Kinbag, John Kokanasinga, Gai Kula, John Lamaris, M. Yves Létocart, Trevor Maeda, Peter Manueli, Honourable Timothy Masiu, Dr Melchior Mataki, Alivereti Naikatini, Jeffrey Noro, Junior Novera, Malik Oedin, Sarah Pene, Douglas Pikacha Jr., Patrick Pikacha, Edgar Pollard, Corzzierrah Posala, Veira Pulckera, Cornelius Qaqara, Mick Raga, Lester Seri, Kevin Sese, Birandra Singh, Myknee Sirikolo, Martika Tahi, Ikuo Tigulu, Marika Tuiwawa and Agnetha Vavekaramui.

Others based outside the region have also contributed greatly, by conducting surveys and research in the field, or through academic, logistic or moral support of such work. Thank you to David Akin, Simon and Joelle Albert, Michael Andersen, Lucas DeCicco, Noé de la Sancha, Anja Divljan, Mark Eldridge, Patricia Ennis, Helen Faddy, Wade Fairley, Adam Ferguson, Chris Filardi, Diana Fisher, Robert Fisher, Paul Flemons, Greta Frankham, Pavel German, Linda Gibson, Corey Howell, Sandy Ingleby, Heather Janetzki, Rebecca Johnson, Tanya Leary, Simon and Anne Le Gassicke, Luke Leung, David Lindenmayer, David MacLaren, Michael McCoy, Rob Moyle, Andrew Olds, Karen Olson, Bruce Patterson, Michael Pennay, Helen Pippard, Jonathan Richmond, Jesse Rowland, Jennifer Seddon, Akiko Shinya, Lauren Smith, Elizabeth Tasker and Luis Verde. Susan Tsang, Gary Wiles and Pat Woolley provided feedback on draft manuscripts or species profiles. Special thanks to Harry Parnaby for providing the encouragement needed to begin compiling this book and for providing constructive feedback on draft manuscripts.

Among those who generously donated their photographs or illustrations were Simon Albert, Thibaud Aronson, Bat Conservation International, M. Berroneau (www.matthieu-berroneau.fr), Evan Bowen-Jones, Julie Broken-Brow, Peter Crowhurst, Lucas DeCicco, Robert Fisher, Angus Fraser, Corey Howell, John Irvine, David Jenkins, Christian Langner, Chien Lee, Caroline Legg, George Madani, Madison E. Mayfield, Michael McCoy, Angus McNab, Marco Messina, Roberto Miguez, Malik Oedin, Keiko and Yushi Osawa, Michael Pennay, Patrick Pikacha, Philippe Pourtallier, Stephen Richards, Jonathan 'Tugi' Richmond, Kate Robertson, Klaus Rudloff, Mark Sanders (EcoSmart Ecology), Velizar Simeonovski, Margaret Sloan, Ben Speare, Charles A. Thomas, Elodie Van Lierde, Roger Wasley, Valter Weijola, Ann Williams, Pat Woolley and Anders Zimny.

Finally, it is our families that have borne the hardship that compiling a work like this entails. We cannot thank them enough, especially for enduring the long absences associated with fieldwork.

Introduction

The islands of the South-west Pacific are remarkable for their diversity. Divided among 16 countries and eight territories, they range from tiny coral cays to large heavily forested oceanic islands, to drifting continental fragments of Gondwana. The islands range in size from the South Island of Aotearoa New Zealand (150 718 km^2), to islets less than 1 km^2 in area. Their human histories, culture, flora and fauna also are extraordinarily diverse. This diversity makes it difficult to deal with their faunas as a cohesive unit but, simply because they are islands in the one region, they do share some similarities.

The precise region covered by this book is rather arbitrary. Deciding a cut-off point to the west of New Guinea was particularly challenging. Previous focus on this region had included the Maluku Islands of Indonesia, making the western limit akin to Weber's Line, a purported halfway point of balance between Asian and Australo-Papuan biotas (Weber 1902; Pelseneer 1904). But Wallacea is an extremely complex zone of spatial overlap and fauna interchange. There have been prolonged, complicated and sometimes contradictory attempts to define biogeographical boundaries in the region, and all are somewhat permeable to certain taxonomic groups (Ali and Heaney 2021). Eventually, we deemed it would be problematic to incorporate a subset of the oceanic islands of Wallacea while overlooking New Guinea and omitting islands between Weber's and Wallace's Lines that still have a strong Melanesian influence. We settled on a delineation that included only those islands to the west of New Guinea that were connected to the mainland during the Last Glacial

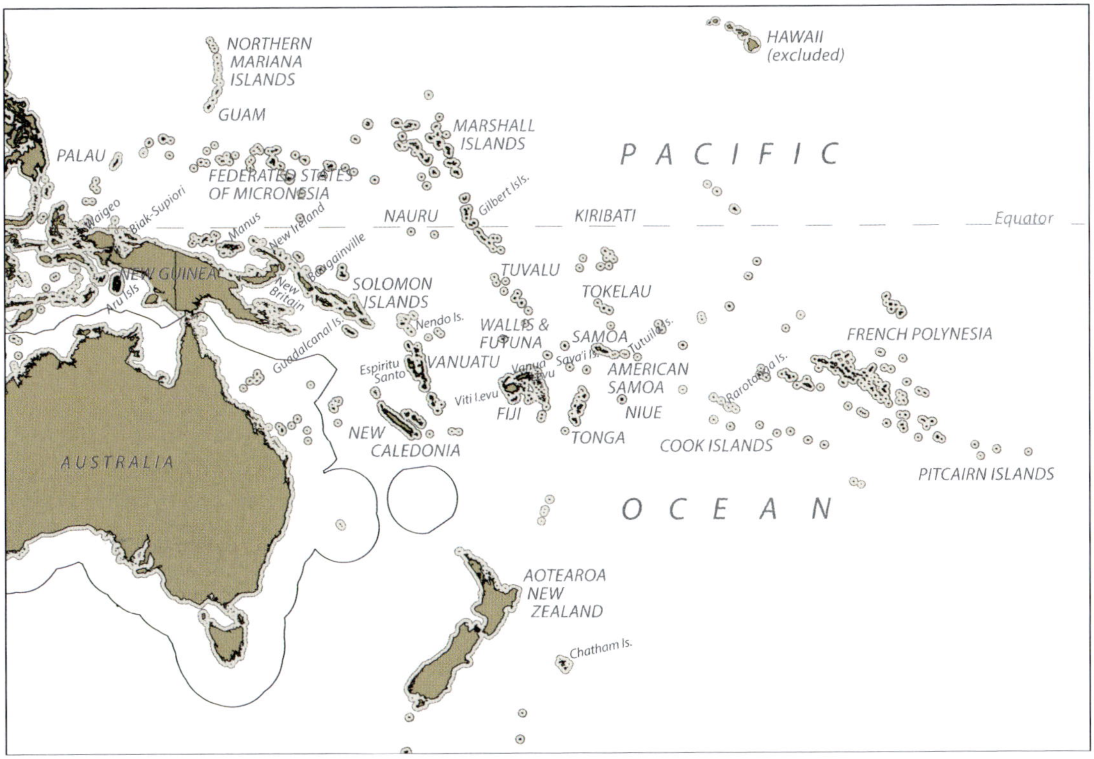

Nations and territories of the South-west Pacific included in this book, and the approximate boundaries of regions regularly labelled as Melanesia, Micronesia and Polynesia.

Maximum, akin to Lydekker's Line (Lydekker 1896). This marks the boundary between exclusively Australian and mixed Australian-Oriental mammal suites. Together, the South-west Pacific region of focus embraces the concepts of Melanesia, Micronesia and Polynesia (e.g. Cox 2008). The islands of these three geographical areas support mammal faunas that are predominantly Australian or Melanesian in origin.

The South-west Pacific comprises four global biodiversity hotspots: East Melanesian Islands; New Caledonia; Aotearoa New Zealand; and Polynesia–Micronesia (Mittermeier *et al.* 2004). These areas are recognised as such because they hold an irreplaceable share of global biodiversity (each containing at least 1500 endemic vascular plants), and because they are threatened (retaining 30% or less of their original natural vegetation). These values of high biodiversity and endemism extend to the mammal faunas. There are 184 described indigenous species recorded from the islands reviewed for this book and 95 of these are endemic to the South-west Pacific. The indigenous mammal fauna includes marsupials, rodents and bats. The bats are the best represented among these, comprising almost 65% of the total mammal fauna. The family Pteropodidae alone comprises over 36% of the entire indigenous fauna, while marsupials and murid rodents make up around 17–19% each.

There has also been widespread prehistoric translocation of indigenous South-east Asian or Melanesian species such as Bismarck common cuscus *Phalanger breviceps*. This Melanesian marsupial is purportedly one of the oldest known human mediated mammal translocations, having arrived on New Ireland some 10 000–20 000 years before present (Heinsohn 2010a). A suite of modern introductions has followed, including many of the most commonly introduced and detrimental species globally (e.g. black rat *Rattus rattus*, feral cat *Felis catus*). At least 40 exotic species have established populations, and virtually every island in the region supports at least one introduced species.

We endeavour to provide an account of every indigenous terrestrial mammal for which modern records are available, as well as the species newly introduced to the region either prehistorically or historically. We have aimed to follow a standard formula for each indigenous species, detailing the available information on distributional ranges, habitat preferences, diet, behaviour, reproduction, major threats and conservation status. In the case of species with extralimital distributions, we aim to restrict accounts to what is known of biology in the region of focus. However, we do incorporate data from Papua New Guinea and Australia where information specific to the South-west Pacific is lacking. Wherever possible, measurements, photographs and illustrations also relate to individuals from the region.

The reader will find this highly variable, because for many species, scientific data, measurements and photographs are simply unavailable. For example, although bats are regularly acknowledged as playing vital ecological roles in tropical rainforests and island ecosystems, much of their basic ecology remains undocumented in the Pacific. However, a lack of information in the literature does not always mean the knowledge of species biology does not exist, for there is usually a deep and intimate understanding held by Indigenous people cohabiting islands. This is particularly the case for the species endemic to single, remote islands, like *Pteropus tuberculatus* on Vanikoro Island in Solomon Islands. On two occasions this species has been considered 'missing' for decades at a time: first, because the type locality had been confused with Guam and, second, because it could not be detected during brief surveys in the 1990s. Yet ask any resident on Vanikoro about *basapine* (the Indigenous name for *P. tuberculatus*) and they can quickly relate observations about current population status as well as many of the important behavioural and ecological traits of the species.

Presenting species distributions is also a challenge given the temporal and spatial shortfalls in fauna surveys for most islands. Individual species vary in their preferences towards particular habitats and are often non-uniformly distributed on any given island. However, for the most part we are able to delineate distributions at the unit of the island only, except on larger islands where more reliable species data are available.

A great deal of work remains to document the diversity of mammals of this region, especially on

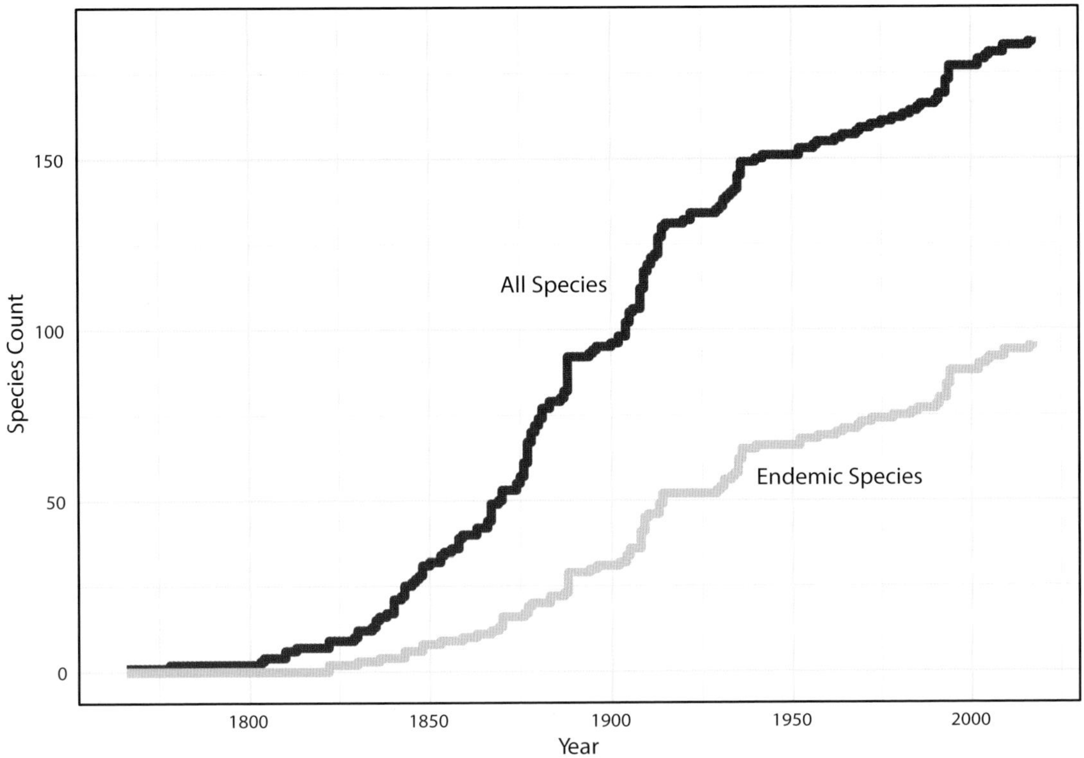

Cumulative taxonomic history of the mammals of the South-west Pacific. Lines indicate the cumulative scientific description of all mammals that occur in the region under consideration (dark brown), and those that are endemic to the region only (light brown).

large, remote islands that have been the focus of few mammal surveys. A recent survey of Mussau, for example, revealed multiple undescribed species there (Aplin *et al.* 2015). It is almost certain that more undescribed species exist in the South-west Pacific, particularly in the remote, high-elevation, mossy forests of large islands such as Bougainville, Guadalcanal, Fergusson and New Britain. Given their inaccessibility and the difficulties associated with surveying mammals in such areas, it could still be some time before the faunas of these regions are fully known. However, emerging generations of Indigenous Pacific Island biologists are best prepared to access these habitats, and with adequate support will continue to chip away at the unknowns of South-west Pacific mammalogy.

Preamble to species accounts

Name

A plethora of Indigenous language names apply to mammals of the South-west Pacific. These remain an incredibly important part of the deep traditional ecological knowledge of Pacific Islanders, and signify the cultural significance held by many of the region's mammals. Species with wide distributions can have numerous local names applied under different languages and dialects. In any single locality, various life stages and sexes of a single Linnaean species can bear numerous names (corresponding to the taxonomic classifications of Indigenous peoples), or multiple Linnean species can be recognised under a single Indigenous name. Indigenous names remain in use to varying degrees and have regularly been documented in field notes and the scientific literature. However, here we have chosen not to include these names for species, to avoid foreign interpretations of pronunciation and spelling, the application of incorrect names to species and the printing of names that may be culturally sensitive.

The New Georgia Group monkey-faced bat (*Pteralopex taki*) occurs on Kolombangara, New Georgia, Rendova, Tetepare and Vangunu, Solomon Islands. The specific epithet *taki* is derived from the name for this bat in the language of the people of Vangunu Island (Tyrone Lavery).

A common name is given in English only and, where possible, this follows the nomenclature used by the International Union for Conservation of Nature (IUCN) Red List. Some common English names are newly formulated here by necessity. A citation for the author and date is provided for each species description. Where author names are given in brackets, the describers placed the species in a different genus to that which it is assigned here.

Size

Where possible, morphological measurements given are field measurements of animals caught in the South-west Pacific. Measurements have been recorded by the authors unless otherwise noted, and where measurements were unavailable from the region, those from individuals in nearby New Guinea or Australia have sometimes been substituted. Weights are in grams and lengths are in millimetres. Hindfoot measurements are given SU (without inclusion of the claw) for murids and marsupials, and CU (with inclusion of the claw) for bats. The ear is measured from the notch to the tip, head–body length from nose tip to cloaca or anus, and tail length from cloaca or anus to tail tip.

Identification

The taxonomy adopted here usually follows that of the most recent reviser, although some changes are adopted as a result of ongoing studies by the authors and others. In such instances, this is made clear in the accompanying text.

Unfortunately, no keys exist to cover the complete list of mammals from this region, and the identification guides provided here are no substitute. We have not attempted to provide full and diagnostic descriptions for species in these profiles. The descriptions provided are intended to be used in conjunction with measurements and distributional data to aid identification of animals in the field. For this reason, readily observed external features are given priority over the sometimes more reliable but difficult to observe cranial and dental features. However, there are very few species for which we have sufficient data on natural history or field observations to allow identification from afar.

The characteristics of calls recorded from species of echolocating bat (using ultrasonic, full-spectrum recording devices) have been described where available. Descriptions of the shape of a single pulse of sound made by a bat are derived from sonograms (sound graphs) that depict time in milliseconds (ms) along the x-axis and frequency in kilohertz (kHz) along the y-axis. Adjustment of the scales of either of these axes will alter the shape and interpretation of a single pulse. Calls are referred to as either frequency modulated (the call covers a range of frequencies) or constant frequency (the call has a flat constant frequency tone at a certain frequency). A single call can consist of harmonics (multiple parallel components of a call at consistent frequency intervals). Other parameters regularly provided are characteristic frequency (the frequency of the characteristic section, the flattest part of the call); peak frequency (the frequency of the maximum amplitude of the call–the loudest part); and duration (the duration of the call in milliseconds).

Synonyms

A substantial proportion of the species included in this work have more than one scientific name (synonyms). In zoological nomenclature, only the earliest name (senior synonym) is to be applied as the scientific name for a species. We have endeavoured to include the relevant scientific names applied to species following their original descriptions (junior synonyms). Occasionally, synonyms are followed by Latin qualifications as follows:

Nomen nudum: a designation intended to be a scientific name that failed rules of the International Code of Zoological Nomenclature because it was not published with an adequate description or reference to such a description.

Nomen dubium: a scientific name that is of unknown or doubtful application either because no specimen was collected, the original description was inadequate, or its name-bearing type lacks important diagnostic features, or has been lost or destroyed.

Conservation status

The governments of South-west Pacific nations and territories apply a range of frameworks and legislation to classify the conservation status of species, regulate their use and trade, and plan for their protection. Some are guided by the IUCN Red List, whereas others adhere to their own pieces of legislation (e.g. Fiji – *Endangered and Protected Species Act 2002*), or a combination of both. For the purposes of consistency, we have provided the IUCN Red List category for each species at the time of publication (2023). Up-to-date classifications for each species can be reviewed at www.iucnredlist.org, along with distribution-wide descriptions of major threats, geographical ranges, habitat requirements and ecology.

Subspecies

Among taxonomists, species concepts and acceptance of the rank of subspecies can vary greatly. However, the geography of South-west Pacific islands, divided into discrete units of land, is somewhat conducive to the delineation of geographically isolated populations and acceptance of subspecies among terrestrial taxa. Most South-west Pacific mammals therefore have recognised subspecies. Described subspecies have been listed chronologically and a brief description of geographical range provided.

Distribution

The names of islands known to form parts of the distribution of each species are provided. Local extinctions at the island scale are denoted by the presence of a dagger symbol (†) next to an island name. Extinct species listed in the annotated faunal list for the nations and territories of the region are identified with the same symbol.

The majority of distribution maps indicate only whether a species has been recorded on a given island, except in cases where more refined distributional data are available for larger islands, or species are known to be restricted to high- or low-elevation areas. Maps incorporate an elevational contour of 1000 m above sea level derived from the Shuttle Radar Topography Mission (SRTM) (Farr *et al.* 2007). The distributions of those species limited to higher or lower elevations have been delineated using this arbitrary 1000 m contour.

Extinct species and those for which a verified record or specimen has not been collected in the last 100 years are shaded light blue. Green shading is used to indicate islands for which the occurrence of a species has not been confirmed, but is presumed based on broader species distributions, adequacy of survey efforts, and island location. Light orange is used to indicate distributions that have been confirmed via the subfossil record, but not by modern surveys.

For some species with an extremely widespread distribution (e.g. Pacific flying fox *Pteropus tonganus*), many small islands appear only as tiny dots on broad-area maps. In this case, the small islands where the species occurs have been given exaggerated shading rather than infill.

Extralimital

General lists of the known distributions outside of the region are briefly listed for the species covered by this work as extralimital distributions. For further information relevant to Australian distributions, readers are directed to Van Dyck and Strahan (2008), and to Flannery (1995a) for New Guinea.

Museum specimens

Specimens from major museum collections are referenced for this work, institutional abbreviations are as follows:

AM Australian Museum, Sydney, Australia
AMNH American Museum of Natural History, New York, U.S.A.

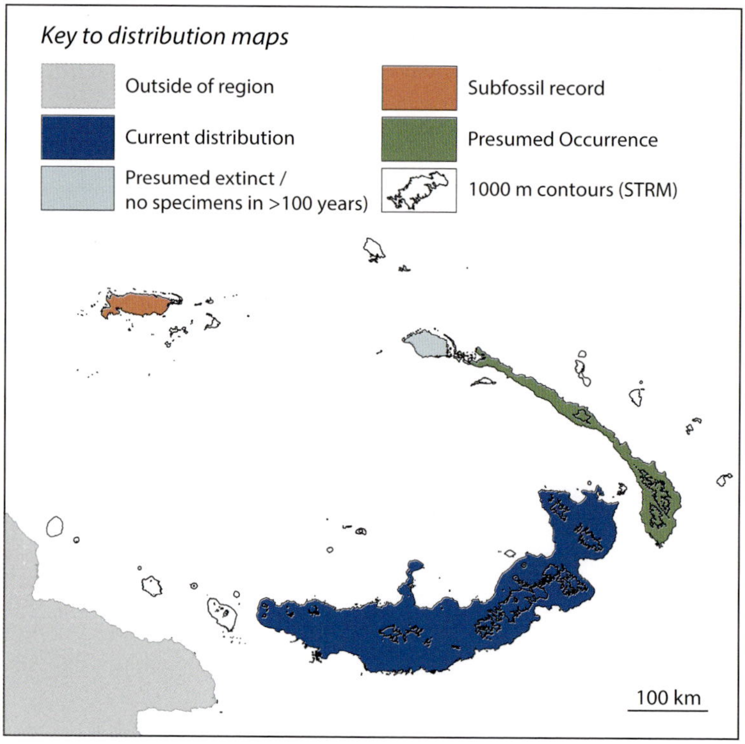

BPBM Bernice P. Bishop Museum, Honolulu, Hawaii, U.S.A.
CM Universitetets Zoologisk Museum, Copenhagen, Denmark
FMNH Field Museum of Natural History, Chicago, U.S.A.
GE Museo Civicio di Storia Naturale, Genoa, Italy
KU University of Kansas Biodiversity Institute and Natural History Museum, Lawrence, U.S.A.
LACM Natural History Museum of Los Angeles County, Los Angeles, U.S.A.
MZB Museum Zoologicum Bogoriense, West Java, Indonesia
NHMUK The Natural History Museum, London, United Kingdom
NMNH Muséum national d'Histoire naturelle, Paris, France
NMV Museum of Victoria, Melbourne, Australia
QM Queensland Museum, Brisbane, Australia
RMNH Naturalis Biodiversity Center, Leiden, Netherlands
UP University of Papua New Guinea, Port Moresby, Papua New Guinea
USNM United States National Museum of Natural History, Smithsonian Institution, Washington, DC, U.S.A.
WAM Western Australian Museum, Perth, Australia
ZMB Museum für Naturkunde, Berlin, Germany

Geology

Introduction

The South-west Pacific comprises areas with some of the most complex geological histories on Earth. A diverse assemblage of continental fragments, volcanic arcs, island chains formed by mantle plumes and large igneous provinces exist today. Consideration of these diverse geological origins and histories is of fundamental importance for understanding the faunas that inhabit the islands. In the South-west Pacific, islands have appeared, subsided and sometimes reappeared, and rising and falling sea levels have reconfigured exposed lands. Many young islands are volcanic and composed entirely of igneous rocks, whereas the largest islands are usually older, with a composite geology derived from subsiding and uplifting. These older landmasses often have limestone overlaying the original igneous rock and include metamorphic rocks from the influences of plate tectonic pressure and heat.

The aim here is to present an overview of regional geology, providing some of the important background necessary to appreciate how geological processes have helped shape the mammal faunas. Wherever possible we aim to provide context and relevant examples of the influences of: (1) the history of subaerial exposure, subaerial extent and elevation through time; (2) the movement of landmasses relative to each other; (3) Pleistocene glaciation, sea level change and island coalescence; and (4) the effect of geology upon soil formation, vegetation and attendant mammal faunas. This is worthwhile, for even a brief overview reveals a very different region in the past from that which exists today.

Regional setting

The break-up of the Australian section of Gondwana from 90 million years ago saw the New Zealand–Norfolk–New Caledonia continental fragment begin to rift from both Australia and Antarctica, to drift north-east (Hall 2002). Various events in the Pacific were initiated as oceanic crust subducted under the continental crust, leading to oceanic trough and island arc formation. Until around 45 million years ago, Meganesia was attached to Antarctica. This was the beginning of a period of northward drift, and collision between the north Australian continental margin and Pacific intra-oceanic island arcs. It resulted in pieces of oceanic plate (ophiolites) being thrusted onto the edge of continental plates at the northern margin of New Guinea, and later in New Caledonia (Hall 2002). An oceanic spreading centre began to develop the deep-water rift that became Wallace's Line, and by the Miocene (15–20 million years ago) the Australian plate had made contact with the Asian plate with profound implications for Wallacea (Hall 2002).

In the Pacific, the development and reactivation of island arcs has provided the main source of new land, as well as the movement of existing lands. By the Middle Eocene (45 million years ago), south-west-directed subduction began beneath the eastern Australian margin, from Papua New Guinea to north of Aotearoa New Zealand (Kroenke 1984; Petterson *et al.* 1997; Hill and Hall 2003). Initial arc volcanism and island-building for present day Melanesia began around this time to the north-east of the Australian craton. It included a broad continuous line of island-building from mainland Papua New Guinea through to the Fiji plateau (Yan and Kroenke 1993). Major development of the Papuan arc was laying the basis of the landmasses of south-eastern New Guinea and produced the older parts of New Britain, while the Rennell arc was laying the basis for Solomon Islands. The modern island arcs of the Solomons and Bismarcks comprising these elements have consistently remained above sea level since the Eocene epoch (40 million years ago), providing a deep time scale for the evolution of unique biotas (Hall 2002).

By the early Oligocene (35 million years ago), the Melanesian arc continued laying the basis for the Bismarck Archipelago, Solomon Islands and parts of Vanuatu, Fiji and the Tonga group. Commencing around the beginning of the Miocene, the Ontong Java plateau collided with the northward moving Melanesian Arc, causing a significant change in plate boundaries (Petterson *et al.* 1997). This set in motion the formation of land configurations important for shaping South-west Pacific mammal assemblages. The island arcs of Melanesia, New Guinea terranes at the southern Caroline margin, and the Halmahera–Philippines arcs were closely associated. Several endemic South-west Pacific lineages (e.g. *Anthops*, *Melonycteris*, *Notopteris*) are estimated to have diverged from sister lineages around this early Miocene period (Almeida *et al.* 2009; Lavery *et al.* 2014). By the late Miocene (around 10 million years ago), the South Solomons–New Hebrides arc had reactivated island building activity in the region of the former North Solomons–Vitiaz arc. Vanuatu began to swing into its present position, and the Loyalty arc (near New Caledonia) and Three Kings arc (to the north of Aotearoa New Zealand) were active (Hall 2002).

The islands of Polynesia and Micronesia vary widely in their geological origins. Many were formed from geological hotspot and fracture zone volcanism and include both older and younger volcanic origins (Allison and Eldredge 1999; Koppers *et al.* 2008). Several islands in Tonga, Samoa and the Northern Mariana Islands are either currently volcanically active or potentially active. Coral atolls, raised limestone islands, mixed volcanic and limestone islands also are common.

During the past two million years, world sea levels have fluctuated as a result of the locking of water (as ice) at the poles during glacial maxima. In the most recent of these events (the Last Glacial Maximum), ice sheets were at their maximum positions from around 26 500 to 19 000–20 000 years ago, resulting in what may have been one of the greatest lowering of sea levels of all time (Clark *et al.* 2009). Australia and New Guinea, along with

Global reconstructions at 40, 20 and 10 million years ago modified from Hall (2002). Green areas are mainly arc, ophiolitic and accreted material formed at plate margins. Dark grey regions represent landmasses that were former Gondwanan connections. Lighter grey areas represent submarine parts of the Australian continental margins.

Tasmania and the continental shelf islands at their margins, coalesced into a single landmass. This period also saw many present-day islands joined into much larger landmasses. Key 'mega islands' formed at this time; their approximate land area and the islands incorporated into them are shown in Table 1.

Other smaller landmasses were formed from the fusion of various islands, while some landmasses simply grew in size. Fiji's Vanua Levu, for example, roughly doubled in size but did not join with any other large islands. In addition to this, some very large and entirely new landmasses were exposed above the sea. Considerable islands appeared off north-eastern Australia, while the Norfolk Rise exposed an enormous area between Aotearoa New Zealand and New Caledonia, as did the Lord Howe Rise to a lesser extent. Smaller islands would have been exposed elsewhere throughout the South-west Pacific.

Table 1. Putative 'mega islands' formed by lowered sea levels of the Last Glacial Maximum, their approximate size and the member islands that were joined by land bridges to create them.

Greater Bukida	46 400 km^2	Buka, Bougainville, Choiseul, Ngella, San Jorge, Shortland and Santa Isabel +/– Guadalcanal
Greater Gatumbangara	5600 km^2	Kolombangara, Vonavona, New Georgia, Vangunu, Gatokae
Greater Trobriands	4000 km^2	Goodenough, Fergusson, Normanby, Kiriwina
Greater Vellonga	2200 km^2	Baga, Ranongga, Vella La Vella
Viti Levu	20 800 km^2	Ovalau, Viti Levu and the Yasawa Group

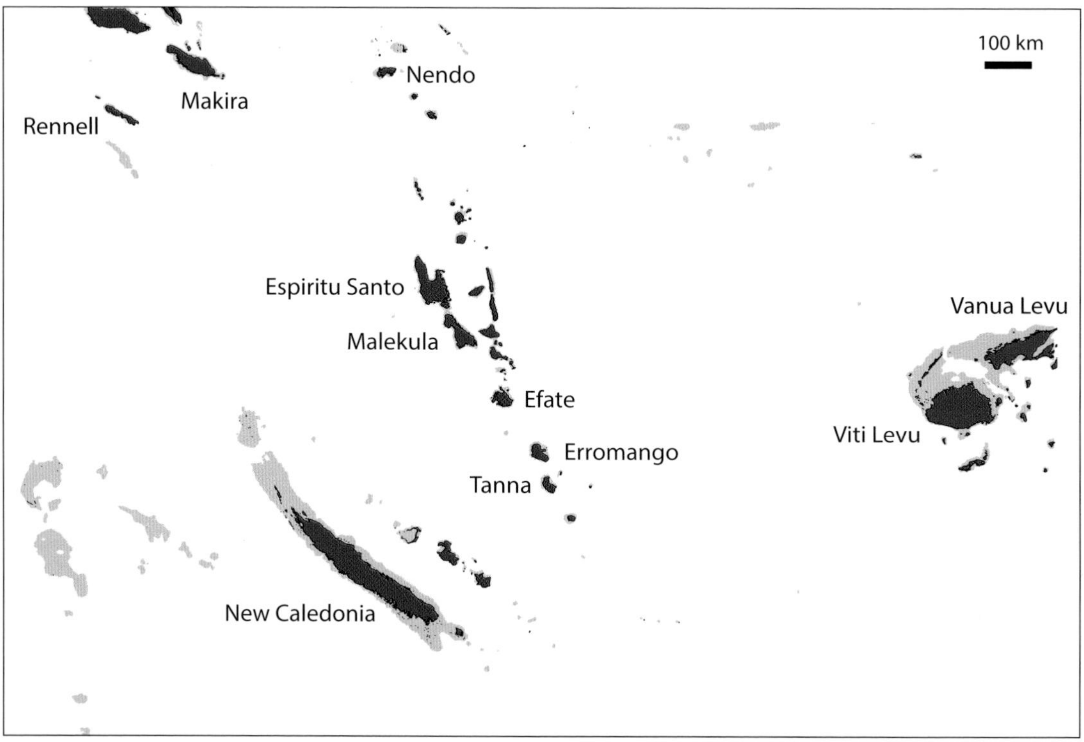

The present-day islands of Fiji, New Caledonia and Vanuatu. Areas of land that were probably exposed during lowered sea levels of the Last Glacial Maximum have been identified in dark grey based on a 120 m bathymetric contour (prepared by Amante and Eakins 2009).

The island groups

Bismarck Archipelago and Manus

The Bismarck Archipelago is an extremely complex region, including up to five separate plates between

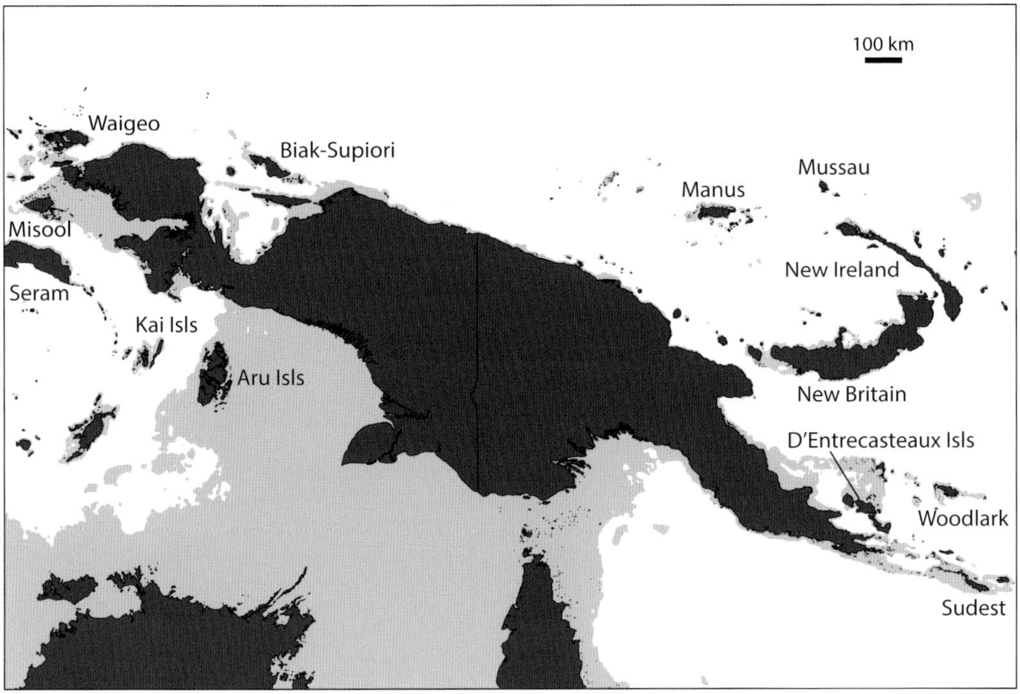

Present-day New Guinea and surrounding islands. Areas of land that were probably exposed during lowered sea levels of the Last Glacial Maximum have been identified in dark grey based on a 120 m bathymetric contour (prepared by Amante and Eakins 2009).

the Manus Trench and the Pocklington Trough (Hall 2002). This region features the large, mountainous islands of New Ireland and New Britain, with peaks over 2000 m in elevation. Despite their proximity, the Bismarck island arc and the mainland of Papua New Guinea are quite independent (Holm *et al.* 2016; Hall 2002). As recently as six million years ago, the Finisterre-Adelbert Terrane (later accreted to New Guinea as the Huon Peninsula) and New Britain probably remained separated from New Guinea by approximately 500 km of open ocean (Holm *et al.* 2016). Collision between the Ontong Java plateau and Australian plate has progressively reduced this distance.

On New Ireland and Manus there is an Eocene or Early Miocene arc sequence, built on presumed older oceanic crust, indicating that these islands were being formed at that time (Francis 1988; Stewart and Sandy 1988; Hall 2002). On New Britain there is also evidence of the Eocene to Miocene arc sequence (Lindley 1988). New Britain, New Ireland and Manus probably all formed a continuous arc with the Solomon Islands archipelago during the Oligocene (Falvey and Pritchard 1982).

Volcanic activity declined in the Early Miocene and the arc sequences were uplifted and faulted, and erosion and sedimentation resumed (Hall 2002). A period of subsidence began on New Ireland and New Britain, continuing until volcanism recommenced in the Pliocene (Hall 2002). Biologically, New Ireland is unusual because it supports lower species diversity and endemism than might be expected for such a large island so close to New Guinea (Flannery and White 1991; Mayr and Diamond 2001; Lavery *et al.* 2016). This is probably due to a period of recent submersion followed by uplift and re-emergence above sea level.

Between approximately 17–10 million years ago Manus began moving into its present position, and at the end of this time the various fragments that would form present-day New Britain had begun to move into place (Hall 2002). A period of very active volcanism resumed on a regional scale in the

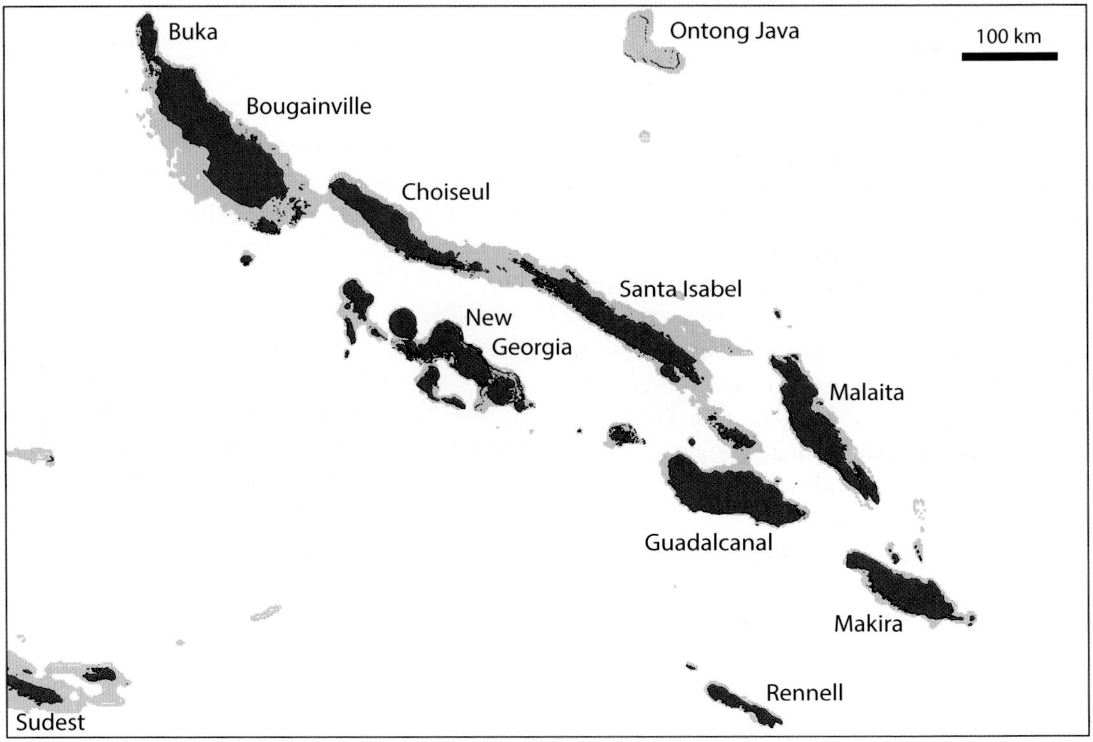

The islands of the Solomon Islands archipelago. Areas of land that were probably exposed during lowered sea levels of the Last Glacial Maximum have been identified in dark grey based on a 120 m bathymetric contour (prepared by Amante and Eakins 2009).

Pliocene. Uplift began on New Britain and New Ireland and both islands reached their present extent and elevation over this time.

The islands of the Tabar-Feni groups are very recent, dating to the Pleistocene and Pliocene (less than three to four million years) (Wallace *et al.* 1983), as are the islands of Papua New Guinea's East Sepik Province (e.g. Vokeo, Blup Blup, Kadovar). Numerous active volcanoes exist in the region on larger islands (e.g. Rabaul, New Britain) and many smaller islands in the archipelago, and along the northern New Guinea coastline there are active volcanoes (e.g. Kadovar, Karkar, Long).

Cook Islands

The Cook-Austral Islands are the summit portions of extinct Tertiary volcanoes. They are a hotspot chain of islands, and the various life stages of volcanic Pacific islands (as they move away from their birthplace) are superbly on display. However, the islands do not exhibit a simple linear age progression, and may have been formed by multiple hotspots (Rose and Koppers 2019). Six of the Northern Group islands (e.g. Palmerston) are atolls formed around submerged sea mounts (Wood 2012). These are thought to have built up and emerged above sea level during the Eocene and Oligocene. Some were probably already extinct and beginning to subside during the Eocene (Wood 2012).

Four of the Southern Group consist of an uplifted island surrounded by a ring formed of atoll or high barrier reef (Makatea-type islands). These are also old, with Mangaia probably extinct and beginning to subside well before the Miocene (Wood 2012). Aitutaki has subsided from its level of formation and is an atoll with an emergent volcanic peak at the northern end. The largest island, Rarotonga, is the emergent summit of a Pliocene–Pleistocene volcanic complex surrounded by a fringing reef. Pleistocene volcanism has been recorded on Aitutaki, Rarotonga and Rurutu in the Austral Islands.

D'Entrecasteaux Islands, Trobriand Islands and Louisiade Archipelago

The rocks forming the D'Entrecasteaux Islands and some of the Louisiades date from the late Palaeocene and are part of the Papuan Ultramafic Belt, which is formed from rocks of the oceanic crust. They are complexes that lie in an area of active continental extension and have been metamorphosed and thrust high above the sea. Goodenough, Fergusson and Normanby consist of complexly deformed high-grade metamorphic basement that is overlain by ultramafic rocks and more recent sedimentary deposits (Hill *et al.* 1992). The D'Entrecasteaux Islands were probably separated from south-east New Guinea when Goodenough Bay formed or expanded in the Pliocene. At the same time, dramatic uplift was building the mountains on the D'Entrecasteaux Islands, which now reach over 2000 m in elevation at their highest peaks.

To the north, the D'Entrecasteaux Islands are joined to Kiriwina by a broad, shallow platform. This was probably partly exposed at times of low sea level during the Pleistocene, connecting Kiriwina with the D'Entrecasteaux Islands. Woodlark is part of a largely submerged ridge that extends eastwards from Kiriwina. Volcanic rocks date to the Eocene and are overlain by limestones (Pigram and Davies 1987). Other volcanic rocks on Woodlark have been dated to 11 million years (Milsom 1985). Until sea-floor spreading began in the Pliocene (five to two million years ago), Woodlark would have been situated more closely to the Louisiade Archipelago.

Fiji

The oldest rocks in Fiji are island arc volcanics from Viti Levu and are late Eocene in age (Neall and Trewick 2008). They were formed by westward subduction of the Pacific plate beneath the Australian plate along the Vitiaz arc at a time when Fiji was much nearer to Vanuatu. It is likely that parts of Viti Levu were exposed above the sea at this time and deposition of sediments continued throughout the Miocene. From around 28 million years ago, relative movement between the plates changed to a more oblique convergence and volcanism associated with mountain building and intrusion of gabbros (igneous rock) began again. These strata are now exposed in southern Viti Levu, and in the Yasawa and Mamanuca Islands (Neall and Trewick 2008). At about this time Viti Levu began to break away from the south-eastern end of the Vanuatu volcanic arc as the arc began to bend.

By the late Miocene (10 million years ago) volcanism was beginning to build Vanua Levu. Spreading was initiated in the North Fiji Basin, and Viti Levu became attached to the Pacific plate (Neall and Trewick 2008). Since seven million years ago, Fiji has rotated anticlockwise with the opening up of the North Fiji Basin, and Viti Levu moved south-eastwards relative to Vanua Levu (Kroenke 1984).

On the Lau Ridge of eastern Fiji, three phases of volcanic activity are recognised: the Lau Volcanic Group was active from 14 to 5.4 million years ago; the Korabasaga Volcanic Group from 4.4 to 2.4 million years ago; and the Mago Volcanic Group less than 2.2 million years ago (Cole *et al.* 1990). By the Pliocene (five to two million years ago) volcanism was building Taveuni and Kadavu. Taveuni is a basaltic volcanic island with a north-east–south-west lineament of over 150 Quaternary vents. Volcanic activity persisted on Taveuni during the Holocene with least 2.7 km^2 of magma erupted from over 100 events (Cronin and Neall 2001).

Loyalty Islands

The Loyalty Rise is a poorly known arc system. It collided with what is now New Caledonia between 38 and 33 million years ago (Neall and Trewick 2008). The islands themselves are mainly composed of Late Miocene to Pleistocene uplifted reef formations covering basement derived from volcanic action in their present position some 9–11 million years ago (Kroenke 1984). Quaternary uplift has taken place over the past two million years due to upwards flexure of the lithosphere caused by subduction of the Indo-Australian plate beneath the New Hebrides arc (Neall and Trewick 2008). The atolls emerged during the Pleistocene (possibly late Pliocene for Mare) and are still rapidly emerging and increasing in size.

Micronesia

Micronesia covers an enormous area and comprises six distinct geological subregions associated with a complex combination of plate subduction-zones and arc systems. The Caroline Islands span two broad geological provinces and include the volcanic islands of Chuuk, Pohnpei and Kosrae, as well as coral atolls. The volcanic islands have been created by the Caroline hotspot that was active between 12 and one million years ago. The islands were formed as the ocean crust passed over it from the east and the oldest islands thus lie in the west. Chuuk is around 13.9–9.9 million years old, Pohnpei around 5.2 million years old, and Kosrae approximately one to three million years old (Duncan and Clague 1985). The Yap Islands are composed mainly of metamorphic rocks of greenschist to amphibolite facies from oceanic tholeiite and lack raised coral reef (Nedachi *et al.* 2001).

The Palau Archipelago is part of the Palau–Kyushu Ridge. The larger islands are volcanic (e.g. Babeldaob, Arakabesan, Malakal, Koroor) and the main island of Babeldaob comprises old-weathered continental rocks (Neall and Trewick 2008). The volcanics are dated between 37.7 and 20.1 million years (Kobayashi 2000). The islands between Koroor and Peleliu are mainly uplifted limestone islands, with atoll-like islands immediately north of Peleliu (Neall and Trewick 2008).

The atolls and islands of Marshall Islands were formed by layers of marine organisms, built on a foundation of now submerged pre-Tertiary volcanoes. Extensive coral reefs developed during warmer interglacial periods of the Pleistocene and were exposed above the ocean around 2000 years ago, when ocean levels dropped (Vander Velde 2009).

The northern Mariana Islands are formed by two subparallel, concave-west arcs that are the upper 2–3 km of the East Mariana Ridge, a result of an active subduction zone where the Pacific plate is passing beneath the Philippine plate (Trusdell 2009). The oldest islands are in the south (e.g. Farallon de Medinilla and Guam) and are Eocene to Miocene volcanic rocks overlain by uplifted Tertiary and Quaternary limestone. Guam was volcanically active from 32–20 and 9–5 million years ago (Reagan and Meijer 1984). The islands to the north of Farallon de Medinilla do not bear limestone caps, and the islands stretching from Anatahan to Uracas are volcanically active (Meijer and Reagan 1981).

New Caledonia

New Caledonia is one of the oldest landmasses in the region, the oldest rocks being Permian or possibly older. They are thought to have been formed on the active continental margin of the Australian section of Gondwana (Kroenke 1984). Before the late Cretaceous, around 95 million years ago, New Caledonia was connected to the Antarctic portion of Gondwana via Aotearoa New Zealand and the Norfolk Ridge. Around 90 million years ago the New Zealand–Norfolk–New Caledonian portion began to drift to the north-east, away from Australia and the rest of the Gondwana supercontinent. By 60 million years ago, New Caledonia had reached a latitude of around 30° South and its position remained relatively stable (around 30–35° South) before beginning to drift northward around 20 million years ago (Hall 2002). Its distance from Australia has remained approximately the same over the past 60 million years.

Geological and biological evidence indicates that, following divergence from Australia, there were long periods during the Palaeocene and Eocene where Grande Tierre was submerged (Grandcolas *et al.* 2008; Giribet and Baker 2019). Since at least 25 million years ago New Caledonia has remained emergent above sea level (Pelletier 2006, Maurizot and Campbell 2020). Although the geological evidence for submergence cannot be disputed, the influence of submersion on the biota of New Caledonia is more controversial (Giribet and Baker 2019). While the origins of most endemic animal and plant lineages can be dated to after the period of submersion, lineages presumed to have persisted on New Caledonia for more than 37 million years (paleoendemics) do exist. These paleoendemics remain to be fully explained, but interactions with nearby landmasses that emerged to the west between 45 and 50 million years ago could have played a role (Sutherland *et al.* 2020). The land area of New Caledonia increased with lowered sea level during the Last Glacial Maximum, but did not incorporate any other significant islands.

A large part of south-western New Caledonia is composed of ultramafic rocks rich in heavy metals, particularly nickel. This has led to the development of a *maquis* vegetation, a low, stunted heath-like

Montagne des sources, South Province, New Caledonia. Large parts of New Caledonia are composed of ultramafic rocks that support a stunted heath-like vegetation of high flora species richness (Malik Oedin).

Ben Lomond, Queenstown, Aotearoa New Zealand (Angus McNab).

formation with extremely high flora species richness. Although the smallest of the world's biodiversity hotspots, New Caledonia is host to approximately 3400 species of vascular plants of which almost 75% are endemic (Morat *et al.* 2012). Few mammals are known to occur in the floristically diverse *maquis*, but rainforest habitats elsewhere provide suitable habitats.

Aotearoa New Zealand

Zealandia began to separate from Gondwana some 90 million years ago (Mortimer 2004). Sea-floor spreading began around 83 million years ago and as Zealandia drifted northwards towards the Coral Sea, the northernmost parts separated and initiated the opening of the Tasman Sea by 62 million years ago (Hall 2002). The subsequent history of Aotearoa New Zealand has aptly been described as a 'changing archipelago', for during the Tertiary, widespread fragmentation and appearance and disappearance of islands occurred. For the earlier part of the Tertiary (until around 25 million years ago) most of Zealandia was an eroded, flat surface. It began to subside from around 52 million years ago and marine transgression began, reaching its maximum between 37 and 25 million years ago (Neall and Trewick 2008). At this time much of today's Aotearoa New Zealand was submerged and widespread limestone, calcareous sandstone, siltstone and greensand was deposited (Fleming 1962).

This process was reversed in the early Miocene when Australia began to separate from Antarctica and major lateral and oblique compressive movements began to shape Aotearoa New Zealand (Neall and Trewick 2008). Convergence between the Australian and Pacific plates subsequently led to the deformation of the Aotearoa New Zealand region and the uplift of the Southern Alps since the onset of the Pliocene (Kamp *et al.* 1989; Whitehouse and Pearce 1992).

Aotearoa New Zealand is the only island group in the region to have been affected by Pleistocene glaciation. Between 25 000 and 15 000 years ago, much of the South Island and small parts of the North Island were covered by glaciers and snowfields.

Samoa Islands

The Samoan islands of Upolu and Savai'i belong to the independent state of Samoa, whereas Tutuila and the Manu'a group are the main islands of American Samoa. This broadly linear chain of subaerial islands is derived from a primary hotspot trail associated with a seafloor mantle plume more than 4000 m deep. The islands form a linear age progression, with the volcanic construction of Savai'i Island commencing as early as five million years ago (Koppers *et al.* 2008). However, the bulk of subaerial volcanics on Savai'i are younger than 0.39 million years. The mean age for the shield-building phase of Upolu was 2.15 million years ago. Much of the island is covered with younger volcanic rocks associated with numerous small youthful volcanic cones along the backbone of the island (McDougall 2010). Tutuila consists of several volcanoes with a mean age estimated at 1.26 million years ago for the shield-building volcanism. The Manu'a Islands are remnants of quite youthful volcanoes, (less than 0.4 million years old) that have lost parts of their structures from major collapse (McDougall 2010).

Solomon Islands and Autonomous Region of Bougainville (Buka to Makira)

The Solomon Islands archipelago is purely oceanic in origin. The island arc formed where the south-western boundary of the Pacific plate contacted the Australian plate. Almost all islands are plate-boundary islands in close proximity to subduction zones and associated deep ocean trenches. The complex central geological province contains old igneous, sedimentary and metamorphic rocks as well as mineral-rich ultramafic intrusions (Hackman 1973). The oldest rocks are lavas beneath limestone, formed along a spreading mid-ocean ridge during the Cretaceous. They are oldest to the east of this arc around Guadalcanal.

Around the early Eocene, a convergent plate boundary formed in the vicinity of Rennell and Bellona, to the south of the main Solomons chain. By the Oligocene this had led to metamorphism and uplift in the northern part of Solomon Islands (Choiseul, Santa Isabel, Ngella).

By the Oligocene, volcanic activity and uplift was occurring in the area of Makira, Guadalcanal and Bougainville. Construction of Bougainville and Buka began as a submarine volcanic pile around 45 million years ago (Davies 2005). By the early Miocene (25 million years ago), volcanic activity had ceased and between around 25 and 20 million years ago the Ontong Java plateau collided with the Melanesian Arc. This caused a major change in the character of plate boundaries (Hall 2002) and uplifted limestones laid down in shallow waters are found on many of the islands.

Volcanic activity recommenced by around seven million years ago, again accompanied by uplift. During this time, Guadalcanal and Bougainville may well have reached their present extent and elevation, the highest points of each being well over 2000 m. Also at this time the volcanoes of New Georgia and other areas in present-day Western Province came into existence. Thus these landmasses are much more recent that the rest of the group (Kroenke 1984). The outlying islands of Rennell and Bellona themselves are intraplate landforms formed by raised coral limestone. Raising of a former atoll associated with Rennell Island formed the largest freshwater lake in the insular Pacific.

The influences of volcanism are evident on the island of Bougainville today, from the large Mt Balbi stratovolcano that forms Bougainville's highest peak, to smaller areas affected by sulphur and denuded of vegetation such as this area at Kagalalo in the south of the island (Tyrone Lavery).

During glacial maxima over the past two million years, various islands have been connected together. Perhaps the most important was the landmass coined Greater Bukida by Diamond (1983). It included all islands between Buka and Ngella, which would have been joined when sea level dropped by over 100 m. Similarly, many of the islands of present-day Western Province would have been joined into

The Kavachi submarine volcano, to the south of Vangunu Island, is an example of ongoing volcanism in the region. Kavachi regularly forms a new island, before waves cause it to subside beneath the ocean again (Simon Albert).

two landmasses at these times, although they would have remained isolated from all other landmasses. Malaita, Makira and probably Guadalcanal would have been joined to small islands but remained isolated from each other and from other major landmasses. Greater Bukida and Guadalcanal would have been separated at most by only a few kilometres of sea during these times.

There are five active volcanoes in the Solomon Arc. Kavachi submarine volcano, south of Vangunu Island in the Western province regularly forms a new subaerial island every few years, before quickly subsiding beneath the ocean again as activity decreases.

Tonga

The Tongan archipelago consists of roughly parallel eastern and western chains of geologically different islands (Burley 1998). They were formed as part of the Tonga–Kermadec subduction zone, initiated from around 45 million years ago. The arc system rifted away from the eastern Australian margin at around 40 million years ago (Hall 2002).

The oldest rocks are in the eastern arc of islands. They are volcanics and conglomerates on 'Eua, formed by the mostly submerged Tonga ridge and date to around 40 million years (Kroenke 1984). 'Eua is the only island in the group that has risen high enough to expose this underlying Eocene volcanic bedrock. The 'Eua ridge was either initially attached to the eastern end of the New Caledonia Arc until detaching at around 40 million years ago, or the entire Vitiaz arc formed close to the eastern Gondwana margin, but geological evidence is lacking to distinguish between these two alternatives (Hall 2002). The remaining islands of the eastern chain are dominated by more recent raised coral limestone formations and emerging sand cays. Tongatapu is the largest of these, with limestones of Pliocene and Quaternary age.

The western chain of the Tongan Volcanic arc comprises a series sea mounts and steep-sided cones of volcanic origin. Activity on the Tofua arc dates from three million years ago. Several active volcanoes are present in the archipelago, including the submarine volcano Hunga Tonga–Hunga Ha'apai that exploded with a destructive eruption in January 2022.

Vanuatu and Temotu Province, Solomon Islands

Vanuatu and the Santa Cruz islands of Temotu Province, Solomon Islands, are a single geological unit forming part of the New Hebrides arc above a convergent boundary between the Australian and Pacific plates. The islands in the north of Vanuatu form an unusual configuration of three parallel chains identified as the western chain (Espiritu Santo and Malakula), central chain (active volcanoes of Aoba and Ambrym) and eastern chain (Maewo and Pentecost) (Mitchell and Warden 1971).

The earliest known phase of arc growth in the Vanuatu archipelago (formed via subduction of the Pacific plate at the Vitiaz Trench) is Oligocene to mid-Miocene in the western belt of islands (Meffre and Crawford 2001). Volcanism had begun by early Oligocene times and from the late Oligocene to early Miocene submarine volcanic activity built up small landmasses. Volcanic rocks of this age are exposed in the western belt of islands on Espiritu Santo, Malakula, Nendö and the Torres Group (Robin *et al.* 1993).

By the early Miocene volcanism had ceased, and remained absent in the western and eastern chains, with the exception of volcanism of probable early Pliocene age in southern Pentecost (Mitchell and Warden 1971). In the middle Miocene there was subduction reversal and development of new arc crust was initiated by arrival of the Melanesian oceanic plateau at the Vitiaz Trench. This began first in the eastern chain, and then continued from six million years ago in the central chain of islands. Uplift and erosion at the end of the middle Miocene was followed by progressive subsidence and marine transgression across all the western chain islands (Buys *et al.* 2014).

During the early Oligocene to Miocene, Vanuatu, Fiji and the Lau–Tonga region formed a roughly linear north-east–facing island arc along the margin of the Australian plate (Hall 2002). This arc lay distant from Australia. Opening of the North Fiji basins and the Lau Basin since the late Miocene further fragmented the Vanuatu (New Hebrides), Fiji and Tonga arc systems. Vanuatu rotated to the south-west, bringing it into its present-day position.

Volcanism had begun again by the Pliocene and volcanics dated to 2.2–2.4 million years old have been found in the Santa Cruz Islands and the central chain (Kroenke 1984). This activity continues in the central chain of islands. Active volcanoes are present from Aneityum to Tinakula and on Tanna.

During the last two million years, uplift has taken place rapidly. Older islands thus have massive late Pliocene and Quaternary reef limestones that form tilted plateaux and terraces overlaying an igneous basement. Today, limestones less than two million years old have been lifted as high as 300 m above sea level on Erromango, 600 m on Efate and 950 m on Pentecost. Rates of uplift during the Holocene uplift in the central chain have been estimated at up to 5.5 mm per year in the south-west of Espiritu Santo and up to 3.4 mm per year in the north of Malakula (Taylor *et al.* 1987).

Flora and vegetation

The flora and vegetation of the South-west Pacific is complex, influenced by past continental connections and island biogeography, and shaped by disturbance regimes and insular environmental patterns (Müeller-Dombois and Fosberg 2013). Melanesia is essentially an attenuating extension of the floristically rich Indo-Malaysian region (which includes New Guinea) (Müeller-Dombois and Fosberg 2013). Geologically complex, large islands in close proximity to one another support diverse floras compared to other parts of the Pacific.

New Caledonia is an isolated relict of Gondwana. Long periods of submersion followed the Cretaceous and the origins of most plant lineages date to after the island's re-emergence from 37 million years ago (Giribet and Baker 2019). Nonetheless, New Caledonia supports levels of floral diversity and endemism far exceeding those expected under biogeographical principles related to island size and isolation. Aotearoa New Zealand also has a high proportion of endemic flora (approximately 75%) but is depauperate in species relative to land area and the range of climates available, more closely resembling an island flora than a land mass with former continental connections (Müeller-Dombois and Fosberg 2013).

Micronesia and Polynesia, consisting of relatively small islands distributed over a vast area of ocean, have floras descended from chance immigrants that arrived via long-distance dispersal. In Polynesia these species are largely Indo-Malaysian in origin, with American and Aotearoa New Zealand elements (Müeller-Dombois and Fosberg 2013). Micronesia also comprises an attenuated Indo-Malaysian flora, but with influences from the Philippines, Asia, Melanesia, New Guinea and Australia.

In an island context, the presence or absence of certain key taxa can result in significant differences in the ecology of ecosystems, vegetation and habitat for fauna. Generally, the most important habitats for mammals in the region are provided by the various closed forest plant associations, and many of the region's endemic species are restricted to such habitats. Throughout Melanesia there exists a widespread suite of lowland forest tree species. These commonly include species of *Alstonia*, *Canarium*, *Ficus*, *Pometia*, *Sterculia*, *Syzygium*, *Terminalia* and *Vitex* among others. Palms are common (*Areca* spp. and *Licuala* spp.), and wild gingers (*Alpinia* spp.) and bananas (*Musa* sp.) are abundant in natural canopy gaps, providing critical foraging resources for a range of nectarivores and frugivores. Such forests support the large array of

Paéoua, Northern Province, New Caledonia. Approximately 74% of New Caledonia's plant species are endemic to the archipelago (Malik Oedin).

Wabu, Viti Levu, Fiji. Lowland rainforests of the South-west Pacific provide complex habitats that support many of the region's endemic mammals (Patrick Pikacha).

lowland forest-dwelling mammals found throughout the region.

Unusual vegetation communities exist on the higher peaks. Transitions into montane forest generally occur around 1000 m above sea level, but these elevational limits vary according to the size of the island (the Massenerhebung effect), with montane transitions occurring at lower elevations on smaller islands (Grubb 1971). High elevation forests are smaller in stature with finer leaves. Cloud cover is ever-present, making moisture abundant and sunlight deficient, and encouraging thick growths of bryophytes. On Goodenough, oak forests (*Castanopsis* sp.) become established at 1200–1800 m, while above this *Nothofagus* dominates. On Guadalcanal above 1200 m, *Metrosideros* and *Syzygium* become dominant tree species, with orchids, *Rhododendron* spp., *Freycinetia* sp. and bamboos (*Nastus* spp.) also becoming common. On Fiji's Taveuni above 1000 m, a dense thicket of monocots and herbs dominate, and often extend above the canopy of short trees and bushes. An endemic palm (*Clinostigma exorrhizum*) is also dominant. Although clear altitudinal zonation is evident in the flora and vegetation of the high islands, in most instances the effect of this upon mammal distributions is poorly studied.

Several islands have endemic species that are restricted to montane forest, and examination of these associations may give an understanding of the effect of altitude, changing climate and vegetation on mammals. Islands and montane endemics of note are Goodenough (*Dorcopsis atrata*, above 1200 m); Fergusson (*Dactylopsila tatei*, above 600 m); Guadalcanal (*Pteralopex pulchra*, above 1200 m); and Taveuni (*Mirimiri acrodonta*, above 800 m). Other species probably await scientific discovery, as the mountainous areas of many islands are remote, difficult to access and almost entirely unsurveyed.

Open forests, woodlands, savannah and grasslands also occur on various islands (e.g. Guadalcanal, New Caledonia, Aotearoa New Zealand and Viti Levu). Although some are natural, many are the long-established results of human activity (either vegetation clearing or periodic burning). Other important vegetation types are mangroves and communities on ultramafic soils. The latter are best developed in New Caledonia

The slopes of Guadalcanal's highest peak, Mt Popomanasaeu, are covered in montane cloud forests with a stunted canopy layer and dense growths of bryophytes and epiphytes. Such habitats are difficult to access and thus poorly surveyed. Numerous mammals of the South-west Pacific are endemic to them (Patrick Pikacha).

Much of the southern half of Guam that was formerly covered in rainforest vegetation has now been converted to grassland via periodic burning (Robert Fisher).

Monu Island, Fiji. Period burning in Fiji can result in the conversion of diverse rainforest vegetation to lower diversity grasslands (Robert Fisher).

Jejevo, Santa Isabel, Solomon Islands. Ultramafic areas on Choiseul, Santa Isabel and San Jorge support a unique forest type dominated by ironwood (*Xanthostemon melanoxylon*) (Tyrone Lavery).

where they are known locally as *maquis*. However, smaller areas occur in Solomon Islands (Choiseul, Santa Isabel and San Jorge). Although mammals do occur in these environments, they are few in number and no species are known to be endemic to them.

At the other end of the vegetational spectrum lie the modern villages, gardens and agroforests. On large islands these punctuate the broader native vegetation communities, but they can almost entirely occupy the small islands inhabited by humans (along with strand vegetation communities). Commonly cultivated species include *Canarium indicum* (ngali or galip nut), *Cocos nucifera* (coconut), *Areca catechu* (betel nut), *Mangifera indica* (mango), *Ceiba pentandra* (kapok), *Carica papaya* (papaya), *Artocarpus altilis* (breadfruit), *Syzygium malaccense* (Malay apple) and *Barringtonia edulis* (cutnut/naveli). Agroforests and gardens support a distinctive suite of widespread bats including various species of *Dobsonia* and *Pteropus*, and *Macroglossus minimus* and *Rousettus amplexicaudatus*. It is possible that some species may have expanded their distributions thanks to human settlement and establishment of such food rich habitats.

Old gardens and fallow can be a focus for foraging bandicoots (e.g. *Echymipera kalubu*), and the edges between open gardens and intact forests form an important foraging niche targeted by a range of insectivorous bats. For example, Davies *et al.* (2016) found that activity of echolocating bats on Makira, Solomon Islands, was greater in garden areas than primary forests (as measured by the number of passes by echolocation recording devices). Abandoned old village sites can be equally important (but less obvious) landscape features. Villages were abandoned in prehistoric times for various reasons and at the beginning of the 20th century more isolated inland villages were vacated in preference for coastal locations. Abandoned sites retain abundant planted fruit and nut trees and, set among regenerated mature forests, provide resource rich habitats for mammals. The South-west Pacific endemic genus *Pteralopex*, for example, is frequently associated with old village sites where roosting and feeding trees are abundant.

Village gardens like this one in the highlands of East Makira, Solomon Islands, are often embedded within expansive areas of native forest and can provide important food resources and habitats for mammals (Tyrone Lavery).

Human presence, translocation and extinction

The tripartite division of the South-west Pacific into the regions of Melanesia, Polynesia and Micronesia is a well-established but also extremely problematic classification. The diverse and heterogeneous variation in language, culture and human biology among people of the Pacific defies categorisation, especially in Melanesia (Kirch 2000). Nonetheless, by virtue of their deeply entrenched acceptance, these labels can provide handy geographical referents that we have sometimes incorporated into this work to describe the distributions of species, rather than cultural or ethnic affinities of people.

Coinciding with various phases of human settlement of the Pacific have been the introductions of both domesticated, and wild species. In the case of translocated wild animals, three principal categories are generally recognised: *stowaways* or smaller species capable of concealing themselves in vessels or cargo; *ethnotramps* that are economically and culturally important species regularly carried around with humans; and *incidentals* or species that

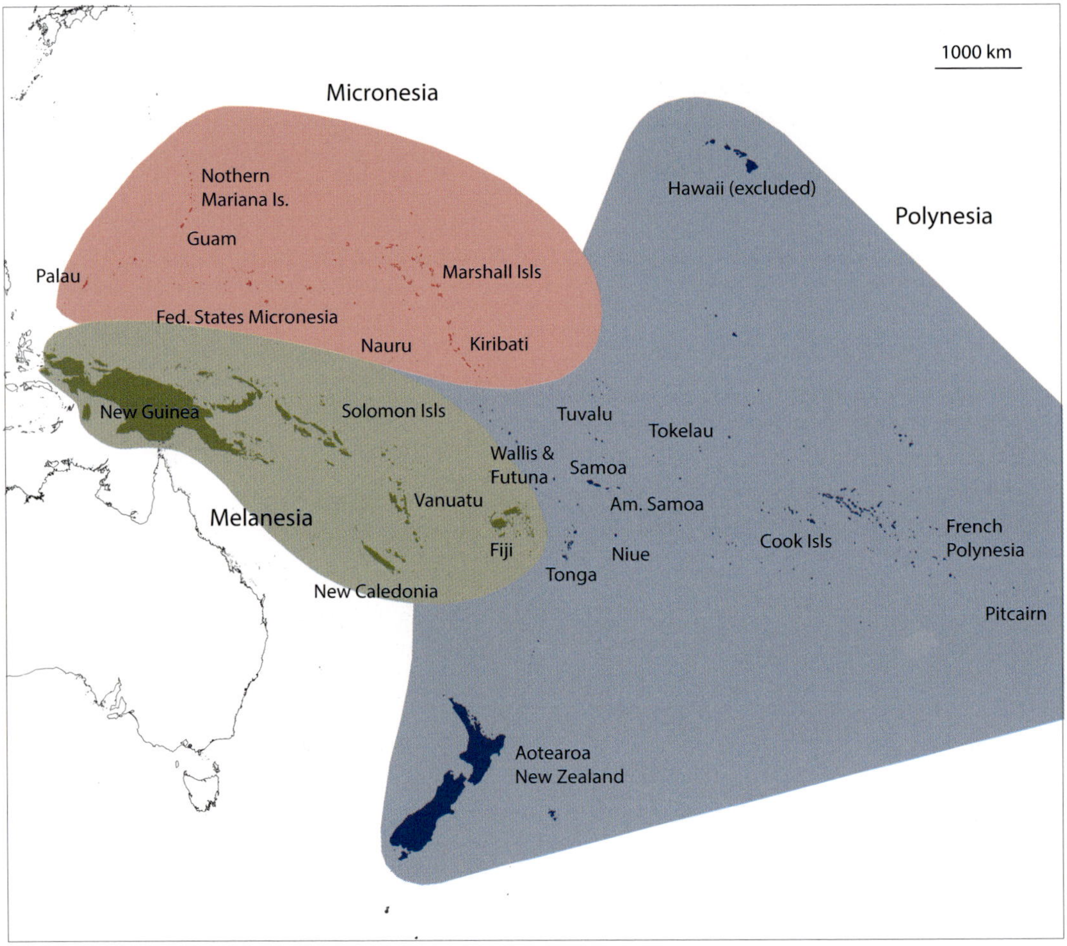

Islands of the South-west Pacific are regularly grouped into the subregions of Melanesia, Micronesia and Polynesia. These groupings have been used as geographical referents to help define the region of focus for this book.

are infrequently transported and introduced (Heinsohn 2010a).

The first phase of settlement of the South-west Pacific is generally considered to reflect one of the oldest passages of modern humans out of Africa, reaching Sahul (Australia, Tasmania, New Guinea, Aru Islands) and adjacent eastward islands at least 50 000 years ago (Wollstein *et al.* 2010; O'Connor and Hiscock 2018). Evidence from New Ireland and New Britain confirms a similar antiquity of human occupation of Melanesia's large islands (45 000 years) (Allen *et al.* 1988; Pavlides and Gosden 1994; Leavesley and Chappell 2004). By 35 000 years ago, these non-Austronesian or Papuan-speaking people had crossed the significant geographical divide between the Bismarck islands and Northern Solomons (Wickler 2001), and by 25 000 years ago reached Manus Island (Fredericksen *et al.* 1993). The Vitiaz Strait forms a significant biogeographical boundary, and lower species richness of mammals and birds in Near Oceania arguably resulted in a heavier reliance on marine resources (Allen 2000). However, following time lags after initial colonisation, trading was established within and between the Bismarck and Solomon Archipelagos and wild mammal translocations took place, bolstering potential game.

Translocation of the Bismarck common cuscus (*Phalanger breviceps*) to New Ireland from New Britain occurred between 10 000 and 20 000 years ago, ranking among the earliest known mammal introductions around the world (Flannery and White 1991; Heinsohn 2010a). The New Guinea pademelon (*Thylogale browni*) was similarly introduced to New Ireland around 7000 years ago (Flannery and White 1991). By 4850 years ago, the Bismarck common cuscus had been introduced to Nissan, and sometime between 6700–2000 years ago it was introduced to Buka and was subsequently dispersed throughout Solomon Islands as far as Makira and Santa Ana (Flannery *et al.* 1988; Flannery and White 1991; Flannery and Wickler 1990; Spriggs 1991). New Guinea pademelon was also carried from the Bismarcks to Buka, Lihir and possibly Emirau, but apparently never established wild populations on Buka or Emirau (Flannery *et al.* 1988, Flannery and White 1991).

These introductions and other anthropogenic impacts presumably resulted in some transformation of indigenous flora and fauna communities. But because of an almost complete lack of fossils which pre-date or date to the initial arrival of humans, the impacts of early species introductions are poorly understood. In most instances it is simply not known whether mammal extinctions occurred some 35 000 years ago.

A second phase of human expansion reached the Bismarck Archipelago, Palau and the Mariana Islands by around 3500 years ago, bringing people associated with a distinct material culture who mostly spoke Austronesian languages (Denham 2018). These people were also linked with the translocation of numerous domestic and commensal species such as the chicken (*Gallus gallus*), dog (*Canis familiaris*), pig (*Sus scrofa*) and Polynesian rat (*Rattus exulans*). Around the same period, the large spiny rat (*R. praetor*) was spread throughout the Bismarcks and Solomons. By around 3200 years ago, people had begun to expand into remote Oceania with a range of plant domesticates derived from the New Guinea region (Denham 2018). Following a time lag after settlement of remote Tikopia and Vanuatu, the large spiny rat was introduced there too, probably via contact with the main Solomon Archipelago (Flannery *et al.* 1988; Anderson 2008). During this second phase of settlement, a pulse of dispersal between around 1190–1290 CE saw humans reach virtually every corner of Polynesia, including Aotearoa New Zealand (Wilmshurst *et al.* 2011). The dispersal of pigs, dogs and chickens occurred at this time but was not ubiquitous; thus the presence of these three domesticates in archaeological sites is variable across Remote Oceania (Denham 2018). However, the remains of Polynesian rats are present in all archaeological sites associated with both Lapita and Polynesian settlements as far as Aotearoa New Zealand (Matisoo-Smith and Robins 2004). This ubiquitous distribution and the large numbers of skeletal remains that are found in archaeological middens indicate this species was intentionally introduced, possibly as a food item (Matisoo-Smith and Robins 2004).

Evidence for prehistoric settlement of the Marianas island chain at 3500 years before present is contemporaneous with the expansion of people into Remote Oceania. On Yap and Palau, human presence is marked by a major period of forest destruction accompanied by fire (Kirch 2000). Central and Eastern Micronesia, including the Caroline and Marshall Islands, were probably settled around 1500 years after Western Micronesia (Athens 2018). Along with this temporal divide, archaeological and linguistic evidence suggests Eastern and Western Micronesia was colonised by Austronesian speaking people with separate origins in South-east Asia and Near Oceania respectively. The remains of what is probably *R. tanezumi* (a South-east Asian human commensal) are present in deposits in Western Micronesia (Ant Atoll, Fais Atoll, Palau, Pagan, Rota, Guam, Ngulu, Chuuk, Nukuoro) from around 1700–1000 years before present, suggesting continued contact with Asia (Anderson 2008). *Rattus exulans* is absent from a 3000 year profile on Palau, but does occur at around 1000 years before present at Utrok Atoll in Marshall Islands (Weisler 2001; Clark 2005).

This second phase of human expansion in the Pacific is marked by extinction, best documented for birds, but also including indigenous terrestrial mammals (Steadman 1995). Buka provides one such example, but a gap in the archaeological record between around 6700–2000 years ago means the dates and causes of extinction are uncertain. The sequence predating 6700 years ago revealed five species of endemic murid rodent were present on Buka. However, in the sediments deposited from around 1860 years before present, only three of these species remained and *R. exulans*, *R. praetor*, *P. breviceps*, *S. scrofa* and *C. familiaris* had all been introduced. On New Ireland, the extinct endemic murid *R. sanila* was found in archaeological deposits dating to between 3000 and 7000 years before present. In younger levels above this *R. praetor* was found, suggesting a potential role for competition or introduced disease driving extinction (Flannery and White 1991).

Extinction, depletion and translocation of flying foxes also occurred during this second phase. At the Teouma Lapita site on Efate, Vanuatu, up to 13 forms of unidentified pteropodid fruit bat were recovered from deposits dated between 3000–2500 years before present (Hawkins 2015). It was apparent that at least nine disappeared from later deposits and today only four pteropodid species are known from the entire Vanuatu archipelago. Kirch *et al.* (1992) identified bones of Pacific flying fox (*Pteropus tonganus*) in archaeological deposits on Mangaia in Cook Islands. The remains become less frequent with time, suggesting a population decline due to hunting. But this species was also apparently translocated by people. A population on Rarotonga in Cook Islands represents the easternmost extension of fruit bats into the Pacific, and was apparently initiated from animals brought from Mangaia in the 1870s (Tiraa 1992). Martin (1981) reported animals being carried as far afield as Tahiti.

From the 16th century, Europeans began to arrive and establish settlements, territories and colonial outposts in the South-west Pacific. Initially these were restricted to Micronesia but by the 19th century they were widespread throughout the region. In Melanesia, European settlements were often tenuous, but permanent settlements were better established on the more temperate islands of Aotearoa New Zealand and New Caledonia. Europeans brought a whole host of new introduced species. For example, the Dutch, British and Germans assisted with the spread of rusa deer (*Cervus timorensis*) into Melanesia (Lever 1985). The small Indian mongoose was introduced to Fiji in the late 1800s as an attempt to control rodents in sugarcane fields (Gorman 1975). The feral cat (*Felis catus*), one of the most devastating introduced species for island faunas, has now been almost ubiquitously introduced to the South-west Pacific. All of these phases of human arrival brought new influences on the indigenous mammals of the South-west Pacific. Recent times have seen the possible extinction of seven more mammal species (two murids on Guadalcanal; a bat species each on Aotearoa New Zealand, Guam, the Palau Islands and the Aru Islands; and two bats in Samoa) as human predation, introduced species and environmental change all take their toll.

Palaeontology

Almost all known sites in the South-west Pacific containing mammal remains are archaeological in nature. This means that the pre-human fauna of most islands is almost entirely unknown. Therefore, species that became extinct rapidly after human arrival, particularly in northern Melanesia, which was largely settled by around 35 000 years ago, are unlikely to be known.

Both the archaeological and few non-archaeological sites provide evidence for richer faunas before human settlement. A deposit of probable Pleistocene age on Woodlark Island has yielded the remains of dugong (*Dugon dugon*) and evidence of longirostrine crocodiles, the nearest relatives of which may be a Miocene South American species (Molnar 1982). Sites on Lord Howe Island, New Caledonia and Walpole Island have yielded the remains of horned tortoises (*Meiolania* spp.) (Balouet 1991). Seven early archaeological sites across Vanuatu and Fiji have similarly produced extinct hornless tortoises (*?Meiolania damelipi*) (Hawkins *et al.* 2016). In Fiji and Vanuatu, these megafaunas vanished soon after human arrival (3000–2700 years before present). A crocodilian and a large varanid have also been reported from New Caledonia (Balouet 1991). A non-human site of uncertain age is known from Balof Cave, New Ireland. It contained an essentially modern fauna, the only extinct species being *Rattus sanila* (White *et al.* 1991).

Sites of note

American Samoa

The site at To'aga on Manu'a in American Samoa documented by Hunt and Kirch (1988) is in an open coastal area and is Lapita to post-Lapita in age. In post-ceramic stratigraphic contexts, bones from pigs and *Rattus* (probably *R. exulans*) were common, and those from dogs were also present. No remains of flying foxes (*Pteropus* spp.) were reported.

Autonomous Region of Bougainville

The prehistory of Buka Island, interpreted across over 20 sites, has been comprehensively described by Wickler (2001). Flannery and Wickler (1990) published on a Pleistocene fauna from the Kilu site where initial human settlement dated to around 29 000 years before present. Apart from humans, the only mammals present in the Pleistocene levels were murids (five species) and Chiroptera. The remains were dominated by rodents with much lower numbers of large to small bats.

Among the murids were two extinct and three extant species. The extant *Melomys bougainville*, *Solomys ponceleti* and *S. salebrosus* were common in the site. The extinct species (*M. spechti* and *S. spriggsarum*) were moderately large, the former probably weighing some 100–150 g, the latter around 1 kg. Both extinct species may have been terrestrial, whereas all surviving species are arboreal, and both became extinct somewhere between 6700–2000 years before present. Conspicuously absent from older levels (dating to before 6700 years ago) are *Phalanger breviceps*, species of *Rattus*, and human commensals such as pigs and dogs. All of these (including *R. praetor*), however, were present on Buka by around 2000 years ago (Flannery *et al.* 1988; Flannery and Wickler 1990). Dogs, pigs and competition with, or diseases from, introduced *Rattus* and the Bismarck common cuscus may have all contributed to the extinction of the two endemic terrestrial murids.

Phalanger breviceps was present on Nissan by around 4850 years ago and remains of a *Pteropus* species (probably *P. admiralitatum*) were also dated to that period (Spriggs 1991). Both survive on the island today. *Rattus exulans* was present by around 3200 years ago, and pigs by 3650 years ago. A single dentary of *R. praetor* from the island was recovered from levels dating to 700–500 years old. This species is no longer found on Nissan, and the bone may represent a single unsuccessful introduction or item of trade (Flannery *et al.* 1988).

Cook Islands

Kirch *et al.* (1992) reported upon fossils and sediments from various locations on Mangaia in Cook Islands. Humans have been present for some 3600 years, but the faunal remains recovered date only to the last 1000 years. The only indigenous mammal recorded is *P. tonganus*, and analysis of remains indicated a major reduction in its frequency in the deposits over the last 1000 years. Remains of pig and *R. exulans* were also present in the sediments.

Steadman (1991) recorded the remains of *P. tonganus* from archaeological sites on Aitutaki in the lower Cook Islands. The species is absent from the island today, and there appear to be no historic records of its occurrence there.

Fiji

The Qaranipuqa rock shelter on Lakeba has been characterised as representing sporadic, short-term occupations in its lowest layers (with light predation of local fauna). The bulk of mammal bones were derived from the A levels, dated to approximately the 14th–17th centuries CE (Best 1984). Many of the long bat bones recovered were broken and burnt, indicating they were food remains rather than *in situ* deaths of animals. *Rattus praetor* and *R. exulans* were both found together in the same layers at the site, overlying layers that contain the abundant remains of extinct land birds (White *et al.* 2000; Worthy and Clark 2009).

At Qaranicagi in the Yasawa group of islands, pig bones that had an estimated age range of 1000–2500 cal. years before present were recovered (Cochrane 2006).

French Polynesia

Weisler *et al.* (2006) recovered bones of Pacific flying fox (*P. tonganus*) from an archaic East Polynesian site at Peva Valley on the east coast of Rurutu, Austral Islands, French Polynesia. The excavation site was situated around 100 m from the coast in an overgrown area of unused private land section atop a coastal sand dune. The *Pteropus* bones were dated to between 1064–1155 CE (Weisler *et al.* 2006). Human transportation of this species to French Polynesia and extirpation were both raised as possibilities as no bones were present in either the pre-human deposits or late prehistoric to historic layers (Weisler *et al.* 2006)

Indonesia

Aplin and Pasveer (2006) completed a systematic review of the prehistoric vertebrate fauna of the Liang Lemdubu and Liang Nabulei Lisa limestone caves on Pulau Kobroor, Aru Islands. The excavations produced a rich record of the late Pleistocene to Holocene vertebrate fauna via substantial quantities of bone and other remains. Around 30 mammal species were identified from the deposits. Several of these are yet to be documented among the living faunas of the Aru Islands, and a suite of others are unknown from the region entirely (e.g. *Isoodon macrourus*, *Parahydromys asper*, *Pseudomys* cf. *nanus*, *R. sordidus*, *Tachyglossus aculeatus*, *Thylogale stigmatica*) (Aplin and Pasveer 2006). Those that are apparently absent from the modern faunas are primarily species adapted to drier grassland or savannah habitats (e.g. *Isodon macrourus*, *Notamacropus agilis*, *R. sordidus*). Their former presence constitutes strong evidence that considerably drier and more open habitat dominated the Aru Islands during Pleistocene times before shrinking to be replaced by wetter forests as sea levels rose (Aplin and Pasveer 2006). A second group of species unknown today are rainforest adapted, but may have been susceptible to hunting or competition in an environment of increasing heterogeneity (e.g. *N. agilis*, *T. stigmatica*). The apparent loss of New Guinea Quoll (*Dasyurus albopunctatus*) is consistent with trends in extinction of dasyurids from other continental shelf islands following sea level rise (Lavery *et al.* 2013b).

New Caledonia and Loyalty Islands

The vertebrate faunas of 17 fossil bearing sites on New Caledonia and the Loyalty Islands were reviewed by Balouet (1991). Some sites (including Walpole Island) had produced the remains of large extinct birds and reptiles, including the horned turtle *Meiolania mackayi*, the crocodile *Mekosuchus inexpectatus* and a large species of *Varanus* (similar to the living *Varanus indicus*). Balouet (1991) mentioned approximately 1000 bat bones from the various deposits, but stated that all represented extant species. Fossil *R. exulans* (dated to the last few thousand years) were also identified. The site known as Lapita is located in a bay on the peninsula of Foué in Koné, on the west coast of Grande Terre.

A stratigraphic context containing cultural materials and a rich charcoal layer there has been dated to 2930 years before present (Sand 1998).

Aotearoa New Zealand

An early Miocene site (19–16 million years ago) at St Bathans, Central Otago, South Island, revealed the modern genus *Mystacina* was established in Aotearoa New Zealand from at least 16 million years ago (Hand *et al.* 2015). Close relatives of the plants and arthropods eaten, inhabited, pollinated or dispersed by extant *Mystacina* were also recorded as fossils in the same area, suggesting the ecological associations of the genus have been long established. Also present was a new bat genus and species (*Vulcanops jennyworthyae*) from the family Mystacinidae (Hand *et al.* 2017). This was a relatively large bat that according to its dentition was probably omnivorous and weighed around 40 g. The same site has produced a relatively plesiomorphic femur and two partial mandibles from a mouse-sized terrestrial mammal with unknown relationships to other mammals (Worthy *et al.* 2006). Its ancestors are thought to have diverged from their nearest relatives more than 125 million years ago. Aotearoa New Zealand's non-volant terrestrial mammal fauna disappeared some time since the Middle Miocene, possibly the result of climatic cooling (Worthy *et al.* 2006).

Late Pleistocene to Holocene fossils of *Chalinolobus tuberculatus*, *Mystacina robusta* and *M. tuberculata* have been recovered from numerous sites in Aotearoa New Zealand (e.g. Hill and Daniel 1985; Flannery 1987; Gill 2002). These species probably experienced range contractions due to environmental changes caused by the arrival of humans, along with the introduction of *R. exulans*.

Papua New Guinea

Kirch (1988) reported on the excavation of the Talepakemelai site on Eloaua in the St Matthias Group. The site was occupied between 3200–2000 years before present. The bones of pig and dog, and at least two species of pteropodids (whose specific identity remains uncertain) and the phalangerid *Spilocuscus maculatus*, were identified. Egloff (1975) also reported bones from the Talepakemelai site and recorded the phalanges of a wallaby (a possible misidentification for *Spilocuscus*), as well as cuscus, pig and dog. Steadman and Kirch (2016) provided an analysis of bird bones present in the deposits of three sites in the St Matthias Group (including Eloaua) and reported a lower percentage of extinct and extirpated species compared to sites in Western Polynesia, possibly related to rugged terrain of the islands, or extirpations having pre-dated the available archaeological records.

Material of Pleistocene age has been recovered from the Pamwak area on Manus. *Rattus detentus* and *Melomys matambuai* are present throughout the deposits and into the most recent layers (Williams 1999). Faunal composition at the site was found to have changed somewhat at 12 400 years before present when the common Echymipera (*E. kalubu*) first appears in the deposits. This corresponded with a shift from the use of stone tools to obsidian. The Admiralty cuscus (*S. kraemeri*) first appeared in the sequence at around 11 000 years before present. However, layers at the site were dated back to just 14 000 years before present (C^{14} calibrated). Hence both species might be natural elements in the Manus fauna, and further assessment of the taxonomic status of *E. kalubu* is warranted (Helgen and Flannery 2004).

Green and Anson (1991) and Smith (2000) reported a single distal humerus fragment of a peramelid bandicoot (likely *E. kalubu*) from 2000 year old deposits on Watom, just off northern New Britain. The remains of pigs were by far the most common, and all material was highly fragmented, with 88% being fragments of less than 3 cm in length. Although it is a large island, New Britain has so far failed to reveal any substantial mammalian faunal assemblages in the archaeological record.

New Ireland has one of the best documented fossil mammal records of any island in the region. Five sites yielding Pleistocene faunas have been recovered, with evidence for human occupation going back to 33 000 years ago. Only the basal levels at Balof 1 and 2 seemed to pre-date the arrival of humans at the site.

Flannery and White (1991) suggested the pre-human mammal fauna of New Ireland comprised around 29 species, including two murids. One was the widespread species *Melomys rufescens*. The second was originally described as a subspecies of *R. mordax* but subsequently elevated to full species (*R. sanila*). This species appeared in an archaeological sequence at Panakiwuk Cave from approximately 15 000 years before present and was continually present until disappearing at approximately 1600 years before present (Marshall and Allen 1991). *Rattus sanila* is now presumed extinct.

Bones of *Phalanger breviceps* at Matenbek represent the earliest evidence for the occurrence of this species in the Bismarck Archipelago, appearing around 20 000 years before present (Allen 1996). The Panakiwuk site was most intensively occupied between around 10 000–8000 years before present and at the end of this period, introduction of *R. praetor* was evident (Allen 1996). Over half the number of identified specimens at Panakiwuk were bats, but much of this material was from a cave roosting species of *Dobsonia*, that may have died *in situ* (Macdonald 2002).

Solomon Islands

Cave sites in the Poha River Valley in northern Guadalcanal have revealed early evidence for the exploitation of primary forest taxa by humans including *Uromys* rodents and the arboreal monkey-tailed skink (*Corucia zebrata*) (Roe 1992). Today, the lower Poha River crosses grasslands, far from the forest habitat preferred by these species. Northern Guadalcanal's grasslands are maintained through a process of periodic burning by humans and pollen records indicate they were established as an integral part of the landscape by 3300 years before present (Roe 1992).

The prehistory of Tikopia has been documented exhaustively by Kirch and Yen (1982). They recorded the island was initially colonised by 2000 years ago, and all remains reported were more recent than this. Pigs, dogs, *R. exulans* and *R. praetor* were among the introduced species, while *Pteropus* cf. *tonganus geddiei* and *P.* cf. *nitendienses* were indigenous. The contribution of flying foxes to the diet was low initially and decreased further over time (apparently culturally restricted). The occurrence of *R. praetor* is interesting. Its remains were not present in the lowest levels of the site, and it has not been recorded from the island in historic times, suggesting that it may now be extinct there. Both pigs and dogs were rare or extinct by the early historic period, although their remains were common in early prehistoric levels.

Tonga

Rainey and Pierson (1992) and Koopman and Steadman (1995) recorded remains of bats similar to *Notopteris macdonaldi*, *Pteropus samoensis*, *P. tonganus*, *Emballonura semicaudata* and *Mops jobensis* from Holocene sediments on the island of 'Eua in the Tongatapu Group. The only bats recorded from the island in historic times are *P. tonganus* and *E. semicaudata*, of which the latter is now probably extinct in Tonga. The extinction of bats on 'Eua, along with at least 23 land birds and two species of lizards was probably related to human impacts (Koopman and Steadman 1995).

Vanuatu

Ifo is an open site situated within modern Ifo Village, Erromango, consisting of five layers dating back to around 3500 years before present (Macdonald 2002). Terrestrial fauna deposits include an extinct imperial pigeon (*Ducula* sp.), rats, pigs and turtles. Unidentified flying foxes made up a significant component of the assemblage. Ponamla, also on Erromango, is slightly younger than Ifo, having been occupied from around 2700–2500 years before present (Macdonald 2002). The remains of *Rattus* sp., *Pteropus* sp. and pig were common throughout the layers. The abundance of pig and *Pteropus* sp. declined in younger layers, but *Rattus*

A subadult Polynesian rat (*Rattus exulans*), East Kwaio, Malaita. This species was prehistorically introduced to many Pacific Islands, is common among archaeological deposits, and remains one of the most widely distributed and commonly encountered mammals of the region (Ben Speare).

sp. remained constant. It was postulated this reflected a decrease in site usage rather than a decline in flying fox numbers.

On Efate, the Teouma site is an early Lapita settlement, and Arapus and Mangaasi sites are immediately post-Lapita. At Teouma (3000–2500 years before present), the remains of fruit bats and large-bodied sea turtles are common in early deposits, with the remains of giant tortoises and pigs increasing as these initial proximal resources were exploited (Hawkins 2015). Following an ecological tipping point, crocodiles, several species of fruit bats, birds and giant tortoises disappear from the deposits. Predation of the introduced large spiny rat (*R. praetor*) also increased in later deposits causing a population decline in that species.

Zoogeography

Introduction

The region considered here is inhabited by some 184 described indigenous terrestrial mammals. Among these are 35 marsupials, 32 murids and 117 bats. Levels of endemism are relatively high. For example, a comparison between the birds and mammals or Northern Melanesia using an endemism index developed by Mayr and Diamond (2001) produces higher values for mammals than for birds (Lavery *et al.* 2016). The South-west Pacific mammal fauna includes an endemic family of insectivorous bats (Mystacinidae), an endemic genus of murid rodents (*Solomys*), and endemic genera of pteropodid bats (*Melonycteris*, *Mirimiri*, *Nesonycteris*, *Notopteris*, *Pteralopex*) and insectivorous bats (*Anthops* and *Mystacina*). There are 95 endemic species of mammal (51.6% of the species presented in this work).

Inadequate fauna inventories continue to present a significant hurdle for developing a zoogeographical synthesis of the mammals of the South-west Pacific. For many islands, fauna inventories are derived from just a few weeks of field surveys that are inadequate to detect even the majority of species from a moderately large island. So great is this lack of survey effort that for some quite large islands (for example, Lavongai, 1186 km^2), there is still not a single record of an indigenous mammal other than pteropodid bats. The regular addition of new, undescribed species to island inventories is further evidence of survey inadequacies. For some 100 years leading up to 1990, many surveys had been completed on Guadalcanal in Solomon Islands, yet an expedition that year found a previously undescribed species of *Pteralopex* – a large bat weighing 280 g. The holotype remains the only specimen of this species and as far as we know, highland Guadalcanal (above 1300 m elevation) is still yet to be the focus of a dedicated mammal survey. The Solomon Islands Western Province is one of the more accessible regions of the country and had been the focus of major expeditions led by the Australian Museum in the 1990s, including a 3-month period of sustained research focused on the endemic monkey-faced bat (*P. taki*) on Kolombangara, New Georgia and Vangunu. Yet in 2015 a new species of endemic giant rat (*Uromys vika*) was uncovered on Vangunu. Similarly, New Britain had been the focus of numerous biological surveys and collecting expeditions including Reverend George Brown (operating from Duke of York Island between 1875 and 1881); Thomas and Margaret Gilliard in the Whiteman Range 1958–1959 (Gilliard and Lecroy 1967); Jared Diamond in the 1970s (reported by Koopman 1979); Smith and Hood (1981) June–August 1979; multiple expeditions by the Australian and South Australian Museums (in the early 1990s); expeditions by the Papua New Guinea National Museum (Bonaccorso 1998); and the New Britain Biological Survey March–April 1999 (Anthony 2001). Despite this body of work, a biological survey in the Nakanai Mountains in 2014 uncovered multiple new species of murid rodent (Aplin and Opiang 2011).

A further problem for the zoogeographer is that human-attributed extinctions and introductions have altered the species compositions of many islands. These factors can severely bias zoogeographical conclusions if not accounted for, but unfortunately relatively few good fossil records have been documented from the South-west Pacific. Those that have been assessed have demonstrated the magnitude of changes that have occurred. On New Ireland, for example, the pre-human non-volant mammal fauna consisted of two murid species (one endemic). The endemic species became extinct, and humans introduced another two murid species and three marsupials, as well as the widespread Asian commensal pigs and dogs.

Marsupials

Marsupials make up 19% of the mammalian fauna of the region, and just one monotreme has been documented. Of these, 15 of 35 species are exclusively limited to New Guinean land bridge

Kabui Bay, Waigeo, on the northern edge of Raja Ampat, Indonesia. The archipelagos of the South-west Pacific offer alluring opportunities for the zoogeographer, but inadequate faunal inventories continue to present significant hurdles across much of the region (Chien Lee).

islands, nine are endemic (25.7%), with two of these endemic to old continental fragments and seven to what are probably oceanic islands. Numerous species have a known record of human translocation and owe at least part of their island distribution to introduction by humans.

These figures clearly indicate that marsupials are generally very poor over-water dispersalists.

This was quantified by Lavery *et al.* (2016) using a non-sharing index. The results demonstrate the significance of the Vitiaz Strait as a biogeographical barrier, with the faunas of New Guinea's closest satellite islands and Bismarck Archipelago highly dissimilar to those of the mainland. Evidently, even the small divide between New Britain and New Ireland (approximately 30 km in width) has been large enough to prevent natural dispersal of marsupials to New Ireland (Flannery 1996). Only among phalangerids is there presumed over-water dispersal to an oceanic island (the Woodlark cuscus *Phalanger lullulae*), although bandicoots and wallabies also may have achieved this.

Long-term survival of certain marsupials on isolated continental fragments is also poor. Australia and New Guinea are surrounded by a multitude of continental shelf islands that were connected to Sahul during the Last Glacial Maximum. Lavery *et al.* (2013b) compared the modern faunas found on 274 of these with the faunas of adjacent mainland source pools and found that following isolation from rising sea levels, species from the order Dasyuromorphia were less likely to persist than those from Peramelemorphia, Diprotodontia and Rodentia. As carnivores and insectivores, dasyuromorphs require larger home ranges relative to body size to acquire sufficient resources, and probably exist at lower densities compared to species from other orders – exposing them to higher likelihood of extinction. Today only six dasyurids are known from the region, several of these by just a single specimen.

The D'Entrecasteaux Islands provide a most fascinating opportunity to consider marsupial zoogeography in the region. The three main D'Entrecasteaux islands have been separate from mainland New Guinea since the Pliocene, but lie relatively close to the Papuan Peninsula (less than 20 km in the case of Normanby). At least two endemic marsupials exist (*Dorcopsis atrata* on Goodenough, and *Dactylopsila tatei* on Fergusson), and all share a bandicoot (*Echymipera rufescens*), a widespread petaurid (*Petaurus papuanus*), a phalangerid (*Phalanger intercastellanus*) and (except for Normanby) a macropodid (*Notamacropus agilis*).

Rodents

Indigenous murid rodents are restricted to Melanesia, and naturally extend no further than Solomon Islands where they are represented by three genera. The genus *Solomys* is endemic to the Solomon Islands archipelago; *Melomys* is widespread in Australia, New Guinea and parts of Indonesia. *Uromys* is similarly widespread and represented in the region both by a Solomon Islands endemic subgenus (*Cyromys*) and the more widespread subgenus *Uromys*. Secondary radiations within these groups indicate over-water colonisation from New Guinea has probably been infrequent. Only six of the 32 species (around 18.8%) are restricted to New Guinean land bridge islands, and 18 (56.3%) are endemic to oceanic or long-isolated continental islands. *Rattus praetor* is the only indigenous species known to have been introduced to islands outside its natural range.

Bats

Pteropodid bats are the only indigenous mammals in large parts of Micronesia and the central Pacific, having dispersed much more widely than either marsupials or murids. Fifty-two described species (or 76.5% of the pteropodid fauna) are endemic to oceanic islands or distant continental fragment islands. *Pteropus* is most diverse, comprising 33 described species, and species of this genus also are the most widely distributed of all members of the family. Sympatry of distantly related pairs of species indicates repeat colonisation by *Pteropus* has taken place in some of the more remote parts of the Pacific (Almeida *et al.* 2014). The species on Guam (*P. mariannus* and *P. tokudae*), New Caledonia (*P. ornatus* and *P. vetulus*), and Samoa (*P. tonganus* and *P. samoensis*), for example, all represent sympatric pairs of species that are derived from two separate species groups.

Among insectivorous bats, only 16 of the 50 species (32%) are endemic to the region, and all are found on islands that required colonisation via over-water dispersal. The reason for so much endemism among pteropodids and so little among insectivorous bats remains unexplained.

Insectivorous bats rank high among the species most difficult to survey in this region, especially in Melanesia. Conventional trapping techniques (harp traps and mist nets) are largely ineffective unless placed at known roost exits. This probably relates to the fact the region's echolocating bats are well accustomed to avoiding objects in the cluttered environments typified by lowland tropical rainforests. The bulk of distributional records are thus derived from searches of caves. This results in significant biases, where the distributions of non-cave-roosting species are less well known, and islands without known or accessible caves have poor species inventories. The known distributional ranges of many insectivorous species are thus often irregular.

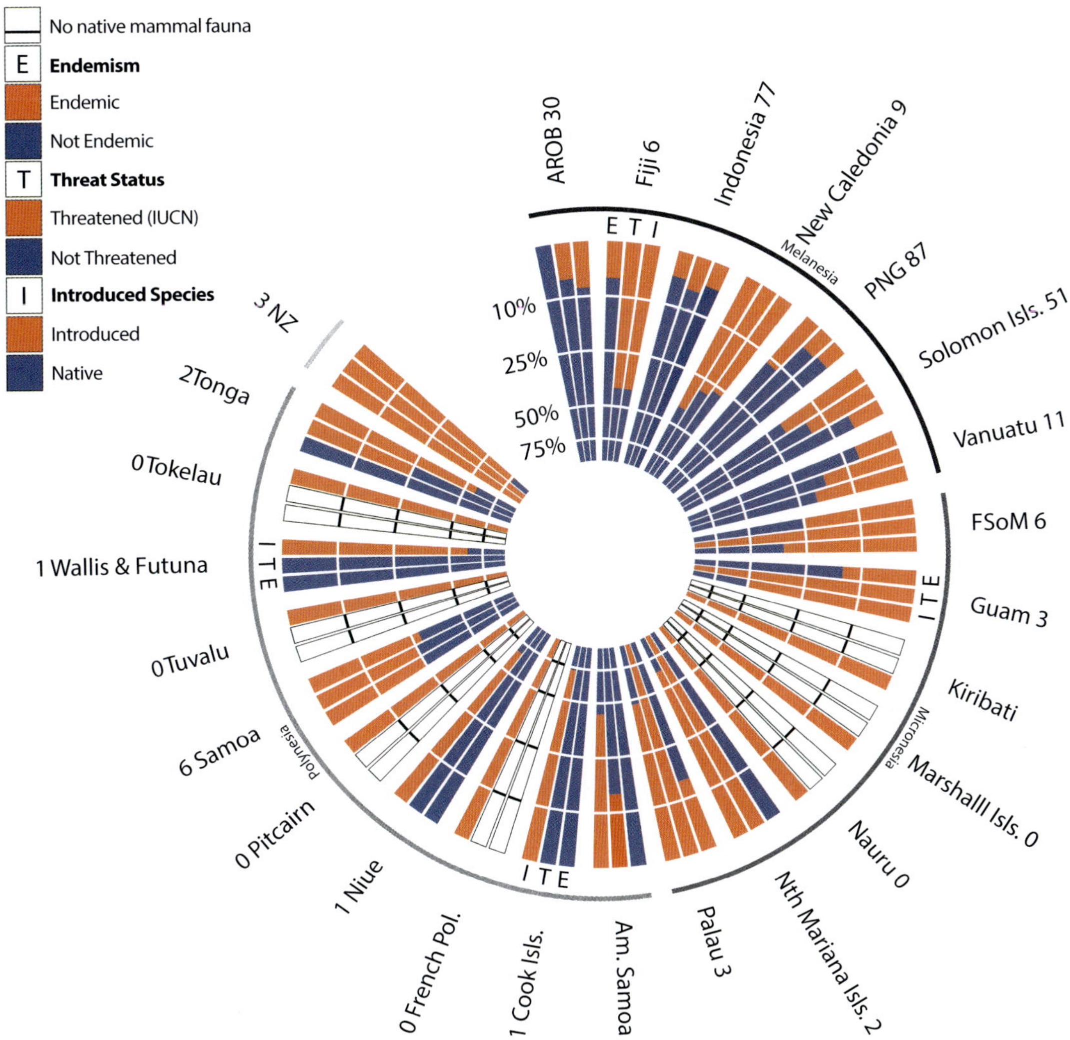

Patterns of indigenous mammal species richness and endemism, and introduced species richness, for the countries and territories of the South-west Pacific. Total indigenous species richness (including modern extinctions) is shown adjacent to each country or territory name. The clockwise order of the bars remains unchanged around the central axis of the plot.

Conservation of South-west Pacific mammals

Introduction

Island biotas, including mammals, have proven extremely vulnerable to extinction. Islands harbour disproportionately high numbers of endemic species and thus island conservation is incredibly important for global biodiversity. A significant hurdle to conservation of South-west Pacific mammals is the fact that basic data on distributions, population sizes and habitat preferences are severely lacking. For many species it is too difficult to understand the potential for impact from a given land use change, let alone implement measures to manage or mitigate those impacts. The paucity of fossil sites in the region further restricts the assessment of past mammal diversity, extinction and decline.

The settlement of Pacific islands by people brought hunting, habitat loss and species introductions that all contributed to declines and extinctions of endemic species. In the case of near Oceania, prehistoric anthropogenic impacts are poorly understood because of a scarcity of fossils that pre-date the arrival of humans. However,

Kwaio Chief Esau Kekeubata, Mt Tolobusu, Malaita, Solomon Islands. A large native rodent, probably a species of *Uromys* or *Solomys*, has been reported from Malaita. Esau recalls seeing this species (known locally as *kwete*) on a hunting trip in East Kwaio in 1996. In the early 1990s in the East Kwaio village of Naufe'e, senior Kwaio man Folofo'u revealed that in his youth he also had seen this large forest-dwelling rat with a tail like a rasp. Sustained efforts to locate *kwete* have sadly failed and it may be that it is now extinct (Tyrone Lavery).

deposits from Buka Island have provided one clear example of the loss of mammal diversity since the early Holocene (two murid rodents) (Flannery and Wickler 1990). Unfortunately, gaps in the archaeological sequence mean the drivers of these two extinctions are unclear. A total of five species of murid were present up until around 6700 years before present, and in the sequence from 2000 years before present, two of these had become extinct and several other mammals (*Rattus exulans*, *R. praetor*, *Phalanger breviceps*, *Sus scrofa* and *Canis familiaris*) had been introduced to the island.

The impacts of later human expansion into remote Oceania commencing some 3600 years before present have been better documented, especially for birds. Steadman (1995) estimated the extinction of birdlife from the tropical Pacific following initial human arrival exceeded 2000 species. The impacts on mammals are less well understood but, in Tonga at least, the pre-human mammal fauna 3500–3000 years before present included at least five bats, three of which subsequently vanished (Koopman and Steadman 1995). Similarly, the Pacific flying fox (*Pteropus tonganus*) was extirpated from Aitutakia and Ma'uke in Cook Islands (Walter 1990; Steadman 1991).

Subsequent colonisation by Europeans initiated secondary phases of extinction on many islands, sparked by increased or more intense habitat modification, new introduced species and further hunting. A fundamental problem for island species, including those of the South-west Pacific, is the fact that they are confined to small land areas surrounded by oceanic barriers. Restricted ranges render island species highly susceptible to environmental, climatic and anthropogenic change. Disturbance on an island at any spatial scale has a high chance of having an impact on an endemic species, whereas continental species are more likely to be more widespread or have access to refuges. Moreover, specialised insular adaptations and ecology (often including naïvety to predators) render many island biotas extremely vulnerable to introduced species (Manne *et al.* 1999; Karels *et al.* 2008).

The ripple effects of decline and extinction on islands can be profound because the loss of species from simple ecological networks can also mean the loss of critical ecosystem processes, functions and interactions between species (Wood *et al.* 2017). A great example of this can be found among mammals of the South-west Pacific. The Pacific flying fox (*Pteropus tonganus*) is a keystone pollinator and seed disperser. When flying foxes are abundant, they will forage by plucking fruits from trees and flying short distances to consume them in locations where they can avoid disturbance from other individuals. Population declines can reduce the potential for conflict between individuals and lead to a shift in this behaviour. Flying foxes instead become more likely to remain feeding in the fruiting tree and less likely to disperse their fruits and seeds. On Tonga, a population drop of *P. tonganus* below a threshold resulted in a shift in the proportion of seeds (from 58% to less than 1%) being distributed away from the tree (McConkey and Drake 2006).

The threatened species

Of the 184 mammals identified from the region under consideration here, 63 are classified as extinct or threatened (Vulnerable, Endangered or Critically Endangered) on the IUCN Red List. One hundred and nine species are classified as Near Threatened (13) or Least Concern (96) (IUCN 2021).

Eight species endemic to the South-west Pacific are feared extinct in historic times. Four species, all flying foxes (*Pteropus* spp.), have been officially classified as such under the IUCN Red List. Two were from Samoa (*Pteropus allenorum* and *P. coxi*), each known only from one or two specimens collected in mid-1800s and apparently not seen again. The other two were Micronesian: *P. tokudae* (Guam) last seen in the 1960s and *P. pilosus* (Palau) known only from a pair of specimens collected before 1880.

Five species presently classified as Critically Endangered are probably also extinct. Three of these are *Uromys* rodents (*U. emmae*, *U. imperator* and *U. porculus*), two of which were collected for the first and last time in the late 1800s and most likely driven to extinction by predation by feral cats. The fourth is the bat *Mystacina robusta* of Aotearoa New Zealand, last seen in the 1960s. The fifth is the Aru flying fox (*P. aruensis*), a poorly known species that is understood to have been last recorded in 1992 (a presumed trophy specimen).

An additional 15 species are classified as Critically Endangered, all but one of which are Melanesian. This includes the western long-beaked echidna (*Zaglossus bruinji*), and three murid rodents – *Uromys vika* and *Solomys ponceleti* (Solomon Islands), and *Uromys boeadii* (Biak). Three are monkey-faced bats (two Solomon Islands *Pteralopex* and *Mirimiri acrodonta* of Fiji). Three are the marsupials *Dorcopsis atrata* of highland Goodenough Island and *Spilocuscus wilsoni* of Biak-Supiori, and the insectivorous bat *Nyctophilus nebulosus* of New Caledonia.

Twenty-five species are classified as Endangered. Five are Melanesian murids: *Melomys matambuai* and *Pogonomys fergussoniensis* (Papua New Guinea); and *Solomys salebrosus*, *S. sapientis* and *Uromys rex* (Solomon Islands). Eight are large-bodied pteropodid bats of the genera *Pteropus* and *Pteralopex*. Three are marsupials from the Milne Bay islands of Papua New Guinea (*Dactylopsila tatei*, *Echymipera davidi* and *Phalanger lullulae*). The remainder are from families of insectivorous bat, and the pteropodid *Notopteris neocaledonica*.

Nineteen species are classified as Vulnerable, eight of which are large-bodied bats of the genera *Pteropus* and *Pteralopex*. Four are marsupials (*Dendrolagus inustus*, *Spilocuscus papuensis*, *Thylogale browni* and *T. brunii*), six are insectivorous bats and a further two are pteropodid bats. Ten species are Data Deficient.

Threats

The status of habitat loss, invasive species and hunting as key threats to South-west Pacific mammals today is clearly evident in data from the IUCN Red List of threatened species. For the indigenous mammals of Micronesia and Polynesia, the threats most commonly identified correspond to invasive and other problematic species, biological

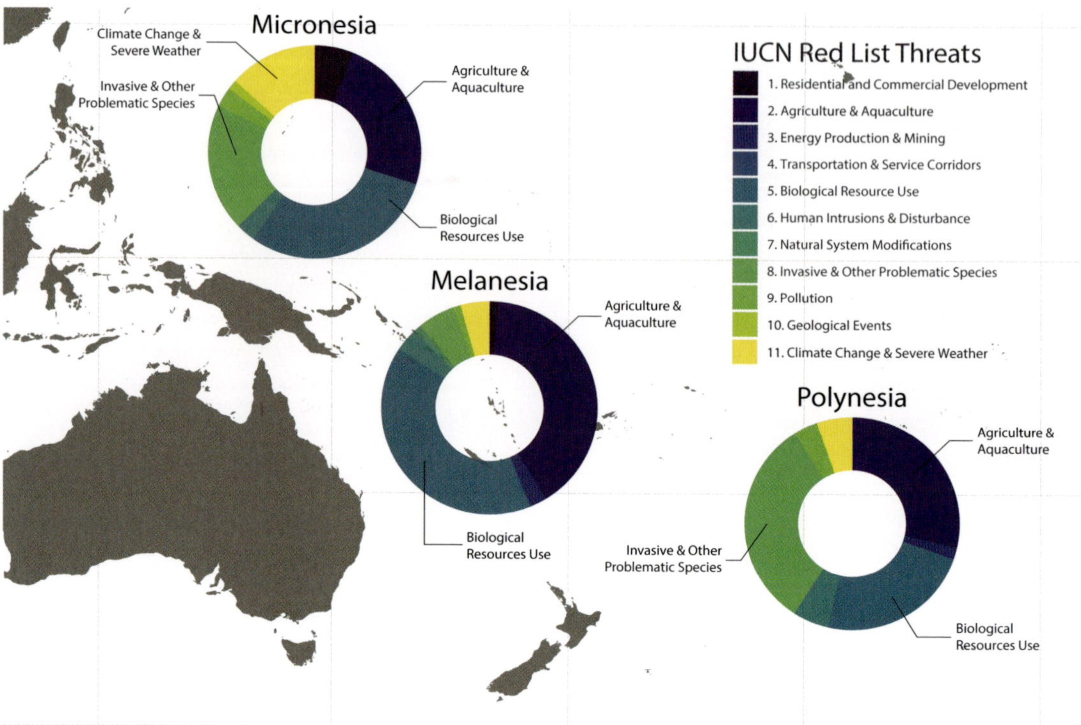

Threats to indigenous mammals of the South-west Pacific, according to data provided by the IUCN Red List (IUCN 2021). Pie charts represent the frequency with which various threat categories have been identified in IUCN Red List assessments for the indigenous terrestrial mammals of Micronesia, Melanesia and Polynesia (as a percentage of mammal species assessed).

resource use (including logging, wood harvesting and hunting) and agriculture (varying scales of habitat modification from shifting cultivation to plantations and industrial farming). In Micronesia, climate change and severe weather have also commonly been identified as threats to mammals (e.g. storms and flooding, droughts). Invasive species have been less commonly identified as an issue for Melanesian mammals. Instead, hunting and trapping (biological resources use), and annual and perennial non-timber crops (agriculture and aquaculture), featured as the two most commonly identified threats.

Forest disturbance and deforestation

Forest loss (reduced quantity) and forest disturbance (reduced quality) are key threats to South-west Pacific mammals. Between 1990 and 2010, deforestation in the South-west Pacific (excluding Aotearoa New Zealand) occurred at a rate of 0.4% loss, over three times the global average of 0.14% (SPREP 2016). At a country or territory level, the Secretariat of the Pacific Regional Environment Programme (2016) showed that remaining forest cover as of 2013 varied from 0% (Nauru and Tokelau) to 92% in the Federated States of Micronesia (Table 2).

Considering habitat loss at a biodiversity hotspot level, Mittermeier *et al.* (2004) estimated that remaining habitat extents in the four hotspots of the South-west Pacific ranged between 5% (New Caledonia) and 30% (East Melanesian Islands) (Table 3).

Commercial logging is a major driver of forest disturbance and loss in Melanesia. In Papua New Guinea and Solomon Islands, companies selectively log viable trees and abandon damaged forests to natural regeneration. This has a limited effect on country-wide estimates of forest cover; for example, an estimated 79% of Solomon Islands remained forested in 2013 (SPREP 2016) despite

Table 2. Estimates of forest cover in square kilometres and as a percentage of land area for countries and territories of the South-west Pacific.

Country	Forest (km²)[1]	% land area[1]
American Samoa	180	89
Cook Islands	160	65
Fiji	10 140	56
Federated States of Micronesia	640	92
French Polynesia	1550	42
Guam	260	47
Kiribati	120	15
Marshall Islands	130	70
Northern Mariana Islands	300	66
New Caledonia	8390	46
Nauru	0	0
Niue	190	72
Aotearoa New Zealand[2]	95 800	36
Papua New Guinea (country wide)	287 260	63
Pitcairn	40	83
Palau	400	88
Samoa	1710	60
Solomon Islands	22 130	79
Tokelau	0	0
Tonga	90	13
Tuvalu	10	33
Vanuatu	4400	36
Wallis and Futuna	60	42

Figures provided by 1. Secretariat of the Pacific Regional Environment Programme (SPREP) (2016); 2. Global Forest Watch (2020)

Table 3. Estimates of original extent (tropical and subtropical moist broadleaf forests, and temperate broadleaf and mixed forests), remaining habitat and percentage of remaining habitat for biodiversity hotspots of the South-west Pacific (Mittermeier *et al.* 2004).

Biodiversity hotspot	Original extent (km²)	Remaining habitat (km²)	Percentage remaining habitat
East Melanesian Islands	99 384	29 815	30
New Caledonia	18 972	5122	5
Aotearoa New Zealand	270 197	59 443	22
Polynesia–Micronesia	47 239	10 015	21

widespread logging there. The majority of lowland primary forests in the country have now been commercially logged at some point in time, with 53% of Solomon Islands land area that is less than 400 m elevation now within 1 km of a logging road (Global Witness 2018). Commercial logging leaves behind a residual stand of severely damaged immature trees, altered microclimates, and

Southern Santa Isabel Island, Solomon Islands. 53% of the land area below 400 m elevation in Solomon Islands is now within 1 km of a logging road (Tyrone Lavery).

deteriorated environmental and soil conditions (Katovai *et al.* 2015). This disturbance has been found to be most detrimental to endemic mammals, because logging is more likely to damage larger proportions of their limited distributions, and because endemic species may be more sensitive to the loss of specialised habitat resources (e.g. tree hollows) (Lavery *et al.* 2020a).

Logging operations regularly revisit logged sites two to three decades later to harvest regenerating trees. This causes further deteriorations in species diversity, forest structure and environmental conditions. A more recent development in parts of Melanesia (such as West New Britain and New Ireland in Papua New Guinea) has seen the complete conversion of forests to oil palm plantations (Buchanan *et al.* 2008). This same type of commercial logging is a less-prominent threat to the islands of Micronesia and Polynesia. Nonetheless, rates of deforestation are high in some countries, driven in part by agricultural activities such as commercial cropping of kava, taro, copra and cocoa.

Hunting

Many countries and territories of the South-west Pacific are undergoing rapid human population growth, especially in the biodiverse nations of Papua New Guinea, Solomon Islands and Vanuatu where annual rates of increase are 2.0, 2.6 and 2.4% per annum respectively (World Bank). Often coupled with increasing populations is an intensification of the use of natural resources, and the erosion of traditional management systems that ensured their sustainable use.

Bismarck common cuscuses (*Phalanger breviceps*) prepared and ready to be cooked at Kainake, south Bougainville Island. This species is an important hunting resource for communities in parts of Papua New Guinea and the Solomon Islands archipelago. Despite the widespread hunting focus, it remains common in most areas, even close to villages (Tyrone Lavery).

Admiralty cuscus (*Spilocuscus kraemeri*), Manus Island, Papua New Guinea. This species is one of very few large-bodied mammals present on Manus, and as such it represents an important source of protein for people on the island and is a primary target for hunters (Elodie Van Lierde).

A prime example of this concerns David's echymipera (*Echymipera davidi*) of Kiriwina Island, Papua New Guinea. Cultural protections for this species prevented the majority of the population on Kiriwina from consuming bandicoots. However, a burgeoning population and influx of outsiders unaware of, or unbound by, cultural protections have apparently led to a significant increase in hunting pressure and anecdotal decline of the bandicoot (D. Mitchell, pers. comm.).

On Manus Island, Papua New Guinea, the endemic Admiralty cuscus (*Spilocuscus kraemeri*) is a species favoured by hunters. Customary conservation (*tambu*) areas continue to be used as a method for regulating hunting. However, long-term population studies have indicated they may not be enough to conserve self-sustaining populations of the species (Lamaris and Whitmore 2018). Breaking of the *tambu* protections is also common (J. Lamaris, unpublished data).

Large-bodied, endemic mammals with low reproductive output face the greatest risk from over-hunting, and many Pacific Island flying foxes possess these traits. Females tend to reach sexual maturity around 18–24 months and have low birth rates, usually of one young per year (Pierson and Rainey 1992). Endemism among Pacific Island *Pteropus* is also extraordinarily high, and many species roost conspicuously in large groups, making them a vulnerable target for hunters. Globally, the Pacific islands region has the highest proportion of bats known to be hunted (Mildenstein *et al.* 2016; Wheatley *et al.* 2020). Bats are hunted in American Samoa, Commonwealth of the Northern Mariana Islands (CNMI), Cook Islands, Federated States of Micronesia, Fiji, Guam, New Caledonia, Niue, Palau, Samoa, Solomon Islands and Vanuatu (Mildenstein *et al.* 2016; Wheatley *et al.* 2020). Accordingly, only six of 35 species of *Pteropus* from the region are classified as Least Concern under the IUCN Red List. Prior to their inclusion under the Convention on International Trade in Endangered Species (CITES), flying foxes in Samoa and Palau were heavily impacted by over-exploitation for the export trade to Guam, where they are considered a delicacy. Poaching continues to pose some threat. An additional dynamic influencing flying fox hunting in parts of Melanesia is the value of their canine teeth for use as traditional currency. On Makira, Solomon Islands, where two species of flying fox exist, the value of teeth apparently results in a hunting preference towards the larger of the two species (*Pteropus tonganus*) that possesses larger canines (Lavery and Fasi 2017). Flying fox teeth also are used for currency or decoration in the Santa Cruz Islands of Solomons Islands and Banks Islands of Vanuatu (T. Lavery, pers. obs.).

Introduced species

The first humans to arrive in Micronesia and Polynesia more than 3000 years ago introduced a range of plants and animals that had impacts on native biotas. Pigs, dogs, and Polynesian rats (*Rattus*

The canine teeth of large flying foxes (*Pteropus* spp.) continue to be used as traditional currency and decoration in parts of Melanesia. On Makira, Solomon Islands, this remains more valuable than modern currency (Tyrone Lavery).

Pigs (*Sus scrofa*) were widely introduced in the South-west Pacific prehistorically. Their digging can cause significant damage to the environment, especially near wetlands and waterways (Angus Fraser).

exulans) in particular, probably had significant implications for the incumbent island mammals. Following European colonisation, hundreds more species have been introduced, both intentionally and accidentally.

Island ecosystems are particularly vulnerable to the effects of invasive species. The mammals of the South-west Pacific evolved in the absence of mammalian predators; thus their introductions (e.g. feral cats, mongoose and stoats) have been devastating. On many of the islands of Polynesia and Micronesia, the diversity of introduced mammals now rivals or outweighs the diversity of indigenous mammals. Aotearoa New Zealand is an obvious example. The incumbent mammal fauna there included just three species of insectivorous bats. Human settlement brought species such as the Polynesian rat (*Rattus exulans*) and dogs (*Canis familiaris*) that wrought initial impacts on native biotas that had evolved in isolation. However, European colonisation greatly overshadowed these initial introductions. At least 31 species of introduced mammal have been reported from Aotearoa New Zealand and species such as brushtail possums (*Trichosurus vulpecula*) and stoats (*Mustela erminea*) have caused irreparable losses of biodiversity (King and Forsyth 2021).

Climate change

Islands of the South-west Pacific are acutely vulnerable to anthropogenic climate change. Greater climate variability, rising sea levels, increasing frequencies of extreme weather events (e.g. cyclones), and elevational shifts in species distributions will all pose great challenges to South-west Pacific mammal conservation in the coming decades.

Extreme weather events have the potential for significant impacts on short-range endemic flying foxes restricted to small islands. The initial impacts of storms themselves can result in animal mortality, and the reduced foraging resources of damaged forests have ongoing impacts. Hunting can also be exacerbated following storms, as animals are forced to relocate to suboptimal roost sites, or become more exposed and obvious among denuded vegetation.

Numerous mammals of the region are high-elevation species limited to island mountain tops. *Mirimiri acrodonta* (Taveuni, Fiji), *Pteralopex anceps* (Bougainville), *P. pulchra* (Guadalcanal, Solomon Islands) and *Dorcopsis atrata* (Goodenough, Papua New Guinea), for example, are all restricted to the uppermost montane forests of their small island distributions. On Taveuni, the cloud forest that provides suitable habitat for *M. acrodonta* exists along a 10 km stretch of the island's main range, with an extent of occurrence of just 99 km^2 (Scanlon 2019). Shifts in temperature and altitudinal distributions of plants, forest types and habitats thus have huge potential to increase the extinction risk of such species by reducing the sizes of montane zones and habitats.

Monotremata

Monotremes are the only egg-laying mammals. They are restricted to Australia and New Guinea and there are five living species, in the family Tachyglossidae under the genera *Tachyglossus* (one species) and *Zaglossus* (three species), and the monotypic genus *Ornithorhynchus* in the Ornithorhynchidae.

Only a single island within the region covered by this book has a voucher-backed record of a monotreme. That is of *Zaglossus bruijnii* from Salawati, which lies immediately west of New Guinea's Vogelkop Peninsula. Other unconfirmed records of monotremes exist for the region, including that of Menzies (2011) for *Tachyglossus aculeatus* from Salawati, and *Zaglossus* from the nearby island of Waigeo. Aplin and Pasveer (2006) revealed that *T. aculeatus* was once present on Kobroor in the Aru Islands, but may have gone extinct there as the climate became wetter since the Last Glacial Maximum. Salawati and the Aru Islands lie on the shallower continental shelf and so would have been part of the Meganesian landmass (combining New Guinea and Australia) during the Last Glacial Maximum. Thus the presence of monotremes there is not surprising. Reports of *Zaglossus* on Biak-Supiori (which was apparently not connected to New Guinea at lower sea levels) are intriguing but yet to be substantiated.

Tachyglossidae

Zaglossus

WESTERN LONG-BEAKED ECHIDNA

***Zaglossus bruijnii* (Peters and Doria, 1876a)**

zah'-glos-us brue'-nee-ee

Thomas (1907) reported that Walter Goodfellow collected two long-beaked echidnas from Salawati, off the Vogelkop Peninsula in October 1906. Thomas used one of these, an old female (NHMUK 7.9.5.1), as the holotype for his proposed species *Acanthoglossus goodfellowi*, now recognised as a synonym. On a visit to Salawati in the late 1950s, Bergman (1961) recorded information that long-beaked echidnas did not occur on the island, but lived in the Tamrau Mountains of the nearby Vogelkop Peninsula of New Guinea. However, Van Deusen and George (1969) considered this information suspect and thought previous records of long-beaked echidnas from Salawati should not be dismissed.

In the mountainous northern part of Salawati in 1992, local people were familiar with *Zaglossus*, but said that it occurred only in the flatter southern part of the island where there was soft soil. The most recent sighting of *Zaglossus* from the island was apparently of an animal kept briefly in captivity near Kampong Walyam, southern Salawati, in 1975. It was reportedly caught in the vicinity of that village.

There are two further reports of *Zaglossus* from the region. Menzies (2011) included it among the faunas of Waigeo, but without further detail. On Supiori in September 1992, a resident originally from the flat-lying northern area of the island indicated that over his lifetime he had three encounters with an animal he called *landuk Irian* (*landuk* being the Bahasa Indonesian word for hedgehog). When shown a photograph of *Zaglossus*, he immediately identified it as the animal he had seen. He had encountered it near the northern slope of the mountains, on the last occasion in 1975. Unfortunately no one else from the island was able to corroborate with similar observations, and the animal seemed entirely unfamiliar.

Information on the biology of this species, and *Zaglossus* species more generally, is sparse. Western long-beaked echidnas occur from primary forests close to the coastline into high montane forests up to 2500 m elevation. Pattiselanno *et al.* (2021) recently documented three individuals from the Teluk Bintuni Regency of West Papua. They were found in lowland forests dominated by merbau (*Intsia bijuga*), pulai (*Alstonia scholaris*), nyatoh (*Palaqium* sp.) and medang (*Litsea timoriana*) close to small creeks. The sites were muddy and the forest floor was covered by scrub, fallen and rotting trees, and hollow logs that were suitable for echidna burrows. The curvature of the rostrum in *Z. bruijnii* is better suited to vertical probing of the soil rather than low-angle feeding and prying as is practised by short-beaked echidnas (*Tachyglossus aculeatus*) (Murray 1978). Where long-beaked and short-beaked echidnas co-occur, their feeding signs can be differentiated by the fact that *Zaglossus* produce holes deeper than 5 cm where they have inserted their rostrum into the soil (Opiang 2009). Similar to *Z. burtoni*, the diet of western long-beaked echidnas probably includes earthworms, beetle larvae and invertebrates found in the leaf litter, soil, and underneath rotting logs and debris (Menzies 2011). A note entered in the records for an American Museum of Natural History *Z. bruijnii* specimen taken from Arfak Mountains in 1913 (AMNH 33949) states: 'Taken at night when this animal goes out to hunt. Apparently rare, local, found singly.'

A captive western long-beaked echidna (*Zaglossus bruijnii*). This is the only monotreme known to occur in the region, apparently recorded from Salawati in 1906. It may be extinct there today (Klaus Rudloff).

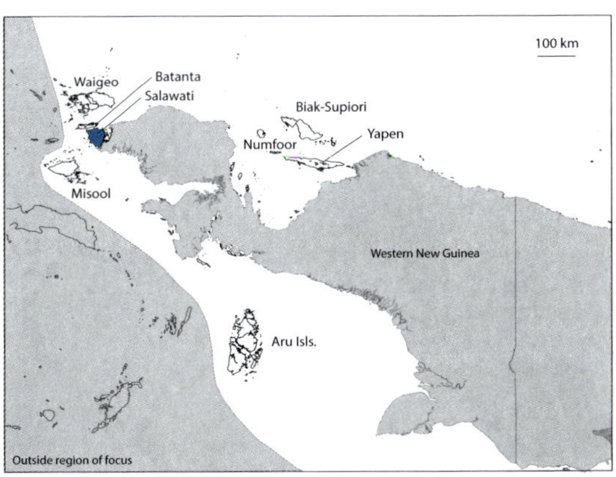

Nothing else is known of reproductive biology or behaviour. A radio-telemetry study focused on *Z. burtoni* in the Crater Mountains of Simbu Province, Papua New Guinea, found that animal home ranges varied in size between 10 and 168 ha (Opiang 2009). Dens were most often in underground burrows, and the animals were never observed foraging during daylight. One female was found to be lactating in April 2002 and again in April 2005.

The western long-beaked echidna is a large-bodied mammal, presumably with a slow rate of reproduction. The species has a limited distribution and is suspected to be in serious decline; the population on Salawati may already be extinct.

Zaglossus bruijnii			
Size*			
	Mean	**Range**	**Sample**
Head–body	674.0	630–730.0	4
Tail	–	–	–
Hindfoot	–	60	1
Ear	–	–	–
Weight	8779.3	6430.0–12 520.0	4
* From Laurie (1952) and Pattiselanno *et al.* (2021)			

Identification The western long-beaked echidna is unlike any other mammal recorded from the region. The fur between the spines on the dorsum is relatively short and pale brown in colour, some of the spines of the dorsum are tipped black, and there are visible spines on the underside of the animal.

Synonyms *Bruynia tridactyla* Du Bois, 1882
Proechidna villosissima Du Bois, 1884
Proechidna nigroaculeata Rothschild, 1892
Acanthoglossus goodfellowi Thomas, 1907
Zaglossus bruijni gularis Rothschild, 1922
Zaglossus bruijni pallidus Rothschild, 1922
Zaglossus bubuensis Laurie, 1952

Subspecies None

Distribution Salawati (?extinct)

Extralimital New Guinea

Conservation status (IUCN) Critically Endangered

Etymology Gen.–very + tongue (Gr.); Sp.–for the Dutch merchant A. A. Bruijn, who sent the holotype to Europe

Dasyuromorphia

Dasyuridae

The Dasyuridae is a family of carnivorous and insectivorous marsupials. The largest living species – the Tasmanian devil *Sarcophilus harrisii* – can weigh up to 14 kg, but the bodyweights of the majority of species lie between 17 and 80 g. Dasyurids reach their greatest diversity in Australia (with around 60 species), but New Guinea also supports a diverse fauna, including endemic genera such as *Murexia* and *Myoictis*. This New Guinea fauna is vastly understudied.

Few species are found in the South-west Pacific and all but one are restricted to the land bridge islands that were connected to New Guinea during the Last Glacial Maximum. The exception is *Murexia murex* from high elevation on the island of Normanby in the D'Entrecasteaux Group. It is only known from a single specimen collected during the Fifth Archbold Expedition in 1956. It has unusual pelage that differentiates it from other *Murexia*, including a narrow white stripe from the chin to the mid-ventral region. The type localities of two Melanesian dasyurid species (*Murexia longicaudata* and *Myoictis wallacei*) lie within the region under consideration here, so the study of South-west Pacific dasyurids is important from a taxonomic perspective. Unfortunately, all species have rarely been documented in the region, and ecological data specific to the South-west Pacific are sparse.

Dasyurus

NEW GUINEA QUOLL

Dasyurus albopunctatus Schlegel, 1880

daz'-ee-yue'-rus al'-boh-punk-tah'-tus

The mandibles of a New Guinea quoll were collected on Yapen by Jared Diamond in 1983 and are held in the collection of the American Museum of Natural History (AMNH 266892). The specimen was collected along with large series of trophy jaws from *Echymipera clara*, *E. kalubu*, *Peroryctes raffrayana*, *Myoictis melas* and *Murexia longicaudata*. The bones of this species were also reported from Holocene deposits on Malai Island (near Umboi) (Lilley 1986). Heinsohn (2010b) was unable to determine whether the remains were derived from human translocation to the island, or a chance arrival via over-water dispersal. Aplin and Pasveer (2006) referred three partial lower jaws and a proximal femur from archaeological deposits in the Aru Islands to this species. The presence of New Guinea quoll there was considered surprising, given strong evidence for a dominance of savannah habitats during the late Pleistocene. It may well be that all cases represent instances of living animals or their bones being translocated from New Guinea to offshore islands.

The New Guinea quoll is widespread on the mainland, most commonly in forested areas above 1000 m, and rarely occurs at lower altitudes (Flannery 1995a). People of New Guinea consider it a fierce predator that will attack prey as large as or larger than itself. Brass (1956), and Flannery (1995a) both reported instances of New Guinea quolls being lured and trapped using the bodies of *Rattus* spp. and Mianmin hunters reported it will enter villages to hunt rats. Breeding occurs throughout the year; litter size is four to six. A subadult female and subadult male were trapped in Papua New

Bones of the New Guinea quoll (*Dasyurus albopunctatus*) have been recorded from several islands close to New Guinea. It may be that these represent animals transported from the mainland by people (P.A.Woolley and D. Walsh).

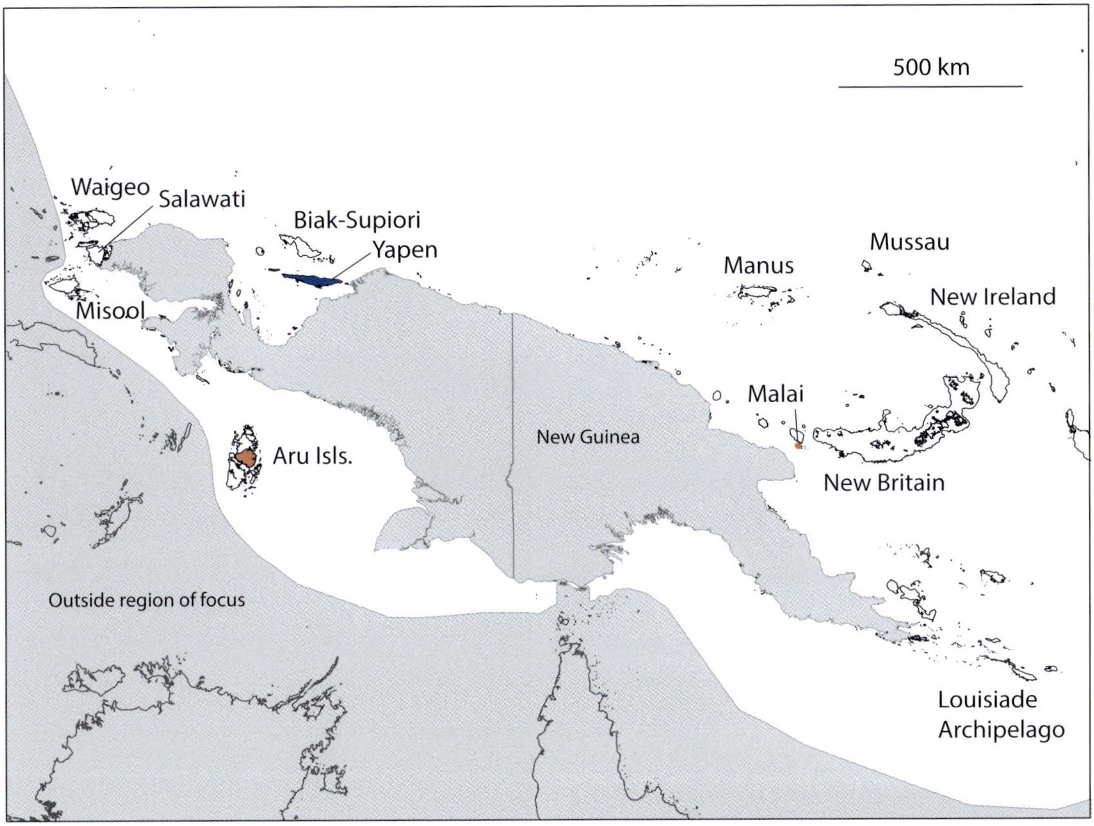

Dasyurus albopunctatus			
Size*			
	Mean	**Range**	**Sample**
Head–body	267.5	231.0–298.0	13
Tail	256.8	221.0–290.0	13
Hindfoot	47.8	43.0–54.0	13
Ear	29.1	27.0–31.9	13
Weight	645.0	580.0–710.0	3
* From AM specimens and Laurie (1952)			
Identification The largest dasyurid in the region, body size and presence of white spots on the dorsum will serve to differentiate this species from all other mammals of the region.			
Synonyms *Dasyurus fuscus* Milne-Edwards, 1880 *Dasyurus daemonellus* Thomas 1904a			
Subspecies None			
Distribution Yapen			
Extralimital New Guinea			
Conservation status (IUCN) Near Threatened			
Etymology Gen.–hairy-tailed (Gr.); Sp.–white-spotted (L.)			

Guinea's Telefomin area in January 1984 and March 1986 respectively (Woolley 1994; Flannery 1995).

The New Guinea quoll is listed as Near Threatened because, despite being widespread and locally abundant, expanding agriculture and hunting with dogs have apparently led to population declines at several localities.

Murexia

SHORT-FURRED DASYURE

Murexia longicaudata (Schlegel, 1866)

mue-rex'-ee-ah lon'-gee-kaw-dah'-tah

The short-furred dasyure is widespread in the lowlands of New Guinea. The holotype was collected in 1865 at Wonumbai (sic Manumbai), Kobroor, Aru Islands by Baron C. B. H. von Rosenberg, a military cartographer. This is a mounted specimen held by the Naturalis Biodiversity Center, Leiden (RMNH 35135, *v.* spec. a) and remains the only known specimen from the Aru Islands. It is unusually small and there are numerous abnormalities, including features of the cranium and dentaries. Woolley (2019) concluded it is either a highly abnormal individual or the skull and dentaries have been mismatched. However, the premolar teeth are spaced and the footpads are fused, consistent with specimens from New Guinea and Yapen Island considered to be conspecific (Woolley *et al.* 2020). Attempts to collect further specimens on the Aru Islands in 1992 were unsuccessful (Woolley 2019), and it was not encountered in the Kobroor fossil deposits examined by Aplin and Pasveer (2006).

The short-furred dasyure is known from Yapen via two specimens. One is the holotype of *Phascogale maxima* (Stein 1932) (ZMB 44328), and the second was collected on the Dawai River (elevation 30 m), 1.5 km north-east of Sumberbawa on 1 November 1962. These Yapen animals, both males, are exceptionally large. Woolley *et al.* (2020) demonstrated *M. longicaudata* comprised three to four distinct genetic clades corresponding to described subspecies. Reinstatement of *M. murex* was recommended, but further sampling was needed to properly resolve the status of *M. aspera* and *M. maxima* (a classification adhered to here).

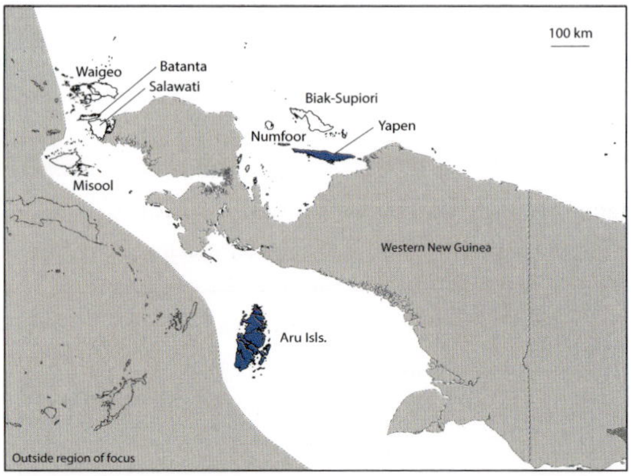

Murexia longicaudata			
Size*			
	Mean	**Range**	**Sample**
Head–body	248.5	234.0–262.0	2
Tail	222.5	215.0–230.0	2
Hindfoot	43.5	40.0–47.0	2
Ear	23.5	23.0–24.0	2
Weight	–	434.0	1
* From Stein (1932) and BPBM 22109 (ex AMNH 221626) (Yapen)			
Identification A large, powerful dasyurid. The fur is short, sleek and rather uniformly coloured brown. It is far larger than the sympatric *Sminthopsis virginiae* and lacks the dorsal striping of *Myoictis melas*.			
Synonyms *Murexia longicaudata parva* Laurie, 1952			
Subspecies *Murexia longicaudata longicaudata* (Schlegel, 1866) Aru Isls. *Murexia aspera* (Thomas, 1913a) New Guinea central cordillera *Murexia maxima* (Stein, 1932) northern New Guinea and Yapen Is.			
Distribution Aru Isls, Yapen			
Extralimital New Guinea			
Conservation status (IUCN) Least Concern			
Etymology Gen.–meaning and derivation unknown; Sp.–long-tailed (L.)			

Short-furred dasyure (*Murexia longicaudata*), mainland New Guinea. This species is widespread in New Guinea's lowland forests and extends to some of the associated land bridge islands (Michael Pennay).

Little information is available on the biology of short-furred dasyure on the Aru Islands or Yapen. A note accompanying the holotype of *M. l. maxima* stated it was trapped on a low ridge in deep jungle. At Mt Missim, Morobe Province, Papua New Guinea, tracking of two adult males found both occupied nests composed of dry leaves located 4.0–4.5 m above the ground. One was positioned in a leaf axil of a *Pandanus* tree and the other in a tangle of lianas (Woolley 1989). A subsequent study at the same site found that the most common dietary items for this species (frequency of occurrence in scat and stomach samples) were coleopterans (94.5%), orthopterans (78.2%), hemipterans (33.6%) and lepidopterans (33.6%) (Grossek *et al.* 2010). Bird eggshell was also occasionally detected.

Reproduction occurs throughout the year and lactating females and juveniles can be found in most months (Woolley 2004). One female had two litters 9 months apart, one in the wild in January and one in captivity the following September. The number of young to a litter ranges between two and four (mean 3.8), and females lactate for a duration of 3.5 months (Woolley 2004). The short-furred dasyure has a wide distribution, no major threats are known, and it is considered unlikely to be declining at a rate that would qualify the species for listing in a threatened category.

The lesser short-furred dasyure (*Murexia murex*) has only been documented from the region on a single occasion, in 1956 at 820 m elevation, Mt Pabinama, Normanby Island (Madison Erin Mayfield)

LESSER SHORT-FURRED DASYURE

Murexia murex (Thomas, 1913b)

mue'-rex

The lesser short-furred dasyure is known from the region by a single specimen caught on 8 May 1956 at Mt Pabinama, 820 m elevation, Normanby Island. The adult male is a skin and skull (M-159473) assigned to *Murexia longicaudata* by Brass (1959) and entered in the AMNH catalogue as such. The teeth of this animal were not greatly worn, suggesting that it was a relatively young animal.

Van Dyck (2002) placed this specimen in *M. melanurus* based on morphological features. However, Woolley *et al.* (2020) found it belonged in the *M. longicaudata* species group which comprised three to four genetically distinct clades. Genetically, specimens corresponding to *M. murex* Thomas 1913, from Mt Missim and Big Wau Creek (Morobe Province, Papua New Guinea) were highly distinct and formed a sister group to all exemplars of *M. longicaudata*, as well as *M. naso* and *M. melanurus*. This *M. murex* form was also distinctly smaller than other specimens included under *M. longicaudata* and reinstatement of the species was recommended. Guided by morphological data, Woolley *et al.* (2020) tentatively placed the single Normanby specimen within this *M. murex* group, but highlighted the unusual pelage (including a narrow, 4–5 mm white stripe running from the

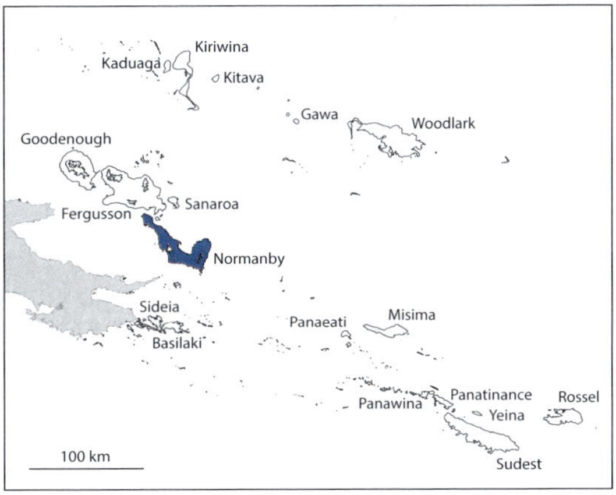

Murexia murex			
Size*			
	Mean	**Range**	**Sample**
Head–body	–	165	1
Tail	–	165	1
Hindfoot	–	31.3	1
Ear	–	18.0	1
Weight	–	113	1
* From AMNH 159473 (Normanby)			
Identification *Murexia murex* is rather uniformly brown with a narrow, 4–5 mm white stripe running from the chin to the mid-ventral region. The premolar teeth are spaced.			
Synonyms None			
Subspecies Monotypic			
Distribution Normanby			
Extralimital Mainland New Guinea			
Conservation status (IUCN) Least Concern (assessed under *Murexia longicaudata*)			
Etymology Sp.–meaning and derivation unknown			

chin to the mid-ventral region). We have followed this tentative classification here; however, additional data including genetic sequencing are needed to confirm the identity of Normanby *Murexia*.

The terrain at Mt Pabinama where the single specimen was collected consisted of steep narrow spurs alternating with deep ravines supporting cloud forests to 18 m in height and dominated by *Xanthomyrtus*, *Calophyllum*, *Schizomeria*, *Ilex* and *Syzygium* (Brass 1959). The understorey was sparse, with heavy growth of bryophytes and gullies lined with ferns including *Cyathea* sp. The conservation status of *M. murex* has not been individually assessed.

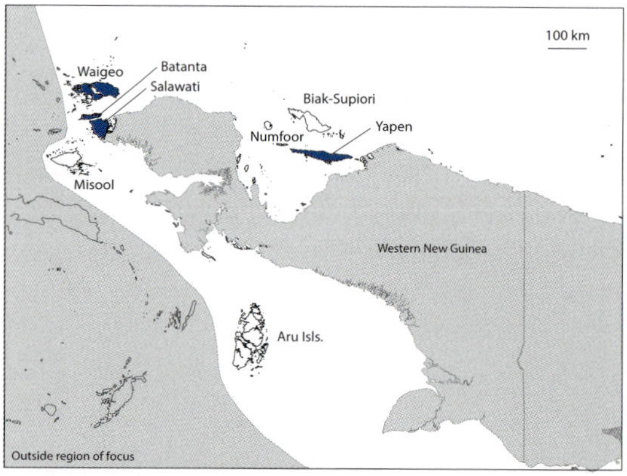

Myoictis melas			
Size*			
	Mean	**Range**	**Sample**
Head–body	206.6	190.0–213.0	5
Tail	–	–	–
Hindfoot	39.7	36.0–44.0	6
Ear	–	–	–
Weight	220.0	172.0–255.0	4
* From Salawati, Waigeo and Yapen (head–body, hindfoot) and New Guinea (bodyweight) (Woolley 2005)			
Identification Differs from other species of mammal in the region in having three dorsal stripes and a well haired, tapering tail with a dorsal crest.			
Synonyms *Phascogalea thorbeckiana* Schlegel, 1866 *Chaetocercus bruijnii* Peters, 1875 *Phascogale melas bürgersi* Stein, 1932 *Phascogale melas senex* Stein, 1932			
Subspecies None			
Distribution Batanta, Salawati, Waigeo, Yapen			
Extralimital New Guinea			
Conservation status (IUCN) Least Concern			
Etymology Gen.–mouse-weasel (Gr.); Sp.–black (Gr.), the holotype was melanistic			

Myoictis

MÜLLER'S THREE-STRIPED DASYURE

Myoictis melas (Müller, 1840)

mie'-oh-ik'-tis me'-las

Müller's three-striped dasyure may include more than a single species and further taxonomic research is needed. There are several records of this species from the region. Thomas (1887a) recorded an adult female specimen in alcohol, collected by Bruijn on Salawati and held in the collections of the Museo di Storia Naturale, Genoa. Four syntypes of *Phascogale thorbeckiana* collected by Bernstein on Salawati are held by the Naturalis Biodiversity Center, Leiden. The Genoa Museum also holds a single male specimen in alcohol, collected by Beccari at Ansus, Yapen. Finally, the Musée d'Histoire Naturelle, Paris, holds a specimen apparently collected on Waigeo by Laglaise in the late 19th century. In October 1992, an animal thought to be this species was spotted in the mid-morning near the village of Yeinanas on Batanta. It scampered across a track that ran through an area of highly disturbed forest and disappeared under a pile of leaf matter into a hole in the ground.

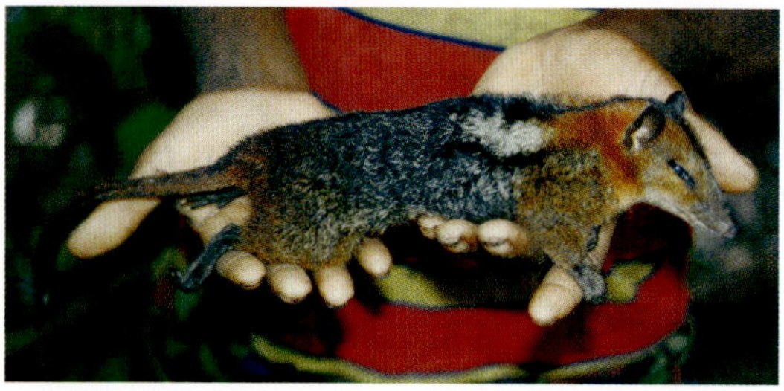

Müller's three-striped dasyure (*Myoictis melas*). This widespread species of *Myoictis* occurs on several islands on the western margin of New Guinea (P. A. Woolley).

Müller's three-striped dasyure is the most widespread species of *Myoictis*, found from sea level to 1800 m elevation (Menzies 2011). Limited information is available on species biology. Menzies (2011) indicated all species of *Myoictis* are diurnal or at least crepuscular, and feed on small vertebrates and insects. The Müller's three-striped dasyure is known to visit village houses to hunt rodents. Females have six nipples and Menzies (2011) postulated reproduction occurs throughout the year. Using a combination of museum and field studies, Woolley (2004) identified pregnant or lactating females present in January, February, March, May, July, September and October. However, the specimens examined were considered to represent more than one species of *Myoictis* and reappraisal was recommended.

Müller's three-striped dasyure is widely distributed, no major threats are known, and it has been deemed unlikely to be declining at a rate that warrants listing under a threatened category.

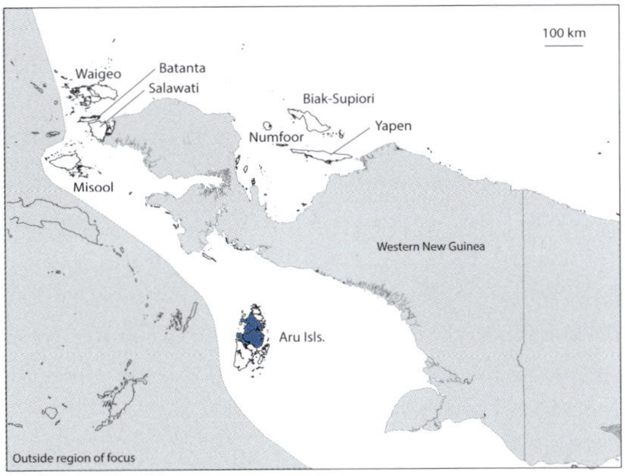

WALLACE'S THREE-STRIPED DASYURE

Myoictis wallacei Gray, 1858

wol'-liss-ee

Wallace's three-striped dasyure is found from low-elevation forests and secondary habitats around 30 m elevation up to 923 m elevation. The holotype was collected by Alfred Russel Wallace in the Aru Islands in 1857. This specimen and another, also collected from the Aru Islands by H. Kühn, are held by the Museum of Natural History, London. The Naturalis Biodiversity Center, Leiden, also holds one specimen collected by C. B. H. von Rosenberg in 1865, and another five specimens from Wokam and Kobroor are held by the Australian National Wildlife Collection, collected in October 1992 by Dr Pat Woolley (Woolley 2005). Wallace's three-striped dasyure is different from other species of *Myoictis* in possessing a bushy tail that is densely long-haired to the tip. Hairs on the top and sides of the tail are up to 20 mm long and the underside is flattened and almost bare (Woolley 2005).

Myoictis wallacei			
Size*			
	Mean	**Range**	**Sample**
Head–body	210.2	200.0–222.0	5
Tail	–	–	–
Hindfoot	38.0	34.0–39.9	7
Ear	–	–	–
Weight	224.6	192.0–245.0	5
* From the Aru Isls (Woolley 2005)			
Identification The combination of a bushy, reddish-haired tail and presence of three dorsal stripes serve to distinguish this species from all other mammals in the region.			
Synonyms *Phascogale pilicauda* Peters and Doria, 1881			
Subspecies None			
Distribution Aru Isls			
Extralimital New Guinea			
Conservation status (IUCN) Least Concern			
Etymology Sp.–for Sir Alfred Russel Wallace, collector of the holotype			

Woolley (2001) was unsuccessful at trapping this species on Kobroor, Aru Islands, in October–November 1992. However, five individuals were caught with the help of local hunters. All individuals had been active in the late afternoon. They were said to live primarily on the ground (retreating to rock piles and tree hollows for refuge), and were active in mornings and evenings when they hunt insects and lizards (Woolley 2005). Animals kept in captivity were similarly active for a period in the morning and again in the evening. They consumed geckos and freshwater shrimps that were offered. The shrimps had been found in large numbers at the edges of shallow streams where Wallace's three-striped dasyures were observed drinking and may contribute to the natural diet.

The holotype of Wallace's three-striped dasyure (*Myoictis wallacei*) was collected by Alfred Russel Wallace in the Aru Islands in 1857 (P. A. Woolley and D. Walsh).

Females possess six nipples. One adult female examined in October 1992 had a pouch area that was flat and lacked skin folds, and the mammary tissue was not enlarged (Woolley 2001). The nipples were small, but appeared as though they had been suckled, indicating the animal may have previously reproduced. Four males captured at the same time were all mature and throughout their time in captivity (up to 7.5 months in duration) they showed spermatorrhoea development (Woolley 2001). Wallace's three-striped dasyure is widely distributed, no major threats are known, and it is unlikely to be declining at a rate that warrants listing under a threatened category.

Sminthopsis

RED-CHEEKED DUNNART

Sminthopsis virginiae (Tarragon, 1847)

smin-thop'-sis ver-jin'-ee-ee

The red-cheeked dunnart apparently remains known from this region only by the holotype of *Sminthopsis viriginiae rufigenis*, collected in the Aru Islands (Thomas 1922a). The specimen label reveals it was obtained by Thomas via the Buitzenborg Museum, Java (Tate 1947).

Little is known of the specific biology of Aru Islands populations. Generally, the red-cheeked dunnart prefers monsoonal savannah, woodlands and grasslands. The diet consists mostly of insects, but vertebrates (lizards) also are killed and eaten (Flannery 1995; Woolley 2008). Females from the New Guinea subspecies (*S. v. rufigenis*) possess six nipples, whereas females from the two other Australian subspecies possess eight. New Guinea animals are thus limited to smaller litter sizes than their Australian counterparts. Females in captivity are capable of breeding throughout the year (Woolley 2008). The red-cheeked dunnart is widely distributed, no major threats are known, and it is unlikely to be declining at a rate that warrants listing under a threatened category.

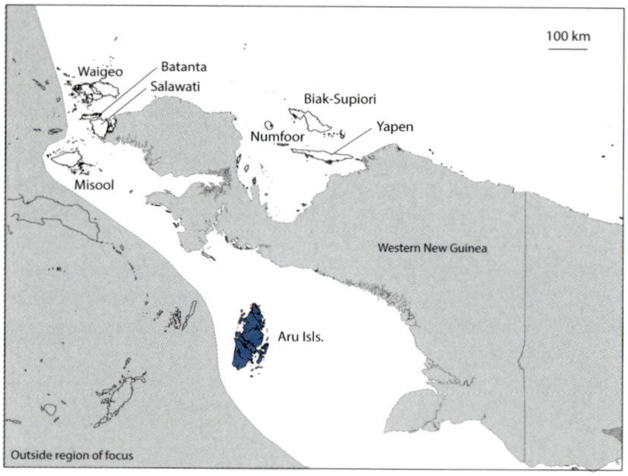

Red-cheeked dunnart (*Sminthopsis virginiae*), northern Australia. This small carnivorous marsupial is shared between New Guinea and Australia and in the region under consideration here has been recorded from the Aru Islands (Anders Zimny).

Sminthopsis virginiae			
Size*			
	Mean	**Range**	**Sample**
Head–body	–	91	1
Tail	–	102	1
Hindfoot	–	21.3	1
Ear	–	–	–
Weight	–	–	–
* From the holotype of *Sminthopsis virginiae rufigenis*, Aru Isls (NHMUK 1922.2.2.76)			
Identification *Sminthops virginiae* is absolutely smaller than other dasyurids of the region, with distinctly red cheeks.			
Synonyms *Sminthopsis lumholtzi* Iredale and Troughton, 1934 *Phascogale rona* Tate and Archbold, 1936			
Subspecies *Sminthopsis virginiae virginiae* (Tarragon, 1847) (north-east Queensland) *Sminthopsis virginiae nitela* Collett, 1897 (north-western Australia) *Sminthopsis virginiae rufigenis* Thomas, 1922a (Aru Isls, New Guinea)			
Distribution Aru Isls			
Extralimital Australia, New Guinea			
Conservation status (IUCN) Least Concern			
Etymology Gen.–mouse-like (Gr.); Sp.–significance unknown			

Peramelemorphia

Peroryctidae

The bandicoots are a relatively small and highly distinctive order of marsupials. The living species in seven genera are distributed between four families, including the predominantly New Guinean Peroryctidae, which are estimated to have diverged from the predominantly Australian family Peramelidae some 13.8 million years ago (Westerman *et al.* 2012). Peroryctidae comprises three genera (*Echymipera, Microperoryctes* and *Peroryctes*), all of which are almost entirely restricted to Melanesia. Just a single species, *Echymipera rufescens*, extends to northern Cape York Peninsula.

Bandicoots are well represented in the region under consideration. Five species have been recorded, including *E. davidi*, endemic to Kiriwina in the Trobriand Islands. Peroryctids are, or have been, present on several oceanic islands in the region including Biak-Supiori and parts of the Bismarck Archipelago. Although debate continues as to whether they have been introduced by humans to parts of this range (e.g. Admiralty Islands: see Williams 1999 and Helgen and Flannery 2004). The presence of well-differentiated subspecies (such as on Biak-Supiori) indicates some populations have been long established. This in turn indicates bandicoots might be better at dispersing across oceanic barriers than other marsupial families.

Echymipera

CLARA'S ECHYMIPERA

***Echymipera clara* Stein, 1932**

ek'-ee-mip'-e-rah klar'-ah

Clara's echymipera is widely distributed in northern New Guinea between elevations of 300–1700 m. The holotype was collected by George Stein on Yapen on 21 March 1931. Between July and October 1983 on Yapen, Jared Diamond collected 25 mandibles identified to this species. They were collected at the same time as 16 mandibles identified as *Echymipera kalubu* and eight mandibles identified as *Peroryctes raffrayana*. This suggests *E. clara* is the more common of the three bandicoots, a favoured or more durable trophy item, or a preferred or easier hunting target on Yapen. Flannery (1995a) stated that on mainland New Guinea this species is usually far less common than *E. kalubu*. However, in some areas (such as the high-elevation Toricelli Mountains) it can be equally, or more common.

Very little is known of species biology. Clara's echymipera is strongly sexually dimorphic, with adult males being at least 50% heavier than adult females. Males also have enlarged canines, larger premolars and an exaggerated sagittal crest

Long-nosed echymipera (*Echymipera rufescens*) (Michael Pennay).

Clara's echymipera (*Echymipera clara*). This species is sexually dimorphic: adult males are much heavier than adult females with enlarged canines, larger premolars and an exaggerated sagittal crest (Madison E. Mayfield).

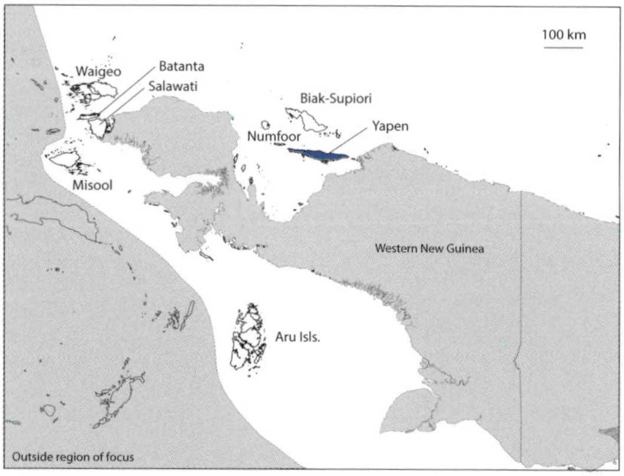

Echymipera clara			
Size*			
	Mean	**Range**	**Sample**
Head–body	339.0	270.0–410.0	5
Tail	95.1	82.3–106.0	5
Hindfoot	61.3	54.0–70.3	5
Ear	33.4	30.0–35.0	5
Weight	1357.2	825.0–2100.0	9
* From Stein (1932), Flannery (1995a), Aplin *et al.* (2010) (Yapen and mainland New Guinea)			
Identification Clara's echymipera can be distinguished from other species of *Echymipera* by the finer spines in the pelage, and the presence of a very large, raised black pad on the hindfoot. Males have a very large head with large canines and premolars. White hair and skin can be present around the lips, and males are much larger than females.			
Synonyms None			
Subspecies None			
Distribution Yapen			
Extralimital New Guinea			
Conservation status (IUCN) Least Concern			
Etymology Gen.–pouched hedgehog; Sp.–for Clara Stein, wife of George Stein and his companion on the 1931 expedition to west New Guinea, who shared with him the full burden of fieldwork (Renata Angermann, pers. comm.)			

(Menzies 2011). The magnitude of difference in appearance is so great that the two sexes can be easily mistaken for different species. Clara's echymipera seems to prefer undisturbed habitats but will occur in secondary forests and village gardens. It is reportedly a strictly nocturnal species that is probably omnivorous and has been recorded eating the fruits of figs (*Ficus* spp.) and *Pandanus* sp. (Flannery 1995a). Females have six nipples, and litters of one or two young have been recorded in January and July. Clara's echymipera is widespread, no major threats are known, and it does not appear to be declining at a rate that warrants listing under a threatened category.

DAVID'S ECHYMIPERA

Echymipera davidi Flannery, 1990a

day'-vid-ee

David's echymipera is known to occur only on Kiriwina Island, in the Trobriand Group of islands, south-east of New Guinea. Its natural history is poorly known, and it is represented by only five specimens in world museum collections. A series in the American Museum of Natural History was collected at Liluta at 10 m elevation by R. F. Peterson during the fifth Archbold Expedition in December 1956. An adult male specimen held in the Australian Museum had been killed by dogs at Nambuso in May 1987.

Kiriwina is a small island (290 km^2) and is heavily cultivated with gardens. This, in combination with the dense human population of the island, suggests David's echymipera could be a species facing significant threat. However, Brass (1959) indicated all arable land was cultivated when *Echymipera* specimens were collected in 1956. From the time the expedition was stationed at Liluta, the only other terrestrial mammals reported were agile wallaby (*Notamacropus agilis*), Polynesian rat (*Rattus exulans*) and common water rat (*Hydromys chrysogaster*). These three species are capable of persisting in disturbed habitats and thus David's

David's echymipera (*Echymipera davidi*) is known only from Kiriwina, and an apparent increase in hunting pressure over the past decade has reduced abundance of the species on the island (Madison Erin Mayfield).

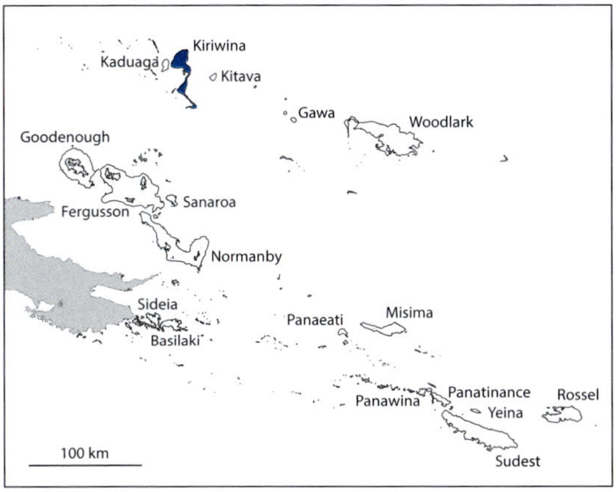

Echymipera davidi			
Size*			
	Mean	**Range**	**Sample**
Head–body	362	327–397	2
Tail	–	108	1
Hindfoot	67.2	62.4–72.0	2
Ear	28.6	28.2–29.0	2
Weight	–	950	1
* From AM specimens (Kiriwina)			
Identification The combination of the lack of a black pad on the hindfoot, the presence of enlarged third premolars in males, and molars with small hypocones (and thus triangular in basal outline) in both sexes distinguishes this species from all other bandicoots in the region. The crown of the head is the same colour as the dorsal surface of the body and the face lacks a contrasting border between the light ventral surface and the crown.			
Synonyms None			
Subspecies None			
Distribution Kiriwina			
Extralimital None			
Conservation status (IUCN) Endangered			
Etymology Sp.–for David Flannery, son of the second author, Tim Flannery			

echymipera also may be tolerant of some degree of human disturbance. Similarly, in 1987 David's echymipera was found in degraded forests and areas of regrowth, where it appeared to be fairly common.

In 2008 there was somewhat of a customary aversion or *tambu* preventing the consumption of bandicoots by the majority of the population on Kiriwina. But this has apparently declined more recently with an influx of outsiders who do not adhere to those cultural preclusions. David's echymipera now seems to be under heavier hunting pressure and has become rare (D. Mitchell, pers. comm.). The species is classified as Endangered on the IUCN Red List, and further studies to assess population trends and conservation status are urgently needed.

COMMON ECHYMIPERA

Echymipera kalubu (Fischer, 1829)

kah-lue'-boo

The common echymipera is a widely distributed species found across much of mainland New Guinea below 2000 m elevation. In the South-west Pacific it has been reported from several islands in the Bismarck Archipelago, and smaller satellite islands along the northern coastline of New Guinea, and on islands to New Guinea's west (Biak, Yapen, Misool, Salawati, Waigeo). However, it probably comprises two or more distinct forms throughout this range.

Under the synonym *Perameles myoides* Günther 1883, this species was reported from New Britain, and the American Museum of Natural History holds several specimens collected by Gilliard in the Whiteman Range area of West New Britain. More recently, Anthony (2001) caught a juvenile in lowland rainforest behind Walindi resort near Kimbe in 1999. Aplin and Opiang (2011) also captured four animals in the Nakanai Range of East New Britain in April 2009. Two of these were females weighing 780 and 690 g and each had two pouch young at the time. The third was a juvenile animal weighing 103 g, and the fourth was a skull recovered from an animal that was recently hunted (Aplin and Opiang 2011). Specimens from the islands of Karkar, Bagabag, Tolokiwa, Umboi and Sakar are skins and skulls, or trophy dentaries collected by Jared Diamond (Koopman 1979).

On Manus, *Echymipera kalubu* bones in the Pamwak archaeological site date back to around 12 400 years, but uncertainty remains as to whether it is a natural occurring element of the fauna there (Williams 1999). Recent surveys on Manus failed to locate the animal or signs of its occurrence (Aplin *et al.* 2015). Landowners indicated they occurred in low numbers in rockier habitats on Mt Sobomu, but were more plentiful at Yeri River and could be snared along runways in the undergrowth (Aplin *et al.* 2015). Two modern specimens are available from Manus: a male collected by Meek and Eichhorn on 23 September 1913 (NHMUK 14.4.1.24); and MZB 15061, a male collected by H. Schoede at Seeadlerhafn. There is also a specimen (unregistered) held in the collections of the Museum für Naturkunde, Berlin, from the nearby island of Lou. Animals from Manus have not been included in molecular studies and detailed comparisons with specimens from the wider range of *E. kalubu* are needed to determine whether it is an endemic taxon.

Ramsay (1877a) described the subspecies *Echymipera kalubu cockerelli* and reported the specimens as having been collected from New Ireland. However, they were collected by James Cockerell, who accompanied the Reverend George Brown to the Duke of York Islands in the 1870s. Many of Brown's specimens were poorly provenanced, leaving it unclear whether they were from New Britain, New Ireland or Mioko in the Duke of York group. Flannery and White (1991) determined no bandicoots occur on New Ireland, and surveys at four sites on southern New Ireland in 1994 similarly did not detect *Echymipera* (Emmons and Kinbag 2001). Thus, it seems likely that the *E. k. cockerelli* specimens were not collected from New Ireland, but some other locality in the Bismarck Archipelago.

Common echymipera is abundant in a range of environments including secondary forests, grasslands, coffee plantations and gardens (Menzies 2011; Cuthbert and Denny 2014). A detailed study in the Eastern Highlands Province of Papua New Guinea revealed the animals were primarily nocturnal, with activity commencing from 18:00–20:00 h and peaking in activity between 20:00 and 02:00 h (Cuthbert and Denny 2014). Home-range sizes of 10 individuals radio-tracked for 8–18 days ranged

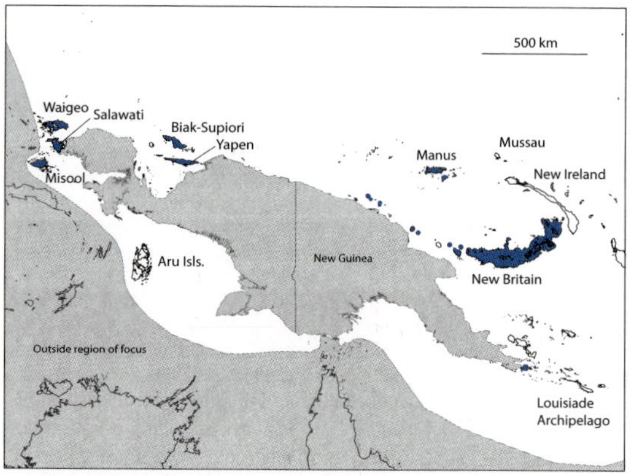

The common echymipera (*Echymipera kalubu*) is one of the more widespread marsupials of the region (Michael Pennay).

Echymipera kalubu			
Size*			
	Mean	**Range**	**Sample**
Head–body	260.0	242.0–345.0	9
Tail	81.6	73.0–92.0	8
Hindfoot	51.8	45.8–58.7	9
Ear	24.9	21.7–27.0	9
Weight	666.3	405.0–1000.0	4
* From AM specimens (Biak-Supiori, Indonesia) (subspecies *E. k. philipi*)			
Identification Can be distinguished from *Echymipera rufescens* and *E. clara* by the lack of any vestige of a black pad on the foot, and from *E. davidi* by its enlarged hypocones on the upper molars, smaller premolars and lack of a hair whorl on either side of the snout.			
Synonyms *Perameles doreyana* Quoy and Gaimard, 1830 *Perameles myoides* Günther, 1883 *Brachymelis garagassi* Miklouho-Maclay, 1884 *Anuromeles rufiventris* Heller, 1897 *Suillomeles hispida* Allen and Barbour, 1909 *Perameles doreyana alticeps* Cohn, 1910 *Perameles doreyana breviceps* Cohn, 1910 (nomen nudum)			
Subspecies *Echymipera kalubu kalubu* (Fischer, 1829) (Most of New Guinea and nearby islands – including Bagabag, Blup Blup, Kadovar, Karkar, Koil and Vokeo) *Echymipera kalubu cockerelli* (Ramsay, 1877a) (Bismarck Archipelago) *Echymipera kalubu oriomo* Tate and Archbold, 1936 (Fly River Plateau, New Guinea) *Echymipera kalubu philipi* Troughton, 1945 (Biak-Supiori and Owi Islands)			
Distribution Batanta, Biak-Supiori, Misool, Owi, Salawati, Waigeo, Yapen (Indonesia); Bagabag, Blup Blup, Kadovar, Karkar, Koil, Lou, Manus, New Britain, Sakar, Tolokiwa, Umboi, Vokeo (Papua New Guinea)			
Extralimital New Guinea			
Conservation status (IUCN) Least Concern			
Etymology Sp.–from an Indigenous name for the animal on Waigeo Island			

between 0.9 and 7.3 ha (average 2.79 ha) (Cuthbert and Denny 2014).

Stomach samples examined by Flannery (1995a) were full of seeds and fruit pulp, indicating this species may be largely frugivorous. Indeed, fruiting trees are often a point of focus when hunters aim to trap this species on mainland New Guinea (Flannery 1995a). However, at Baiyer River Sanctuary, Western Highlands Province, a study revealed common echymipera were most active around secondary forests and tree fall sites (Anderson *et al.* 1988). At these sites, decaying logs (and presumably invertebrates) formed the focus for foraging.

Common echymipera are highly fecund. Hide *et al.* (1984) noted that of 85 females caught, 56 had pouch young. Between March–May 2002 and May–June 2004, 10 of 15 adult females (less than 400 g in weight) captured were carrying one or two young (average 1.53) (Cuthbert and Denny 2014). Between one and three pouch young is typical (Hide *et al.* 1984). Breeding occurs throughout the year and the average birth interval is probably around 120 days (Flannery 1995a). The common echymipera has a large distribution, is tolerant of disturbed habitats, and no major threats have been identified.

LONG-NOSED ECHYMIPERA

Echymipera rufescens (Peters and Doria, 1875)

rue-fes'-enz

The long-nosed echymipera has a wide distribution that extends across several oceanic islands, and the taxonomy of this species is very poorly understood. Aru and Kai Islands populations need revision in order to determine whether or not they are distinct subspecies. The status of populations on the D'Entrecasteaux Islands also are particularly uncertain. They are tentatively retained with *Echymipera rufescens* here, but there are similarities with *E. davidi* (Flannery 1990a) and individuals from Goodenough and Normanby were assigned to the latter species by Elliott *et al.* (2022) without rationale.

The long-nosed echymipera occurs in grassland, closed heath, lowland primary and secondary forests and lower montane forest (Gordon and Lawrie 1978). In the Iron Range of Cape York Peninsula, Australia, this species was found to have

Long-nosed echymipera (*Echymipera rufescens*), Iron Range National Park, Australia. This bandicoot is common and widespread in northern Australia and New Guinea and is found on the D'Entrecasteaux Islands, Aru Islands, Misool and Yapen (Anders Zimny).

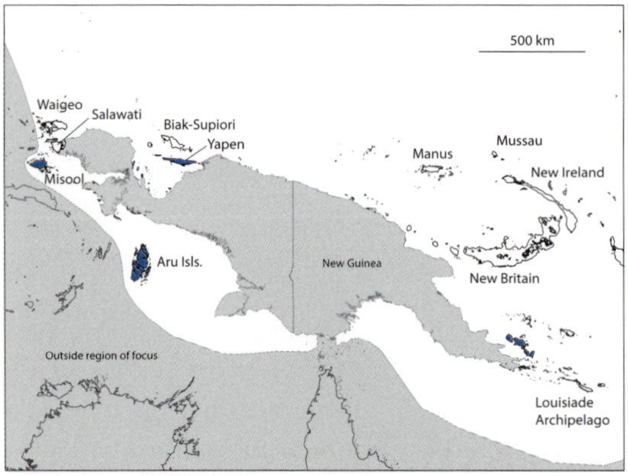

Echymipera rufescens			
Size*			
	Mean	**Range**	**Sample**
Head–body	473.0	393.0–545.0	9
Tail	96.7	84.0–115.0	9
Hindfoot	67.8	54.0–83.0	8
Ear	29.8	25.0–40.0	10
Weight	1186.1	595.3–2721.6	5
* From Peters and Doria (1875), AMNH specimens (Milne Bay and Goodenough Is., Papua New Guinea) and Flannery (1995a)			
Identification The presence of a small black pad on the hindfoot serves to distinguish this species from *E. davidi* and *E. kalubu*.			
Synonyms *Perameles aruensis* Peters and Doria, 1875 *Peramelopsis welsianus* Heude, 1896 *Perameles keiensis* Cohn, 1910 *Echymipera gargantua* Thomas, 1914a			
Subspecies *Echymipera rufescens rufescens* (Peters and Doria, 1875) (New Guinea, Misool Is, Yapen Is, Aru Isls, Kai Isls) *Echymipera rufescens australis* Tate, 1948 (Cape York, Australia)			
Distribution Aru Isls, Misool, Yapen (Indonesia); Goodenough, Fergusson, Normanby (Papua New Guinea)			
Extralimital Australia, Kai Isls, New Guinea			
Conservation status (IUCN) Least Concern			
Etymology Sp.–becoming red (L.)			

an omnivorous and flexible diet, consuming a wide variety of foods, including fruits, invertebrates, fungi and dicot plants (Shevill and Johnson 2008). High proportions of the material identified in faeces were seeds from *Pandanus zea* and native ginger (*Alpinia caerulea*). Females produced one or two litters per year and reproduction was seasonal, with births occurring between December and March (Shevill and Johnson 2008). Females produced an average of 4.9 young per year. The long-nosed echymipera is hunted in New Guinea and this could pose a threat in insular parts of its range including the D'Entrecasteaux Islands.

Peroryctes

RAFFRAY'S BANDICOOT

Peroryctes raffrayana (Milne-Edwards, 1878)

pe'-roh-rik'-tayz raf'-ray-ah'-nah

This very distinctive species is widely distributed in New Guinea. The Natural History Museum, London, holds two specimens (NHMUK 1946.687–8) from Yapen Island collected by Evelyn Cheesman on 8 and 12 September 1938. They were taken at approximately 300 m elevation in mossy forest at 'Camp 1' on Mt Baduri. NHMUK 1946.687 was recorded as 'feeding on fallen *Pandanus* fruit at night'. Between July and October 1983, Jared Diamond collected eight sets of mandibles from this species on Yapen (presumably trophy jaws). Also collected in the same period were the mandibles of 16 *Echymipera kalubu* and 25 *E. clara*.

Raffray's bandicoot prefers undisturbed forests and is rarely encountered in gardens or secondary forests (Menzies 2011). It is a widespread species, occurring up to 2000 m elevation on mainland New Guinea. In the Eastern Highlands Province, Papua New Guinea between March–May 2002 and May–June 2004, 11 Raffray's bandicoots were examined as part of an ecological study (Cuthbert and Denny 2014). Eight individuals were caught in mature secondary forest and two from old garden sites. A single animal that was radio-tracked frequented old garden areas adjacent to mature secondary forest and primary forest. The animal was entirely nocturnal, and active between 19:00 and 02:00 h. Of

A juvenile Raffray's bandicoot (*Peroryctes raffrayana*). This is a distinctive and widespread species. On Yapen Island it occurs with two other species of bandicoot from the genus *Echymipera* (Pavel German).

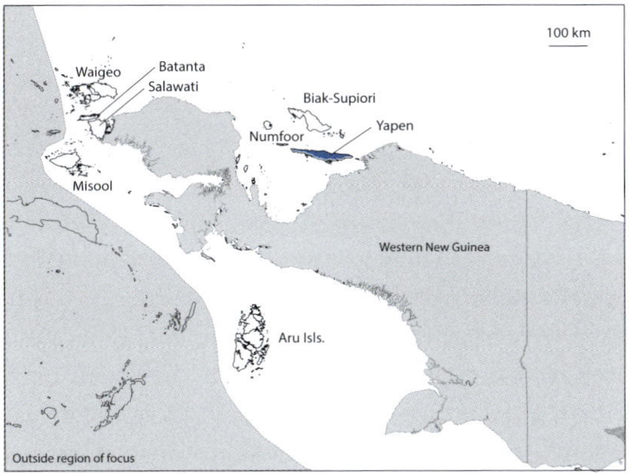

Perorytes raffrayana			
Size*			
	Mean	**Range**	**Sample**
Head–body	316.3	217.0–392.0	9
Tail	154.3	115.0–197.0	9
Hindfoot	71.7	64.0–80.0	9
Ear	28.2	20.0–32.0	9
Weight	827.6	550.0–1038.0	7
* From NHMUK 1946.687 (Yapen), Flannery (1995a) and Cuthbert and Denny (2014) (mainland New Guinea)			
Identification The relatively soft pelage, brownish dorsum and pure white venter will distinguish this species from other bandicoots of the region.			
Synonyms *Perameles mainois* Förster, 1913			
Subspecies *Peroryctes raffrayana raffrayana* (Milne-Edwards, 1878) (most of New Guinea, Yapen Is.) *Peroryctes raffrayana rothschildi* (Förster, 1913) (Huon Peninsula)			
Distribution Yapen			
Extralimital New Guinea			
Conservation status (IUCN) Least Concern			
Etymology Gen.–pouched digger (Gr.); Sp.–for Mr Raffray, collector of the holotype			

two females captured, neither was carrying young; the larger female weighing 880 g had reproduced, but the smaller weighing 220 g had not (Cuthbert and Denny 2014).

Females have eight nipples and have been recorded with one or two pouch young (Menzies 2011). Pouch young have been observed in March, June, August, September, November and December, indicating reproduction may occur throughout the year (Flannery 1995a). Raffray's bandicoot is widely distributed, lacks known major threats, and is unlikely to be declining at a rate required for listing in a threatened category.

Diprotodontia

Acrobatidae

The family Acrobatidae consists of just three species, two of which are restricted to Australia and belong to the genus *Acrobates*, and one restricted to New Guinea and Waigeo belonging to the genus *Distoechurus*. All three possess a fringe of long lateral hairs along the tail. Species of *Acrobates* also possess a well-furred patagium that acts as a gliding membrane, a feature that is absent in *Distoechurus*.

Feather-tailed possum (*Distoechurus pennatus*) (illustration by Gustav Mützel in Peters and Doria 1881).

Distoechurus

FEATHER-TAILED POSSUM

Distoechurus pennatus (Peters, 1874)

dist'-ee-kyue'-rus pen-ah'-tus

Feather-tailed possum was a relatively recent addition to the faunas of the South-west Pacific, observed on Waigeo in August 2019 (Chien Lee, pers. comm.). It may be present on other land bridge islands associated with the Vogelkop Peninsula. The diversity of forms across the range indicates it might be a species complex.

Feather-tailed possums occur in a wide variety of habitats. On mainland New Guinea, they are common inhabitants of regrowth and disturbed forests as well as gardens and villages (Flannery 1995a). However, they can also be encountered in lowland swamp forests, and in undisturbed primary forests from sea level up to 2000 m elevation (Aplin 2015). They are arboreal mammals that seem to

Feather-tailed possum (*Distoechurus pennatus*), Waigeo Island, Indonesia. This small possum is a recent addition to the list of mammals from the South-west Pacific. It is widespread on mainland New Guinea where it can be found in regrowth, disturbed forests, and secondary and primary forests (Chien Lee).

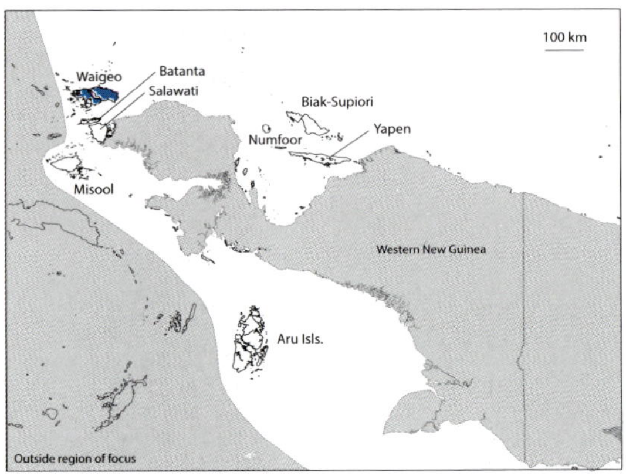

Distoechurus pennatus			
Size*			
	Mean	**Range**	**Sample**
Head–body	116.0	106.0–132.0	18
Tail	146.0	131.0–155.0	18
Hindfoot	20.3	18.9–21.2	18
Ear	11.4	9.7–13.8	18
Weight	49.1	38.0–62.0	10
* From Flannery (1995a) (females, Chimbu Province, Papua New Guinea)			
Identification Feather-tailed possum can be immediately distinguished from all other mammals in the region by the distinctive tail that possesses lateral lines of elongated hairs.			
Synonyms *Distoechurus neuhassi* Matschie, 1916 *Distoechurus pennatus amoensis* Thomas, 1920 *Distoechurus pennatus dryas* Thomas, 1920			
Subspecies None			
Distribution Waigeo			
Extralimital New Guinea			
Conservation status (IUCN) Least Concern			
Etymology Gen.–distally spined tail (Gr.); Sp.–feathered (L.), a reference to the tail			

rarely venture to the ground. Feather-tailed possums nest in tree hollows, usually individually, but sometimes in small communal groups. Repeated studies at Mt Karimui, Chimbu Province, Papua New Guinea, indicated breeding there occured over an extended season from April to August (Aplin 2015). Females are slightly larger than males. Nothing has been reported of the diet. Feather-tailed possums appear resilient to disturbance, and can be found in a range of vegetation types that provide some form of tree cover within their habitat.

Macropodidae

The family Macropodidae includes the familiar kangaroos and wallabies of Australia, New Guinea and nearby islands. It is a very widespread and diverse group which reaches its greatest diversity (46 species) in Australia. New Guinea has 17 endemic species, some of which belong to less well-known endemic macropodid genera such as *Dorcopsis* and *Dorcopsulus*. Four genera (*Dendrolagus, Lagorchestes, Notamacropus* and *Thylogale*) are shared between Australia and New Guinea. *Dendrolagus* and *Thylogale* have their greatest diversity in New Guinea. Just three species (*Lagorchestes conspicillatus, Notamacropus agilis* and *T. stigmatica*) are shared between Australia and New Guinea.

New Guinea pademelon (*Thylogale browni*) (illustration by Joseph Smit in Alston 1877).

Of the six species recognised as indigenous to the region discussed here, three (*Dendrolagus inustus*, *Dorcopsis meulleri* and *T. brunii*) are restricted to land bridge islands, and a further two (*T. browni* and *Notamacropus agilis*) have been introduced by humans to at least part of their insular distribution. Only *Dorcopsis atrata* has presumably crossed oceanic barriers, being endemic to Goodenough in the D'Entrecasteaux group.

Dendrolagus

GRIZZLED TREE KANGAROO

Dendrolagus inustus Müller, 1840

den-drol-ah'-gus in-us'-tus

The grizzled tree kangaroo is a highly distinctive species which (surprisingly given its northern New Guinea distribution) shares features with the two Australian species of tree kangaroo. These include the position of the hair whorl (on the shoulders rather than the back), long hindfoot, short side toe (fifth digit) of the foot, and extensive contact between the two lower leg bones (tibia and fibula). In all of these features, the grizzled and the Australian tree kangaroos resemble typical ground-dwelling kangaroos.

Groves (1982) provided a list of grizzled tree kangaroo specimens examined by him during his extensive revision of the genus *Dendrolagus*, including two specimens from Mt Eiori on Yapen and one from the island of Waigeo. The specimen from Waigeo was considered dubious. Suspect records also exist from the Kai and Aru Islands, both associated with specimens held in the collections of the Museum für Naturkunde, Berlin. Rothschild and Rothschild (1898) reported that a Mr Frost, who collected mammals in the region, stated that no species of *Dendrolagus* occurred in either the Kai or Aru Islands and concluded it was likely both specimens were in fact from mainland New Guinea. Although not confirmed, a tree kangaroo (possibly this species) may also occur on

A dubious record of grizzled tree kangaroo (*Dendrolagus inustus*) exists from Waigeo to the west of New Guinea. Field surveys are needed to better understand the distribution of this species in the region (Pavel German).

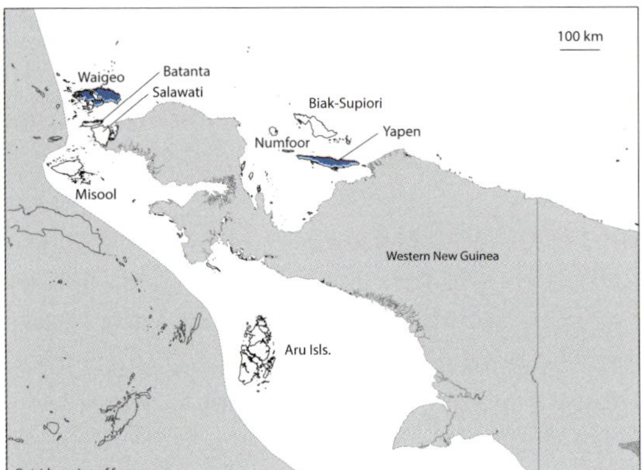

Dendrolagus inustus			
Size*			
	Mean	**Range**	**Sample**
Head–body	696.7	650.0–750.0	5
Tail	780.0	750.0–830.0	5
Hindfoot	148.8	142.5–154.0	5
Ear	55.3	50.0–60.0	5
Weight	12 500	10 500–14 000	5
* From AM specimens (North Coast Range, New Guinea)			
Identification A large, grizzled grey tree kangaroo, with distinct black ears and a long grey tail. It has a longer hindfoot than any other Melanesian species of *Dendrolagus*.			
Synonyms *Dendrolagus maximus* Rothschild and Rothschild, 1898 *Dendrolagus keiensis* Matschie, 1916 *Dendrolagus shoedei* Matschie, 1916 *Dendrolagus sorongensis* Matschie, 1916			
Subspecies *Dendrolagus inustus inustus* Müller, 1840 (?Waigeo, Yapen, Vogelkop Peninsula) *Dendrolagus inustus finschi* Matschie, 1916 (northern New Guinea)			
Distribution ?Waigeo, Yapen			
Extralimital New Guinea			
Conservation status (IUCN) Vulnerable			
Etymology Gen.–tree hare (Gr.); Sp.–dusky, or imprint, crimp or curl (L.), a reference to the pelage			

Salawati. In 1992 people there regarded it as rare, reporting that the only time it was possible to catch it was in the dry season when animals come out from the mountains to drink at streams. Grizzled tree kangaroo is sometimes kept as a pet, and captive animals have been encountered as far from their natural range as Halmahera. Further verification of the species' range outside mainland New Guinea is needed.

Nothing is reported of the grizzled tree kangaroo's biology within the region. In the Toricelli Mountains of New Guinea, Flannery (1995a) obtained information that the species feeds on a variety of young leaves and fruits. Young have been recorded in March, June and December. Flannery (1995a) surmised that breeding is non-seasonal, and females will give birth soon after a young has left the pouch and well before the older young becomes independent. Twin births have been recorded, highly unusual for a macropodid.

Grizzled tree kangaroo is listed as Vulnerable on the IUCN Red List due to suspected population declines over the last generation, and projected declines over the next two generations. Hunting, habitat loss and degradation from agriculture are major threats.

Dorcopsis

BLACK DORCOPSIS

Dorcopsis atrata **Van Deusen, 1957**

dor-kop'-sis ah-trah'-tah

The black dorcopsis is the only macropodid species endemic to the region under consideration. It is very distinctive, undoubtedly a relict species that has probably been isolated on Goodenough for several million years and may be only distantly related to the *Dorcopsis* of mainland New Guinea.

Black dorcopsis was first scientifically documented during the fourth Archbold Expedition in September–November 1953. Van Deusen (1957) based his species description on seven specimens all collected in *Castanopsis* (oak) forest between 1200 and 1800 m elevation near the expedition's 'Top Camp' on Goodenough Island. He estimated these were the limits of the altitudinal range for the species. In 1987 an Australian Museum team was able to interview local people extensively and came to similar conclusions, suggesting that, at least on the northern (rain shadow) side of the island, black dorcopsis is restricted to between 1000 and 1800 m. The northern lowlands are predominantly savannah, but tongues of grassland (the result of burning) extend into the forest to the upper altitudes of the island. On the southern and western sides, forest cover is more or less continuous to the coast. The few enquiries made of people living on the south side of the island indicated this species was restricted to higher elevations there also.

This is the only member of the genus known to inhabit montane forests. The trees in the cloud forests of Goodenough were described by Van Deusen (1957) as being heavily covered with long, wispy moss. The forest had a canopy height of approximately 25 m, with a thin sapling storey including *Pandanus* sp. which was more common in gullies. The forest floor was a maze of rotting tree trunks and broken ground, with little herbaceous growth, but there were thick growths of hepatics, mosses and lichens. A creeper of the genus *Vaccinium* was common, ascending into the canopy. The predominant canopy tree was a species of *Castanopsis* (Van Deusen 1957). Species of *Begonia*, *Elatostema* and *Medinilla* grew along creeks where more light penetrated, and clouds sat against the mountain for large parts of the day. Only around 100 km^2 of Goodenough lies between the altitudes of 1000 and 1800 m. Of this, an even smaller area retains this forest type with a large proportion of the northern side having been converted to grassland not inhabited by *Dorcopsis atrata*.

Black dorcopsis may be a seasonal visitor to gullies at lower elevations (D. Mitchell, pers. comm.). At a site on the Boitutu River in 1987, the trackways of wallabies were present, and landowners indicated that the animals congregated in valleys during the dry season. Information on reproductive biology is scarce. The single female specimen collected by Van Deusen sometime between September and November 1953 had a single elongated teat, but no pouch young was found. Five of the individuals caught by the Australian Museum in August were female, of which two were subadults. Of the adults, one caught on 21 August 1987 weighed 4 kg, and had a single hairless pouch young weighing 38 g. A female caught on 22 August weighed 4.25 kg and was lactating, but no young was seen. The remaining adult female (also caught on 22 August) weighed 4.75 kg and had two enlarged teats but was not lactating and no young was observed. People of Goodenough believe that the black dorcopsis breeds year round and never has more than a single young.

The claws of the hands and feet are blunt and almost worn flat in older animals, possibly as a result of digging for truffles in Goodenough's rocky substrates. Vernes and Lebel (2011) identified 12 fungal taxa present in a single stomach sample from specimen AM19260. The authors suggested this indicated that the black dorcopsis performs an important role as a fungal disperser on Goodenough Island. The black dorcopsis also exhibits considerable diversity in forepaw colour. Out of six study skins held in the collections of the Australian Museum, two have both paws white, two have one white and one black paw, one has both paws black and one is missing a paw, the remaining one being black. The significance of this is unknown, but white paws could plausibly serve as a tool to signal other individuals in the dark cloud forests of Goodenough.

Within the small area of suitable habitat, the black dorcopsis was reasonably common during a visit by the Australian Museum in the late 1980s. However, the species was being heavily hunted with dogs at the time. Eight specimens were captured in this manner as part of the expedition, and one man from the southern side of the island related that he and his family had recently camped up high on the mountain for a week and killed 50 wallabies. This level of hunting pressure appeared as though it could be decreasing because younger people did not seem to frequently venture into higher elevation forests.

Brass (1956) made mention of the possibility of this species existing on Fergusson Island, stating, 'Kim, also out before dawn, came across something in the forest which must have been either a dog or *Dorcopsis*, but being a cook instead of an accomplished hunter, he had no shell ready in his gun so the beast got away.' Brass (1959) expanded on this in his report from the high-elevation Fergusson camp, stating, 'One glimpse of it by the cook we thought probably was *Dorcopsis atrata*, a small black wallaby of the mountain forests discovered on Goodenough by us in 1953.'

Dorcopsis atrata			
Size*			
	Mean	**Range**	**Sample**
Head–body	615.6	550.0–693.0	5
Tail	347.0	255.0–423.0	5
Hindfoot	105.4	96.8–118.0	5
Ear	32.8	28.0–35.0	5
Weight	5125.0	4000.0–7500.0	4
* From AM and AMNH specimens (Goodenough)			
Identification The only *Dorcopsis* with a black dorsum and contrasting white underfur. The ears are shorter than in other *Dorcopsis* species and forepaws can be white.			
Synonyms None			
Subspecies None			
Distribution Goodenough			
Extralimital None			
Conservation status (IUCN) Critically Endangered			
Etymology Gen.–gazelle-faced (Gr.); Sp.–black (L.)			

The black dorcopsis (*Dorcopsis atrata*) is the most distinctive of the species of *Dorcopsis*. It is known only from Goodenough Island and has probably been isolated on the high mountains of its island home for a million years or more (Madison Erin Mayfield).

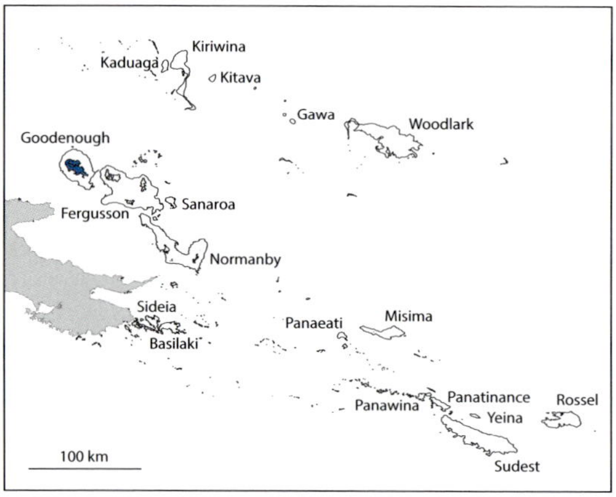

BROWN DORCOPSIS

Dorcopsis muelleri (Schlegel, 1866)

mule'-er-ee

The brown dorcopsis is a poorly known forest wallaby that is restricted to lowlands of western New Guinea and adjacent islands. Island populations on Misool and Yapen show several distinctive features and have been described as distinct subspecies. Animals from Misool differ from populations on other islands in having a conspicuous parting of the hair from the armpit to the shoulder. Animals from Yapen are similar in appearance to those from Misool, but the transverse hair parting over the shoulder fails to reach the midline of the shoulders, and the dorsum is a richer brown colour. Shallow seas lie between these islands and New Guinea; they were probably connected during the Last Glacial Maximum and may have remained so until as recently as 10 000 years ago.

The brown dorcopsis was common in riverine habitats in northern Salawati in October 1992.

A captive brown dorcopsis (*Dorcopsis muelleri*). This poorly known forest wallaby is restricted to lowlands of western New Guinea and adjacent islands (Klaus Rudloff).

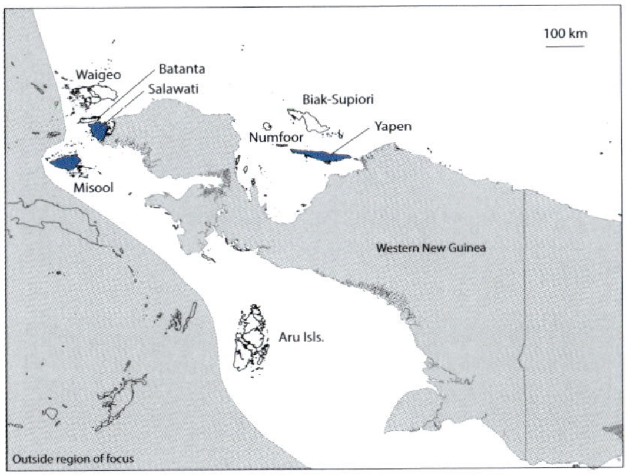

Dorcopsis muelleri			
Size*			
	Mean	**Range**	**Sample**
Head–body	632.7	54.0–708.0	3
Tail	372.3	320.0–417.0	3
Hindfoot	136.3	134.0–140.0	3
Ear	42.3	37.0–47.0	3
Weight	6000	5200–6800	3
* From NHMUK and AM specimens, Yapen and Salawati			
Identification This species differs from the only other *Dorcopsis* known from the region (*D. atrata*) in being light brown rather than blackish dorsally, having longer ears and thinner fur.			
Synonyms *Didelphis brunii* Quoy and Gaimard, 1830 *?Hypsiprymnus dorsocephalus* Owen, 1839 (nomen nudum) *Halmaturus asiaticus* Gray, 1843b (nomen nudum) *Dorcopsis rufolateralis* Rothschild and Rothschild, 1898			
Subspecies *Dorcopsis muelleri muelleri* (Schlegel, 1866) (Salawati, Vogelkop Peninsula) *Dorcopsis muelleri lorentzii* Jentink, 1908 (southern West Papua) *Dorcopsis muelleri mysoliae* Thomas, 1913c (Misool) *Dorcopsis muelleri yapeni* Groves and Flannery, 1989 (Yapen)			
Distribution Misool, Salawati, Yapen			
Extralimital New Guinea			
Conservation status (IUCN) Least Concern			
Etymology Sp.–for Müller, collector of the holotype			

Seven individuals were captured in a single afternoon by hunters and their dogs. A common strategy to catch the animals is to cut the branches of a favoured food tree and lay them on a riverbank. At dusk the animals come to feed and are easily captured either by hand or with dogs. Breeding may be continuous, as young weighing 300, 1500 and 1900 g were identified on 20–21 October 1992 on Salawati. The brown dorcopsis has a wide distribution and a presumed large population.

Notamacropus

AGILE WALLABY

Notamacropus agilis **(Gould, 1842a)**

no-tah'-mak-rop'-oos a-jil'-is

The agile wallaby is the largest of Melanesian macropods and one of two species from the region to also occur in Australia, where it can be very abundant. It is a large wallaby (up to 27 kg in weight; Merchant 2008), and a favoured game species of south-eastern New Guinea. It may have been prehistorically introduced to the D'Entrecasteaux Islands from mainland New Guinea via captive animals carried for food, trade, ceremony, pets or stock purposes (Heinsohn 2010a).

The agile wallaby prefers grasslands and the insular distribution of this species is thus limited to places where extensive areas of this habitat occur. On Goodenough, it is found on the low-lying savannah on the northern side of the island known as Vivigani Plains. Brass (1959) reported the collection of trophy skulls at Bolu Bolu, Goodenough, and an individual at Wakonai during the Fourth Archbold Expedition in 1953. A single juvenile (weight 2.5 kg) was recorded near Wabuwabuna Village on the Vivigani Plains on 25 August 1987 by an Australian Museum team. Additional records were made at Agamoia on Fergusson during the Fifth Archbold Expedition (Brass 1959), and near Iamalele Village, Fergusson (by the Australian Museum).

The agile wallaby also inhabited open country around Bwebweso on Normanby (Brass 1959), although no specimens were obtained by expedition members and the population became locally extinct there in 1987 (D. Mitchell, pers. comm). It is also now

Agile wallaby (*Notamacropus agilis*) Herberton, Australia. This widespread species inhabits grasslands of the D'Entrecasteaux Islands (Michael McCoy).

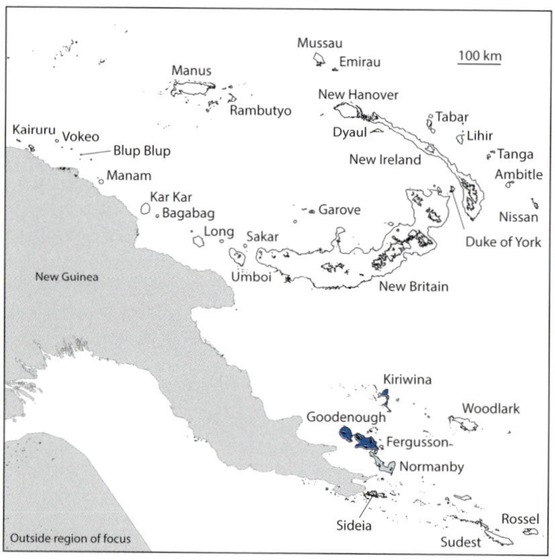

Notamacropus agilis

Size*

	Mean	Range	Sample
Head–body	655.8	490.0–850.0	16
Tail	616.9	433.0–840.0	16
Hindfoot	194.8	118.0–228.0	12
Ear	68.1	60.0–75.0	7
Weight	16 800.0	9000.0–27 000.0	4

* From AMNH specimens (Goodenough Is. and Milne Bay, Papua New Guinea), Peterson (1956b), Husson (1958) and Merchant (2008)

Identification A large sandy-brown wallaby, paler underneath. The head can have a darker median stripe between the ears and eyes and there are pale stripes present on the cheek and hip.

Synonyms *Halmaturus binoe* Gould, 1842b
Macropus papuensis Peters and Doria, 1875
Halmaturus crassipes Ramsay, 1876
Dorcopsis auranticus Rothschild and Rothschild, 1898
Macropus agilis aurescens Schwarz, 1910

Subspecies *Notamacropus agilis agilis* (Gould, 1842b) (Northern Territory)
Notamacropus agilis papuanus Peters and Doria, 1875 (New Guinea, D'Entrecasteaux Isls)
Notamacropus agilis jardinii (De Vis, 1884) (Queensland)
Notamacropus agilis siva (De Vis, 1895) (fossil only, eastern Australia
Notamacropus agilis nigrescens Lönneberg, 1913 (Western Australia)

Distribution Goodenough, Fergusson, ?Sideia, Kiriwina, ?New Ireland

Extralimital Australia, New Guinea

Conservation status (IUCN) Least Concern

Etymology Gen.–stripe (L.) a reference to distinct facial stripes found in the genus, big foot (Gr.); Sp.–agile (L.)

extinct from Baniara Island, but was present on Sideia Island in 1990 and may still exist there (D. Mitchell, pers. comm). Suspicions that this species occurs on New Ireland were based on reports of a large wallaby inhabiting grassland in the southern Namatani area in 1988. More recent consultations with biologists from New Ireland have similarly indicated the agile wallaby possibly exists there, but confirmation is needed (J. Lamaris, pers. comm).

Few studies have examined the natural history of the agile wallaby within its island distribution. *Themeda triandra* grasslands on Goodenough support approximately one individual per hectare (Liem 1977). Young (weighing 2.4 and 4 kg) have been encountered on Goodenough in August–September, and in August on Fergusson (weights 2.4 and 4 kg). This is a widespread and abundant macropod; nonetheless, evidence indicates island populations can be susceptible to local extinction.

Thylogale

NEW GUINEA PADEMELON

Thylogale browni **(Ramsay, 1877b)**

thie'-loh-gah'-lay brown'-ee

New Guinea pademelon was formerly considered a subspecies of *Thylogale brunii*. However, it was elevated by Flannery (1992) as a distinct species with quite different ecological requirements.

Molecular evidence does not support this distinction, but further work is needed (Macqueen *et al.* 2010). The New Guinea pademelon is a species that has been widely translocated and introduced to much of its current insular distribution. Archaeological evidence (some corroborated by Macqueen *et al.* 2010) shows that this species was introduced from New Guinea to New Ireland some 7000 years ago, and that animals were carried from there to Buka, Lihir and possibly Emirau (Flannery *et al.* 1988; Flannery and White 1991). It appears to have never established on Buka or Emirau. The distribution also included numerous smaller islands in the Tabar, Tanga and Feni groups, and Eloaue, Lavongai, Dyual, Tome, Kaut and Mangai (Heinsohn 2005).

Sometime around the 1940s the distribution of New Guinea pademelons on New Ireland contracted

New Guinea pademelon (*Thylogale browni*), central New Britain. This small macropod was prehistorically translocated to much of its insular range and has become extinct on numerous smaller islands to which it was introduced (Peter Crowhust).

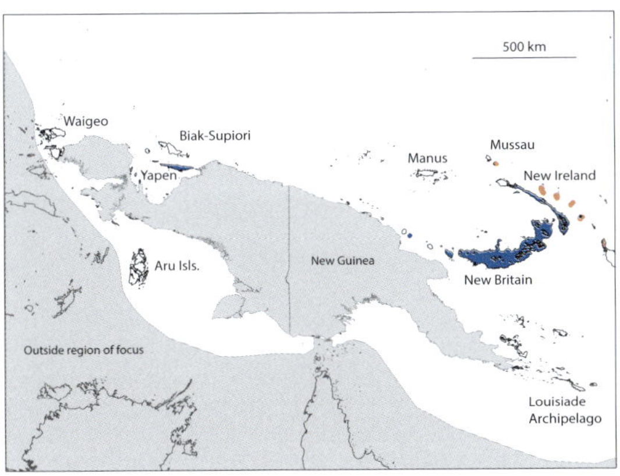

Thylogale browni

Size*

	Mean	Range	Sample
Head–body	586.0	495.0–667.0	5
Tail	421.6	365.0–464.0	5
Hindfoot	143.2	127.0–167.0	5
Ear	49.7	48.0–53.0	3
Weight	6000	5000–7000	2

* From AMNH (New Britain) and AM (New Ireland) specimens

Identification A small brown wallaby with fur of the belly white tipped with grey bases. It lacks hip and face stripes and is smaller and darker than *Notamacropus agilis*.

Synonyms *Macropus lugens* Alston, 1877
Macropus tibol Miklouho-Maclay, 1885a
Macropus keysseri Förster and Rothschild, 1914
Thylogale lauterbachi Matschie, 1916

Subspecies *Thylogale browni browni* (Ramsay, 1877b) (northern and eastern New Guinea, Bismarck Archipelago)
Thylogale browni lanatus (Thomas, 1922b) (Huon Peninsula, New Guinea)

Distribution Yapen (Indonesia); Bagabag, Buka†, Emirau†, Feni†, Lavongai, Lihir†, New Britain, New Ireland, Tabar†, Tanga†, Umboi (Papua New Guinea)

Extralimital New Guinea

Conservation status (IUCN) Vulnerable

Etymology Gen.–pouched weasel (Gr.); Sp.–for Reverend George Brown, who collected the syntypes

from the lowlands into more mountainous areas (Heinsohn 2005). Similarly, the last sightings of the species on Feni, Lihir, Tabar and Tanga groups date to this same period around the Second World War. Heinsohn (2005) postulated that seizure of local food by Japanese troupes and a breakdown of ethnozoological translocation regimes led to overhunting and the failure of restocking that was essential to sustain small island populations.

In 1994 New Guinea pademelon was recorded at all four sites surveyed between 165 and 1800 m elevation on southern New Ireland, and tracks and sightings showed it occurred above 2000 m there (Emmons and Kinbag 2001). From the high densities of animals observed at lower elevations it was presumed that hunting pressure at the time was low. On New Britain it has been recorded at elevations over 1000 m in the Whiteman Range. Field surveys in the Nakanai Mountains in 2009 found it to be present in moderate densities in relatively undisturbed forests above 200 m elevation (Aplin and Opiang 2011). Populations of this species are suspected to be suffering ongoing declines due to increasing hunting pressure.

DUSKY PADEMELON

Thylogale brunii (Schreber, 1778)

brue'-nee'-ee

Thylogale brunii			
Size*			
	Mean	**Range**	**Sample**
Head–body	556.0	458.0–600.0	6
Tail	454.0	315.0–570.0	5
Hindfoot	123.8	115.0–143.0	5
Ear	64.7	60.0–70.0	6
Weight	–	–	–
* From NHMUK 1910.3.2.24 (Aru Islands) and Flannery (1995a) (mainland New Guinea)			
Identification Differs from other New Guinea species of *Thylogale* in having the fur of the belly white to the roots, and from *T. browni* in having a pale hip-stripe and longer ear.			
Synonyms *Macropus jukesii* Miklouho-Maclay, 1885b *Macropus gracilis* Miklouho-Maclay, 1885b			
Subspecies None			
Distribution Aru Isls			
Extralimital Kai Isls, New Guinea			
Conservation status (IUCN) Vulnerable			
Etymology Sp.–for Le Brun, who first reported upon this species			

The dusky pademelon has a restricted distribution in southern New Guinea and co-occurs with agile wallaby and red-legged pademelon in parts of its range. It is apparently restricted to the rainforest-savannah interface in areas with a highly seasonal, monsoonal climate. A population near Port Moresby is probably extinct, but discrete populations survive in the Trans-Fly plains and Aru Islands.

There are very few records of the dusky pademelon from within the region, and nothing has been recorded of biology from the Aru Islands. Aplin and Pasveer (2006) identified well-preserved remains of this species from late Pleistocene to Holocene deposits on Kobroor and suggested it can be locally abundant. It apparently remained a favoured game animal for hunters on Kobroor Island in October–November 1992 when 13 specimens (trophy jaws or skulls) were collected from 'kitchen middens' by Pat Woolley (Aplin and Pasveer 2006). One held by the Australian Museum is a juvenile (the third molar erupting). Dusky pademelon is sensitive to hunting (having apparently been extirpated from the south-eastern portion of its range near Port Moresby).

Dusky pademelon (*Thylogale brunii*) occurs in southern New Guinea, the Aru and Kai Islands. It is a species that is probably sensitive to overhunting, having apparently disappeared from parts of its range around Port Moresby, Papua New Guinea (Chien Lee).

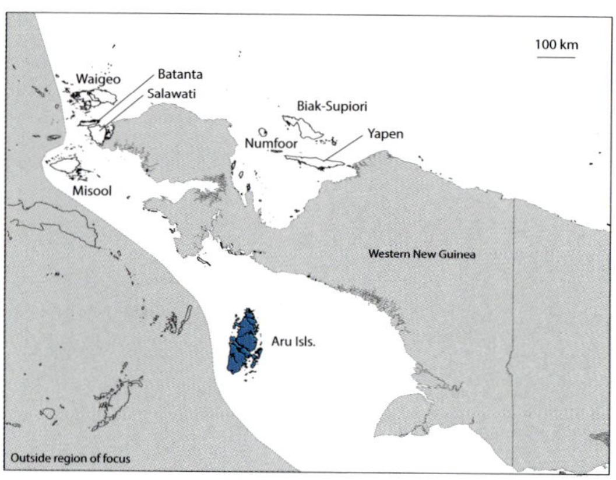

Phalangeridae

The family Phalangeridae includes the cuscuses of New Guinea and surrounding islands, and the brush-tail and scaly-tailed possums of Australia. Encompassing 29 species in all under six genera, it is a widespread family, with representatives extending from southern Tasmania at almost 44° South, to the Talaud Islands at nearly 5° North. Phalangerids are arboreal herbivores (usually more than 2 kg in weight) that take a variety of leaves, fruit, flowers and bark. Some will eat meat (e.g. eggs, nestlings, invertebrates) when available. The region under consideration has particularly rich and fascinating assemblage of phalangerids, including five endemic species.

Phalangerids are better dispersers than many other marsupials (as the presence of endemic species on oceanic islands such as Woodlark and Biak-Supiori shows). Phalangerids have also been translocated by humans more frequently than any other group. The distributions of some species were expanded via prehistoric introductions to locations well removed from New Guinea, including New Ireland, Solomon Islands, the Maluku Islands and Timor. In historic times, the brush-tailed possum has been introduced to Aotearoa New Zealand. This has led to a rich mosaic of old endemic and recently introduced populations within the region. It also means developing a good understanding of natural geographical distributions and taxonomic relationships between species is challenging. For example, *Spilocuscus kraemeri* is endemic to Manus, but sporadic records have also been made from New Britain and the adjacent Bali-Witu. It remains uncertain whether these were derived from a New Britain population or single individuals carried from Manus.

Grey cuscuses of the *Phalanger orientalis* complex were all until recently treated as a single species that had been translocated widely from Timor in the west as far as Solomon Islands in the east. However, this complex has now been separated into the northern common cuscus (*Phalanger orientalis*), eastern common cuscus (*P. intercastellanus*) and southern common cuscus (*P. mimicus*). *Phalanger orientalis* and *P. intercastellanus* at least apparently have their own translocation histories, the timing and distribution of which still need to be fully resolved. A fourth species, the Bismarck common cuscus (*P. breviceps*), is recognised here based on recent genetic data. All but the latter species are known members of the indigenous fauna of New Guinea. The Bismarck common cuscus was apparently confined to New Britain before its extensive translocation to New Ireland and Solomon Islands. However, further sampling of New Guinea cuscuses may reveal that it too has origins on the New Guinea mainland.

Bismarck common cuscus (*Phalanger breviceps*) New Britain, Papua New Guinea (Stephen Richards).

Phalanger

BISMARCK COMMON CUSCUS

***Phalanger breviceps* Thomas, 1888a**

fa-lan'-jer breh-vee'-seps

Phalanger breviceps was long considered a subspecies of *P. orientalis* of northern New Guinea. However, genetic data have shown the two are quite distinct (Kealy *et al.* 2019; Olson and Lavery, upubl. data). It may yet prove that this species was prehistorically transferred to New Britain from an unknown source population on New Guinea. Wider geographical sampling of New Guinea *Phalanger*, or the discovery of archaeological sites on New Britain, are needed to better understand its evolutionary origin.

This is the most widespread marsupial in the region, owing much of its insular distribution to prehistoric translocations. It is probably well suited to introduction by humans because it is an abundant lowland species and has a generalised diet. Juveniles are commonly kept as pets, but it is rare to see fully grown animals, presumably a large proportion die in captivity before reaching adulthood. The earliest known introduction of this species was to New Ireland, between 10 000 and 19 000 years before present (Flannery and White 1991). Its introduction to the Solomon Archipelago followed sometime after 6700 years before present (Flannery and Wickler 1990).

The external and cranial morphology of the Bismarck common cuscus varies across its distribution, presumably as a result of new populations having been established by translocations of small numbers of individuals. Animals in the eastern Solomon Islands (Makira) and Nissan have the smallest body size, where weights of approximately 1 kg upon maturity are common. The species is predominantly grey, but a range of colour morphs from near white to dark brown (almost black) can be encountered within any given population in the wild. Certain islands are known for the predominance of a particular colour morph, especially white. All females possess a white tip to the tail.

The Bismarck common cuscus will inhabit primary and secondary forests and garden areas close to human habitation. In fact, it tends to be more common in village environs, gardens and disturbed areas because of high concentrations of edible garden plants and fruiting pioneer species. It dens by day in tree hollows, especially the cavities formed by strangler figs (*Ficus* spp.). In high-elevation montane forests where hollows are lacking, Bismarck common cuscuses will conceal themselves in dense epiphytic ferns and orchids. Limited dietary information is available, although it appears to be a generalist folivore and frugivore. On Mono Island in July 2019, numerous individuals were seen on consecutive nights feeding on the fruits of a large *Sterculia* sp. Breeding is aseasonal and the incidence of twin births is high. Younger twins are more commonly encountered than older ones, presumably due to high mortality among the young.

Based on information from local residents, Heinsohn (2002) reported the apparent extinction of the Bismarck common cuscus at the extreme eastern end of its distribution on Santa Ana adjacent to Makira, Solomon Islands. Local knowledge indicated it became extinct there around 1969 when

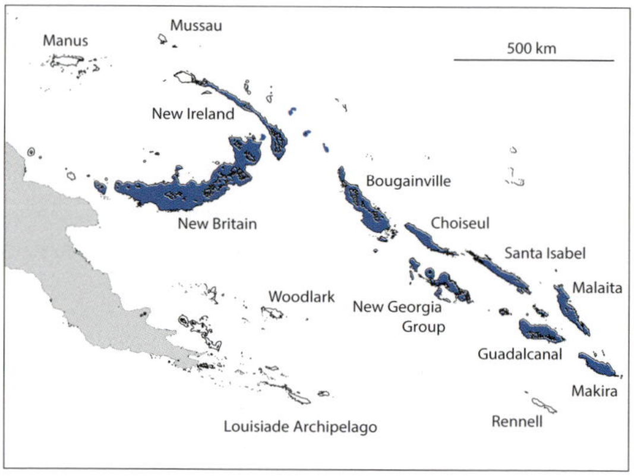

Phalanger breviceps			
Size*			
	Mean	**Range**	**Sample**
Head–body	353.9	276.0–428.0	16
Tail	284.4	221.6–340.0	16
Hindfoot	49.9	38.8–59.6	16
Ear	23.7	18.9–27.3	14
Weight	1536.5	620.0–2300.0	13
* From AM (New Ireland), QM and KU (Solomon Isls) specimens			
Identification Highly variable in colouration, but males are usually buff or grey and females a darker reddish-brown. There is usually a distinct dark dorsal stripe running from the head to the lower back and in females the tail tip is invariably white.			
Synonyms *Phalanger orientalis ducatoris* Thomas, 1922c *Phalanger orientalis albidus* Feiler, 1978a *Phalanger orientalis kori* Menzies and Parnetta, 1986			
Subspecies None			
Distribution Boang, Buka, Bougainville, Duke of York, Lihir, Mioko, New Britain, New Ireland, Umboi, Nissan, Tabar (Papua New Guinea); Choiseul, Fauro, Gatokai, Guadalcanal, Makira, Malaita, Mono, New Georgia, Ngella Sule, Ngella Pile, Ranongga, Rendova, Russell Isls, Santa Isabel, Shortland, Tetepare, Vella La Vella (Solomon Isls)			
Extralimital None			
Conservation status (IUCN) Least Concern (assessed under *Phalanger orientalis*)			
Etymology Sp.–a combination of brevi (L.) meaning 'short' and ceps (L.) meaning 'headed'			

Bismarck common cuscus (*Phalanger breviceps*) Kainake, Bougainville Island. This species is widespread on islands in Melanesia, having been prehistorically translocated to most inhabited islands (Tyrone Lavery).

a major cyclone struck and damaged much of the forest. In Solomon Islands, Sanford's sea eagle (*Haliaeetus sanfordi*) is adept at hunting beneath forest canopies and is a predator of the Bismarck common cuscus.

The Bismarck common cuscus was included in an IUCN Red List assessment with *P. orientalis* and classified as Least Concern. It remains common across most of its range, even where it is heavily hunted, and retention under the category of Least Concern is probably appropriate.

GROUND CUSCUS

Phalanger gymnotis (Peters and Doria, 1875)

jim-noh'-tis

The type locality for the ground cuscus is recorded as Gialnhegen Island, Aru Islands (Peters and Doria 1875). It is a large and common species that is widespread in New Guinea and usually favours rocky areas in forested habitats from sea level up to an elevation of around 2700 m. Menzies (2011) noted this species survives quite well in captivity. Schwartz (1934) discussed one animal from Kabroor Island that lived in the London Zoological Gardens for some time, and another two specimens from the same island collected by Mr Wilfred Frost (NHMUK 1933.11.4.1–2, a female and juvenile). He noted that Mr Frost found the ground cuscus lived in caves there, was well known to residents and was a ground-dwelling animal unable to climb trees like other species of cuscus. The ground cuscus is nocturnal and will den in holes in the ground, among rocks, among tree roots, thickets of vegetation, and even inside disused mine tunnels (Menzies 2011).

The ground cuscus possesses a smaller caecum and proximal colon than other species of the genus, suggesting a different diet less reliant on leaf matter (Menzies 2011). Nonetheless, it is primarily a folivorous-frugivorous species. The fruits of plants from the genera *Elaeocarpus*, *Ficus*, *Pandanus*,

A captive ground cuscus (*Phalanger gymnotis*). This is a large and common cuscus that usually favours rocky areas in forested habitats (Klaus Rudloff).

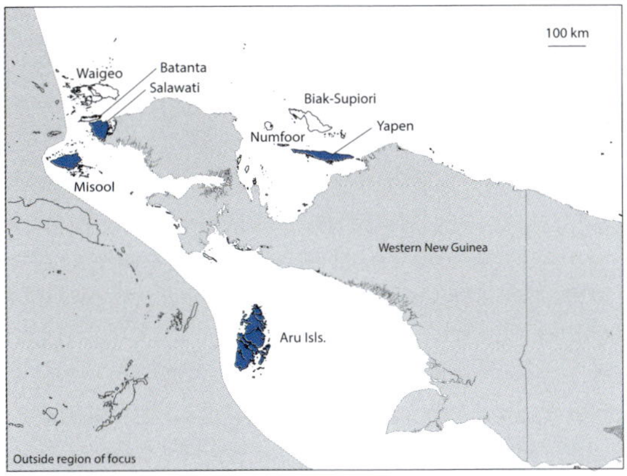

Phalanger gymnotis			
Size*			
	Mean	**Range**	**Sample**
Head–body	437.5	310.0–539.0	6
Tail	331.8	290.0–395.0	5
Hindfoot	57.8	39.0–71.2	6
Ear	28.9	24.4–34.0	5
Weight	2787.5	1500.0–4850.0	4
* From NHMUK specimen (Aru Isls) and Flannery (1995a)			
Identification The ground cuscus can be distinguished from other species of *Phalanger* in the region by its coarsely tuberculate tail and very large third premolars, which are set at an acute angle to the molar row.			
Synonyms None			
Subspecies *Phalanger gymnotis gymnotis* (Peters and Doria, 1875) (Aru Isls, ?lowland New Guinea) *Phalanger gymnotis leucippus* Thomas, 1898a (New Guinea)			
Distribution Aru Isls, Misool, Salawati, Yapen			
Extralimital New Guinea			
Conservation status (IUCN) Least Concern			
Etymology Sp.–naked ear (Gr.)			

Syzygium and *Garcinia* are taken. It will also consume parts of cultivated species of *Rungia*, *Oenanthe*, *Ipomoea* and *Musa*.

Reproduction is non-seasonal and, as with other species of *Phalanger*, the mating system is polygynous, with males playing no part in rearing of the young (Menzies 2011). One offspring is produced at a time and, in captive animals, juveniles begin to leave the pouch after around 108 days. By 161–200 days they have left the pouch permanently. The ground cuscus has a wide distribution and a large population, and is tolerant of a variety of habitats including degraded habitats.

EASTERN COMMON CUSCUS

Phalanger intercastellanus Thomas, 1895

in'-ter-kas-tel-ah'-nus

The eastern common cuscus is the common cuscus of the lowlands of south-eastern New Guinea and the D'Entrecasteaux, Louisiade and Trobriand Islands of Milne Bay. The species in monotypic, with *Phalanger mimicus* of southern New Guinea and Cape York Peninsula having been elevated to species level (Norris and Musser 2001). However, *P. intercastellanus* is a highly variable species, and it is possible subspecies will be recognised.

Surprisingly, limited information is available on the ecology of this species, including diet and reproduction. The eastern common cuscus inhabits both primary and secondary forests and is a popular game animal for hunters across its distribution. It is probably a generalist frugivore and folivore, similar to *P. orientalis*. A female with a single pouch young was recorded on 5 August 1987 on Kiriwina, and on Normanby island dependent young have been documented in August. The eastern common cuscus is an adaptable species able to live in a range of habitats, and threats are not considered severe enough to place the species in a threatened category.

Eastern common cuscus (*Phalanger intercastellanus*), Fergusson Island, D'Entrcasteaux Group. This is the common cuscus of many of the islands of Milne Bay (Pavel German).

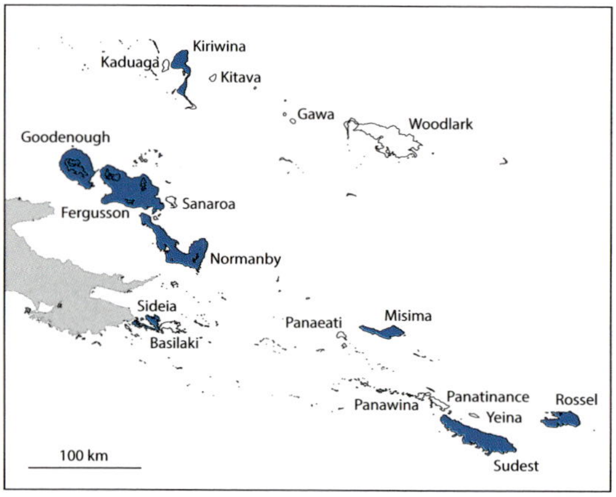

Phalanger intercastellanus			
Size*			
	Mean	**Range**	**Sample**
Head–body	386.0	354.0–440.0	4
Tail	306.8	277.0–325.0	4
Hindfoot	51.1	46.5–55.0	4
Ear	24.0	19.6–27.3	4
Weight	1800.0	1400.0–2200.0	4
* From AM specimens (Goodenough, Ito, Normanby)			
Identification The eastern common cuscus is highly variable in coat colour. It always has a narrow dorsal stripe and lacks mottling on the dorsum as seen in *P. lullulae*. Females have a uniformly dark tail rather than a white tail tip as in *P. orientalis*. Males are difficult to distinguish from *P. orientalis* and *P. breviceps*.			
Synonyms *Phalanger orientalis kiriwinae* Thomas, 1896 *Phalanger orientalis meeki* Thomas, 1898b *Pseudochirus vulpecula* Förster, 1913 *Phalanger brevinasus* Tate and Archbold, 1935a *Phalanger matsika* Tate and Archbold, 1935a			
Subspecies None			
Distribution Goodenough, Fergusson, Ito, Kiriwina, Misima, Normanby, Rossel, Sariba, Sideia, Sudest			
Extralimital New Guinea			
Conservation status (IUCN) Least Concern			
Etymology Sp.–a combination of inter (L.) meaning 'between' and *castellanus*, presumably deriving from D'Entrecasteaux Isls, the type locality			

WOODLARK CUSCUS

Phalanger lullulae Thomas, 1896

lul'-ue-lee

The Woodlark cuscus may have the most unusual colour pattern of all marsupials. The upper surface is formed of a mixture of dark brown, ginger and white patches, giving an almost brindle appearance. The undersurface is predominantly white, but irregular brown patches intrude onto it. No two individuals are alike, with colour varying between predominantly dark brown with whitish spots to predominantly ginger with a few paler and darker spots.

This species occurs on both Woodlark and Alcester Islands, which are separated by around 70 km. No differences have been detected between the two populations, and it is possible that humans have introduced it from Woodlark to Alcester. It also reportedly occurs on the island of Madau which lies just off the western end of Woodlark (Gerry Maynes, pers. comm.).

In 1987, an Australian Museum team found it to be common on Alcester. Surveys on Woodlark in the same period also found it moderately common, even close to human settlements (Norris 1999). The Woodlark cuscus is found in both secondary and primary forests, but is not known to exploit village gardens. The animals are most abundant on the drier eastern side of Woodlark rather than denser forested areas of the west. The 1956–57 Archbold Expedition visited the western end of Woodlark and encountered only four individuals. This led them to believe that it was a rare species overall.

Radio-tracking studies found home ranges are localised and contain a small number of denning

Woodlark cuscus (*Phalanger lullulae*) is an unusually coloured species endemic to remote Woodlark and two nearby islets (Madison Erin Mayfield).

Phalanger lullulae			
Size*			
	Mean	**Range**	**Sample**
Head–body	377.5	328.0–407.0	4
Tail	309.8	276.0–343.0	4
Hindfoot	49.1	40.2–55.0	4
Ear	20.1	17.0–22.2	4
Weight	1687.5	1500.0–2050.0	4
* From AM specimens (Alcester and Woodlark)			
Identification The fur colour pattern of this species is unique, consisting of an irregular mixture of mottled brown, ginger and white. The facial skin is black and contrasts with a pink rhinarium. The ventrum is white with irregular darker spots.			
Synonyms None			
Subspecies None			
Distribution Alcester, Madau, Woodlark			
Extralimital None			
Conservation status (IUCN) Endangered			
Etymology Sp.–woodlark, a bird (L.), for the name of the island where this species was first discovered			

sites (Norris 1999). According to people on Woodlark, dens are typically under epiphytes or in tree hollows. The Woodlark cuscus is nocturnal and solitary, and intraspecific interactions are often aggressive. When animals come into contact with each other during foraging, they can produce a range of vocalisations including snarls, barks and a cry similar to a baby (Norris 1999).

When examined on 7 August 1987, a 1.6 kg female from Guasopa was not lactating but had two enlarged teats. A 2.05 kg female caught there on the same day had a single enlarged nipple and a well-grown young on its back weighing 440 g. Two more females were examined at Guasopa on 12 August 1987. One of these weighing 1.9 kg had a hairless pouch young weighing 23 g, while the second weighed 1.6 kg and was parous but lacking young. On Alcester, two females were examined on 14 August 1987: a subadult weighing 1.2 kg and an adult weighing 1.5 kg, which had a single naked pouch young weighing 10 g. These data suggest that breeding probably takes place over an extended period. In what was a very dry year in 1987, females still had young in August (the dry season).

On Woodlark Island this species is hunted, and hunting pressure varies seasonally, peaking during the 'garden season' or when seas are too rough for fishing. A potential threat to the Woodlark cuscus would be the introduction of the more widespread *P. orientalis* or *P. intercastellanus*. *Phalanger* cuscuses have been widely introduced elsewhere, and interspecific competition, hybridisation or pathogens could represent a problem for this long-isolated species.

SOUTHERN COMMON CUSCUS

Phalanger mimicus Thomas, 1922c

mim-ee'-kus

Norris and Musser (2001) revised the lowland cuscuses of New Guinea and Australia and distinguished *Phalanger mimicus* as a species found from northern Australia to the Trans-Fly area of the New Guinean mainland, west to the Mimika River. In the region under consideration here, southern common cuscus has only been recorded with certainty from Kobroor in the Aru Islands via the collection of trophy specimens by Dr Pat Woolley in 1992 (Aplin and Pasveer 2006). This material was not included in the study by Norris and Musser (2001), but it is consistent with their description of *P. mimicus* and examples of this species from Iron Range, Australia (Aplin and Pasveer 2006). The southern common cuscus was also consistently present throughout the late Pleistocene to Holocene excavation sites in Liang Lemdubu and Liang Nabulei Lisa limestone caves on Kobroor (Aplin and Pasveer 2006). It is probably widespread on other islands of the Aru Group.

In Australia, the southern common cuscus is more or less confined to rainforest habitats but does penetrate into fringing *Acacia* forests (Winter and

Southern common cuscus (*Phalanger mimicus*), Iron Range, Cape York Peninsula, Australia. This is a common species found on Cape York Peninsula, southern New Guinea, and the Aru Islands (Mark Sanders).

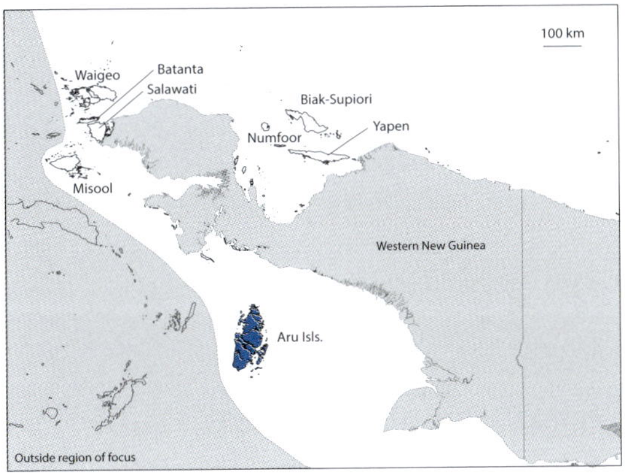

Phalanger mimicus			
Size*			
	Mean	**Range**	**Sample**
Head–body	371.9	245.0–467.0	13
Tail	300.3	210.0–353.0	13
Hindfoot	50.5	37.0–60.0	13
Ear	25.2	16.0–37.0	11
Weight	–	–	–
* From AMNH specimens (north Queensland, Australia and Southern New Guinea)			
Identification Distinguished from *P. orientalis* and *P. intercastellanus* by its smaller size, and females lack the presence of a white tail tip found in *P. orientalis* and *P. breviceps*.			
Synonyms *Phalanger microdon* Tate and Archbold, 1935a *Phalanger orientalis peninsulae* Tate, 1945a			
Subspecies None			
Distribution Aru Isls			
Extralimital Australia, New Guinea			
Conservation status (IUCN) Least Concern			
Etymology Sp.–to mimic, imitate (Gr.), perhaps an allusion to perceived similarities to other species of the *P. orientalis* complex			

Leung 2008). At sites where this species has been recorded on southern New Guinea, the predominant vegetation types have varied from classic lowland rainforest to scrub savannah with 100–200 m wide fringing strips of rainforest along creeks and streams. The savannahs were dominated by *Melaleuca* and on ridges this genus was mixed with *Tristania*, *Eucalyptus* and *Acacia* to form savannah forests (Rand and Brass 1940; Norris and Musser 2001).

The southern common cuscus was found to be solitary and strictly nocturnal in Australia, where it shelters by daytime inside dens (Winter and Leung 2008). It is primarily folivorous and has been observed feeding on red cedar (*Toona australis*), and the buds and flowers of corky bark (*Carallia brachiate*). The stomach of one animal was filled with a paste derived from black bean (*Castanospermum australe*) (Winter and Leung 2008). Females have four teats in the pouch and can give birth to twins. One female captured at Dogwa on the Oriomo River in Papua New Guinea's Western Province had twins on 27 February 1934. Another caught on the same day had a large pouch young with a hindfoot length of 39.3 mm.

The southern common cuscus is widely distributed, occurs in protected areas and shows some tolerance towards habitat modification. However, information specific to the conservation status of this species in the Aru Islands is needed.

NORTHERN COMMON CUSCUS

Phalanger orientalis (Pallas, 1766)

o'-ree-en-tah'-lis

Cuscuses of the Bismarck Archipelago were formerly considered a subspecies of this taxon (*Phalanger orientalis breviceps*). However, genetic data have shown these populations are different from the *P. orientalis* of mainland New Guinea and should be recognised as a distinct species (Kealy *et al.* 2019; Olson and Lavery, upubl. data.). The distribution of *P. orientalis* in the region is thus smaller than previously understood. Nonetheless, it is widespread and has probably been introduced to numerous islands along New Guinea's northern coastline, and further west beyond the region as far as Seram,

Northern common cuscus (*Phalanger orientalis*), Waigeo Island, is widespread on mainland New Guinea and was probably prehistorically translocated by humans to numerous islands (Chien Lee).

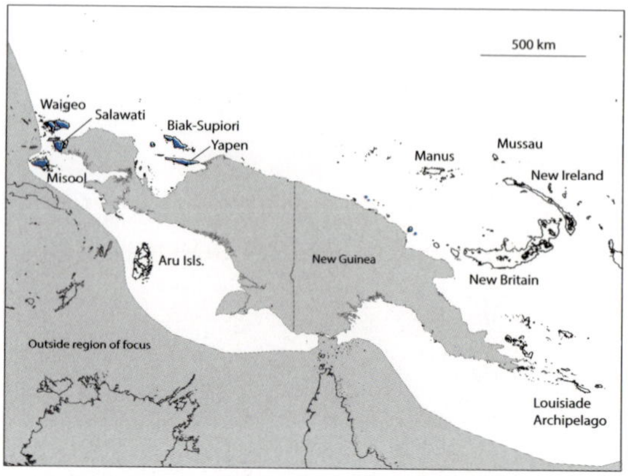

Phalanger orientalis			
Size*			
	Mean	**Range**	**Sample**
Head–body	353.9	276.0–428.0	16
Tail	284.4	221.6–340.0	16
Hindfoot	49.9	38.8–59.6	16
Ear	23.7	18.9–27.3	14
Weight	1536.5	620.0–2300.0	13
* From AM (New Ireland), QM and KU (Solomon Isls) specimens			

Identification Males are usually grey and females a darker reddish-brown. There is usually a distinct dark dorsal stripe running from the head to the lower back, the ear is unfurred internally and in females the tail tip is invariably white.

Synonyms *Didelphis indica* Müller, 1776
Didelphis molucca Gmelin, 1789
Coescoes amboinensis Lacépède, 1799
Phalangista alba Geoffroy, 1803
Phalangista rufa Geoffroy, 1803
Balantia fusca Oken, 1816
Balantia minor Oken, 1816
Balantia moluccensis Oken, 1816
Phalangista rufa Desmarest, 1818
Phalangista rufa var. *alba* Desmarest, 1820 (*nec*. É. Geoffroy)
Phalanger cavifrons Temminck, 1824
Cuscus albus Lesson and Garnot, 1826
Phalanger orientalis timoriensis Menzies and Parnetta, 1986

Subspecies None

Distribution Batanta, Biak-Supiori, Misool, Numfoor, Salawati, Waigeo, Yapen (Indonesia); Bagabag, Karkar, Koil, Vokeo (Papua New Guinea)

Extralimital Ambon, Banda Isls, Buru, Kai Isls, Sanana, Saparua, Seram, Su Mios, Timor-Leste

Conservation status (IUCN) Least Concern

Etymology Sp.–eastern (L.), a reference to the East Indies. At the time of discovery, it was thought to be the eastern counterpart of the American possums.

Buru, Sanana, the Kai Islands and Timor (Glover 1986).

The northern common cuscus is widely distributed and abundant in northern New Guinea, up to an elevation of around 1500 m (Flannery 1995a). It is an inhabitant of primary and secondary lowland forests and higher elevation mossy forests. On mainland New Guinea it reportedly feeds on the fruits of *Diospyros* sp., *Eugenie* sp., *Pandanus* sp. and *Gnetum gnemnon* (Liem 1983).

Spilocuscus

ADMIRALTY CUSCUS

Spilocuscus kraemeri (Schwarz, 1910)

spie'-lo'-kus-kus krem'-er-ee

Spilocuscus kraemeri is a small spotted cuscus endemic to the Admiralty group of islands. It occurs on Manus, Los Negros, Ponam, Rambutyo, Lou, Ninigo Group, Luf and Wuvulu, and probably other nearby small, remote islands. Subfossil deposits at the Pamwak site on Manus reveal the presence of this species around 11 000 years ago. Two records exist from outside the Admiralty Islands, on New Britain and the adjacent Bali-Witu (Anthony 2001). The Admiralty cuscus is a culturally significant species on Manus and continues to be traded beyond the Admiralty Islands (Lamaris 2021). Until established populations are documented on New Britain or Bali-Witu, it is presumed these animals were exported from Manus as inter-island trade.

The Admiralty cuscus is sexually dimorphic and juvenile colours vary from those of mature animals. Males are predominantly grey in colour when

Admiralty cuscus (*Spilocuscus kraemeri*) Manus, Papua New Guinea. This large and spectacularly coloured cuscus remains abundant on Manus (Elodie Van Lierde).

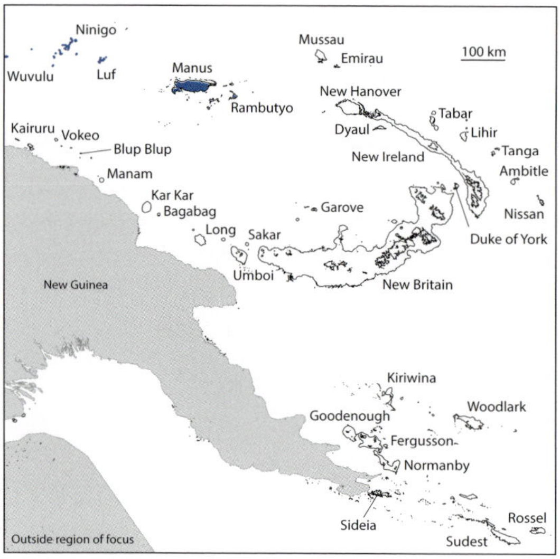

Spilocuscus kraemeri			
Size*			
	Mean	**Range**	**Sample**
Head–body	415.0	310.0–654.0	60
Tail	425.9	289.0–707.0	60
Hindfoot	64.5	47.5–85.0	16
Ear	23.9	20.0–30.0	16
Weight	2944.4	1500.0–4000.0	142
* From AM, AMNH and USNM specimens and Lamaris (2021) (Manus)			
Identification Smaller than *S. maculatus*. The females have a black saddle with a reddish head, limbs and tail; males have small blackish spots on a white background, a reddish head and limbs, and are grey as juveniles.			
Synonyms *Phalanger maculatus minor* Cohn, 1914			
Subspecies None			
Distribution Baluan, Los Negros, Lou, Luf, Manus, Ninigo Group, Pak, Ponam, Rambutyo, Wuvulu			
Extralimital None			
Conservation status (IUCN) Near Threatened			
Etymology Gen.–spotted (Gr.) + cuscus; Sp.–for Herr Professor Dr Kraemer, who helped collect the original series			

juveniles and, as they mature, they develop small blackish-brown spots on a white background with a reddish head and limbs. Adult and juvenile females are similar in colour, with an extensive black saddle on the back. The head, limbs and tail are initially dark, shifting towards more of a rusty brown with maturity.

Spilocuscus kraemeri has been recorded on Manus in a range of habitat types, from mature primary forests to disturbed village gardens. Estimated home-range sizes for 13 animals living in a traditional conservation (*tambu*) zone on northern Manus varied between 0.5 and 18.3 ha (mean = 2.9 ha), and there were intra- and inter-sex home range overlaps (Lamaris and Whitmore 2018).

Females are larger and more numerous than males (Cohn 1914; Lamaris 2021). Of 274 animals hand captured between 2016 and 2020, the number of females was approximately double that of males (181:93) (Lamaris 2021). Cohn (1914) suggested the species may breed seasonally; on a visit to Manus in September–December 1912 he caught many juveniles with a head–body length of approximately 9.5 cm and a tail of 8 cm, but none smaller. However, the results of a long-term monitoring program focused on this species on northern Manus does not support this conclusion. Breeding instead seems to occur in at least two seasonal peaks. The greatest numbers of females with pouch young have been recorded between October and January, with smaller numbers recorded in April and May (J. Lamaris, pers. comm.).

On Manus, the Admiralty cuscus is the largest game animal and is considered a staple food source. It is also culturally significant, and is an expected offering at ceremonial occasions such as weddings and mourning ceremonies (Lamaris and Whitmore 2018). It continues to be traded to other islands and is classified as Near Threatened on the IUCN Red List.

COMMON SPOTTED CUSCUS
Spilocuscus maculatus (Geoffroy, 1803)

mak'-yue-lah'-tus

This large and rather beautiful cuscus is very common in a range of lowland habitats across its range. There are several recognised subspecies but the status of these are uncertain. Ongoing taxonomic revisions may see some subspecies elevated to species level.

The common spotted cuscus is possibly an indigenous component of the faunas of nearshore islands along the coastline of New Guinea. However, it has probably been introduced throughout much of its insular distribution via captive animals carried as food, pets, trade and deliberate introductions. Bones found at the Talepakemelai archaeological

Common spotted cuscus (*Spilocuscus maculatus*) female, mainland New Guinea. This species is widely distributed on Melanesian islands; many populations are the result of translocation by humans (Elodie Van Lierde).

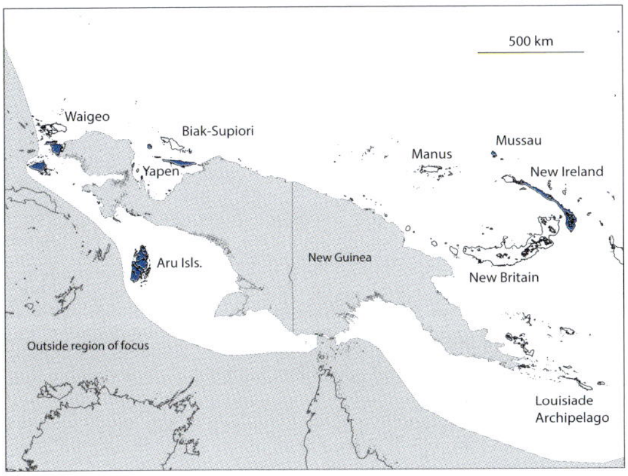

Spilocuscus maculatus			
Size*			
	Mean	**Range**	**Sample**
Head–body	518.3	376.0–630.0	10
Tail	495.0	371.0–630.0	10
Hindfoot	80.5	64.5–100.0	10
Ear	23.0	18.0–31.0	5
Weight	3572.3	1190.7–6123.5	9
* From AM specimens (New Ireland) and AMNH specimens (Milne Bay, Papua New Guinea)			
Identification A very large phalangerid, variable in colour, usually either all white, or with brown or grey mottling, or a saddle on the back. The pupils form vertical slits like those of a cat.			
Synonyms *Balantia variegata* Oken, 1816 *Cuscus brevicaudatus* Gray, 1858 *Cuscus maculatus ochropus* Gray, 1866a			
Subspecies *Spilocuscus maculatus maculatus* (Geoffroy, 1803) (northern New Guinea, Mussau, New Ireland) *Spilocuscus maculatus chrysorrhous* (Temminck, 1824) (south-western New Guinea, Maluku Is.) *Spilocuscus maculatus goldiei* (Ramsay, 1877c) (south-eastern New Guinea)			
Distribution Aru Isls, Batanta, Numfoor, Misool, Salawati, Yapen (Indonesia); Bona Bona, Karkar, Mussau, ?New Britain, New Ireland, Samarai, Walis (Papua New Guinea)			
Extralimital Buru, Kai Isls, Kur, Maluku Isls, New Guinea, Num, Seram, Su Mios, Tioor, Wammer			
Conservation status (IUCN) Least Concern			
Etymology Sp.–a spot (L.), in reference to the coat of the males, which is usually spotted			

site on the island of Eloaua were considered evidence that common spotted cuscus was introduced to the St Matthias Group, but it has been present there for at least 2000–3000 years (Kirch 1987). A modern specimen of a spotted cuscus collected from Boland on Mussau Island is held by Museums Victoria (NMV C787). Flannery and Calaby (1987) remarked upon the completely white fur of this female specimen and identified that it shared cranial similarities with *Spilocuscus maculatus* from northern New Guinea. Helgen (2007) similarly assigned Mussau populations to a spotted cuscus of the northern New Guinea lowlands, but postulated this was an undescribed species.

The common spotted cuscus was reportedly established on New Ireland from founding individuals carried from Mussau (Flannery and White 1991). They were accidentally released in the Kavieng region of northern New Ireland in 1929. The species has spread southward along the island with the main invasion frontier reaching as far south as the Lemakot area by 2010 (Heinsohn 2010b). Secondary introductions have also occurred further southward beyond the invasion frontier (e.g. the Weiten River Valley area in far southern New Ireland). Surveys in this region documented the species at two lowland sites in 1994 (Emmons and Kinbag 2001). Populations had been established on northern New Britain in the late 20th century and early 21st century via individuals that were probably carried across the St George Straight from New Ireland (Heinsohn 2010b). The status of New Britain populations is unclear.

Limited information is available on the biology of island populations, but the common spotted cuscus is generally less common than species of *Phalanger*, with which it is broadly sympatric. Individuals from northern New Guinea, Mussau and New Ireland are small and usually reach 3–4 kg in size (Flannery 1995a). Common spotted cuscus is reportedly common on Mussau. Several were detected by their odour during surveys on the island in 2014, but none was seen or captured (Aplin *et al.* 2015).

Females have a 28-day oestrous cycle, with the end marked by a period of calling for one or two nights, which is likened to the braying of a donkey (Menzies and Dennis 1979). The female pouch contains four teats, but normally only one young is carried, and this vacates at 6–7 months of age (Menzies and Dennis 1979). The young is carried on the mother's back for many months after. The common spotted cuscus has a wide distribution and a large population, and appears resilient to habitat disturbance and hunting.

WAIGEO CUSCUS

Spilocuscus papuensis (Desmarest, 1822)

pah'-pue-en'-sis

Although described in the early 19th century, the Waigeo cuscus has remained very poorly known. The coat colour pattern is remarkable. The adult male has a background of almost pure white, with irregular black splotches covering the back. The female is similar, but the black spots are smaller. Juveniles have blackish-grey tips to the fur, obscuring the pattern of blotches.

The Waigeo cuscus is found in primary and secondary forests. Feiler (1978b) reported on a series of this species held in the Museum für Naturkunde, Berlin, collected in May–June 1931 by George Stein. There are 14 individuals suggesting it may have been common at the time. More than a century before, Lesson and Garnot (1826) had noted that it was then abundant. Modern citizen science records of this species also remain relatively common today. Nothing has been recorded of diet, but animals have been photographed in the branches of fruiting *Ficus* sp., in the upper branches of breadfruit (*Artocarpus altilis*) and clutching the fruits of *Pandanus* sp. Waigeo cuscuses presumably produce one offspring at a time. The species has a small range limited to one island and is thus susceptible to a range of potential threats such as habitat loss and hunting.

Waigeo cuscus (*Spilocuscus papuensis*) is a strikingly coloured species endemic to Waigeo Island in the Raja Ampat group of islands to the west of New Guinea (Chien Lee).

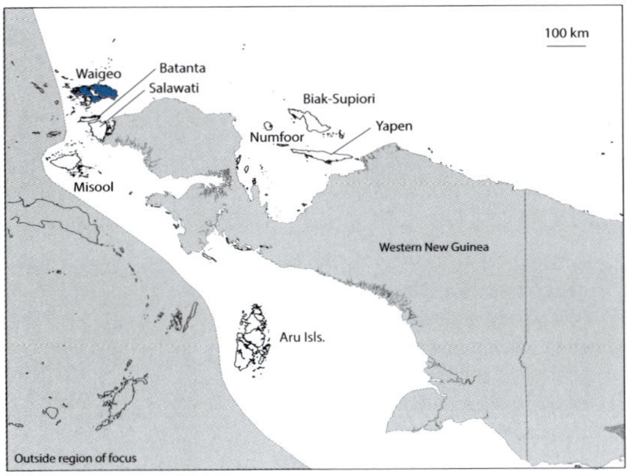

Spilocuscus papuensis			
Size*			
	Mean	**Range**	**Sample**
Head–body	509.7	472.0–560.0	3
Tail	507.7	476.0–555.0	3
Hindfoot	71.0	70.0–72.0	2
Ear	25.7	20.0–29.0	3
Weight	–	2650.0	1
* From ZMB specimens (Waigeo)			
Identification Differs from other spotted cuscuses in that both sexes are spotted, and the individual spots are often smaller.			
Synonyms *Phalangista quoy* Gaimard, 1824 *Cuscus macrourus* Lesson and Garnot, 1826			
Subspecies None			
Distribution Waigeo			
Extralimital None			
Conservation status (IUCN) Vulnerable			
Etymology Sp.–from Papua (L.)			

Blue-eyed spotted cuscus (*Spilocuscus wilsoni*) is endemic to Biak-Supiori. It is poorly known and has been recorded by biologists on very few occasions. This individual was an immature animal kept as a pet on Biak (Tim Flannery).

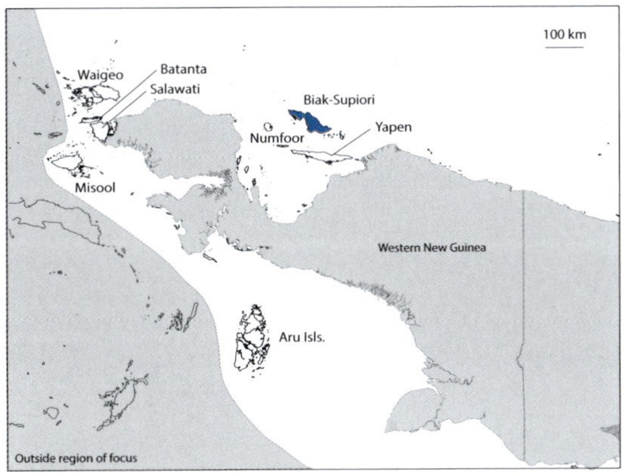

BLUE-EYED SPOTTED CUSCUS

Spilocuscus wilsoni Helgen and Flannery, 2004

will-son'-ee

The blue-eyed spotted cuscus, with its blue-green eyes (unique among the cuscuses) is one of the most striking species of *Spilocuscus*. It is a very poorly known species, rare in museum collections. The holotype, a juvenile, was collected in 1955 by L. D. Brongersma and is held in the collections of Naturalis Biodiversity Center, Leiden. An adult paratype with an imprecise location was registered with the museum in 1906. The blue-eyed spotted cuscus was not encountered in the wild during a 2-week intensive mammal survey on Supiori in September 1992. However, an immature individual

Spilocuscus wilsoni			
Size*			
	Mean	**Range**	**Sample**
Head–body	407.5	335.0–480.0	2
Tail	–	–	–
Hindfoot	–	55.0	1
Ear	–	–	–
Weight	1612.5	1125.0–2100.0	2
* From Dahruddin *et al.* (2014) and Helgen and Flannery (2004)			
Identification a small-bodied species of spotted cuscus, the blue-green eyes are unique in the genus			
Synonyms None			
Subspecies None			
Distribution Biak-Supiori			
Extralimital None			
Conservation status (IUCN) Critically Endangered			
Etymology Sp.–for Don E. Wilson of the Division of Mammals at the United States National Museum of Natural History (Smithsonian Institution), Washington, DC.			

kept as a family pet on Biak was photographed as part of the same expedition. Blue-eyed spotted cuscus is also absent among the significant series of nonvolant mammals collected on Biak in 1962 (AM and AMNH specimens collected by P. Temple and N. Dixon) or during the NAMRU-2 expeditions to Biak and Owi in 1976 (AMNH, MZB, and WAM specimens) (Bergmans and Sarbini 1985; Helgen and Flannery 2004). It is therefore understood to be relatively rare on the island.

A study by Dahruddin *et al.* (2014) provided observational data for this species. The animals were primarily nocturnal; however, the diet and denning behaviour reported were for cuscuses generally (both *S. wilsoni* and the sympatric *Phalanger orientalis*) making it difficult to identify observations that were specific to *S. wilsoni*. Nothing else has been reported of the life history of this species.

Blue-eyed spotted cuscus is believed to have undergone a population decline of more than 80% (inferred the rate and extent of deforestation on Biak-Supiori in the last 10 years). Hunting is also likely to be a major threat.

Petauridae

The family Petauridae contains 14 living species in three genera. The greatest generic diversity is found in Australia (where the family probably evolved), for all three genera – and eight species – are present there. In the New Guinea region it is represented by six species, but only two genera. The distinctive subfamily Dactylopsilinae is most diverse in New Guinea where five species of *Dactylopsila* exist. One species *Dactylopsila trivirgata* is shared between New Guinea and Australia's Cape York Peninsula and the monotypic genus *Gymnobelideus* is endemic to Australia.

Four species of petaurids – two each of the genera *Dactylopsila* and *Petaurus* – are found in the region under consideration here. Two of these are endemic – *D. tatei* (Fergusson Island) and *P. biacensis* (Biak-Supiori). The sugar glider (*P. breviceps*) was formerly regarded as a single species, broadly distributed across much of eastern and northern Australia, New Guinea and surrounding islands. However, recent taxonomic research has seen Australian representatives split into three distinct species. Similar undescribed

Papuan glider (*Petaurus papuanus*) (Angus Fraser).

diversity way well exist in New Guinea and surrounding islands.

Gliders are widespread in the region under consideration, and it is possible that at least some island populations are the results of human introductions. It is evident; however, that other populations are indigenous. For example, on Karkar a distinctive and presumably long-isolated population exists. The striped possum (*Dactylopsila trivirgata*) is largely restricted to land bridge islands in the region, but also occurs in New Guinea and Australia's Cape York Peninsula.

Dactylopsila

TATE'S TRIOK

Dactylopsila tatei **Laurie, 1952**

dak'-til-op'-sil-ah tayt'-ee

Tate's triok is one of the world's least studied species of possum. It was described from nine specimens collected by Fred Shaw Mayer in 1935. These were from the mountains above Taibutu Village in the Faralulu District on the western part of Fergusson Island, at 600–900 m elevation. Another held in the Natural History Museum, London (NHMUK 95.5.8.10), is an adult male preserved entire in alcohol collected by Albert Meek. Expeditions to Fergusson from the 1950s to the 1980s failed to find the species again. Remarkably, the Fifth Archbold Expedition also failed to encounter it during their long sojourn in the D'Entrecasteaux Islands in 1956, and it was subsequently feared extinct or at best extremely rare. However, in 1992 a single animal was captured, providing proof that is still survived on Fergusson. A tail of Tate's triok was reported in the 1980s, and then held by the Nature Conservation Section of the Department of Environment, in Port Moresby. It had been used as a body ornament on Fergusson and may have already been quite old when acquired.

The record of this species on 4 September 1992 was made by an Australian Museum team near Tutumvasaie Village, some 2-3 h walk from Saibutu Village (the same locality referred to by Shaw Mayer as Taibutu). The site was in an area of primary forest at an elevation of at least 1000 m. When located by a hunter, the animal was sleeping in the hollow of a tree known locally as *Kili-Kili*. The resting place was apparently detected when a rolled leaf was noticed protruding from the hollow. Such leaves are reportedly used by the possums for urination and are considered by hunters as a sure sign that an animal is in residence. The tree was cut down to obtain the possum – a common practice in the D'Entrecasteaux Islands for obtaining hollow-dwelling species. Doubtless the method leads to a severe reduction of suitable roost sites.

Field work on Goodenough in 1987 established that Tate's triok does not occur there. However, some island residents who had travelled to west Fergusson were familiar with it, as they had seen or heard of it during visits. Most recently, in 2019 there were reports of hunters taking a Tate's triok on Orthona peak, east Fergusson (D. Mitchell, pers. comm.). Apart from this, all localised records are from the mountains above 600 m elevation near Saibutu in west Fergusson, and the species may be restricted to this small area of high-elevation habitat.

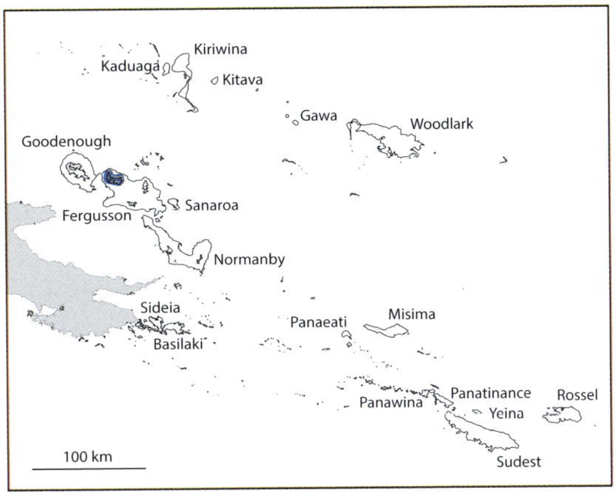

Dactylopsila tatei			
Size*			
	Mean	**Range**	**Sample**
Head–body	193.8	173.0–213.0	8
Tail	272.5	262.0–286.0	8
Hindfoot	44.1	42.0–46.0	8
Ear	23.2	22.0–24.0	8
Weight	–	252.0	1
* From AM specimen (male, weight only) and Laurie (1952)			
Identification smallest member of the genus *Dactylopsila*, the tail is sleek and with a white tip, hands and feet are white. The fourth digit of the hand is elongated, there is no large dark chin spot as is found in *D. trivirgata*.			
Synonyms None			
Subspecies None			
Distribution Fergusson Is.			
Extralimital None			
Conservation status (IUCN) Endangered			
Etymology Gen.–digit (finger or toe) + naked (Gr.), a reference to the sparse fur on the long digits; Sp.–probably for George Tate, mammalogist at the American Museum of Natural History, who died the year the species was described			

Tate's triok (*Dactylopsila tatei*), Fergusson Island, D'Entrecasteaux Islands, Papua New Guinea. This is one of the most poorly known of all marsupials. This photo was the first scientific record for 60 years, made in 1992. Records since are again similarly lacking, but it may have recently been captured by hunters in east Fergusson Island (Pavel German).

STRIPED POSSUM

Dactylopsila trivirgata Gray, 1858

trie'-ver-gah-tah

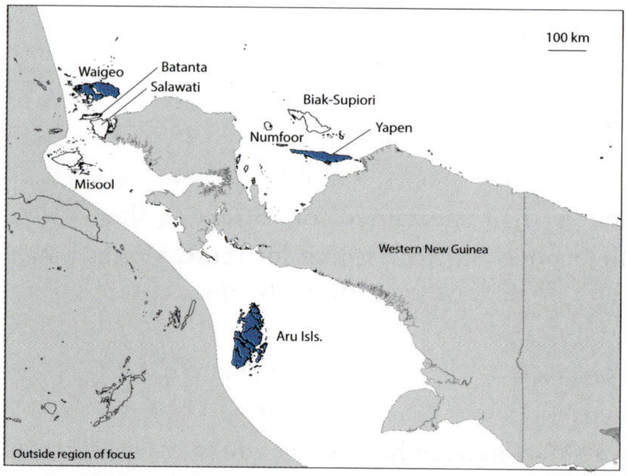

In the region under consideration here, striped possum is restricted to several land bridge islands of western New Guinea. The type locality is in the Aru Islands (presumably Wokam or Kobroor), first recorded there by Alfred Russel Wallace. Excavations that have provided a rich record of the Pleistocene to Holocene vertebrate fauna confirm a long prehistoric occurrence on Kobroor (Aplin and Pasveer 2006). However, few modern records exist from the Aru Islands. One specimen was apparently among the 'kitchen midden' collections made on Kobroor by Pat Woolley in 1992 (Aplin and Pasveer 2006). Matschie (1916) named *Dactylopsila occidentalis*, based on specimens collected by the explorer Luigi Maria D'Albertis on Waigeo. This form is now regarded as a synonym of *D. trivirgata*. Laurie and Hill (1954) recorded it as occurring on Yapen.

Dactylopsila trivirgata			
Size*			
	Mean	**Range**	**Sample**
Head–body	245.7	202.0–278.0	12
Tail	352.9	315.0–390.0	12
Hindfoot	47.2	41.0–51.0	12
Ear	26.3	22.2–31.0	12
Weight	371.8	226.8–527.3	12
* From AM specimens (Chimbu Province and Sandaun Province, Papua New Guinea) and AMNH specimens (Collingwood Bay, Papua New Guinea)			
Identification Differs externally from *D. tatei* in its larger size, longer tail (usually greater than 300 mm and pale at the base), darker hands and feet, presence of a dark chin spot and smaller wrist tubercle.			
Synonyms *Dactylopsila albertisii* Peters and Doria, 1875 *Phalangista (Dactylopsila) angustivittis* Peters and Doria, 1881 *Dactyopsila hindenburgi* Ramme, 1914 *Dactyopsila arafakensis* Matschie, 1916 *Dactyopsila biedermanni* Matschie, 1916 *Dactyopsila occidentalis* Matschie, 1916 *Dactyopsila trivirgata infumata* Tate, 1945b			
Subspecies *Dactyopsila trivirgata trivirgata* Gray, 1858 (most of New Guinea, Aru Isls, Waigeo, Yapen) *Dactyopsila trivirgata melampus* Thomas, 1908b (southern Papua) *Dactyopsila trivirgata picata* Thomas, 1908b (north-eastern Australia) *Dactyopsila trivirgata kataui* Matschie, 1916 (lower Fly River area)			
Distribution Aru Isls, Waigeo, Yapen			
Extralimital Australia, New Guinea			
Conservation status (IUCN) Least Concern			
Etymology Sp.–three striped (L.), in reference to the boldly striped head and body			

Striped possum (*Dactylopsila trivirgata*), Atherton Tablelands, Australia. This common and widespread possum found in Australia and New Guinea has also been recorded from several land bridge islands to the west of New Guinea (Mark Sanders).

Striped possum has a broad ecological tolerance and is the most widespread species of the genus *Dactylopsila*. In Australia it occurs in rainforests and adjacent woodlands in two discrete patches from the

Iron Range to the Paluma Range north of Townsville. In New Guinea it is ubiquitous in lowland and lower montane forests.

Wood-boring beetle larvae are a primary food item for this species, and it possesses an elongated fourth digit specialised for extracting them from their holes. In their search for these, animals may tap on the wood with their claws causing the concealed insects to move and reveal their presence (Menzies 2011). They also feed on a wide variety of insects, flowers and fruits. Striped possum makes leaf-lined nests in hollow trees or among epiphytes. These can be occupied by pairs or same sex groups (Menzies 2011).

Females give birth to one or two young (Handasyde 2008). In captivity, one young remained in the pouch for 90 days, was then carried on the mother's back for 2 weeks and appeared independent by 171 days. Reproduction appears to be non-seasonal. Striped possum has a wide distribution, a presumed large population and appears somewhat tolerant of habitat modification.

Petaurus

BIAK GLIDER

Petaurus biacensis Ulmer, 1940

pet-or'-us bee'-ak-en'-sis

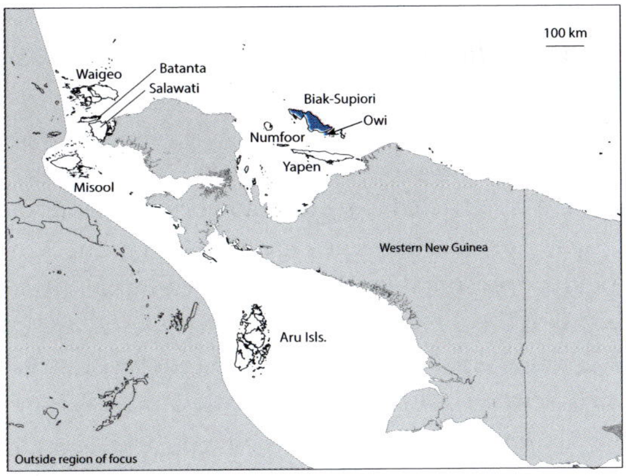

Petaurus biacensis			
Size*			
	Mean	**Range**	**Sample**
Head–body	145.6	140.0–150.0	5
Tail	161.0	145.0–175.0	5
Hindfoot	24.4	22.0–25.9	5
Ear	21.0	19.8–22.5	5
Weight	88.3	79.0–100.0	5
* From AM specimens (Biak-Supiori)			
Identification A relatively small species of glider (head–body length less than 150 mm) with rather uniform tawny or chocolate-brown fur unlike populations of *P. papuanus*.			
Synonyms *Petaurus kholsi* Troughton, 1945			
Subspecies None			
Distribution Biak-Supiori, Owi			
Extralimital None			
Conservation status (IUCN) Least Concern			
Etymology Gen.–tightrope dancer, tightrope walker, tumbler (Gr.); Sp.–from Biak (L.), the island name of the type locality			

This small gliding possum is endemic to Biak-Supiori and Owi in Cenderawasih Bay. The original scientific description was based on four specimens collected on or around 9 December at Korido, Biak (Ulmer 1940). A single specimen collected from Owi by members of the US Typhus Commission was described by Troughton (1945) as *Petaurus kholsi.*

In the 1990s Biak glider remained common on Biak-Supiori and was observed in the evenings climbing on coconut trees in villages. It was also a common inhabitant of gardens and village surrounds in the Korido area of Supiori, and in eastern Biak. Several individuals were captured when feeding on bananas (*Musa* sp.), while a group of six were located in a hollow in a mangrove tree (*Bruguiera* sp.) growing near a village. Three of these escaped and could not be examined, while the remaining three were an adult female (with a pouch young), an adult male and a subadult male. Two females with pouch young were examined on 25 and 27 September. The young (one set of twins and one singleton) all weighed around 7 g. Although limited to a small distribution, the Biak glider is common and tolerates habitat disturbance, and no major threats have been identified.

PAPUAN GLIDER

Petaurus papuanus Thomas, 1888a

pah'-poo-ah'-nus

The sugar glider (*Petaurus breviceps* Waterhouse 1838a) was until recently treated as a single species distributed from north of the equator in the

Halmahera Group of islands, Indonesia, all the way to southernmost mainland Australia, and Tasmania (where it was apparently introduced in the 19th century). However, morphological and genetic studies had suggested it was probably a species complex (Malekian *et al.* 2010). A revision of Australian specimens has now resulted in the recognition of three distinct species, with *P. breviceps* restricted to south-eastern coastal regions of Australia (Cremona *et al.* 2020). The earliest available name for New Guinea is *P. papuanus*; however, taxonomic revisions of Melanesian members of the genus would probably see the recognition of additional new species, including some which may be confined to the region under consideration here (Flannery 1994a; Malekian *et al.* 2010; Colgan and Flannery 1992). The genetic study by Malekian *et al.* (2010), for example, identified five divergent clades in New Guinea. One of these comprised samples from Normanby Island and levels of genetic divergence from mainland clades was indicative of long-term isolation and potential species status. Another distinct clade comprised samples from Karkar Island and Tifalmin (western Papua New Guinea).

Certain populations of Papuan gliders in the South-west Pacific (such as on Karkar) are clearly long-established. However, populations on other oceanic New Guinea satellites (e.g. New Britain) could be due to long-term human agency (Heinsohn 2010b). Jared Diamond recorded this species on Bagabag (Koopman 1979), and it was documented by the Australian Museum in the late 1980s and early 1990s on Karkar, the D'Entrecasteaux Islands (Goodenough, Fergusson and Normanby) and Sudest. Field notes published by Brass (1959) describe regular encounters with the Papuan glider among the D'Entrecasteaux Islands during the Fifth Archbold expedition in 1956. It was present at many of the sites visited on Normanby, Fergusson, Misima, Woodlark and Sudest islands, as well as the mainland near Samarai Island. Although known from New Britain, Aplin and Opiang (2011) did not detect it at high elevation in the eastern Nakanai Mountains in 2014. Searches of New Ireland in 1988 established that it does not occur there (T. Flannery, pers. obs.), as did surveys in the south of the island in 1994 (Emmons and Kinbag 2001).

Papuan gliders are adaptable and on New Guinea they occupy a range of habitats including primary lowland and montane forests, floodplain forests, young secondary regrowth and disused garden areas (Flannery 1995a). The diet is omnivorous and on New Guinea it has been recorded eating the fruits of *Ficus* spp. and beetle larvae (Hide *et al.* 1984). At an elevation above 800 m on Mt Pabinama, Normanby Island, this species was seen feeding on the white fruits of a *Syzygium* species in early May (alongside a species of *Nyctimene*) (Brass 1959). At lower elevation, along the littoral fringe of Normanby it was observed feeding on the flowers of tamanu (*Calophyllum inophyllum*) (again alongside a species of *Nyctimene*).

Papuan gliders live in social groups and individuals are known to share a single nest in the hollow cavities of trees. A group of five animals was encountered in a hollow tree on Normanby (Brass 1959). On 26 December 1988 in Guleguleu Village, east Normanby, Lester Seri noticed a hole in the side of a coconut palm. When he climbed the palm to check whether it was inhabited, three Papuan

The Papuan glider (*Petaurus papuanus*) is common in both disturbed and primary forest on many of the Melanesian Islands that lie close to New Guinea (Chien Lee).

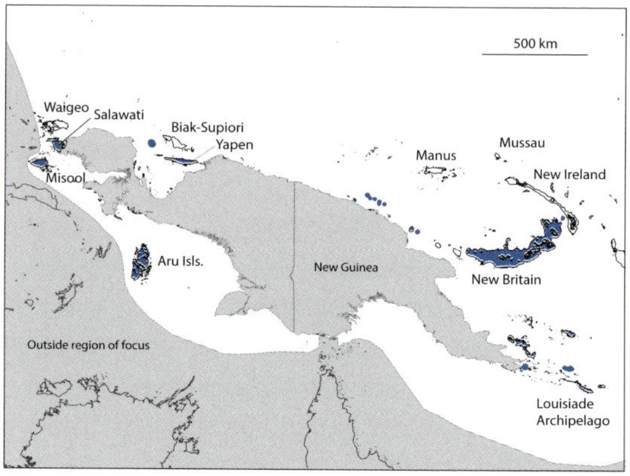

Petaurus papuanus			
Size*			
	Mean	**Range**	**Sample**
Head–body	155.5	117.0–183.0	21
Tail	185.5	145.0–221.0	21
Hindfoot	27.9	24.5–32.0	21
Ear	19.5	17.0–23.0	20
Weight	97.1	39.7–144.6	21
* From AM specimens (Normanby), AMNH specimens (Goodenough and Milne Bay)			
Identification A small greyish or brownish possum with a dorsal stripe and a gliding membrane extending from the wrists to the ankles.			
Synonyms *Petaurus* (*Petaurella*) *papuanus tafa* Tate and Archbold, 1935a			
Subspecies None			
Distribution Misool, Numfoor, Salawati, Yapen (Indonesia); Bagabag, Bam, Blup Blup, Duke of York, Fergusson, Goodenough, Kadovar, Karkar, Koil, Misima, New Britain, Normanby, Sudest, Vial, Vokeo, Woodlark (Papua New Guinea)			
Extralimital Maluku Isls, New Guinea			
Conservation status (IUCN) Least Concern (assessed under *P. breviceps*)			
Etymology Sp.–of Papua (L.), a name for the south-eastern part of New Guinea			

gliders flew out. Two were females, each carrying two very small pouch young. The only other animals caught around the same time also had pouch young. A female collected on 24 September 1956 on Goodenough Island was carrying two pouch young. Little more is known of the specific reproductive biology of this species within its insular distribution. The Papuan glider has not been assessed for the IUCN Red List, and specific assessments will be required following taxonomic revisions and the clarification of species distributional limits.

Pseudocheiridae

The ringtail possums (family Pseudocheiridae) comprise a small family of around 20 species in six genera. They are a group of slow-moving, arboreal mammals with short limbs and woolly fur. The family is entirely confined to Australia, Tasmania, New Guinea, and a few of the larger Meganesian land bridge islands. All pseudocheirids are specialised folivores with complex molars used to shear leaves. They all possess a large caecum to allow for the digestion of such a diet. There also are a large number of extinct fossil species and genera.

Two genera (*Pseudochirops* and *Pseudochirulus*) occur in the region under consideration here. These

Lowland Ringtail (*Pseudochirulus canescens*) (by G. Mützel in Peters and Doria 1881, Annali del Museo Civico di Storia Naturale di Genova, 16, 664–707).

genera occur in both New Guinea and Australia, but no species are shared between the two landmasses. Molecular data indicate that the ancestors of extant members of *Pseudochirops* and *Pseudochirulus* dispersed from Australia to New Guinea between approximately 12.1–6.5 and 6.0–2.4 million years ago respectively (Meredith *et al.* 2010).

The two pseudocheirid species found in the region inhabit Meganesian land bridge islands off western New Guinea. *Pseudochirulus canescens* has been recorded from Yapen and Salawati, and *Pseudochirops albertisii* only from Yapen. Most pseudocheirids prefer montane forests. However, they are absent from the D'Entrecasteaux Group and other mountainous old continental fragments that lie close to New Guinea and support endemic marsupials from other families. This restriction to larger land bridge islands and absence from large proximal islands with mountainous habitats suggests pseudocheirids are poor over-water dispersers and may not be able to survive on the smaller of the land bridge islands that have long been isolated.

Pseudochirops

D'ALBERTIS'S RINGTAIL

Pseudochirops albertisii (Peters, 1874)

sue'-doh-kie'-rops al-ber'-tis-ee-ee

The D'Albertis's ringtail is patchily distributed in northern New Guinea (Arfak, Cyclops, Foja and Weyland Ranges) and Yapen Island. It prefers high elevation moss forests between 1000–1900 m elevation but has also been recorded from secondary forests. The Natural History Museum, London, holds a series of 11 specimens collected by Evelyn Cheesman on Yapen from late August–October 1938. The series includes five adult males and five adult females. All but one were collected on Mt Eiori, and those with elevational data listed were collected at around 600 m. The exception was obtained while Cheesman was camping at around 300 m elevation, although this individual may well have been hunted by locals in higher forests. A note accompanying this specimen states 'in mountain forests, not seen near the coast'. It may be that on the smaller landmass of Yapen, suitable habitat extends to below the 1000 m threshold identified as an elevational limit for the species on the mainland.

D'Albertis's ringtails are nocturnal and the diet includes leaves, fruits and the buds of fig trees (*Ficus* sp.). They are not known to use tree hollows, instead resting in the forks of trees or on the ground under logs (Flannery 1995a). Two females collected on 26 and 30 August on Yapen were

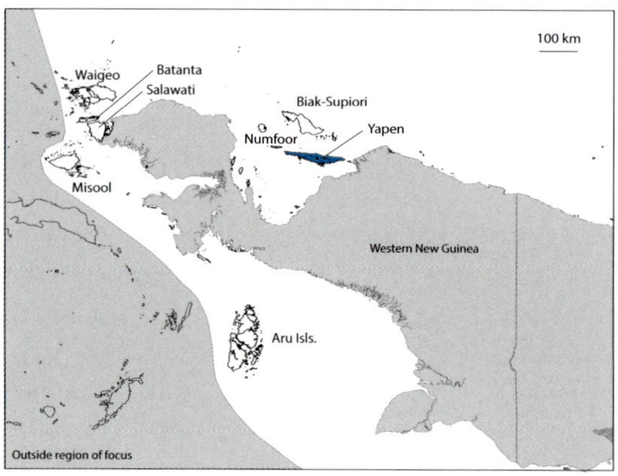

Pseudochirops albertisii			
Size*			
	Mean	**Range**	**Sample**
Head–body	330.7	289.0–347.0	12
Tail	298.5	267.0–322.0	11
Hindfoot	39.2	35.7–46.0	10
Ear	19.3	15.2–25.0	11
Weight	779.0	640.0–875.0	5
* From NHMUK specimens (Yapen), AMNH specimens (West Papua, Indonesia), and Flannery (1995a) (Sandaun Province, Papua New Guinea)			
Identification The only *Pseudochirops* in the region, differing from all phalangerids in its well-furred tail and from *Pseudochirulus canescens* in its golden-tipped fur.			
Synonyms None			
Subspecies *Pseudochirops albertisii albertisii* (Peters, 1874) (Vogelkop Peninsula, West Papua) *Pseudochirops albertisii schultzei* Matschie, 1915 (north coast ranges, Papua New Guinea) *Pseudochirops albertisii insularis* Stein, 1933 (Yapen)			
Distribution Yapen			
Extralimital New Guinea			
Conservation status (IUCN) Near Threatened			
Etymology Gen.–*Pseudocheirus* (sic) + face (Gr.); Sp.–for Luigi Maria D'Albertis, the eminent Italian naturalist and collector of the holotype			

recorded as having single young in the pouch. On mainland New Guinea, females have been found lactating in July, with single newborn young in March, and with more mature but unfurred young (head–body length 98 mm) in September (Flannery 1995a).

D'Albertis's ringtail seems to have suffered a population decline over the last 10 years. In the Torricelli Mountains in the 1990s, the species was limited to small areas of the higher peaks and populations that were close to human settlements were apparently threatened by hunting. Habitat loss also may pose a major threat.

Pseudochirulus

LOWLAND RINGTAIL

Pseudochirulus canescens (Waterhouse, 1846)

sue'-doh-kie'-ru-lus kah-nes'-enz

This widespread species of *Pseudochirulus* inhabits lowland forests of mainland New Guinea and its presence on land bridge islands is therefore not unexpected. However, the records are few and it may be that undocumented populations remain on other islands connected to New Guinea during the Last Glacial Maximum.

An adult male specimen collected from Yapen on 13 March 1931 (MZB 91735) is held by the Museum für Naturkunde, Berlin. No other information accompanies the specimen. Schlegel (1866) described two specimens from Salawati as *P. bernsteini*. The specimens on which he based his description are an adult female and juvenile male, collected by Mar Bernstein, and held in the Naturalis Biodiversity Center, Leiden. This form is still recognised here as a subspecies along with four other described subspecies. However, the status of these needs revision.

On mainland New Guinea, lowland ringtail occurs up to an elevation of 1300 m (Flannery 1995a). Hide *et al.* (1984) reported three animals captured from near this upper limit in Papua New Guinea's Simbu Province. All were in secondary forest: one was inside a tree hollow, and another in the top of a tree fern. The anatomy of the gut indicates it is a primarily folivorous species, and stomach contents have been found to contain plant leaves, mosses, ferns, lichens and fungi (Hume *et al.* 1993). The diet is better quality than that of other ringtails, being higher in nitrogen. Flannery (1995a) postulated this may be an adaptation to the extremely small body size for a folivore.

Lowland ringtails probably reproduce throughout the year. Females have been recorded in New Guinea with pouch young in February, March and October and one subadult (seemingly independent) was recorded in March. Between one and three pouch young have been recorded, the latter an unusually high number for pseudocheirids. The lowland ringtail has a wide distribution, and does not seem to be declining at the rate required to qualify for listing in a threatened category.

The lowland ringtail (*Pseudochirulus canescens*) is common in the lowland forests of mainland New Guinea and is also present on land bridge islands such as Salawati and Yapen (Angus Fraser).

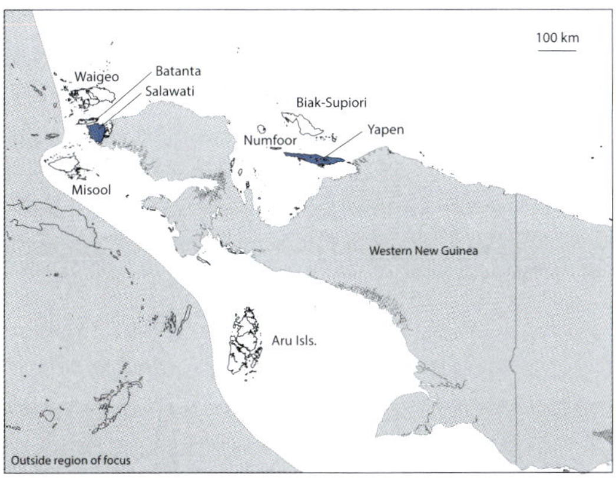

Pseudochirulus canescens			
Size*			
	Mean	**Range**	**Sample**
Head–body	246.4	201.0–282.0	11
Tail	207.9	173.0–268.0	11
Hindfoot	31.9	28.0–35.0	12
Ear	18.0	16.0–18.0	12
Weight	32.1	235.0–380.0	4
* From MZB 91735 (Yapen), Flannery (1995a) (Sandaun Province, Southern Highlands Province, Papua New Guinea), and AMNH specimens (Central Province, Papua New Guinea)			
Identification The only *Pseudochirulus* in the region, differing from all other possums in the region in its possession of a dark patch behind the ears, and from phalangerids in having a well-furred tail.			
Synonyms *Phalangista grisescenti* Peters, 1874			
Subspecies *Pseudochirulus canescens canescens* (Waterhouse, 1846) (West Papua) *Pseudochirulus canescens bernsteini* (Schlegel, 1866) (Salawati) *Pseudochirulus canescens gyrator* (Thomas, 1904a) (north-east Papua) *Pseudochirulus canescens avarus* (Thomas, 1906) (south-east Papua) *Pseudochirulus canescens dammermani* (Thomas, 1922a) (West Papua)			
Distribution Salawati, Yapen			
Extralimital New Guinea			
Conservation status (IUCN) Least Concern			
Etymology Gen.–*Pseudocheirus* (sic) + diminutive (L.); Sp.– becoming light, or becoming old or hoary (L.), presumably a reference to coat colour			

Rodentia

Muridae

The family Muridae is well represented in the region, with 33 indigenous living species described. Of these, 18 are endemic, with many restricted to a single island or island group. Despite the large number of species present, the degree of evolutionary radiation in the group is relatively modest in the region. Perhaps one of the most interesting regions is Solomon Islands, where a rich assemblage of nine endemics can be found. Among them are four species of *Solomys* and two *Uromys* that have evolved into large or very large arboreal murids, probably the evolutionary counterparts of the possums of Meganesia. There also are two species of *Solomys* and two *Uromys* (all now apparently extinct), which were, or probably were, terrestrial. These may have been the ecological equivalents of bandicoots and small forest wallabies found in other parts of Melanesia.

Almost all genera have a predominantly New Guinean or Australopapuan distribution, indicating this has been the major source area for the region. The endemic genus *Solomys* is most closely related to the widespread genus *Melomys* (Bryant *et al.* 2011). New Guinea has also acted as a major secondary centre of radiation for the genus *Rattus* (Rowe *et al.* 2011).

King rat (*Uromys rex*), Guadalcanal, Solomon Islands. This was one of two animals captured from forested areas near Kakambona, west Honiara, and kept for a period by Michael McCoy. In captivity *Uromys rex* accepted shelled ngali nuts, semi-ripe papaya (*Carica papaya*) and some coconut (*Cocos nucifera*) (Robert Fisher).

Five murid species (*R. rattus*, *R. tenzumi*, *R. exulans*, *R. norvegicus* and *Mus musculus*) have been introduced into the region and are now well established. *Rattus exulans* has by far the widest distribution, while *R. norvegicus* is the most restricted, being found only in and around major towns. In addition, the New Guinean *R. praetor* has become established on New Ireland and in Solomon Islands through human agency within the last 3500 years (White *et al.* 2000).

At least three murids (*R. sanila*, *Solomys spriggsarum* and *Melomys spechti*) have become extinct in the region prehistorically, while a further three (*Solomys salamonis*, *Uromys imperator* and *U. porculus*) have probably become extinct in historic times.

Chiruromys

FORBES'S TREE MOUSE

***Chiruromys forbesi* Thomas, 1888b**

kie-rue'-roh-mis forb'-zee

Forbes's tree mouse, with its long fur, long black tail, black eye rings and short face, is a very attractive murid. It is also a very docile species and usually does not attempt to bite when handled. It is a common species throughout its range and is probably one of the most commonly obtained game animals.

The subspecies *Chiruromys forbesi pulcher*, which inhabits the D'Entrecasteaux Islands, is distinctive (being larger than the mainland subspecies), but variable in its morphology. Individuals from high elevation on Goodenough Island have very dense, brown fur with a variable amount of white on the belly. Those from Normanby have shorter, often greyish fur. It seems to be relatively abundant, particularly in primary forests, although there is at least one record of a group being found in a hollow coconut palm in a plantation near disturbed forest on Fergusson Island.

Elliott and Vernes (2021) examined the stomach contents of three Australian Museum specimens collected on Goodenough Island in 1987. The contents were highly digested, but seemed to be primarily composed of insect exoskeletons and plant material. Forbes's tree mouse nests together in groups, usually comprising four or five individuals. At 1300 m elevation on Goodenough Island it was found in hollows around 30–60 cm in diameter that were high up the trunk of straight, tall trees. In two instances the entrance consisted of a small natural hollow that had been enlarged (on the inside only) by chewing. A tunnel led downwards inside the trunk to a nesting chamber 30–40 cm from the entrance. This small nesting chamber was filled with rotting leaves, which were very wet.

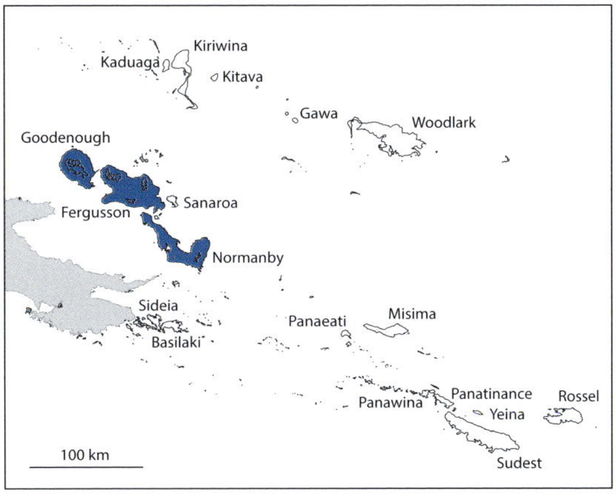

Brass (1959) reported that on Normanby up to four individuals were found in any single hollow. The species was rarely trapped; instead most individuals were encountered by cutting down the trees in which they were nesting. Animals encountered by the Australian Museum on Fergusson Island between 21 and 24 August 1992 were in average-sized groups of four, with the groups shown in Table 4 found in various tree hollows.

Forbes's tree mouse (*Chiruromys forbesi*), Fergusson Island, D'Entrecasteaux Islands, Papua New Guinea. This beautiful murid nests communally in tree hollows (Pavel German).

Few accounts of the reproductive biology of Forbes's tree mouse exist. Menzies and Dennis (1979) examined two colonies in December and found that the first contained four adult animals and one newborn young. The second contained two males and two females, one of which was pregnant with three embryos. Average litter size seems to be two or three and most groups could conceivably consist of a pair, along with one or two of their litters. If this is so, it may well be that the young stay with their parents until they are almost adult-sized, and that the female gives birth to another litter while the earlier litter is still sharing the nest. If this scenario is correct, either breeding is non-seasonal or more than one litter is raised per year. Forbes's tree mouse has a wide distribution and a presumably large population, and there do not appear to be any major threats causing population decline.

Table 4. Composition (age, class and sex) of six groups of Forbe's tree mouse encountered on Fergusson Island between 21 and 24 August 1992. Weights of immature animals in grams are shown in parentheses.

	Group 1	Group 2	Group 3	Group 4	Group 5	Group 6
Adult male	1	1		1	1	
Adult female	1	1	1	1		1
Young adult / subadult male	1 (106 g)		2 (88, 84 g)	1 (80 g)	1 (90 g)	2 (98, 96 g)
Young adult / subadult female			1 (74 g)	1 (86 g)	1 (92 g)	
Juvenile male	1 (56 g)				1 (44 g)	1 (24.5 g)
Juvenile female						1 (25 g)
Total	4	2	4	4	4	5

Chiruromys forbesi			
Size*			
	Mean	**Range**	**Sample**
Head–body	148.5	134.0–161.0	10
Tail	216.5	210.0–222.8	10
Hindfoot	31.7	30.0–32.7	10
Ear	19.1	17.0–22.2	10
Weight	108.3	100.0–122.0	10
* AM specimens from Goodenough Island			
Identification Species of *Chiruromys* can be separated from *Pogonomys* by the possession of raised and strongly overlapping, rather than non-overlapping scales. These black, overlapping and diamond-shaped scales on the tail, the long fur, and very well-developed prehensile tail tip can distinguish this species from other rodents found in the region.			
Synonyms *Pogonomys forbesi vulturnus* Thomas, 1920 *Pogonomys forbesi mambatus* Thomas, 1920 *Pogonomys forbesi satisfactus* Tate and Archbold, 1935b *Pogonomys pulcher major* Tate and Archbold, 1935b *Pogonomys shawmayeri* Laurie, 1952			
Subspecies *Chiruromys forbesi forbesi* Thomas, 1888b *Chiruromys forbesi pulcher* Thomas, 1895			
Distribution Fergusson, Goodenough, Normanby			
Extralimital New Guinea			
Conservation status (IUCN) Least Concern			
Etymology Gen.–hand-tailed mouse (Gr.), a reference to the prehensile tail; Sp.–for Mr H. O. Forbes, collector of the holotype.			

SUDEST TREE MOUSE

Chiruromys sp.

Menzies and Dennis (1979) referred to the presence of a population of *Chiruromys* on Sudest Island. This was substantiated by Musser and Carleton (2005), who identified specimens held in the American Museum of Natural History as a distinct, unnamed species. The series was collected by Russell F. Peterson between late August and early September 1956 at three locations: Joe Landing and Rambuso at sea level, and Mt Rui at 400 m elevation (Peterson 1956b). Head–body and tail length measurements of the largest specimens are slightly larger than those of *C. forbesi*. Nothing is recorded of the ecology of this species. The few specimens reported by Brass (1956) were cut from tree hollows by local guides.

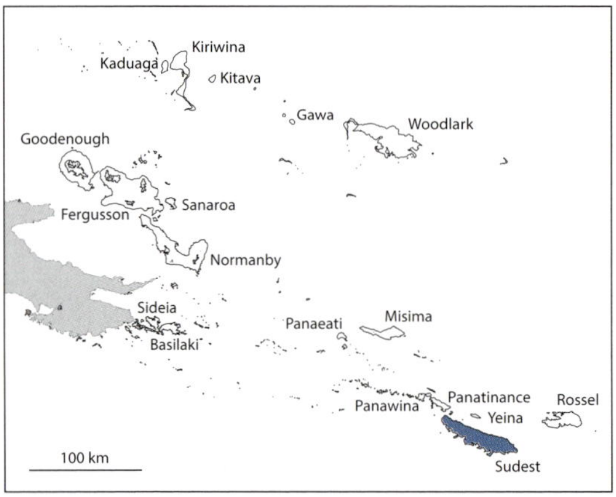

***Chiruromys* sp.**			
Size*			
	Mean	**Range**	**Sample**
Head–body	181.7	165.0–203.0	15
Tail	247.6	229.0–274.0	15
Hindfoot	36.1	34.0–38.0	15
Ear	–	–	–
Weight	–	–	–
* From Russell F. Peterson field notes (Peterson 1956b)			
Identification Species of *Chiruromys* can be separated from *Pogonomys* by the possession of raised and strongly overlapping, rather than non-overlapping, scales			
Synonyms None			
Subspecies None			
Distribution Sudest			
Extralimital None			
Conservation status (IUCN) Not Assessed			

Hydromys

COMMON WATER RAT

Hydromys chrysogaster Geoffroy, 1804

hie'-droh-mis kris'-oh-gas'-ter

The common water rat is a widespread and common rodent, most often associated with littoral and estuarine habitats across Australia, New Guinea and adjacent islands. Given its preference for habitats at the interface of oceanic barriers, one would expect it to be more widespread on the islands surrounding New Guinea. However, it can be cryptic, often going undetected in surveys in parts of Melanesia, and may be more widespread than presently understood. The taxonomy of the common water rat is in need of revision and it probably comprises more than one species. There is a long list of subspecies and synonyms available; all have been listed as synonyms here pending further assessment.

Genetic studies by Bettink (2016) found evidence for Australian and New Guinean clades of this species, as well as an additional sister New

Guinean clade that potentially contained two new taxa paraphyletic with *Parahydromys asper*.

In New Guinea and its satellite islands, the common water rat is usually, but not always, associated with water, and can be found in almost any environment. During the Fifth Archbold expedition in 1956, the species was trapped on Normanby and Kiriwina islands. It was found at two sites on Normanby, one of which was a coconut plantation adjacent to the sea at Waikaiuna. There were several open springs and small, slow-flowing freshwater streams that were bordered by tall reeds (*Phragmites karka*) (Brass 1959). The second location was at around 820 m elevation on Mt Pabinama. The site there consisted of steep narrow spurs, alternating with deep ravines that carried small rocky streams (Brass 1959). The Australian Museum holds specimens taken from Fergusson Island in 1992, similarly collected near rivers or creeks near villages.

The common water rat will den in burrows during the day and is primarily a nocturnal species in the northern extents of its distribution. It is carnivorous and will feed on a wide variety of invertebrates such as crustaceans, and vertebrates such as fish and frogs (Menzies and Dennis 1979). In captivity, females gave birth to litters of one to

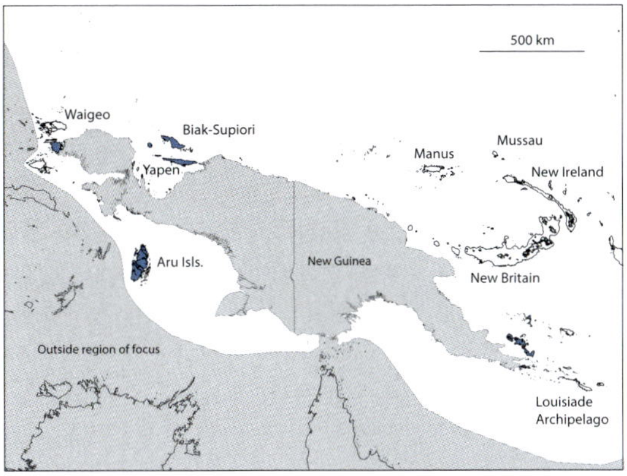

The common water rat (*Hydromys chrysogaster*) is very widespread in Australia, New Guinea and the islands of Milne Bay, Cenderawasih Bay, and the Raja Ampat group. It is usually associated with littoral and estuarine habitats (David Jenkins).

Hydromys chrysogaster			
Size*			
	Mean	**Range**	**Sample**
Head–body	254.0	195.0–282.0	8
Tail	232.9	206.0–260.0	8
Hindfoot	51.6	46.0–55.3	8
Ear	18.2	16.4–20.8	7
Weight	386.4	210.0–540.0	8

* From AM specimens (Sandaun Province, Southern Highlands Province and Fergusson Is.)

Identification The common water rat varies from almost all other murids in the region in having webbed hindfeet. The tail is black and densely furred and has a white tip but lacks a white subterminal band as in *H. neobritannicus* (although whether the subterminal band is present in all *H. neobritannicus* individuals remains uncertain).

Synonyms *Hydromys leucogaster* Geoffroy, 1804
Hydromys fulvogaster Jourdan, 1837
Hydromys fulvoventer Cuvier, 1837
Hydromys fuliginosus Gould, 1853
Hydromys fulvolavatus Gould, 1853
Hydromys lutrilla Krefft, in Gould, 1863
Hydromys beccarii Peters, 1874
Hydromys esox Thomas, 1906
Hydromys chysogaster caurinus Thomas, 1909a
Hydromys chysogaster reginae Thomas and Dollman, 1909
Hydromys melicertes Thomas, 1921
Hydromys nauticus Thomas, 1921
Hydromys esox illuteus Thomas, 1922a
Hydromys longmani Thomas, 1923a
Hydromys grootensis Troughton, 1935a
Hydromys lawnensis Troughton, 1935a
Hydromys moae Troughton, 1935a
Hydromys oriens Troughton, 1937b

Subspecies The number of subspecies is unresolved. All names are listed here as synonyms.

Distribution Aru Isls, Biak-Supiori, Numfoor, Waigeo, Yapen (Indonesia); Fergusson, Goodenough, Kiriwina, Normanby (Papua New Guinea)

Extralimital Australia, Kai Isls, Maluku Isls, New Guinea, Obi

Conservation status (IUCN) Least Concern

Etymology Gen.–water mouse (Gr.); Sp.–golden-bellied (Gr.)

seven (average 3.29) between September and March (Olsen 1982). The young were weaned at around 29 days and most females reached sexual maturity at approximately 240 days. The common water rat is classified as Least Concern on the IUCN Red List because it is widespread, common and occupies a range of habitats of varying levels of human disturbance.

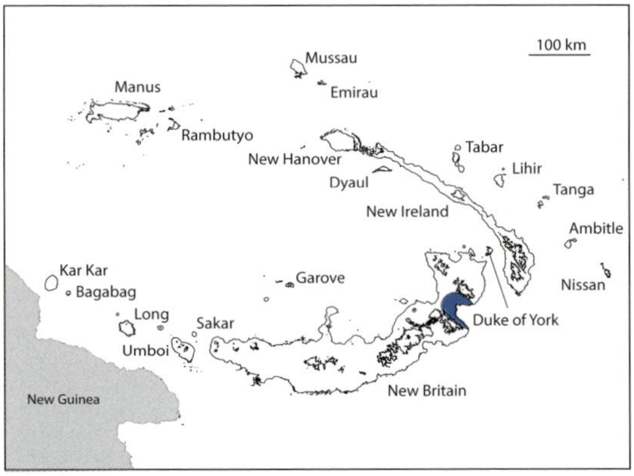

NEW BRITAIN WATER RAT

Hydromys neobritannicus Tate and Archbold, 1935b

nee'-oh-brit-an'-ik-us

The holotype of New Britain water rat (AMNH 99867) was collected by W. F. Coultas during the Whitney South Seas Expedition. It was a female collected on 4 March 1933 at Balayang, Baining Mountains, East New Britain, at an approximate elevation of 760 m (2500 feet). The expedition team at that time was working along the lowlands each side of Wide Bay and along the Mavola River (Gilliard and Lecroy 1967).

A second specimen (ZMB 7960), possibly this species, is held at the Museum für Naturkunde, Berlin. It is an adult female reportedly collected on the north coast of New Britain by E. Hartert on 18 June 1886. It differs from the holotype in having a red belly and in lacking the white tail band anterior to the white tail tip. It is either a mislabelled *Hydromys chrysogaster*, or *H. neobritannicus* is variable in body colouration.

Hydromys neobritannicus			
Size*			
	Mean	**Range**	**Sample**
Head–body	–	288	1
Tail	–	289	1
Hindfoot	–	63	1
Ear	–	25	1
Weight	–	–	–
* Original measurements from the holotype			
Identification Similar to *H. chrysogaster* but differing in its blackish-brown belly and in having a white band around the tail anterior to the white tip.			
Synonyms None			
Subspecies None			
Distribution New Britain			
Extralimital None			
Conservation status (IUCN) Data Deficient			
Etymology Sp.–of New Britain (L.)			

The New Britain water rat (*Hydromys neobritannicus*) has seldom been recorded and its taxonomic status is uncertain (Madison Erin Mayfield).

Extensive, high-elevation field surveys in the Nakanai Mountains in 2009 (above Wide Bay where the holotype was collected) failed to record the New Britain water rat and it may be that it is restricted to larger lowland waterways (Aplin and Opiang 2011). Large-bodied *Hydromys* are semi-amphibious and are typically associated with larger rivers and streams. Nothing else is known of the species and it is presently classified as Data Deficient on the IUCN Red List.

Mammelomys

LOWLAND MAMMELOMYS

Mammelomys rattoides (Thomas, 1922c)

mam-mel'-oh-mis rat-oy'-dayz

This is a large and attractive rodent rarely recorded outside of mainland New Guinea. The

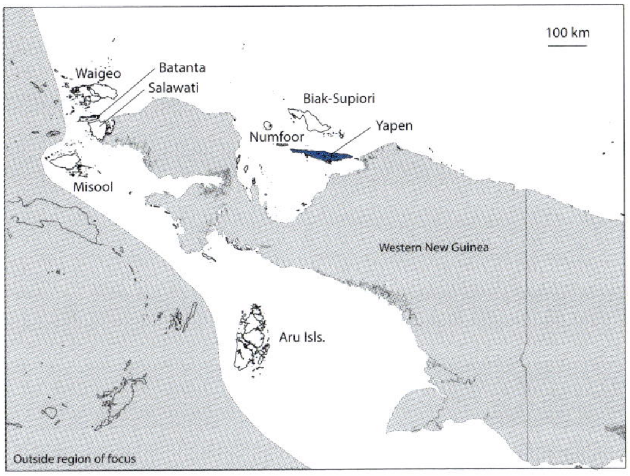

Mammelomys rattoides			
Size*			
	Mean	**Range**	**Sample**
Head–body	193.6	159.0–224.4	22
Tail	147.1	112.0–172.0	22
Hindfoot	42.3	38.0–45.0	22
Ear	20.9	15.0–23.9	22
Weight	205.0	172.0–236.0	4
* From NHMUK 1946.643 (Yapen), Flannery (1995a) (mainland New Guinea), and AMNH specimens (West Papua, Indonesia)			
Identification A large and attractive rodent; females have only one pair of mammae. The hindfoot is long (more than 38 mm), relatively narrow, and white. The tail appears to be slightly hairy to the naked eye, in contrast with the normally naked appearance of species of *Melomys*.			
Synonyms None			
Subspecies None			
Distribution Yapen			
Extralimital New Guinea			
Conservation status (IUCN) Least Concern			
Etymology Gen.–mammae (abbreviated, referring to the single pair of mammae, unique among Papuan rodents) + Melanesian (abbreviated) + mouse (Gr.); Sp.–rat like (L.)			

Natural History Museum, London, holds one specimen collected by Evelyn Cheesman on Yapen (NHMUK 1946.643). It is an adult female preserved as a skin and skull, collected at an elevation of around 1000 m on the 'central range' on 11 November 1938.

On mainland New Guinea, the lowland mammelomys is found from 320 to 1400 m elevation (Flannery 1995a). It is a species that nests in burrows and is usually solitary. Females seem to normally give birth to just a single young at a time. Two females caught in Sandaun Province in May were lactating, another had a single hairless foetus and a fourth had a lightly furred near-term foetus (Flannery 1995a). A pregnant female was also caught in the Torricelli Mountains in late July. The lowland mammelomys has a wide distribution and presumed large population, and no major threats are known.

ROSSELL ISLAND MELOMYS

Melomys arcium (Thomas, 1913d)

mel'-oh-mis ark-ee'-um

The Rossell Island melomys was formerly recognised as a subspecies of *Melomys leucogaster*. However, the two are distinct species with distributions that are widely separated. A series of specimens collected on Rossell by Russell F. Peterson in 1956 is held at the American Museum of Natural History. This remains the only island from which the Rossell Island melomys has been confirmed, a surprising distribution given Rossell is the most isolated of all the Louisiade islands.

The Rossell Island melomys (*Melomys arcium*) is a poorly known species endemic to Rossell Island in the Louisiade Archipelago, Papua New Guinea (Madison Erin Mayfield).

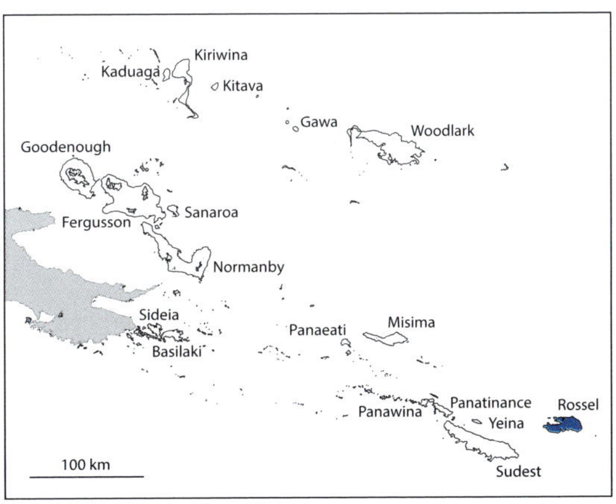

Melomys arcium			
Size*			
	Mean	**Range**	**Sample**
Head–body	–	138	1
Tail	–	127	1
Hindfoot	–	28	1
Ear	–	18	1
Weight	–	–	–
* Recorded from the holotype (Thomas 1913d)			
Identification General colour of the dorsum is rufous brown, hairs of the belly are white to their bases, and there is no sharp demarcation between the dorsum and ventrum colours. Tail is shorter than head–body and practically naked with approximately 14 rings of scales per centimetre.			
Synonyms None			
Subspecies None			
Distribution Rossell			
Extralimital None			
Conservation status (IUCN) Data Deficient			
Etymology Gen.–Melanesian (abbreviated) + mouse (Gr.); Sp.–chest or box (L.), allusion unclear			

Menzies and Dennis (1979) suggested that the Rossell Island melomys also occurred on Misima and Tagula off southern New Guinea. However, subsequent surveys have failed to confirm this. It is possible that Menzies and Dennis (1979) referred to another species of *Melomys*, that additional populations of *M. arcium* do continue to exist outside Rossell, or that range contraction has occurred as a result of extinctions on some islands.

Rossell Island melomys is known to occur between 50 and 700 m elevation. Nothing has been recorded of habitat, diet or reproduction. The tail is unusually short for a species of *Melomys* and this may indicate terrestrial rather than arboreal habits. If this is the case, range declines and local extinctions are possibilities, as suffered by other species of ground-dwelling rodents of the region. The Rossell Island melomys is classed as Data Deficient on the IUCN Red List and an updated assessment of conservation status should be a priority.

BOUGAINVILLE MELOMYS

Melomys bougainville Troughton, 1936a

bue'-gan-veel

The Bougainville melomys is a relatively common rodent previously considered a subspecies of *Melomys rufescens*. Flannery and Wickler (1990) recorded this species in fossil deposits at a site on Buka and reported six specimens from Bougainville and one from Buka held in the Australian Museum. A further two specimens from Bougainville are held in the Musée Royal d'Histoire Naturelle, Belgium, and another individual (an adult female) was encountered at Siwai in southern Bougainville in 2016 and is held at the Australian Museum.

From Choiseul, two specimens are held at the Australian Museum, and a third collected by P. Pikacha in 2014 is in the collections of the Queensland Museum. The latter individual was caught in a trap baited with coconut and set on the floor of ultramafic forest at an elevation of around 400 m. It was the only individual captured despite hundreds of traps being laid over several nights. Researchers from the University of Kansas also spotlighted what may have been this or a closely related species on Santa Isabel in June 2018 (K. Olson & I. Tigulu, pers. comm.). Its possible existence on Santa Isabel is not unexpected as Buka, Bougainville, Choiseul and Santa Isabel were all interconnected during the Last Glacial Maximum. In July 2019, Bougainville melomys was recorded on Mono Island by P. Holland and T. H. Lavery. Its occurrence there is more surprising as Mono is separated from Bougainville by deep trenches and oceanic or human-mediated dispersal would have been essential for Mono to be colonised.

The Bougainville melomys occurs in primary lowland forests, ultramafic forest, secondary forests, and village edges adjoining forests. They are rarely captured using common rodent trapping techniques (e.g. terrestrial traps baited with peanut butter or coconut). This presumably is because they are mostly arboreal, with a diet consisting largely of forest fruits. Bougainville melomys are agile climbers, able to rapidly ascend trees and traverse thin lianas to move through the canopy.

On Mono Island in July 2019, this species was regularly spotlighted in areas of lowland forest. On consecutive nights they were found in the upper branches of a *Sterculia* sp. approximately 20 m in height. The animals were feeding on fruits of the tree alongside Bismarck common cuscuses (*Phalanger breviceps*). Other individuals were encountered feeding on the fruits of wild betel nut (*Areca* sp.), and one was spotlighted feeding on the

The Bougainville melomys (*Melomys bougainville*) is highly arboreal and consumes a variety of native forest and cultivated fruits including bananas (Stephen Richards).

end of a ripe banana (*Musa* sp.) at around 9 p.m. on the edge of Falamai Village. Another individual was seen in a large strangler fig (*Ficus* sp.), also in the vicinity of Falamai Village. At Siwai on Bougainville, a camera trap set in the branches of a small fruiting fig tree (*Ficus copiosa* or *F. wassa*) at the edge of Kainake Village recorded this species repeatedly visiting over several nights to extract the ripe figs.

The reproductive biology of Bougainville melomys is not well known. Two animals recorded at Himandara, Choiseul, on 15 March 1990 were lactating females with young (weighing 26 g).

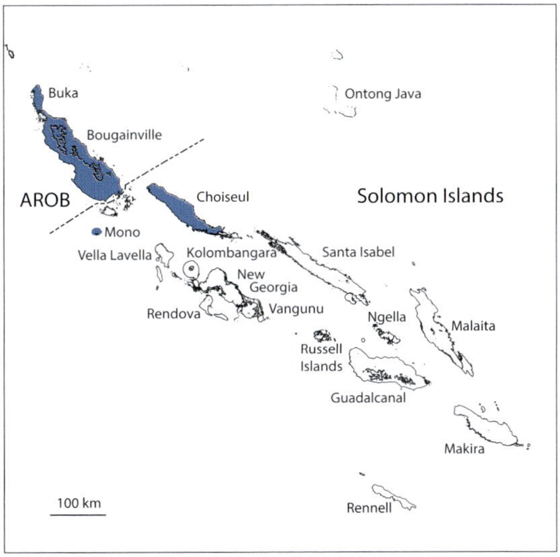

Melomys bougainville			
Size*			
	Mean	**Range**	**Sample**
Head–body	141.5	134.7–151.1	5
Tail	147.0	135.0–165.0	5
Hindfoot	26.8	23.9–29.2	5
Ear	14.3	12.9–15.6	5
Weight	105.8	82.5–139.8	4
* From AM (Bougainville, Choiseul) and KU and QM (Mono) specimens			
Identification A mosaic-tailed rat, with orange dorsal fur contrasting strongly with a pure white underbelly.			
Synonyms None			
Subspecies None			
Distribution Bougainville, Buka, Choiseul, Mono			
Extralimital None			
Conservation status (IUCN) Data Deficient			
Etymology Sp.–for the island of Bougainville, the type locality			

Given the apparent tolerance of disturbed habitats, and the arboreal habits of the Bougainville melomys, it is probably capable of withstanding the major threats posed by commercial logging and feral cats.

GRASSLAND MELOMYS

Melomys burtoni (Ramsay, 1887)

bur'-ton-ee

Populations of *Melomys* from the Milne Bay Islands (Woodlark, Sudest, Misima) have a confused taxonomy and require further assessment. They exhibit similarities with *M. lutillis* from mainland New Guinea and *M. burtoni* from Australia. The results of molecular sequencing by Bryant *et al.* (2011) concluded *M. lutillus* and *M. burtoni* were conspecific. Fabre *et al.* (2018) similarly found *M. burtoni* and *M. lutillus* to be closely related, but concluded further molecular sequencing across a wider taxonomic sample was needed to resolve their relationships. Here we tentatively refer specimens discussed here to *M. burtoni* (after the earliest named member of this species complex – *sensu* Fabre *et al.* 2018).

Melomys are known from Woodlark by three specimens (one of which is a subadult), from Sudest by three specimens (two subadult) and from Misima by three specimens (two subadult). All except an adult specimen from Woodlark (collected during a 1987 Australian Museum expedition) are held in the American Museum of Natural History and were collected during the Fifth Archbold Expedition to New Guinea. Brass (1959) provided further detail on the circumstances under which these animals were encountered. One from Misima was taken at a camp on the north slopes of Mt Sisa in July 1956. The site was in an area previously disturbed by gold mining operations, but with good primary forest a short distance from the camp (with *Schizomeria*, *Metrosideros*, *Elaeocarpus*, *Lophopetalum*, *Buchanania*, *Ternstroemia* and other genera present). Two subadults from Sudest were taken from a camp at 'Joe Landing' where a mixture of primary rainforest and *Themeda*-dominated grasslands were present. The animals were trapped on the ground, and other rodents taken from the same site included *Rattus exulans*, *R. praetor*, and an arboreal rodent listed as '*Pogonomys* of the *forbesi* group'. The mature individual captured on Sudest was taken at Rambuso on the north coast and was trapped in primary forest on lower slopes (Brass 1959).

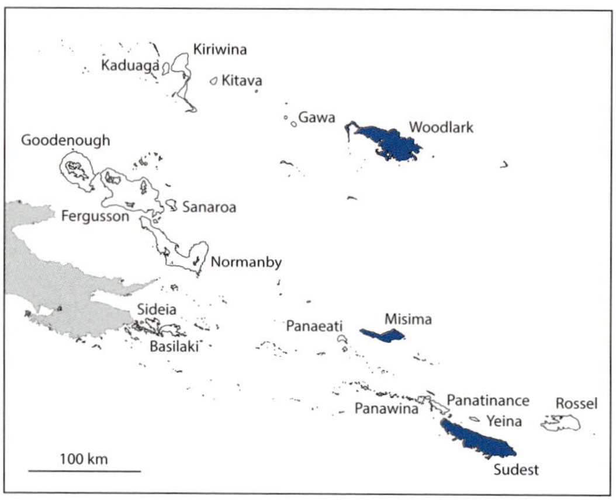

Melomys burtoni			
Size*			
	Mean	**Range**	**Sample**
Head–body	112.8	98.0–145.8	27
Tail	108.5	72.0–131.7	27
Hindfoot	24.1	21.0–30.0	27
Ear	14.9	13.3–17.3	8
Weight	–	91.5	1
* From AM and AMNH specimens (Milne Bay Isls), and AMNH specimens (Central and Western Provinces, Papua New Guinea)			
Identification A medium-sized melomys with a reddish-grey dorsum, and an irregularly mottled red and white belly. The tail scales are flat and the tail is bicoloured.			
Synonyms *Uromys muscalis* Thomas, 1913d *Melomys lutillus hintoni* Rümmler, 1935 *Melomys muscalis froggatti* Troughton, 1937b			
Subspecies Uncertain, all names listed as synonyms			
Distribution Misima, Sudest, Woodlark			
Extralimital Species complex in Australia, Indonesia, New Guinea			
Conservation status (IUCN) Least Concern (*M. lutillus*)			
Etymology Sp.–little dirty or little yellow (L.)			

On Woodlark this species inhabits rainforests on both the drier eastern and wetter western parts of the island. An Australian Museum specimen (M.20406) was taken in an Elliott trap baited with cooked taro and set near a well adjacent to a village near Guasopa. The site was surrounded by disturbed secondary growth. These surveys were undertaken during an exceptionally dry season, and extensive trapping in forest away from water failed to capture any additional specimens.

MANUS MELOMYS

Melomys matambuai Flannery *et al.*, 1994

mat-am'-bu-i

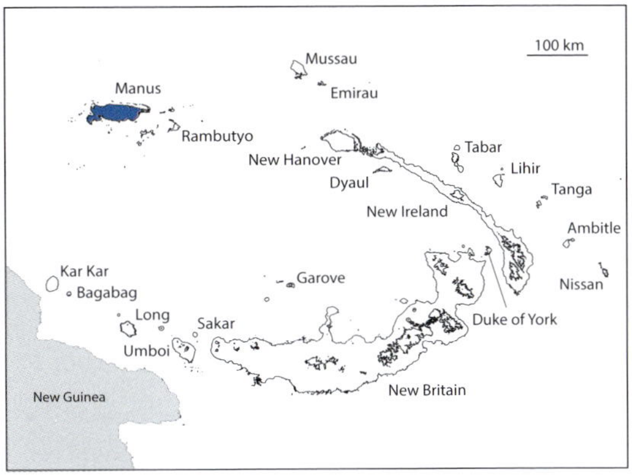

Manus melomys is the largest known member of the *Melomys rufescens* group, which are spread widely throughout mainland New Guinea, the Bismarck Archipelago and northern Solomon Islands. It is very distinct from other *Melomys* and presumably has long been isolated on Manus (Flannery *et al.* 1994).

Melomys matambuai has been recovered from the Pamwak archaeological collection on Manus (Williams 1999). However, it remains known from only a handful of modern records, although it has been observed more recently during ecological research focused on Admiralty cuscus (*Phalanger kraemeri*) (J. Lamaris, pers comm.). The holotype was a non-lactating female collected near Polomou DPI Station (200 m elevation); and the paratype was collected in a forestry area in west Manus. The Polomou individual was encountered climbing in a cacao bush some 1–1.5 m above the ground in a plantation area, and the other was found climbing on a sago palm (*Metroxylon* sp.). Efforts to trap this species on the ground at the same time and locality using live rodent traps failed despite over 100 trap-nights. Similarly, surveys in 2014 using rodent traps (397 trap-nights) and automated cameras failed to record Manus melomys (Aplin *et al.* 2015). However, a small orange rat believed to be this species was spotlighted climbing in the forest subcanopy at Yeri River on the northern coast. Two individuals were captured in north central Manus in August 2002. One was caught in a trap set in a large dead tree buttress and the second was caught in a trap set on the ground under rattan in sparse undergrowth (Pine *et al.* 2017). Camera traps set in trees for *Spilocuscus kraemeri* have been more successful at detecting this species and it seems it is mostly arboreal, similar to *M. bougainville* (J. Lamaris, pers comm.). Very little is known of the reproductive biology. One of two females captured in August 2002 was carrying a single foetus (Pine *et al.* 2017).

Melomys matambuai			
Size*			
	Mean	**Range**	**Sample**
Head–body	159.4	151.3–165.0	3
Tail	177.5	175.0–180.0	2
Hindfoot	34.3	31.0–36.0	3
Ear	18.7	18.0–19.0	3
Weight	171.7	145.0–200.0	3
* From AM specimen (Manus) and Pine *et al.* (2017)			
Identification The tail is dark brown and somewhat paler below and proximally (possibly tipped with white), with roughly circular to squarish to roughly hexagonal scales. The tail of the holotype was damaged when collected. The fur is relatively shorter than in *M. rufescens*, the belly is pure white, hands and feet pale, and the back is reddish with dark grey underfur.			
Synonyms None			
Subspecies None			
Distribution Manus			
Extralimital None			
Conservation status (IUCN) Endangered			
Etymology Sp.–for Karol Matambuai Kisokau, a Manus man and first Permanent Secretary of the Papua New Guinea Department of Environment and Conservation			

Manus melomys (*Melomys matambuai*), an adult female, Manus Island, Papua New Guinea (Ann Williams).

BLACK-TAILED MELOMYS

Melomys rufescens (Alston, 1877)

rue-fes'-enz

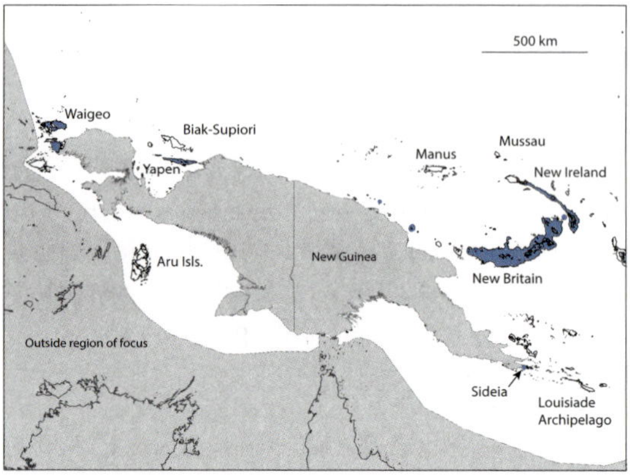

Melomys rufescens is widespread, comprising numerous described subspecies, some of which may eventually be recognised as distinct species. The type locality is Duke of York Island and it has also been documented on adjacent Mioko, New Britain and New Ireland. Within the region under consideration here it has also been documented from small satellite islands off north-eastern New Guinea (Blup Blup, Karkar), adjacent to the Papuan Peninsula (Sideia), and off New Ireland (Lamassa). The archaeological record indicates *M. rufescens* is indigenous to New Ireland (Flannery and White 1991).

Black-tailed melomys is a common species often encountered in forest edges, clearings, disturbed forests and regrowth and is often seen climbing in low vegetation in gardens or near villages at night. The species seems to prefer these more open habitats and is less common in primary forests. It is often found in houses, particularly where introduced rodents are rare or absent. Unlike the related species *M. bougainville* and M. *matambuai*, black-tailed melomys can be readily trapped on the ground using Elliott traps, suggesting its habits may not be as arboreal as those species. It does, however, build nests of woven grass at some height above the ground (Menzies and Dennis 1979).

Melomys rufescens			
Size*			
	Mean	**Range**	**Sample**
Head–body	131.6	113.0–148.0	7
Tail	141.6	131.0–155.0	7
Hindfoot	26.9	15.5–28.0	7
Ear	16.2	15.4–17.0	7
Weight	81.0	78.0–84.0	2
* From AM specimens (New Ireland) and AMNH specimens (New Britain, north-west Cenderawasih Bay)			
Identification Differs from other *Melomys* (except *M. bougainville* and *M. matambuai*) in having a uniformly dark tail. The small tail scales are arranged in more definite rings, and it lacks the up-curled, more prehensile tip of *M. bougainville*. The belly fur is uniform white.			
Synonyms *Mus musavora* Ramsay, 1877d			
Subspecies *Melomys rufescens rufescens* (Alston, 1877) (New Britain and New Ireland) *Melomys rufescens stalkeri* (Thomas, 1904b) (lowlands of northern and south-eastern New Guinea *Melomys rufescens gracilis* (Thomas, 1906) (New Guinea Cordillera from Mt Bosavi east to the Owen Stanley Range) *Melomys rufescens calidior* (Thomas, 1911a) (lowland south-west New Guinea) *Melomys rufescens hageni* Troughton, 1937b (New Guinea highlands from central West Papua to the Kratke Range) *Melomys rufescens niviventer* Tate, 1951 (Digul and Fly River basins southern New Guinea) *Melomys rufescens wisselensis* Menzies, 1996 (Wissel Lakes, western New Guinea)			
Distribution Salawati, Waigeo, Yapen (Indonesia); Blup Blup, Duke of York, Karkar, Lamassa, Mioko, New Britain, New Ireland, Sideia (Papua New Guinea)			
Extralimital New Guinea			
Conservation status (IUCN) Least Concern			
Etymology Sp.–becoming red (L.)			

Black-tailed melomys (*Melomys rufescens*) (illustration by Joseph Smit in Alston 1877).

Melomys rufescens is classified as Least Concern on the IUCN Red List. Its predilection for human modified habitats means its status is probably secure; it may be more widespread that its current patchy distribution suggests.

Paramelomys

MONCKTON'S PARAMELOMYS

***Paramelomys moncktoni* (Thomas, 1904a)**

munk'-tun-ee

This is a widespread species in eastern New Guinea. In the region under consideration here it has only been recorded from Sideia Island adjacent to the Papuan Peninsula. Its presence there has been verified by just a single specimen, a female collected on 25 December 1988 by G. Hangay near Sideia Mission (AM M20166). It was taken in a snap trap and damaged by land crabs after capture.

Menzies and Dennis (1979) stated this species is fully terrestrial and most common in hill forest at low altitudes. One female kept in captivity bore two young, and one of these (a male) had reached adult size by the time it was 6 months old. Little else is known of ecology.

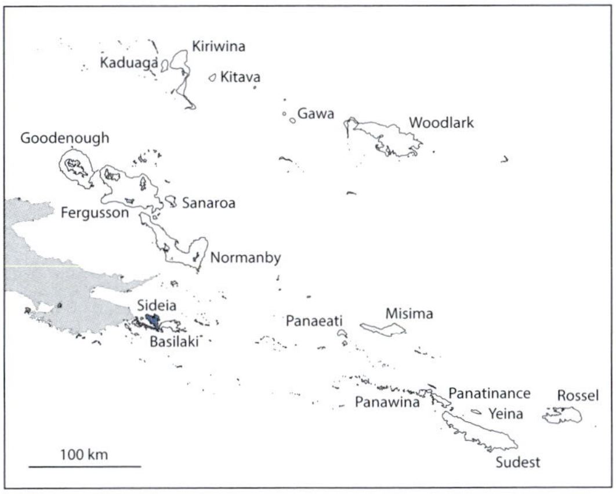

Paramelomys moncktoni			
Size*			
	Mean	**Range**	**Sample**
Head–body	142.5	125.0–170.0	8
Tail	124.6	113.0–145.0	8
Hindfoot	28.0	24.8–33.8	10
Ear	19.4	18.3–20.5	5
Weight	115.4	78.5–18.0	4
* From AM specimens (Marobe, Papua New Guinea), AMNH specimens (Central Province, Papua New Guinea)			
Identification Tail possesses three hairs per scale (rather than one as in other *Paramelomys* species). It is a plump, rotund animal with a broad hindfoot, a tail shorter than head–body, and may possess a dark eye ring. Females have two pairs of mammae.			
Synonyms None			
Subspecies None			
Distribution Sideia			
Extralimital New Guinea			
Conservation status (IUCN) Least Concern			
Etymology Sp.–for the collector of the holotype, Mr C. A. W. Monckton			

LONG-NOSED PARAMELOMYS

***Paramelomys naso* (Thomas, 1911a)**

nah'-soh

Paramelomys naso was considered a synonym of *P. levipes* but Menzies (1996) reviewed taxonomy of the genus and determined *P. naso* was a distinct species distributed across south-western New Guinea and the Aru Islands. It was collected from Wokam, Aru Islands, by Dr Pat Woolley in 1992. Two specimens in the Pleistocene to Holocene deposits at Lemdubu, on Kobroor, Aru Islands,

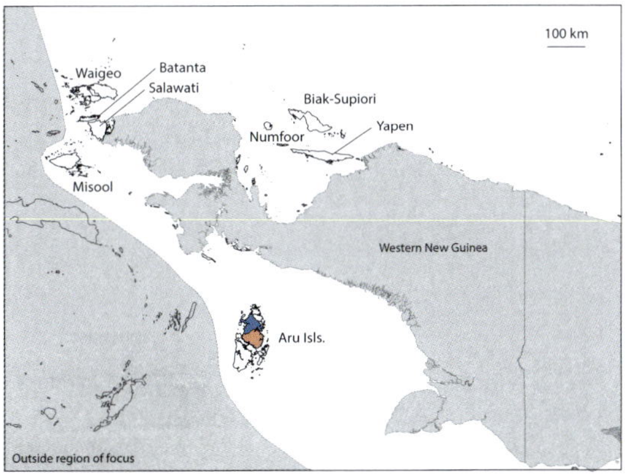

Paramelomys naso			
Size*			
	Mean	**Range**	**Sample**
Head–body	178.5	158.0–191.0	4
Tail	137.0	132.0–142.0	2
Hindfoot	35.5	33.0–38.0	4
Ear	20.0	19.0–21.0	2
Weight	–	–	–
* From the holotype (Thomas 1911a), AMNH specimen (Weyland Range, New Guinea) and Menzies (1996)			
Identification Larger bodied and with larger teeth than *P. platyops* and *P. moncktoni*. The tail is bicoloured, being paler below.			
Synonyms None			
Subspecies None			
Distribution Aru Isls			
Extralimital New Guinea			
Conservation status (IUCN) Least Concern			
Etymology Sp.–nose (L.), a reference to the long rostrum			

were referred to this species by Aplin and Pasveer (2006).

Long-nosed paramelomys is an inhabitant of lowland rainforest habitats and has been documented from just a handful of localities in the lowlands and foothills of south-west New Guinea. Little else has been reported of its ecology. Although there are few records of long-nosed paramelomys, these are widely distributed, and no major threats have been identified.

COMMON LOWLAND PARAMELOMYS

Paramelomys platyops (Thomas, 1906)

plat'-ee-ops

The common lowland paramelomys is widespread on mainland New Guinea and may include multiple undescribed subspecies. One Australian Museum specimen is a male trapped on Goodenough Island in August 1987. It was caught in shrubs along a creek at 1300 m elevation and occurred in the same area as *Rattus exulans* and *R. mordax*. It was locally common, being taken in Elliott traps baited with cooked taro. Another was recorded at 900 m elevation during the fourth Archbold Expedition on Goodenough in 1953. The common lowland paramelomys was also recorded in the D'Entrecasteaux Islands during the Fifth Archbold Expedition in 1956. On Normanby, a male was caught at 820 m elevation on Mt Pabinama, and a female at Waikaiuna (site of a small coconut plantation at 20 m elevation). On Fergusson Island it was caught at 900 m elevation at the expedition's 'Mountain Camp' between Agamoia and Ailuluai. High elevation sites were dominated by Oak forest

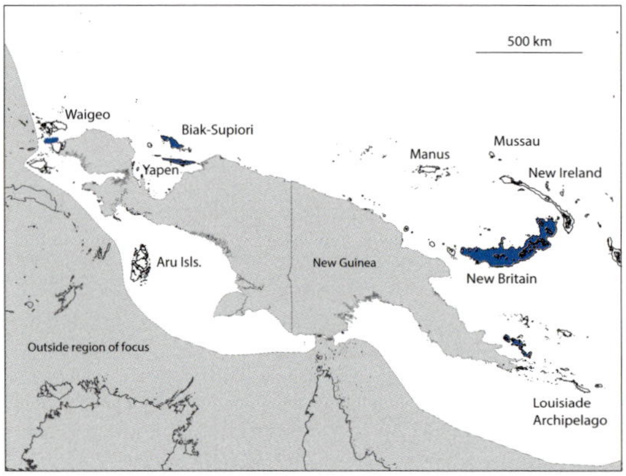

Paramelomys platyops			
Size*			
	Mean	**Range**	**Sample**
Head–body	156.0	123.0–184.8	29
Tail	120.8	100.0–133.0	28
Hindfoot	32.9	29.0–36.0	29
Ear	18.0	15.0–21.5	29
Weight	131.5	125.0–138.0	4
* From AM specimens (Goodenough) and AMNH specimens (Biak, north-west Cenderawasih Bay)			
Identification A plump, sleek-furred species, the tail possesses one hair per scale (rather than three as in *Paramelomys moncktoni*).			
Synonyms *Melomys moncktoni fuscus* Rümmler, 1935 *Melomys moncktoni intermedius* Rümmler, 1935 *Melomys moncktoni jobiensis* Rümmler, 1935			
Subspecies None			
Distribution Batanta, Biak-Supiori, Yapen (Indonesia); Fergusson, Goodenough, ?New Britain, Normanby (Papua New Guinea)			
Extralimital New Guinea			
Conservation status (IUCN) Least Concern			
Etymology Sp.–flat face (Gr.)			

(*Castanopsis* sp.), and on Fergusson there was a dense, low undergrowth of large ferns, herbs and shrubs of the genera *Elatostema* and *Pilea*, and the stinging shrub *Laportea decumaria* (Brass 1959). The existence of common lowland paramelomys on New Britain is uncertain and could instead relate to a similar undescribed species. It was not detected in surveys by Aplin and Opiang (2011).

Common lowland paramelomys seems to favour habitats with a dense understorey, particularly along streams. On mainland New Guinea it is common in low-altitude hill forests and is apparently entirely terrestrial. Examination of the stomach contents sampled from four Australian Museum specimens taken on Goodenough Island in 1987 showed the diet is omnivorous (Elliott and Vernes 2021). Among the samples were the remains of insect exoskeletons (including lepidopteran scales), plant material and fungal spores belonging to the Glomeraceae/Endogonaceae. The only available reproductive data are from mainland New Guinea. A female and a single newborn young (bodyweight 10 g) were caught together in an Elliott trap in Sandaun Province, mainland New Guinea in January 1984. A second lactating female was also collected in the same area (Flannery 1995a). The common lowland paramelomys has a wide distribution and appears tolerant to habitat modification.

Pogonomelomys

BRUIJN'S POGONOMELOMYS

Pogonomelomys bruijnii (Peters and Doria, 1876b)

poh-gon'-oh-mel'-oh-mis brue'-nee

The single known record of Bruijn's pogonomelomys from this region is that of the holotype, held in the Museum of Genova (MSNG 3677). It is unique among specimens of *Pogonomelomys* (being considerably larger in cranial and dental size) (Helgen *et al.* 2008). Bruijn's pogonomelomys is very poorly known and brief surveys on Salawati in 1992 were unsuccessful at finding it. It is apparently an arboreal inhabitant of lowland rainforest, and has been collected from tree hollows.

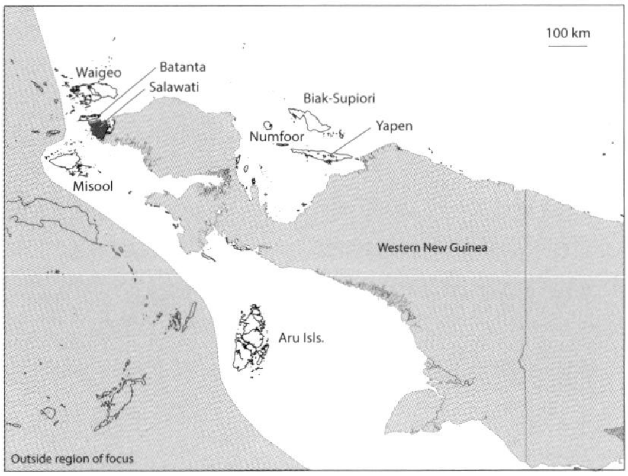

Pogonomelomys bruijnii			
Size*			
	Mean	**Range**	**Sample**
Head–body	–	180.0	1
Tail	–	190.0	1
Hindfoot	–	34.0	1
Ear	–	9.0	1
Weight	–	–	–
* From the holotype in Tate (1951)			
Identification The very long tail covered in brownish hexagonal scales, bright red dorsum, and large size serve in combination to distinguish this species from all other murids in the region.			
Synonyms None			
Subspecies None			
Distribution Salawati			
Extralimital New Guinea			
Conservation status (IUCN) Least Concern			
Etymology Gen.–a combination of *Pogonomys* and *Melomys*, as the species of this genus combine features seen in both *Pogonomys* and *Melomys* species; Sp.–for the Dutch merchant A. A. Bruijn			

Bruijn's pogonomelomys is listed as Least Concern on the IUCN Red List because the population is presumably not severely fragmented and nothing is known about number of locations. There is an apparent need for taxonomic revision of the genus *Pogonomelomys* specimens and the severe lack of information on the ecology and distribution of this species. A Data Deficient category may be more appropriate.

Pogonomys

D'ENTRECASTEAUX ARCHIPELAGO TREE MOUSE

Pogonomys fergussoniensis Laurie, 1952

poh-gon'-oh-mis fer-guss'-oan-ee-en'-sis

Previously considered a subspecies of *Pogonomys loriae*, *P. fergussoniensis* is now recognised as a species endemic to the D'Entrecasteaux Islands. The original description was based on two specimens (one a skull only) collected at Taibutu, Faralulu District, west Fergusson Island. It remains rarely recorded.

Two specimens are held at the American Museum of Natural History. One is an adult female skin and skull collected at 900 m elevation on Goodenough by Van Deusen in 1953 (AMNH 157613). This specimen bears a note on the label indicating it was caught at the transition between rainforest and oak (*Castanopsis* sp.) that occurs as pure stands in the lowland hill forests of the D'Entrecasteaux Archipelago (Paijmans 1976). Another is an adult female collected on Fergusson at Agamoia on moderate south slopes of the Lilai Creek watershed (200 m elevation) by R. Peterson on 24 June 1956. A species of *Dillenia* was common in the intact primary forests on ridges and valleys and mammals collected at the same site included *Phalanger intercastellanus*, *Petaurus papuanus*, *Macroglossus minimus* and *Nyctimene major* (Brass 1959). It was described as being a rare species on Fergusson (Brass 1959). An adult male held by the Australian Museum (AM M20309) was collected at Wamla Village, near Guleguleu, east Normanby. It was discovered at night, climbing in vegetation 1–4 m above the ground and approximately 20 m from the village margin. The D'Entrecasteaux Archipelago tree mouse has a small extent of occurrence and there is a

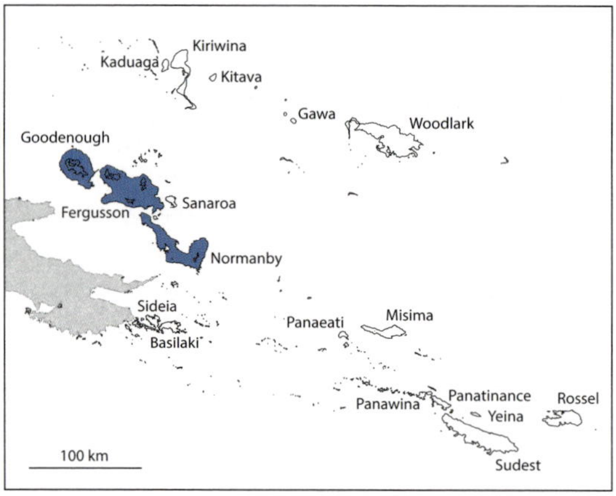

Pogonomys fergussoniensis			
Size*			
	Mean	**Range**	**Sample**
Head–body	179.8	167.0–205.0	4
Tail	251.8	248.0–262.0	4
Hindfoot	30.5	29.0–33.0	4
Ear	14.9	14.0–15.7	3
Weight	146.5	135.0–158.0	2
* From AM and AMNH specimens (Fergusson, Goodenough and Normanby)			
Identification The general colour is rusty brown, belly is whitish-buff with patches of rust coloured hairs and is not sharply demarcated from the dorsum. The tail is prehensile, long and dark with light patches, and covered in rectangular scales.			
Synonyms None			
Subspecies None			
Distribution Fergusson, Goodenough, Normanby			
Extralimital None			
Conservation status (IUCN) Endangered			
Etymology Sp.–for Fergusson Is., the type locality			

continuing decline in the extent and quality of habitat due to deforestation and conversion to grassland.

CHESTNUT TREE MOUSE

Pogonomys cf. *macrourus* (Milne-Edwards, 1877)

mak'-ruh-yue-rus

Pogonomys was long known from New Britain by a single specimen collected at Talasea by P. R. Schumm in October 1927 and held in the collections of the Museum für Naturkunde, Berlin (ZMB 40274). The specimen (probably a male) consists of a skull and a skin that is now in terrible

Chestnut Tree Mouse (*Pogonomys macrourus*). A tree mouse conferred to this species occurs on New Britain Island in the Bismarck Archipelago and has been observed near an underground burrow system (Angus Fraser).

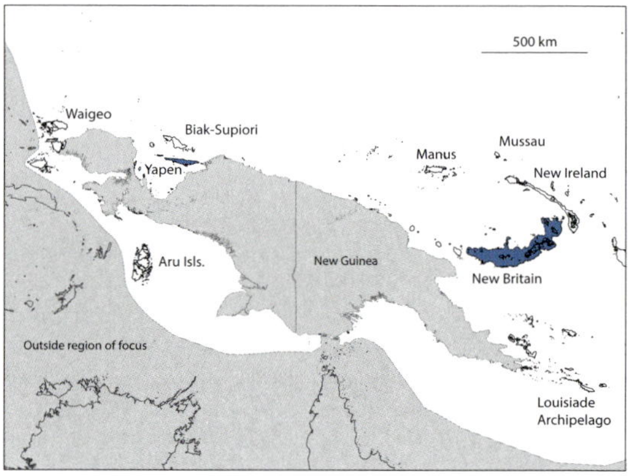

condition, lacking most of the fur. Yet the tail, the remains of a white venter and red dorsum, as well as the morphology of the skull, leave no doubt that it is a species of *Pogonomys*. Menzies and Dennis (1979) suggested this record needed confirmation, and in 1999 a further two specimens were collected at 1000–1200 m elevation in the Nakanai Range of East New Britain (Anthony 2001). A possible sighting of this species was also made on 6 April 2009 in the Nakanai Range at 200 m elevation, in a lowland mosaic of alluvial and hill communities (Aplin and Opiang 2011).

Nothing has been recorded of the biology of *Pogonomys* cf. *macrourus* on New Britain. The animal seen in 2009 was close to a multi-branched burrow system that ran just below the ground surface. On mainland New Guinea, chestnut tree mouse rests in burrows such as this during the day

Pogonomys* cf. *macrourus			
Size*			
	Mean	**Range**	**Sample**
Head–body	121.8	92.0–150.0	4
Tail	156.5	126.0–187.0	4
Hindfoot	20.9	19.1–22.5	4
Ear	13.2	11.0–15.5	4
Weight	44.8	28.0–60.0	4
* From Chimbu and Sandaun Provinces, mainland New Guinea (Flannery 1995a)			
Identification Fur colour is reddish on the dorsum with a pure white ventrum. The tail possesses non-overlapping squarish scales and has a scaleless prehensile area on the dorsal side near the tip.			
Synonyms (of *P. macrourus*) *Mus mollipilosus* Peters and Doria, 1881; *Pogonomys lepidus* Thomas, 1897; *Pogonomys lepidus derimapa* Tate and Archbold, 1935b; *Pogonomys lepidus huon* Tate and Archbold, 1935b			
Subspecies None			
Distribution New Britain, Yapen			
Extralimital Australia, New Guinea			
Conservation status (IUCN) Least Concern			
Etymology Sp.–large tail (Gr.)			

and emerges at night to feed on fruits and leaves in the trees. Males are solitary but females will share burrows (Flannery 1995a). Litters of up to three young have been recorded, and Flannery (1995a) observed one female with two almost independent young on 6 July 1987 in the Telefomin area of New Guinea. Limited information is available for the species on Yapen. Chestnut tree mouse is widely distributed, with a large population and no major threats have been identified.

Rattus

ADMIRALTIES SPINY RAT

***Rattus detentus* Timm *et al.*, 2016**

rat'-us deh-tent'-us

The remains of Admiralties spiny rat were first identified among archaeological deposits. At the Pamwak archaeological site on the southern side of Manus, this large rat persisted into the most recent levels of the site, indicating it probably remained extant on the island (Williams 1999; Timm *et al.* 2016). In August 2002, this was confirmed via the collection of a nearly complete skeleton and an isolated mandible, obtained from hunters on the north-western coastline. Finally, a complete adult

The holotype of Admiralties spiny rat (*Rattus detentus*) collected by Valter Weijola on Manus in 2012 (Valter Weijola).

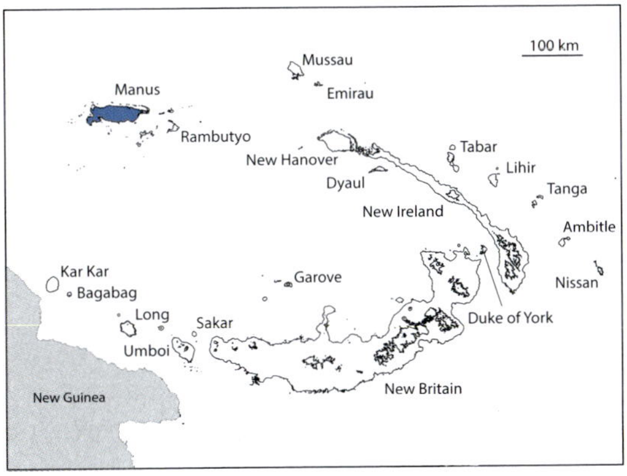

Rattus detentus			
Size*			
	Mean	**Range**	**Sample**
Head–body	–	260.0	1
Tail	–	150.0	1
Hindfoot	–	43.2	1
Ear	–	23.4	1
Weight	–	–	–
* Recorded from the holotype (Timm *et al.*, 2016)			
Identification A large *Rattus*, exceeding the size attained by all other Melanesian *Rattus* except *R. jobiensis*, and possibly *R. praetor coenorum*. The tail is short (approximately 58% of head–body); the mammary formula is 1 + 1 + 2. Pelage is exceptionally coarse and spiny, with prominent black guard hairs.			
Synonyms None			
Subspecies None			
Distribution Manus			
Extralimital None			
Conservation status (IUCN) Data Deficient			
Etymology Gen.–rat (L.); Sp.– detained (L.) a reference to the species' isolation on Manus and use of the island to detain people seeking political and/or economic asylum in Australia			

female specimen (the holotype) was captured in August 2012 by Valter Weijola in the central hill forests of Manus. At that time, local residents indicated Admiralties spiny rat also occurred on adjacent Los Negros Island.

Rattus detentus seems to be at least partially tolerant of disturbance. On Manus it coexists with two introduced rats (*R. exulans* and *R. rattus*), and vegetation in the vicinity of the type locality consisted of a mosaic of secondary forest, village gardens, and agroforestry with sago palm (*Metroxylon* sp.), palms, and cassava (*Manihot esculenta*). Based on the Aplin *et al.* (2015) discovery of a system of tunnels in use by *R. exulans* (a species that does not typically construct burrows), Timm *et al.* (2016) postulated Admiralties spiny rat may be capable of burrowing. Gnaw marks on *Canarium indicum* nuts discovered on Manus indicated Admiralties spiny rat may be capable of eating these hard-shelled nuts. However, *R. rattus* has also been shown to gnaw holes through the smaller nuts of *Canarium* species in other parts of Melanesia (Lavery *et al.* 2020b).

Although apparently tolerant of disturbance, the rarity of modern records suggests Admiralties spiny rat may be worthy of a threatened conservation status. It is actively hunted by people on Manus and all modern records are derived from the efforts of hunters.

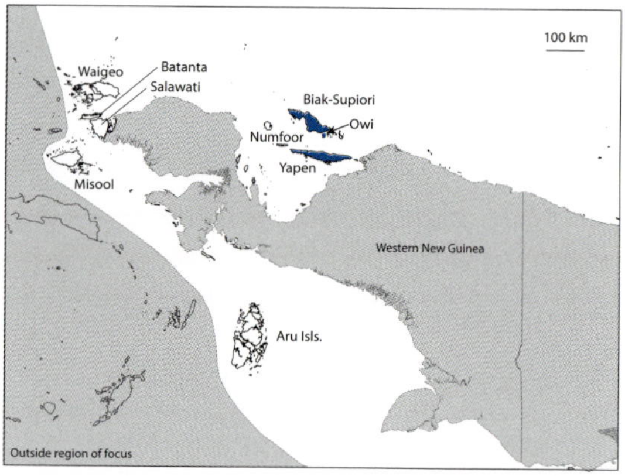

Rattus jobiensis			
Size*			
	Mean	**Range**	**Sample**
Head–body	229.2	170.0–260.0	5
Tail	203.0	166.0–227.0	5
Hindfoot	44.7	36.0–51.0	5
Ear	22.3	20.5–23.3	3
Weight	–	500.0	1
* From AM specimen (Supiori), (Troughton 1945) (Owi) (Troughton 1946) (Biak) and (Taylor *et al.* 1982)			
Identification An enormous species of *Rattus* with a very course, spinous pelage. The tail is 87–97% of the head–body length. The back is medium to dark brown and the belly is grey. The tail appears to be naked but is sparsely haired. Mammary formula is 1 + 2 = 6 (Yapen and Owi) or 2 + 2 = 8 (Biak).			
Synonyms *Rattus owiensis* Troughton, 1945 *Rattus biakensis* Troughton, 1946			
Subspecies None			
Distribution Biak-Supiori, Owi, Yapen			
Extralimital None			
Conservation status (IUCN) Least Concern			
Etymology Sp.–from Jobi (L.), an old name for Yapen			

YAPEN RAT

Rattus jobiensis Rümmler, 1935

joh'-bee-en'-sis

The Yapen rat is immediately recognisable for its very large size, which approaches that of the *Uromys* with which it is sympatric on Biak-Supiori. No other spiny-furred murids within the distribution of this species are comparable in body size. Pattiselanno and Arobaya (2012) listed this species among their inventory of mammals recorded at the Tamrau Nature Reserve, Papua, but no discussion of diagnostic features used to identify the species was provided and so it has not been included in the extralimital distribution here. In the islands of Cenderawasih Bay it is sympatric with *R. exulans* and *R. rattus*.

The mammary formula is either 1 + 2 = 6 (Yapen and Owi Islands) or 2 + 2 = 8 (Biak Island) (Troughton 1946, Taylor *et al.* 1982). Taylor *et al.* (1982) reported trapping Yapen rats in a garden at the edge of native forest, and in a village. In September 1992, two individuals were trapped in gardens using traditional snares. Aluminium box traps set primarily along creeks in the same area failed to catch it, although *Paramelomys platyops* was obtained. An adult female was neither lactating nor pregnant when examined in September. An immature male (weight 260 g) was caught in the same area.

The Yapen rat is a reasonably common species within its limited distribution. However, the effects of habitat loss occurring within its range are largely unknown.

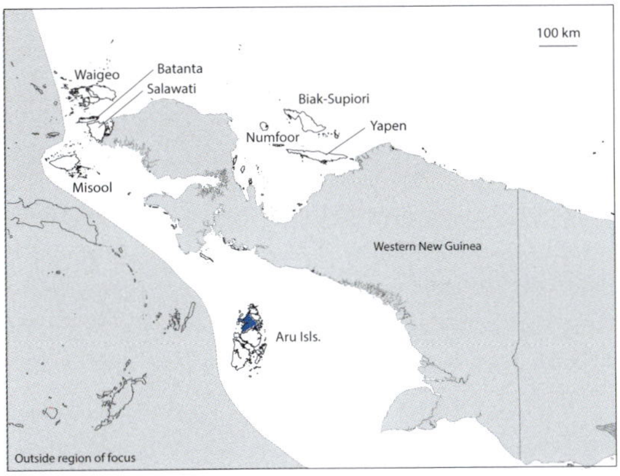

CAPE YORK RAT

Rattus leucopus (Gray, 1867)

leu'-koh-poos

This large, spiny rat is widespread and common in the drier lowland forests of southern New Guinea and rainforest patches of the eastern Cape York Peninsula, Australia. Despite being an easily detected species, it is a relatively recent addition to the mammal faunas of the Aru Islands, first identified on Kobroor by Dr P. A. Woolley in October 1992.

In New Guinea and Australia, the Cape York rat breeds year-round and in subspecies *R. l. ringens* litter size varies from two to five (Taylor *et al.* 1990). Much of what is known of the ecology of this species has been reported from north Queensland, Australia, by Moore and Leung (2008). The Cape York rat is a terrestrial and largely nocturnal species that is omnivorous. The diet is known to include rainforest fruits and seeds, green vegetation and fungi. Insects also feature prominently. Cape York rats occupy hollow logs or burrows with multiple entrances in the forest floor or at the bases of trees. These are usually well hidden under the leaf-litter layer. Animals will move between burrows and burrows can be shared by multiple individuals (several adults and juveniles). Breeding occurs throughout the year, increasing during the wet season, and they can produce three litters per year. Litter size is normally between two and five (Moore and Leung 2008). Cape York rat has a wide distribution, large population, a degree of tolerance for habitat modification, and lacks major threats.

Rattus leucopus			
Size*			
	Mean	**Range**	**Sample**
Head–body	179.0	173.0–184.0	2
Tail	144.0	142.0–146.0	2
Hindfoot	35.1	34.6–35.5	2
Ear	20.7	19.5–21.9	2
Weight	177.5	161.0–194.0	2
* From AM specimens (Aru Islands)			
Identification A relatively large, light-coloured spiny rat. Most animals have a white tail tip and the belly is white. For all subspecies the mammary formula is 0 + 1 + 2 = 6 (Timm *et al.* 2016), but in individuals from the Aru Islands it is 0 + 2 + 2 = 8.			
Synonyms *Hapalotis personata* Krefft, 1867 *Mus terraereginae* Jentink, 1908 *Rattus leucopus mcilwraithi* Tate, 1951			
Subspecies *Rattus leucopus leucopus* (Gray, 1867) (northern Cape York) *Rattus leucopus ringens* (Peters and Doria, 1881) (Fly River area, New Guinea) *Rattus leucopus ratticolor* (Jentink, 1908) (southern West Papua, Aru Islands) *Rattus leucopus dobodurae* Troughton, 1946 (south-eastern New Guinea) *Rattus leucopus cooktownensis* Tate, 1951 (southern Cape York)			
Distribution Aru Isls			
Extralimital Australia, New Guinea			
Conservation status (IUCN) Least Concern			
Etymology Sp.–white foot (L.)			

Cape York rat (*Rattus leucopus*) Iron Range, Cape York Peninsula, Australia. This widespread and common species of Cape York Peninsula and southern New Guinea has also been documented from the Aru Islands (Mark Sanders).

EASTERN RAT

Rattus mordax (Thomas, 1904a)

mor'-dax

Rattus mordax is a distinctive species and the only large *Rattus* indigenous to the islands of Milne Bay. Genetic sequencing by Timm *et al.* (2016) showed this species belongs to a clade of endemic New Guinea *Rattus* and is sister to *R. niobe* of eastern New Guinea.

The eastern rat is very widespread in Milne Bay and can be encountered in a variety of habitats, including gallery rainforests, grassland, *Eucalyptus* savannah and gardens. On Woodlark it occurs in dry secondary forests on limestone near human habitation and on Goodenough its distribution extends up to 1200 m elevation where it has been trapped along creeklines in montane forest. Brass (1959) mentioned the collection of *R. ruber* (certainly *R. mordax*) in low mossy scrub on Mt Pabinima on Normanby in 1956, and among oak forest (*Castanopsis*) and savannah on Fergusson.

This species is known to occur in villages where they will eat stored rice and sweet potatoes (Taylor *et al.* 1982). Stomach samples of two Australian Museum specimens collected from Goodenough Island in 1987 contained plant material, insect exoskeletons and fungal spores belonging to Glomeraceae and/or Endogonacea (Elliott and Vernes 2021). Breeding can occur in any month in both wet and dry seasons, and litters consist of between two and four young (Taylor *et al.* 1990). Given the apparent tolerance of disturbed environments by the eastern rat, its conservation status appears to be secure.

Rattus mordax			
Size*			
	Mean	**Range**	**Sample**
Head–body	184.8	167.0–200.0	4
Tail	148.9	142.0–160.5	4
Hindfoot	36.6	33.0–39.7	4
Ear	20.0	19.0–21.0	3
Weight	221.7	190.0–225.0	3
* From AM specimens (Goodenough, Woodlark)			
Identification A large spiny rat with broad hindfeet. The tail is longer than in *R. praetor* and *R. detentus* (> 80% of head–body length), and it is less rufescent than *R. praetor* and with broader molars. The mammary formula is 2 + 2 = 8.			
Synonyms None			
Subspecies *Rattus mordax mordax* (Thomas, 1904a) (eastern New Guinea and some Milne Bay islands) *Rattus mordax fergussoniensis* Laurie, 1952 (D'Entrecasteaux Isls)			
Distribution Conflict Group, Fergusson, Goodenough, Misima, Normanby, Sideia, Sudest, Woodlark			
Extralimital New Guinea			
Conservation status (IUCN) Least Concern			
Etymology Sp.–prone to biting (L.)			

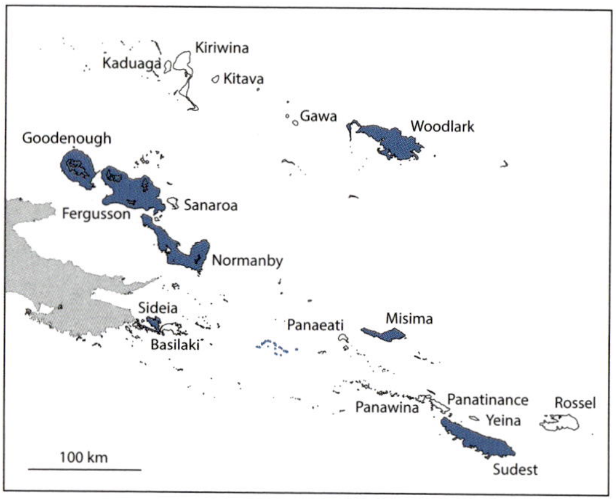

LARGE SPINY RAT

Rattus praetor (Thomas, 1888c)

pree'-tor

The large spiny rat is a natural element of the faunas of Salawati. However, it is a species that is widely distributed in the South-west Pacific today, owing much of this distribution to prehistoric translocation. On New Ireland it appears in late Pleistocene deposits in Panakiwuk cave from 13 000 years before present (Marshall and Allen 1991), but is only in the upper horizon of Balof cave dated 3500 years ago (Flannery and White 1991; White *et al.* 2000). It was detected in deposits on Buka dated to 1860 years before present, but only via a single dentary (Flannery *et al.* 1988). The large spiny rat remains relatively uncommon in Solomon Islands today and, when encountered, is most often found in disturbed habitats and smaller offshore islands. However, it can also be encountered occasionally in primary forests. For example, one individual was captured in one of the most remote regions of Makira, at 1000 m elevation in the Bauro highlands. It was deep in a limestone cave that was inhabited by a large colony of *Dobsonia inermis*.

Curiously, fossils have also been found on Tikopia and Nissan (Flannery *et al.* 1988) although living animals have never been recorded there. White *et al.* (2000) postulated this species may have been introduced intentionally to Vanuatu (Malakula), Fiji and several other islands and, from the large numbers of animals identified in archaeological deposits, surmised that there must have certainly been established breeding populations in the archipelagos. The populations on these islands are presumed extinct, but greater survey effort in remote habitats may reveal it persists (as per Makira).

The large spiny rat is reportedly a strong burrower and will use burrow systems to shelter

The large spiny rat (*Rattus praetor*) is a species commonly encountered near coastlines and villages, and on small islands. However, the individual pictured here was inside a limestone cave at 1000 m elevation on Makira Island, far from villages (Tyrone Lavery).

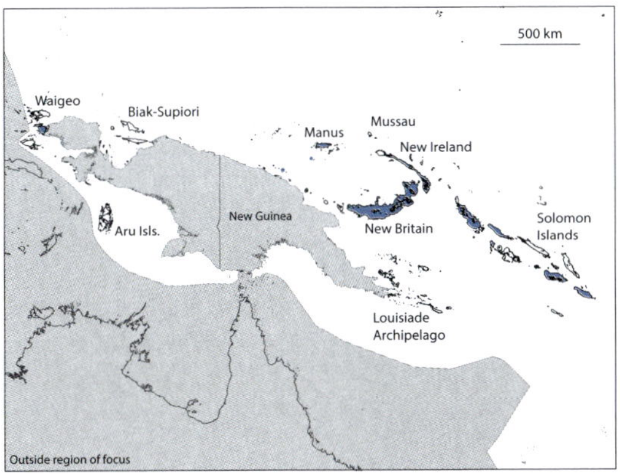

Rattus praetor			
Size*			
	Mean	**Range**	**Sample**
Head–body	200.0	179.0–221.0	6
Tail	167.2	144.0–204.0	6
Hindfoot	35.9	34.1–39.0	6
Ear	20.8	18.6–24.6	6
Weight	200.9	160.0–272.5	6
* From Western Province, Solomon Islands, and Sandaun Province, New Guinea (Flannery 1995a)			

Identification A variable, large spiny rat with narrower hindfeet and smaller molars than *R. mordax*, and more rufous in colouration. In *R. praetor praetor* the tail is approximately 89% of head–body length (Timm *et al.* 2016). The mammary formula is 2 + 2 = 8.

Synonyms *Rattus mordax tramitius* Thomas, 1922a
Rattus bandiculus Thomas, 1922a
Rattus leucopus utakwa Rümmler, 1935
Rattus praetor mediocris Troughton, 1936a
Rattus purdiensis Troughton, 1946
Rattus sansapor Troughton, 1946

Subspecies *Rattus praetor praetor* (Thomas, 1888c) (Manus, Bismarck Archipelago, Solomon Islands)
Rattus praetor coenorum Thomas, 1922a (New Guinea)

Distribution Bougainville, Nissan (fossil) (AROB); Lakeba (fossil) (Fiji); Salawati (Indonesia); Bat, Blup Blup, Karkar, New Britain, New Ireland (Papua New Guinea); Choiseul, Guadalcanal, Makira, Tikopia (fossil), Vella La Vella (Solomon Isls); Efate (fossil) (Vanuatu)

Extralimital Gebe, New Guinea

Conservation status (IUCN) Least Concern

Etymology Sp.–leader, chief, one who goes before (L.)

the young (Troughton 1946). On mainland New Guinea, juveniles and breeding adults have been recorded in all months, suggesting it may breed throughout the year there. Litters contain between two and seven young (average 4.5) (Taylor *et al.* 1990). From the region under consideration here, firm evidence of breeding has been recorded only in July (lactating females) and juveniles have been found in January and March (Taylor *et al.* 1990).

Solomys

PONCELET'S GIANT RAT

Solomys ponceleti (Troughton, 1935b)

sol'-oh-mis pon'-sel-et'-tee

Poncelet's giant rat was described by Troughton (1935b), who was sent a young adult female (in spirit) and two skulls by Fr J. B. Poncelet from Buin, Bougainville. Fr Poncelet subsequently sent a further two specimens to the Australian Museum, and four more to the Musée Royal d'Histoire Naturelle in Belgium (Frechkop 1948). All were taken in the Buin area of Bougainville. Flannery and Wickler (1990) recorded an additional specimen, a partial skin held in the Bishop Museum, Honolulu, collected by J. H. Waterhouse in 'Solomon Islands' in 1930. Remains of this species were moderately common in the Kilu rockshelter subfossil deposits of Buka (Flannery and Wickler 1990), but no modern specimens exist from that island.

A preferred forest type for Poncelet's giant rat is *Terminalia brassii* swamp forests, in the interior of Choiseul (Leary 2003). In February 1990 a female and juvenile male were documented by T. Leary and H. Parnaby of the Australian Museum. The rats were in swamp forest near Vudutaru, and secured by a hunter from a large, loose stick nest (resembling that built by an eagle). The nest was on the outer limb of a huge forest tree that had a girth of more than a metre at ground level. Residents of Choiseul revealed this species will also nest inside the hollows of large trees (Pikacha 2005). The diet reportedly includes the nuts of *Canarium indicum* and *C. salomonense*, fruits of native figs (*Ficus* spp.), and the young shoots and budding leaves of sago palm (*Metroxylon salomonense*) (Pikacha 2005).

In a 1930s report to Troughton, Fr Poncelet stated that Poncelet's giant rat was very rare on southern Bougainville. In 7 months of fieldwork for research focused on the mammals of south-east Choiseul, a survey team encountered Poncelet's giant rat on only one occasion (Pikacha 2005). Questionnaire surveys as part of this work revealed no resident under the age of 20 had ever seen the species and only a small number of elderly men had ever encountered it. Hunters leading the Australian Museum team similarly indicated this species was rare on northern Choiseul. The animals discovered in 1990 represent the most recent scientific record of this species.

Solomys ponceleti			
Size*			
	Mean	**Range**	**Sample**
Head–body	330.0	330.0	2
Tail	352.5	340.0–365.0	2
Hindfoot	74.3	71.5–77.0	2
Ear	24.5	23.0–26.0	2
Weight	–	> 1000	1
* From AM specimens (Choiseul and Bougainville)			
Identification The largest of murids found in the Solomon Isls archipelago. The fur is black and coarse, forming a crest over the head and back. The fur is thin making the pinkish white skin visible beneath.			
Synonyms None			
Subspecies None			
Distribution Bougainville, Buka (fossil), Choiseul			
Extralimital None			
Conservation status (IUCN) Critically Endangered			
Etymology Sp.–for Father J. B. Poncelet, a Belgian Catholic missionary who collected the holotype			

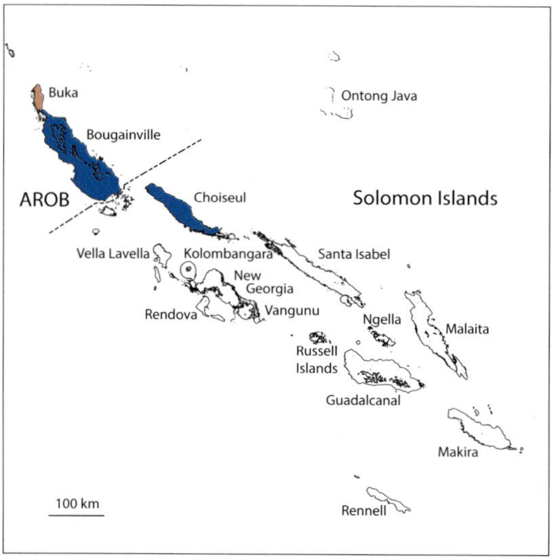

Given its large body size, apparent natural rarity and low fecundity (producing only a single young), Poncelet's giant rat is undoubtedly a species with limited ability to withstand large-scale disturbance. Its preference for building nests in large trees on better soils places the species under severe threat from logging, for such trees are precisely those targeted by logging companies. More research is urgently needed into the biology and conservation of this remarkable murid.

UGI GIANT RAT

Solomys salamonis (Ramsay, 1882)

sal'-ah-mon-is

Solomys salamonis is known only from the holotype collected by Alexander Morton of the Australian Museum in 1881. The type locality was originally given as Ugi Island but later changed in an erratum to Ngella (Florida Island). Lavery (2013) reviewed the repeated confusion of the type locality in the literature and concluded that Ugi is correct.

The Australian Museum register indicates the skin was considered worthless and discarded in 1913, thus only the skull remains (Troughton 1936b; Parnaby *et al.* 2017). Ramsay (1882) described it as light ashy-grey with long black guard hairs and

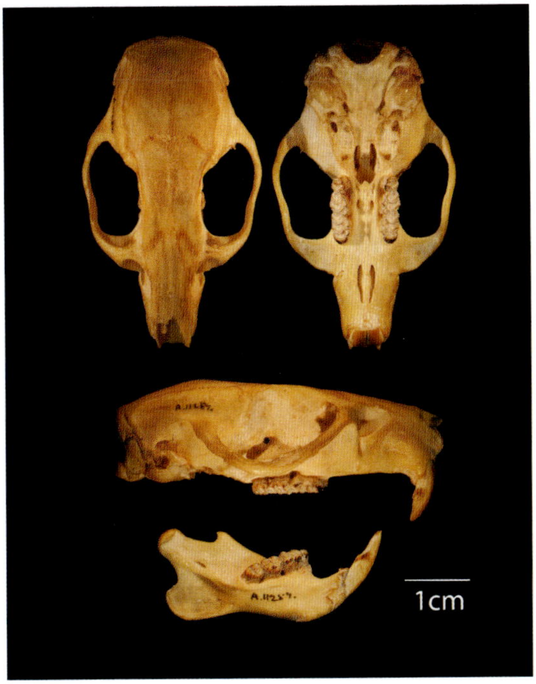

The Ugi giant rat (*Solomys salamonis*) is known only from the holotype skull collected in 1881 (Australian Museum A.11257). The skin has since been lost (Tyrone Lavery).

Solomys salamonis			
Size*			
	Mean	**Range**	**Sample**
Head–body	–	216	1
Tail	–	224	1
Hindfoot	–	44.3	1
Ear	–	13	1
Weight	–	–	–
* Recorded from the holotype (Ramsay 1882)			
Identification According to Ramsay (1882) this species is ashy grey in colour, a characteristic that contrasts sharply with the typical brown-red or brown-black of other *Solomys* species.			
Synonyms None			
Subspecies None			
Distribution Ugi			
Extralimital None			
Conservation status (IUCN) Data Deficient			
Etymology Sp.–relating to Solomon Isls			

small ears. The tail was slightly longer than the head–body length. According to the illustration provided in the species description, the hindfoot was wide with long digits and large plantar pads. This indicates it is an arboreal animal.

The Ugi giant rat could still be present on Ugi and targeted surveys there are urgently needed. Ugi is a small island (42 km^2) with a relatively high human population density, and it supports a patchwork of primary and secondary forests and village gardens. During spotlight surveys on the island in 2012, no rodents were observed. However, residents did indicate rats are regularly seen at night high in coconut palms (*Cocos nucifera*) eating the young fruits (possibly *Rattus rattus*).

A rodent similar in appearance to *S. sapientis* was collected at Hagaehu village on nearby Makira in 2006, but the body was taken by a dog before it could be preserved (R. James, pers. comm.). It is possible this was *S. salamonis*, although the animal's pelage was reddish-brown rather than grey.

BOUGAINVILLE GIANT RAT

Solomys salebrosus Troughton, 1936a

sal'-eb-roh'-sus

The Bougainville giant rat is the most commonly encountered native rodent of the Solomon Islands archipelago. Major collections of this species were made by Reverend J. B. Poncelet around Buin on

Bougainville in the 1930s. Eleven of these were sent to the Australian Museum and an additional four to Musée Royal d'Histoire Naturelle, Belgium. Museums Victoria and American Museum of Natural History hold specimens from Bougainville and Choiseul. In 1990 Harry Parnaby and a team from the Australian Museum recorded an additional 11 individuals on Choiseul Island. In 2016, another two individuals were recorded at Siwai on southern Bougainville as part of research associated with the Australian Museum.

The Bougainville giant rat is an arboreal and nocturnal species recorded from primary forests, the margins of villages and gardens up to 300 m elevation. One animal was discovered dwelling in a nest of leaves inside the hollow of a large tree some 15 m from the ground. Two others, caught at different times, were found in leaf nests made on a branch. At Kainake, southern Bougainville, the Bougainville giant rat was found nesting in a large galip or ngali nut tree (*Canarium indicum*) that had been struck by lightning and was badly damaged. Two individuals (an adult and a subadult) were captured from the tree, and it seemed there were other individuals in residence. On part of the tree's broken trunk, a platform consisting of vines, epiphytes and rotting debris had developed. There, the animals had created separate middens of exocarps, and nut shells from which the kernels had been extracted via large holes gnawed in their centres. The tree was situated in a village garden, isolated from the forest, and to reach it the rodents must have had to traverse the open ground. This

Bougainville giant rat (*Solomys salebrosus*), Kainake, Bougainville. This individual was found living in a *Canarium indicum* tree, where it appeared to be feeding exclusively on the seasonal nuts (Tyrone Lavery).

The Bougainville giant rat is apparently accustomed to exploiting agroforest habitats that exist around villages. This individual was photographed on the edge of Sassamungga Village, southern Choiseul (Patrick Pikacha).

At Kainake, southern Bougainville, holes were gnawed in galip or ngali nuts (*Canarium indicum*) by *Solomys salebrosus* to extract the kernel. The empty nuts were discarded in middens on platforms of rotting vines and detritus in the canopy (Tyrone Lavery).

suggests seasonal *Canarium* nuts are powerful attractant for Bougainville giant rats. In the same region, a camera trap set in a small understorey fig tree (*Ficus copiosa* or *F. wassa*) located on the edge of a village recorded Bougainville giant rat on numerous occasions during the month of November. The rats removed fruits from the tree and carried them away. A dedicated study of this species at Kainake, southern Bougainville, identified 18 important food trees that contribute either fruit or leaves to the diet of the Bougainville giant rat (J. Lamaris, pers comm.). Two individuals that were tracked for a short period using radio-telemetry spent much of their time in riparian vegetation during nocturnal hours and were also observed foraging on the ground. The animals were found

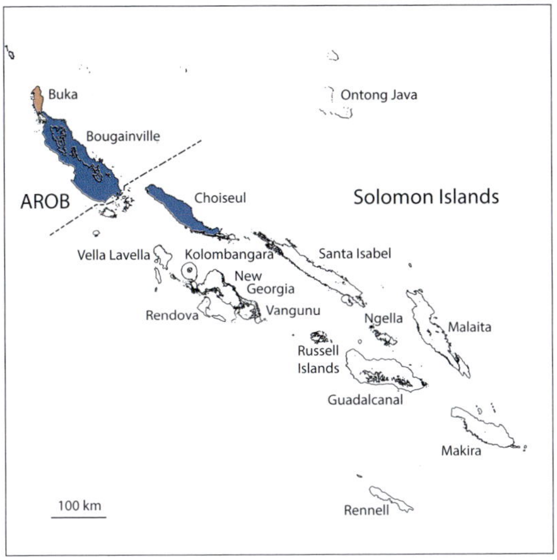

Solomys salebrosus			
Size*			
	Mean	**Range**	**Sample**
Head–body	250.5	225.0–328.0	6
Tail	237.2	200.0–250.0	6
Hindfoot	48.4	46.9–50.5	6
Ear	19.5	18.4–20.7	6
Weight	396.5	290.0–480.0	6
* From AM specimens (Choiseul and Bougainville)			
Identification Can be differentiated from other *Solomys* in being smaller than *S. ponceleti* and larger than *S. salamonis*. It has a shorter hindfoot than *S. sapientis* and brownish rather than reddish back.			
Synonyms None			
Subspecies None			
Distribution Bougainville, Buka (fossil), Choiseul			
Extralimital None			
Conservation status (IUCN) Vulnerable			
Etymology Sp.–rugged, rough (L.), allusion unclear but possibly refers to diagnostic coarser fur cited in the original diagnosis			

denning in dense epiphytes and shifted locations every few nights.

Female Bougainville giant rats give birth to just a single young. A pregnant female carrying a foetus with crown–rump length of 47 mm was recorded on Choiseul in March 1990. Young weighing 69 and 22 g were documented at the same time. On Bougainville, a subadult captured in October 2016 weighed 190 g.

Although Bougainville giant rat has been the most encountered Solomon Islands native murid, very little information is available on population sizes, population trends or major threats. Feral cats are common on Bougainville and Choiseul and undoubtedly represent some degree of threat. Deforestation for gardening, agriculture and commercial logging also probably threatens this species.

ISABEL GIANT RAT

Solomys sapientis (Thomas, 1902a)

sap'-ee-ent'-is

This is the only giant rat known from Santa Isabel. Data on this species are far rarer than for the similar species, *Solomys salebrosus*, which occurs on Choiseul and Bougainville. The Natural History Museum, London, holds three specimens, one of which is a young animal in spirit collected by Albert Meek, and two of which are skulls purchased in London from M.A. Hilton. The Bishop Museum, Honolulu holds two specimens, collected on 12 September 1964 at Kologatoga, at an elevation of around 30 m. The Australian Museum holds a specimen collected by N. S. Heffernan in 1924, as well as specimens collected in 1988 (two) and 1991 (one).

Scientific data on species biology are limited. Troughton (1936a) recorded the observations of N. S. Heffernan regarding this species:

> *a big and large-toothed rat which must be almost entirely arboreal, as it cracks the Ngali* (Canarium) *nuts and gnaws coconuts, and is found in trees felled by the natives, by whom it is eaten.*

The Isabel giant rat (*Solomys sapientis*) is a large arboreal rat endemic to Santa Isabel, Solomon Islands (Pavel German).

Solomys sapientis			
Size*			
	Mean	**Range**	**Sample**
Head–body	203.5	188.0–215.0	4
Tail	216.8	190.0–257.0	4
Hindfoot	56.5	50.0–67.4	4
Ear	18.0	17.3–19.0	4
Weight	–	–	–
* From AM specimens (Santa Isabel)			
Identification Closest to *S. salebrosus* of Choiseul and Bougainville. The molars are lighter, fur softer, and hindfoot longer when compared with that species.			
Synonyms None			
Subspecies None			
Distribution Santa Isabel			
Extralimital None			
Conservation status (IUCN) Endangered			
Etymology Sp.–thoughtful rat or sentient rat (L.), allusion unclear			

In 1991, people in south-western Santa Isabel told Dr Harry Parnaby that this species had become rare. This was before the onset of widespread logging on Santa Isabel and thus the conservation status of this species may have worsened.

Uromys

BIAK GIANT RAT

Uromys boeadii Groves and Flannery, 1994

yue'-roh-mis bu-ah'-dee-ee

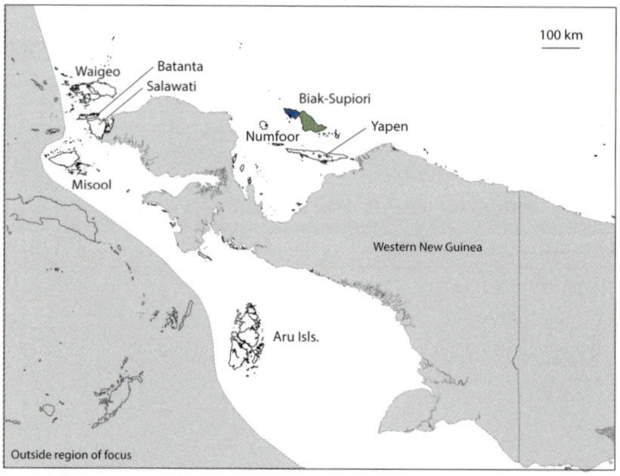

Uromys boeadii			
Size*			
	Mean	**Range**	**Sample**
Head–body	–	255.0	1
Tail	–	235.0	1
Hindfoot	–	62.0	1
Ear	–	25.0	1
Weight	–	–	–
* From AMNH 222242 (Biak)			
Identification This species is most similar to *U. neobritannicus* but differs in having a predominantly brown venter with a small white patch on the chest. It also lacks the golden guard hairs and darker eye rings.			
Synonyms None			
Subspecies None			
Distribution Biak			
Extralimital None			
Conservation status (IUCN) Critically Endangered			
Etymology Gen.–mouse tail (Gr.); Sp.–for the eminent Indonesian mammalogist, Boeadi			

The Biak giant rat is known only from the holotype, a young adult male collected by P. Temple on 22 March 1963. It was caught 25 km north-east of Biak at an elevation of 65 m. Many features of the specimen are plesiomorphic, but restriction of the white zone to a pectoral streak and the absence of any white in the inguinal region are traits unique to this species in the genus *Uromys* (Groves and Flannery 1994). Nothing is known of its biology and nothing could be learned during brief field surveys in September 1992.

The Biak giant rat is listed as Critically Endangered on the IUCN Red List. An estimated population decline of more than 80% in the past 10 years has been postulated, due to habitat loss and hunting. Additional research is urgently needed for this species to provide an assessment of population size and conservation status.

WHITE-TAILED GIANT RAT

Uromys caudimaculatus (Krefft, 1867)

kaw'-dee-mak'-yue-lah'-tus

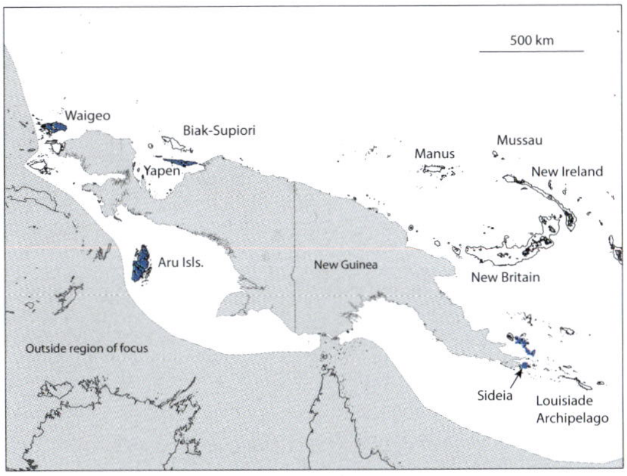

The white-tailed giant rat is a widespread and abundant species in north-eastern Australia and lowland New Guinea. It has also been recorded from numerous islands in the region under consideration here, including the D'Entrecasteaux Islands, Aru Islands, Waigeo and Yapen.

The American Museum of Natural History holds a specimen collected by H. M. Van Deusen from Goodenough Island on the Fourth Archbold Expedition in 1953. Brass (1959) also reported records of this species made during the Fifth Archbold Expedition in 1956 at low elevation near Waikaiuna, Normanby Island; and at Gomwa Bay and the Ukaiokaio coast, Fergusson Island. The Australian Museum holds a female specimen collected by Pavel German from Kokogadi Village on Fergusson Island in 1992. This species was the

White-tailed giant rat (*Uromys caudimaculatus*), Cape York Peninsula, Australia. It is common in northern Australia and New Guinea. It has also been documented from the D'Entrecasteaux Islands, Aru Islands, Waigeo and Yapen. The white-tailed giant rat is a generalist that forages in the canopy and at ground level (Mark Sanders).

Uromys caudimaculatus			
Size*			
	Mean	**Range**	**Sample**
Head–body	–	223.0	1
Tail	–	206.0	1
Hindfoot	–	50.7	1
Ear	–	25.3	1
Weight	–	350.0	1
* From AM specimen (Fergusson)			

Identification This large murid rodent is very distinctive by its large size and long tail, which is black at the base, white at the tip and mottled in between.

Synonyms *Mus macropus* Gray, 1866a
Uromys aruensis Gray, 1873
Uromys papuensis Meyer, 1876 (nomen nudum)
Uromys validus Peters and Doria, 1881
Uromys ductor Thomas, 1913d
Uromys nero Thomas, 1913a
Uromys prolixus Thomas, 1913d
Uromys scaphax Thomas, 1913a
Uromys sherrini Thomas, 1923a
Uromys siebersi Thomas, 1923b
Uromys exilis Troughton and Le Souef, 1929
Uromys waigeuensis Frechkop, 1932
Uromys lamington Troughton, 1937b

Subspecies *Uromys caudimaculatus caudimaculatus* (Krefft, 1867) (Southern New Guinea, Aru Is. and north-eastern Australia)
Uromys caudimaculatus papuanus (Ramsay, 1883) (Eastern New Guinea and the D'Entrecasteaux Isls)
Uromys caudimaculatus multiplicatus (Jentink, 1907) (western New Guinea and Waigeo Is.)

Distribution Aru Isls, Waigeo, Yapen (Indonesia); Fergusson, Normanby, ?Sideia (Papua New Guinea)

Extralimital Australia, Kai Isls, New Guinea

Conservation status (IUCN) Least Concern

Etymology Sp.–spotted tail (L.), a reference to the clearly mottled tail subject of an extensive taxonomic revision (Groves and Flannery 1994). However, specimens from the D'Entrecasteaux Islands were not included and may need closer evaluation.

White-tailed giant rat has been well studied in Australia, but very little has been recorded of its biology from the islands of the South-west Pacific. On mainland New Guinea, it is common in lowland forests as well as more disturbed secondary forests, clearings, forest margins and coconut plantations (where it is a pest capable of opening young coconuts) (Menzies and Dennis 1979; Flannery 1995).

White-tailed giant rats are nocturnal and agile tree climbers that forage in terrestrial and arboreal locations. They are generalist omnivores known to feed on a variety of nuts and fruit, as well as vertebrates, crustaceans, invertebrates and bird eggs. White-tailed giant rats have extremely strong jaws and in northern Australia are notorious for their ability to chew through aluminium small mammal traps. These normally solitary animals build nests in hollow trees and underground burrows beneath logs and overhanging banks. Females normally rear two or three young, which become independent at around 3 months of age (Moore 2008). The white-tailed giant rat is common, widespread and tolerant of habitat disturbance.

EMMA'S GIANT RAT

Uromys emmae Groves and Flannery, 1994

em'-ay

Emma's giant rat is known from a single adult female collected on the island of Owi, West Papua by

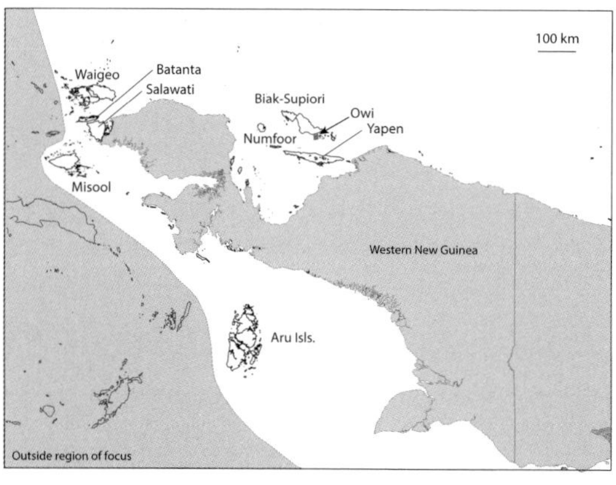

Uromys emmae			
Size*			
	Mean	**Range**	**Sample**
Head–body	–	232.0	1
Tail	–	258.0	1
Hindfoot	–	50.5	1
Ear	–	20.5	1
Weight	–	–	–
* From holotype AM M7200 (Owi)			
Identification Resembles a small *U. caudimaculatus*, but the white mottling is restricted to the tip of the tail and the feet are shorter and broader.			
Synonyms None			
Subspecies None			
Distribution Owi			
Extralimital None			
Conservation status (IUCN) Critically Endangered			
Etymology Sp.–for Emma, daughter of author Tim Flannery			

C. B. Philip of the U.S. Army Typhus Commission in July 1946. It possesses several features (short, broad feet and a short snout) indicating that it is highly arboreal. The enlarged teats suggest it was lactating when caught.

Emma's giant rat is only known from a single, tiny island, much of which is now deforested. On eastern Biak and southern Supiori in 1992, people were familiar with an arboreal, white-tailed rat that may have been this species (the tail of the *E. boeadi* holotype is uniformly black). However, repeated efforts to locate Emma's giant rat have failed, so it may well be extinct.

EMPEROR RAT

Uromys imperator (Thomas, 1888c)

im-pe'-rah-tor

The emperor rat is a rare species, known only from three specimens collected by C. M. Woodford at Aola, northern Guadalcanal, between 1886 and 1888. All three are spirit specimens, some with the skull removed. It is an extremely large species, probably exceeding 1 kg in weight, with a short tail and sparse greyish fur that is somewhat similar to the colouration of the sympatric *Uromys rex*.

Unusually for the genus *Uromys*, its short tail and the morphology of the feet suggest emperor rat may have been terrestrial (Thomas 1888c). The adult male (NHMUK 88.1.5.32), kept entire in spirit, has clay adhering to the muzzle and the claws of the hind and forefeet, indicating it was probably dug from a burrow.

During survey work on Guadalcanal in 1988 and 1990, people knew of a large terrestrial murid that could have been this species. Most sightings had been made 30 or more years earlier in montane mossy forest. Similar traditional ecological knowledge persists among people today, describing a large rodent that is restricted to steeper rocky areas. In the past, emperor rats inhabited Guadalcanal's lowlands and Woodford's specimens were probably collected close to the coast at Aola (Groves and Flannery 1994). This region has now been logged.

The emperor rat (*Uromys imperator*) has not been seen since the late 1800s and is probably extinct as a result of predation by feral cats (Madison Erin Mayfield).

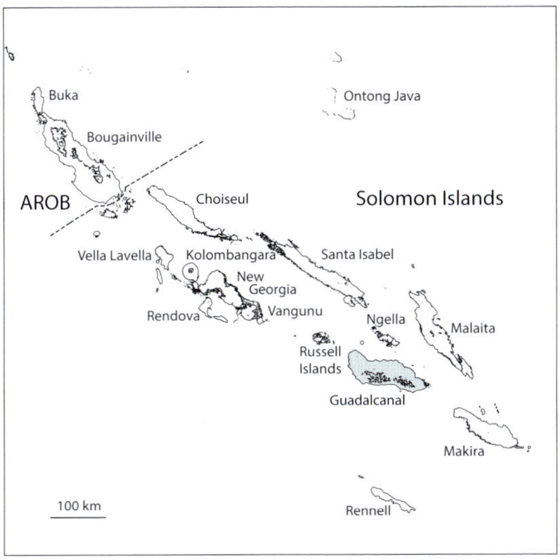

Uromys imperator			
Size*			
	Mean	**Range**	**Sample**
Head–body	–	350	1
Tail	–	258	1
Hindfoot	–	66	1
Ear	–	19	1
Weight	–	–	–
* From Thomas (1888c)			
Identification A very large murid, with a tail less than head–body length, and sparse greyish fur. The tail is naked and the scales relatively small.			
Synonyms None			
Subspecies None			
Distribution Guadalcanal			
Extralimital None			
Conservation status (IUCN) Critically Endangered			
Etymology Sp.–emperor (L.), for its large size			

Jaws of this species have also been found in an archaeological site in the Poha River Valley, close to the northern coastline of Guadalcanal (Flannery 1991a). This area is now surrounded by grassland, maintained through a process of periodic burning by humans that began around 3300 years before present (Roe 1992). These fossil records show that in the distant past the emperor rat was present in the rain shadow of northern Guadalcanal's lowlands as well as more heavily forested higher rainfall areas. Thus, its reported restriction to higher elevation montane, rocky habitats is apparently a more recent phenomenon.

Camera trap surveys have demonstrated feral cats are extremely abundant from sea level to at least 1300 m on Guadalcanal (Lavery *et al.* 2020c). This new threat may have caused the retreat of emperor rats into more topographically complex, higher elevation habitats where some reprieve from predation was available. The emperor rat is classified as Critically Endangered on the IUCN Red List, but is most likely extinct.

BISMARCK GIANT RAT

Uromys neobritannicus Tate and Archbold, 1935b

nee'-oh-brit-an'ik-us

The Bismarck giant rat is a very attractive murid. The tail and fur around the eyes are black, and the head and back have a coppery appearance due to the large number of shiny copper guard hairs over the brownish coat. This contrasts with a bright yellow belly.

Like many of the region's large murids, limited scientific information is available for *Uromys neobritannicus* and few specimens exist in world museum collections. The holotype is a skull collected during the Whitney South Seas Expedition, held at the American Museum of Natural History, with a second collected by F. R. Schumm from the Willaumez Peninsula (Tate 1951). Gilliard and Lecroy (1967) listed two specimens collected by M. Gilliard in 1959 in the Whiteman Range of south-western New Britain. Museums Victoria holds one specimen collected by Clive Champion on the Willaumez Peninsula near Talasea in July 1965.

Uromys neobritannicus			
Size*			
	Mean	**Range**	**Sample**
Head–body	276.7	252–298	3
Tail	256.3	243–279	3
Hindfoot	54.3	51–58	3
Ear	27.5	26.5–28	3
Weight	626.7	570–730	3
* From AM specimens (New Britain)			
Identification The combination of a yellow belly, coppery back with shining golden guard hairs, black tail and black rings around the eyes distinguish this species.			
Synonyms None			
Subspecies None			
Distribution New Britain			
Extralimital None			
Conservation status (IUCN) Near Threatened			
Etymology Sp.–of New Britain (L.), the island habitat of the species			

As with many of the endemic murids from the region, limited scientific information is available for the Bismarck giant rat (*Uromys neobritannicus*) and it has been seldom documented (Madison Erin Mayfield).

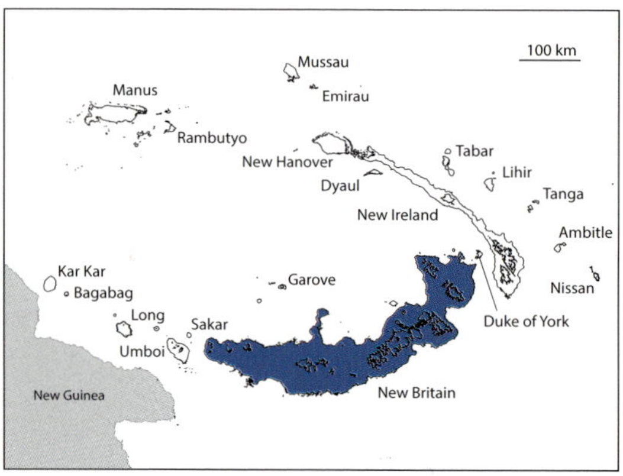

In 1988, an Australian Museum team documented three individuals at Melei Village near Fulleborn on the southern coastline. All were caught as solitary animals, and two were discovered at night by local hunters at 30 m elevation (unfortunately habitat type was not recorded). The third was caught by dogs in regrowth beside the Lofun River at 500 m elevation, suggesting it may have been on the ground at the time. During surveys in the Nakanai Mountains of East New Britain in 2009, a very large, dark-furred rat was seen at 200 m elevation and was likely this species (Aplin and Opiang 2011). The site there was a mixture of lowland alluvial and hill, *Pometia*-dominant forests with depauperate understories. Extensive field surveys on New Ireland have failed to find Bismarck giant rat and it is yet to be detected there in subfossil deposits (Leavesley 2005).

Nothing has been documented of diet or reproduction for the Bismarck giant rat. It has been classified as Near Threatened on the IUCN Red List due to a continuing decline in the extent and quality of habitat.

GUADALCANAL RAT

Uromys porculus Thomas, 1904a

pork'-ue-lus

The Guadalcanal rat is known only from the holotype collected by C. M. Woodford at Aola,

Guadalcanal, presumably between 1886 and 1888. It is an adult male specimen not described until 1904, at which time it was removed from spirit and skinned. It is strikingly different from Guadalcanal's other native rodents, having reddish rather than woolly grey fur and a very short tail.

Information relating to the Guadalcanal rat is extremely rare. The very short tail suggests it was terrestrial and probably a poor climber, making it vulnerable to predation by feral cats. Camera trap surveys from sea level to 1300 m elevation have demonstrated just how pervasive feral cats are in Guadalcanal's forests (Lavery *et al.* 2020c). Feral cats were by far the species most commonly detected, and they are widely regarded as a major driver of extinction in island endemic faunas worldwide.

Traditional knowledge of medium to large ground-dwelling murids persists on Guadalcanal today, including one species said to inhabit caves. However, most encounters with such species are said to have occurred around the 1960s. Just two more recent accounts made by biologists offer some slim hope that Guadalcanal rats may survive today. In the early 1970s, before the forests of Mt Austen had been heavily denuded, Michael McCoy regularly made nocturnal visits to collect reptiles. On one occasion, he came across a large rat perched on a small tree stump a couple of metres above the ground. Spotlights dazzled the rodent, and McCoy was able to pin its tail against the stump and pick it up. It was dazed and didn't struggle at all, being held and viewed for 10 s or more, before twisting to bite his thumb and escape. It was large (approximately the size of *Uromys rex*) but more gracile, slender, and with light reddish fur and a somewhat short snout (M. McCoy, pers. comm.). In November 1989, Tim Flannery saw a medium-sized, reddish murid while spotlighting in the Poha River valley west of Honiara. The animal was climbing among canes on the side of a track in relict forest.

The observation of a climbing habit in both these more recent accounts does not accord with the presumed terrestrial habits of *U. porculus.*

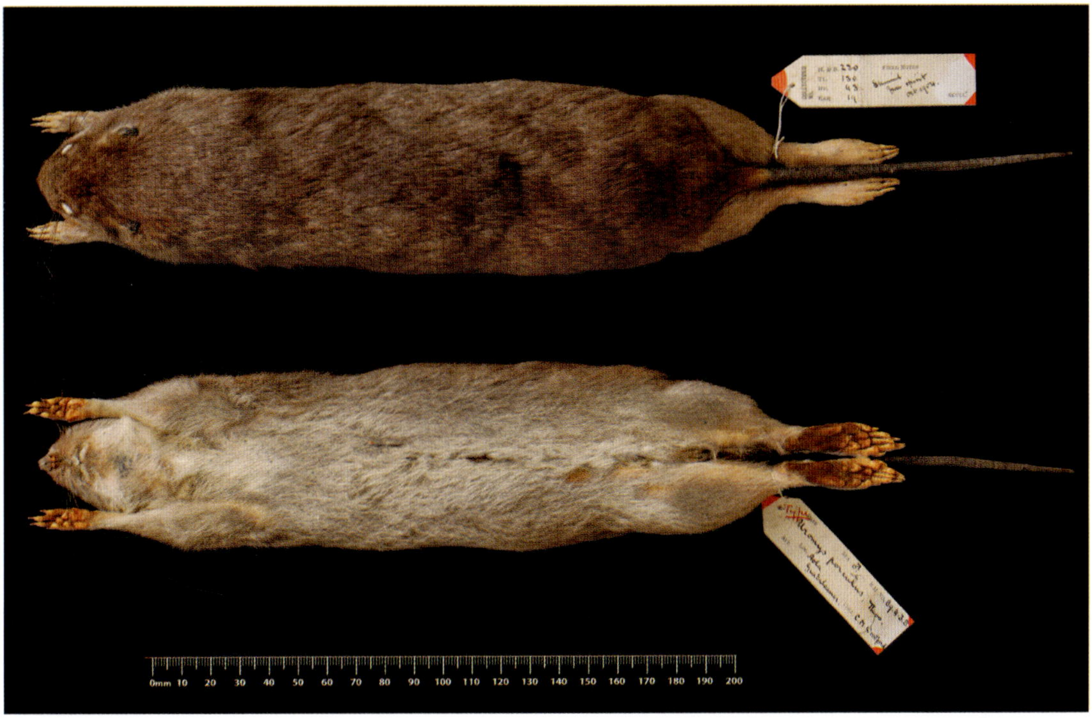

The holotype of the Guadalcanal rat (*Uromys porculus*) is the only known specimen of this species. This animal has probably been driven to extinction by feral cat predation (Photograph by Christian Ching. Courtesy of the Trustees of the Natural History Museum, London).

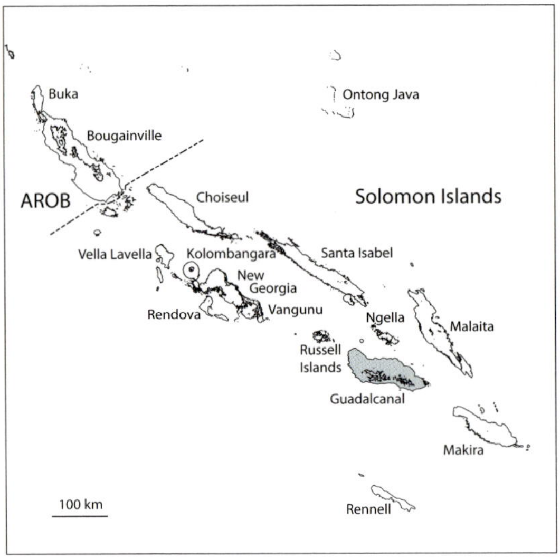

Uromys porculus			
Size*			
	Mean	**Range**	**Sample**
Head–body	–	220	1
Tail	–	130	1
Hindfoot	–	43	1
Ear	–	19	1
Weight	–	–	–
* From Thomas (1904a)			
Identification A reddish rodent with grey venter and a hairless tail approximately two-thirds of the head–body length			
Synonyms None			
Subspecies None			
Distribution Guadalcanal			
Extralimital None			
Conservation status (IUCN) Critically Endangered			
Etymology Sp.–little pig (L.)			

Alternately, both animals may have been a species of *Melomys*, somewhat like *M. bougainville* that exists on Bougainville and Choiseul. The Guadalcanal rat is classified as Critically Endangered on the IUCN Red List, but is most likely extinct.

KING RAT
Uromys rex (Thomas, 1888c)
reks

The king rat is the only one of three species of *Uromys* endemic to Guadalcanal that still survives today. It is also the only arboreal species, and this may account for its persistence, as South-west Pacific terrestrial murids have proven more vulnerable to extinction than their arboreal counterparts. Nevertheless, the King rat is considered Endangered. Modern records are rare, and it now seems absent from large parts of the island.

The type series, collected by Charles Woodford between 1886 and 1888, consisted of four animals from Aola on the north-west coast of Guadalcanal. This area has now been heavily logged. The Bernice P. Bishop Museum, Honolulu, holds two specimens collected by P. Shanahan on 9 and 10 June 1964 at Tabila, at an elevation of 20 m. The Australian Museum holds three specimens: AM M19739 is a skin only, collected in March 1933, and possibly represents *U. imperator* (see Groves and Flannery 1994); AM M13549 is a spirit body and skull probably collected in the 1930s; and the final specimen was collected by T. Flannery in 1987. It is an aged male from Poha River Valley, west of Honiara on northern Guadalcanal. When encountered at around 8:30 p.m., it was climbing on a liana some 10–15 m above the ground in a patch of relict forest. Museums Victoria also holds a single female spirit specimen NMV C35333 (new catalogue no.).

For a short period in 1988, Michael McCoy kept a male and female in captivity in Honiara. Both were captured from forested areas behind Kakambona, west of Honiara. The juvenile male

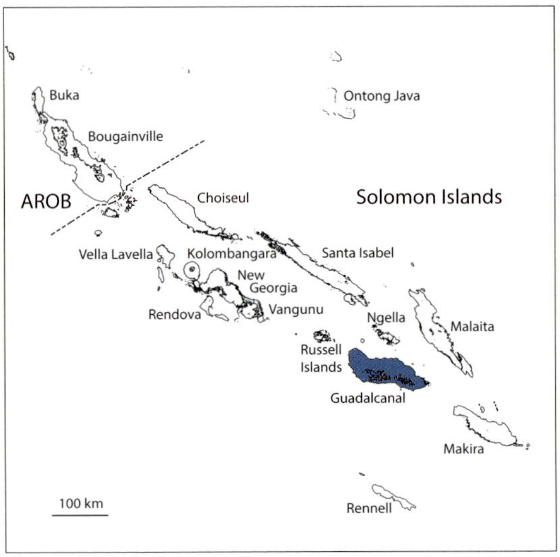

King rat (*Uromys rex*), Guadalcanal, Solomon Islands. This is probably the only one of Guadalcanal's three endemic rodents that still survives today (Michael McCoy).

was a very gentle animal that could be easily handled. However, it was a finicky eater that would take shelled ngali nuts, semi-ripe papaya (*Carica papaya*) and some coconut (*Cocos nucifera*), but refused sweet potato, and any dried fruit or trail mix (M. McCoy, pers comm.).

Uromys rex			
Size*			
	Mean	**Range**	**Sample**
Head–body	268.5	257–290	4
Tail	277.8	232–298	4
Hindfoot	53.5	50–55	4
Ear	18.5	16–20	4
Weight	–	420	1
* From AM and NMV specimens and Thomas (1888c)			
Identification A large, silvery, arboreal murid. It differs from *U. imperator* by its enlarged foot pads and longer tail with larger scales.			
Synonyms None			
Subspecies None			
Distribution Guadalcanal			
Extralimital None			
Conservation status (IUCN) Endangered			
Etymology Sp.–king (L.)			

The king rat probably persists in low numbers in areas where large tracts of primary forest remain. More recent searches in forested areas south of Honiara have revealed gnawed ngali nuts (*Canarium indicum*) that are believed to indicate the presence of this species. Ongoing searches using camera traps will hopefully provide clear evidence that king rat survives in these areas.

VANGUNU GIANT RAT

Uromys vika Lavery and Judge, 2017

vick'-ah

Revealed to the scientific world in 2017, the Vangunu giant rat was the first new species of Solomon Islands murid described in over 80 years. Vangunu's Indigenous people long knew of it by the language name *vika*, and for decades mammalogists had been aware of this knowledge. Nonetheless, several years of targeted surveys using camera traps, rodent traps and active searches failed to produce any evidence of the existence of this species. Somewhat ironically, a specimen was finally procured when a logging company operating on southern Vangunu felled a large tree containing a Vangunu giant rat. The holotype was heavily degraded by the time it was accessioned to the collections of the Queensland Museum, and the skull and partial skeleton remain the only known specimen of this species.

Uromys vika differs from *U. imperator* and *U. rex* of Guadalcanal by its reddish-brown rather than grey pelage. Its mosaic tail is almost entirely hairless, with minute scales of central prominence surrounded by large fleshy areas. It is an arboreal animal with broad hindfeet and prominent plantar pads. A tarsal pad, similar in size to the hypothenar and interdigitals, is present on the lateral heel of the hindfoot, isolated from the hypothenar pad.

The Vangunu giant rat is apparently restricted to the lowland forests of Vangunu, and is reportedly common in large *Dillenia salomonensis* trees, which are a primary target for logging companies. *Dillenia* are said to provide preferred habitat because they readily form hollows that can be used as dens. The Vangunu giant rat is said to feed on coconuts (*Cocos nucifera*) and ngali nuts (*Canarium indicum*). A camera trap survey near Zaira village on southern Vangunu in 2021 recorded at least three individuals of this species at two sites (at elevations of approximately 50 and 250 m). The cameras were set in arboreal positions and baited with canisters filled with sesame oil. Nothing else has been documented of diet, reproduction or habitat preferences.

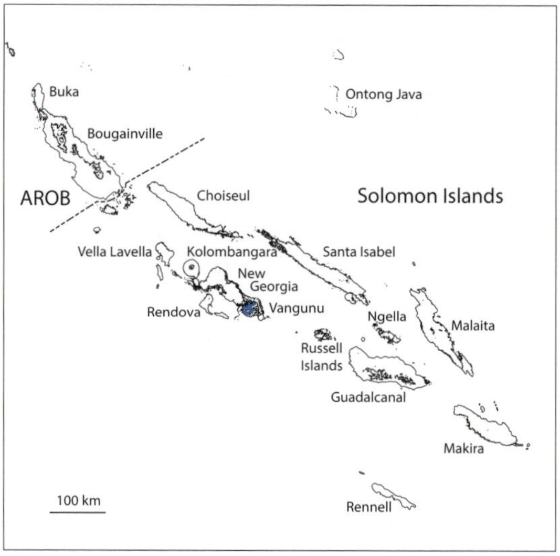

Uromys vika			
Size			
	Mean	**Range**	**Sample**
Head–body	–	–	–
Tail	–	–	–
Hindfoot	–	–	–
Ear	–	–	–
Weight	–	–	–
Identification A large rodent with reddish-brown fur, broad hindfeet, short ears, and a tail with minute scales surrounded by large fleshy areas.			
Synonyms None			
Subspecies None			
Distribution Vangunu			
Extralimital None			
Conservation status (IUCN) Critically Endangered			
Etymology Sp.–*vika* is the Vangunu language name for this rodent			

The Vangunu giant rat (*Uromys vika*) is Critically Endangered and probably limited to the remaining primary forests of southern Vangunu Island (Velizar Simeonovski).

Vangunu giant rat is facing significant threats from commercial logging, with only small areas of suitable primary forest remaining on Vangunu. The largest of these forested areas in the south of the island have been dedicated for conservation by the village of Zaira, but there remains an ever-present threat from logging companies attempting to foil this protection.

Chiroptera

Pteropodidae

The Old World fruit bats (family Pteropodidae) are by far the most species-rich family in the region with some 68 described species. Furthermore, the level of endemism is extremely high, with 78% (53 species) endemic to the South-west Pacific. In the far east of the region, pteropodids are the only indigenous mammals present. Five of the 12 genera are endemic (*Nesonycteris*, *Notopteris*, *Melonycteris*, *Mirimiri* and *Pteralopex*). The genera *Dobsonia*, *Nyctimene* and *Pteropus* are all mostly distributed on islands, and have centres of diversity in Wallacea and Melanesia (Simmons 2005; Tsang *et al.* 2019). Many are old genera – for example, *Syconycteris* and *Macroglossus* diverged from other extant pteropodid genera more than 22 million years ago (Almeida *et al.* 2020). The most recent common ancestor of *Notopteris*, *Nyctimene* and *Paranyctimene* was estimated to have occurred more than 20 million years ago. The genera *Melonycteris* and *Nesonycteris* were estimated to have split around 10 million years before present (Almeida *et al.* 2020).

The diversity of pteropodids is greatest at low latitudes and on moderately isolated, large, old islands. The Solomon Islands archipelago has provided the ideal setting for radiations of pteropodid diversity. This cluster of large, tropical islands has remained close to other landmasses over an extended geological period, providing ample opportunity for colonisation. However, their isolation has been sufficient, and colonisation events probably infrequent enough, to promote secondary radiations. No less than 10 species of *Pteropus* are found, along with five *Pteralopex*, four *Nesonycteris*, three *Nyctimene*, and one species each from *Dobsonia*, *Macroglossus* and *Rousettus*.

The ecology of most South-west Pacific pteropodids remains poorly known. Nonetheless, they are widely regarded as keystone species of island ecosystems, and the ecological niches exploited by this diverse group are equally diverse. For example, *Pteralopex* (Solomon Islands) and *Mirimiri* (Fiji) are medium to large species with large complex teeth and the most robust skulls of all bats. They are capable of breaking materials as hard as a ngali nut (*Canarium indicum*), and strong enough to carry green coconuts (*Cocos nucifera*) short distances. But they also take a variety of other fruits, flowers and leaves, and may even partake in gouging tree bark to feed on exudates (Fisher and Tasker 1997; Helgen 2005). Many of the smaller pteropodids such as species from the endemic genus *Nesonycteris* are understood to be entirely nectarivorous.

Choice of roost sites is variable. Some species are obligate cave-roosters (e.g. *Rousettus*, *Notopteris* and some species of *Dobsonia*), while others roost individually concealed among foliage. The larger

Solomons flying fox (*Pteropus rayneri*) (illustration by Joseph Smit in Thomas 1887b).

species of *Pteropus* often roost in large camps, sometimes numbering in the thousands of individuals, and *Pteralopex* and several smaller species of *Pteropus* will occupy tree hollows.

Many pteropodid species are declining due to human impacts, and five species that once occurred in the region are now possibly extinct. Anecdotal data suggest that many others are in decline, but firm evidence is available in only a few cases. Many of the large *Pteropus* are under threat from over-hunting; smaller species are primarily threatened by habitat destruction.

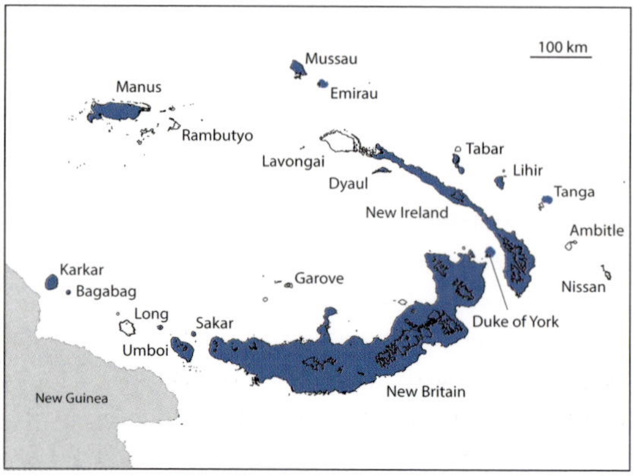

Dobsonia

ANDERSEN'S BARE-BACKED FRUIT BAT

Dobsonia anderseni **Thomas, 1914a**

dob-so'-nee-ah and'-er-sen-ee

Dobsonia anderseni			
Size*			
	Mean	**Range**	**Sample**
Head–body	157.8	130.0–212.0	9
Tail	27.4	15.7–40.0	9
Forearm	123.1	109.0–135.0	9
Tibia	56.6	48.0–61.5	9
Hindfoot	29.5	23.8–38.0	4
Ear	25.6	22.0–28.5	5
Weight	211.2	135.0–325.0	13
* Including data from Bonaccorso (1998) and Aplin *et al.* (2015)			
Identification Slightly larger than the sympatric species *D. praedatrix* and has white, rather than brown claws. The fur is brownish grey, rather than greenish.			
Synonyms None			
Subspecies None			
Distribution Bagabag, Boang, Duke of York, Dyual, Emirau, Karkar, Lihir, Manus, Mussau, New Britain, New Ireland, Sakar, Tabar, Tolokiwa, Umboi			
Extralimital None			
Conservation status (IUCN) Least Concern			
Etymology Gen.–for G. E. Dobson, British Museum authority on bats; Sp.–for Knud Andersen, Danish authority on bats			

This is a relatively common species of *Dobsonia* found in the Bismarck and Admiralty Islands. When seen in flight it can easily be confused with the smaller *D. praedatrix*. However, on closer inspection it can be clearly recognised by its white, rather than brown, claws. During a survey of the bats of the Bismarck Archipelago, Smith and Hood (1981) found *D. anderseni* to be the most commonly recorded species after *Rousettus amplexicaudatus*.

Andersen's bare-backed fruit bat roosts in the twilight zones of caves (Smith and Hood 1981; Bonaccorso 1998). Nearly all caves on New Ireland visited by Smith and Hood (1981) contained large colonies. More recently, Aplin *et al.* (2015) encountered smaller groups on Manus and Mussau. A group of around 20 individuals was found underneath a shallow overhang on a steep, densely forested slope at approximately 310 m elevation on Manus. Landowners indicated a much larger colony could be found in a cleft-shaped cave at lower elevation on the south-western side of Mt Sabomu. Groups of 30–40 individuals were found in two caves on Mussau.

The wings of *Dobsonia* join close to the midline of the back, allowing them great manoeuvrability through the forest. A secondary result of this is that their wing beats produce a hollow clapping noise. This is a useful characteristic for detecting their presence in the upper canopy of complex forests at night. Andersen's bare-backed fruit bat feeds among the canopy and subcanopy, and figs are an important part of the diet (Bonaccorso 1998). An adult male was captured on Mussau Island in a mist net set among a grove of papaya (*Carica papaya*) trees on a limestone terrace at 96 m elevation. Bonaccorso (1998) also reported this species feeding on rambutans (*Nephelium lappaceum*) and juice extracted from the seed pods of rain trees (*Pithecelobium saman*).

Aplin *et al.* (2015) caught a pregnant female and subadult female on Mussau in October 2014. Smith and Hood (1981) encountered lactating females, young and subadults on New Ireland between June and August. Lactating females were frequently mist-netted outside the cave but were never carrying young. However, those mist-netted inside caves were often carrying young. Presumably the females leave them at the roost while they forage but carry them if disturbed at the roost.

This species is Least Concern on the IUCN Red List, and its predilection for disturbed habitat suggests its status is secure. However, as a large colonial species, it is probably under pressure from hunting. Monitoring of colony sizes will better establish population trajectories and determine whether there is significant threat.

BEAUFORT'S BARE-BACKED FRUIT BAT
Dobsonia beauforti Bergmans, 1975
boh'-fort-ee

Beaufort's bare-backed fruit bat is an attractive species with yellow-green fur. It is clearly sexually dimorphic, with males being larger in body size and bright yellow-green in colour. They have large, heavy heads with white cheek glands and pale, somewhat naked skin on the rest of the face (although the intensity of these glands may change with reproductive condition). Females are much more brownish and have a much more gracile head that is blackish and better furred than in males. Young animals lack the yellowish colour so evident in adult males.

Beaufort's bare-backed fruit bat is common on islands of Cenderawasih and the Vogelkop Peninsula, but is largely replaced by *D. magna* and *D. minor* and possibly other pteropodids on Yapen and Salawati. It is not entirely excluded from such islands though. In 1992, a single young adult was mist-netted on a beach on the northern coast of Salawati adjacent to Batanta. It was almost certainly a vagrant that had crossed the few kilometres of sea separating the two islands.

In 1991–1992 this species was found to be common on Gebe, Biak-Supiori and Batanta. Interestingly, it seems to be absent from Numfoor, where it could be reasonably expected to occur. Several other pteropodid species were mist-netted there in different habitats, but Beaufort's bare-backed fruit bat, common on other islands, was not detected. All individuals on Gebe were mist-netted in forest, one of which was carrying a cluster fig (*Ficus* sp.) in its mouth. On Biak-Supiori, it was common and widespread, particularly in secondary forest. One was observed carrying an entire fruit of *Piper aduncum* while in flight, while others defecated the seeds of this plant when captured. Others were mist-netted when foraging around a fruiting bilimbi (*Averrhoa bilimbi*) on the edge of a village, and local people indicated many other cultivated fruits were eaten by this species.

None of the caves visited on Biak-Supiori in 1991–1992 were occupied by Beaufort's bare-backed fruit bats. But a single individual was seen roosting under a fallen tree trunk in a creek near Korido Village, Supiori. It is possible that the similar-sized *D. emersa* excludes this species from caves where both co-occur. On Batanta it was mist-netted in secondary forest, but also observed roosting in caves. In a large inland cave it was found roosting with *D. magna*, while a group of around 30 animals (including females with young) were the sole inhabitants of a shallow, wave-washed sea cave in eastern Batanta.

The type series was collected on 25 December 1909. Of the 14 animals examined, seven were lactating or pregnant females. The embryos had forearm lengths of 19.2, 23.7, 29.3, 32.8 and 34.8 mm, and newborn young had forearm

Beaufort's bare-backed fruit bat (*Dobsonia beauforti*) is common on islands of Cenderawasih Bay and the Vogelkop Peninsula, but is largely replaced by *D. magna* and *D. minor* and possibly other pteropodids on Yapen and Salawati (Chien Lee).

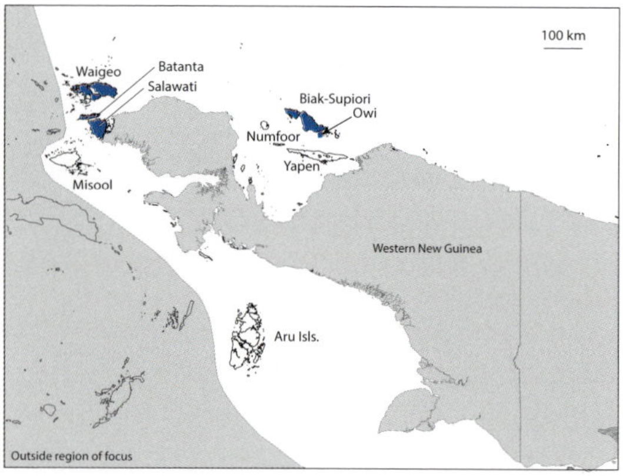

Dobsonia beauforti			
Size*			
	Mean	**Range**	**Sample**
Head–body	144.9	126.0–161.0	18
Tail	19.0	15.0–23.0	18
Forearm	101.5	89.0–112.0	18
Tibia	–	–	–
Hindfoot	26.4	24.0–28.0	18
Ear	25.6	24.0–27.0	18
Weight	117.1	75.0–168.0	18
* From AMNH specimens (Biak)			
Identification A small species of *Dobsonia*, smaller than the sympatric *D. emersa* and *D. magna* and yellowish-green rather than grey-brown in colour.			
Synonyms None			
Subspecies None			
Distribution Batanta, Biak-Supiori, Owi, Salawati, Waigeo			
Extralimital Gebe			
Conservation status (IUCN) Least Concern			
Etymology Sp.–named in memory of Professor Lieven Ferdinand de Beaufort (1879–1968), former director of the Zoölogisch Museum, Amsterdam			

lengths of 42.7 mm. This suggests births take place in December on Waigeo. Females on Biak-Supiori were in an early stage of pregnancy in September. On Batanta many pregnant females were observed in October, but a few females had already given birth. Four of seven adult females examined on Gebe on 29 November were lactating, while another was pregnant. This indicates the timing of reproduction may be relatively consistent across the species distribution. Beaufort's bare-backed fruit bat has a relatively wide distribution and large population and is unlikely to be declining at a rate to qualify for a threatened category.

BIAK BARE-BACKED FRUIT BAT

Dobsonia emersa Bergmans and Sarbini, 1985

ay-mer'-sah

The Biak bare-backed fruit bat is a calm and easily handled species of *Dobsonia*. When captured in a mist net they do not struggle and indeed rarely move if they are gently disentangled. When released, they show a reluctance to fly from the perch on which they are placed. This behaviour contrasts markedly with that of Beaufort's bare-backed fruit bat (with which it is sympatric), which bites and struggles strenuously when captured, and when released usually takes flight after a few moments.

The Biak bare-backed fruit bat roosts in caves. It is easily disturbed at the roost and seems to prefer to fly off rather than seek refuge in the cave crevices as do other species of *Dobsonia*. The earliest scientific records of this species were from a garden at an elevation of approximately 3 m near the coast at Sorido, Biak, and from secondary forest at an elevation of around 30 m on Owi (Bergmans and Sarbini 1985). In September 1992 on Biak-Supiori, Biak bare-backed fruit bat was most common around villages. Large numbers were mist-netted foraging around a fruiting bilimbi (*Averrhoa bilimbi*) growing in the Korido Village, Supiori. It was rarely encountered in secondary forest, a habitat favoured by Beaufort's bare-backed fruit bat.

The population of *Dobsonia* on Numfoor attributed to this species may represent a distinct undescribed species. Beaufort's bare-backed fruit bat is absent there, and *D. emersa* was encountered in a greater range of environments and was common in

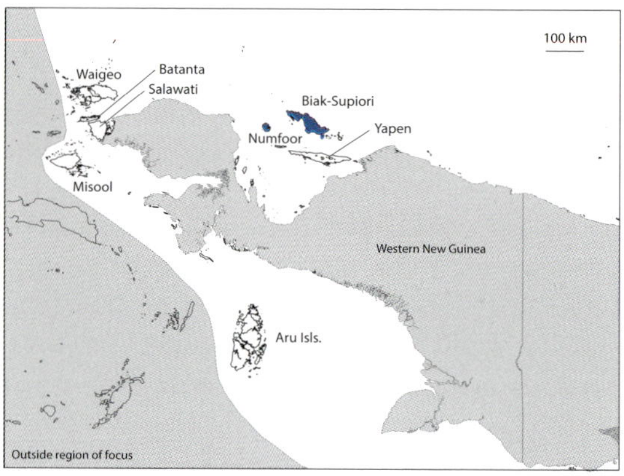

Dobsonia emersa			
Size*			
	Mean	**Range**	**Sample**
Head–body	165.8	158.0–170.0	7
Tail	54.1	53.0–55.1	7
Forearm	113.8	109.1–119.4	7
Tibia	54.1	53.0–55.1	7
Hindfoot	–	–	–
Ear	26.6	25.3–28.1	7
Weight	170.8	104.0–218.0	7
* From AM specimens (Biak-Supiori) and Bergmans and Sarbini (1985)			
Identification A small member of the *D. moluccensis* group, distinguished by its smaller size, distinctly curved rostrum, and small narrow teeth. It can be distinguished from the sympatric *D. beauforti* by its larger size and brownish-black, rather than yellow-green colour.			
Synonyms None			
Subspecies None			
Distribution Biak-Supiori, Numfoor, Owl			
Extralimital None			
Conservation status (IUCN) Vulnerable			
Etymology Sp.–from emergere, to emerge (L.), an allusion to the species remaining known but unrecognised for more than 80 years			

secondary forest. The absence of *D. beauforti* may allow *D. emersa* to expand its ecological niche.

Little is known of reproductive biology. A small sample caught on Numfoor comprised five males and two females. Two Numfoor individuals caught on 29–30 September 1992 were juveniles (weights 30 and 98 g). There was a bias in the sex ratio of captured animals on Biak-Supiori. Of 14 captured, only four were female.

The Biak bare-backed fruit bat has a limited distribution, is dependent on cave roosts, and seems easily disturbed at roost locations. The species appears to tolerate habitat disturbance (being common around villages), but further studies are needed to determine whether populations are in decline, and the category of Vulnerable is appropriate.

SOLOMONS BARE-BACKED FRUIT BAT

Dobsonia inermis Andersen, 1909a

in-er'-mis

This is the only species of bare-backed fruit bat known to occur in the Solomon Islands archipelago. It is an extremely widespread and common species and is one of the most frequently mist-netted bats in the archipelago. It is uncertain whether any subspecies should be recognised. Phillips (1968) named the subspecies *minimus* from Choiseul and recognised an intergrade population from Santa Isabel. Examination of more material is needed before the status of these populations can be fully resolved, and subspecies names are listed tentatively here.

The Solomons bare-backed fruit bat roosts in caves or large hollow trees and strangler figs (*Ficus* spp.). Large caves can contain hundreds of individuals. It is a species encountered in both primary and secondary forests and it will readily venture into gardens and disturbed areas to feed on cultivated banana (*Musa* sp.), papaya (*Carica papaya*), mango (*Mangifera indica*) and Malay apple (*Syzygium malaccense*). Animals caught in mist-nets almost always excrete the seeds of figs (*Ficus* spp.). The prevalence of Solomons bare-backed fruit bats decreases with increasing elevation. On Mt Makarakomburu, Guadalcanal, several individuals were captured in mist nets set below 700 m elevation, but only a single individual was caught at 950 m, and none at 1230 m despite several nights mist-netting and the capture of several other bat species. It was present at Chupukama (elevation 1300 m) on the northern slopes of Popomanaseau, Guadalcanal in 2015. In 2016 it was recorded at an elevation of 950 m close to the summit Mt Tolobusu, Malaita, but in low numbers. Like most large bare-backed fruit bats, the wing beats of this species create a characteristic *pok pok* noise as they move around the forest.

In the Solomon Islands archipelago, the Solomons bare-backed fruit bat (*Dobsonia inermis*) is one of the most common species of bat and can be encountered roosting in large numbers inside caves (Tyrone Lavery).

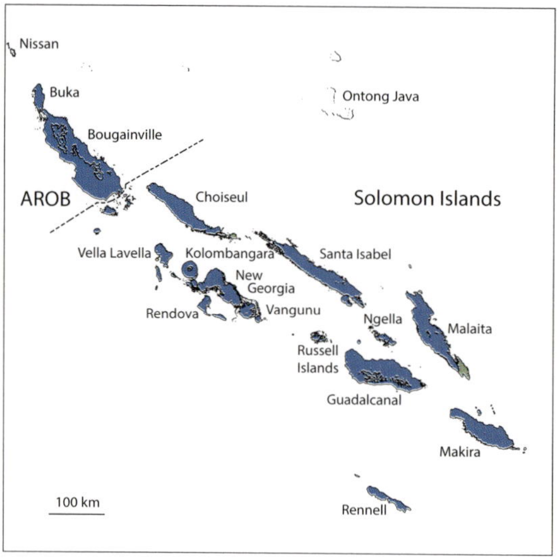

Dobsonia inermis			
Size*			
	Mean	**Range**	**Sample**
Head–body	126.2	87.8–159.3	64
Tail	28.4	20.2–35.4	38
Forearm	104.1	86.7–115.0	83
Tibia	48.7	36.8–55.0	79
Hindfoot	25.2	20.0–29.3	7
Ear	20.5	12.5–25.0	79
Weight	150.2	72.0–203.0	80
* From AM, KU and QM specimens and field measurements (Bougainville, Guadalcanal, Kolombangara, Malaita, Makira, Ngella, Ranongga, Rennell, Tetepare, Vella La Vella)			
Identification The back is entirely hairless – this is only species of *Dobsonia* in the Solomon Islands archipelago and thus the only species with this characteristic.			
Synonyms *Dobsonia inermis nesea* Andersen, 1909a			
Subspecies *Dobsonia inermis inermis* Andersen, 1909a (Solomon Isls excluding Choiseul and Santa Isabel) *Dobsonia inermis minimus* Phillips, 1968 (Choiseul, Santa Isabel)			
Distribution Bougainville, Choiseul, Fauro, Gizo, Guadalcanal, Kolombangara, Makira, Malaita, Mono, New Georgia, Ngella Pila, Ngella Sule, Ranongga, Rendova, Rennell, Russell Isls, Santa Isabel, Shortland, Tetepare, Ugi, Vangunu, Vella La Vella			
Extralimital None			
Conservation status (IUCN) Least Concern			
Etymology Sp.–unarmed (L.), allusion unclear			

Pregnant females have been recorded on Vangunu in February, on Choiseul in October, and on Gatokae in November. Lactating females have been recorded on Vangunu, Malaita and Bougainville in September–October, and on Kolombangara in February. Of eight females examined in the Su'u Harbour area of Malaita between 15 and 22 March 1991, two were lactating and one was carrying dependent young weighing 54 g. Females carrying dependent young were recorded on 15 November (Makira), 17 December (Guadalcanal) and 5 July (New Georgia). Subadult animals seem to be common captures in mist nets at almost any time of year.

Solomons bare-backed fruit bats are hunted to varying degrees across their range. Their habit for roosting in large groups in caves makes them a vulnerable target, and local extirpations do occur. Hunters enter caves and typically bring the animals to ground using slings, or by throwing stones or lengths of wood. Traditional traps also continue to be used – these consist of the barbed tendrils of lawyer cane (*Callamus* spp.), which are strung above the cave entrance to snare the wings of the bats as they flee.

GREAT BARE-BACKED FRUIT BAT

Dobsonia magna Thomas, 1905

mol'-uk-en'-sis

There has been continued disagreement in the literature as to whether *Dobsonia magna* is a distinct species or a subspecies of *D. moluccensis*. Taxonomic revision by Bergmans and Sarbini (1985) suggested *D. moluccensis* be reserved for central Maluku populations. Here we have retained *D. magna* and *D. moluccensis* as species pending further taxonomic and genetic studies.

The great bare-backed fruit bat has been recorded on satellite islands along the northern

The great bare-backed fruit bat (*Dobsonia magna*) has been recorded on satellite islands along the northern coastline of New Guinea (Karkar and Kairuru), and Sideia in the Louisiade Archipelago (Angus McNab).

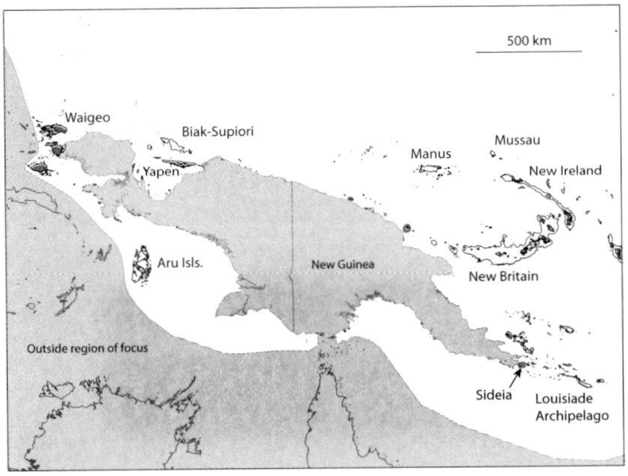

Dobsonia magna			
Size*			
	Mean	**Range**	**Sample**
Head–body	215.0	185.0–254.0	11
Tail	31.1	21.0–42.0	4
Forearm	147.0	135.0–162.0	12
Tibia	68.8	63.0–74.4	9
Hindfoot	39.3	30.0–47.0	4
Ear	32.7	27.0–41.0	11
Weight	461.5	325.0–600.0	10
* From AM specimens from Sideia and mainland New Guinea, Flannery (1995a), and Bonaccorso (1998)			
Identification The largest species of *Dobsonia*, it is dark brown or black in colour with ivory-coloured claws. The wings join in the midline of the back giving it a naked appearance.			
Synonyms None			
Subspecies None			
Distribution Batanta, Misool, Salawati, Waigeo, Yapen (Indonesia); Kairuru, Karkar, Sideia (Papua New Guinea)			
Extralimital Australia, New Guinea			
Conservation status (IUCN) Not Assessed			
Etymology Sp.–large (L.)			

coastline of New Guinea (Karkar and Kairuru), and Sideia in the Louisiade Archipelago. An Australian Museum team on Sideia Island recorded a male and female in a coconut tree roost on 24 December 1988. On New Guinea it occurs up to 2700 m elevation, occupying primary and secondary forests as well as heavily disturbed gardens and village environs (Flannery 1995a).

The great bare-backed fruit bat is a cave dwelling species, but is a visual navigator and thus normally inhabits the dimly lit twilight zones. It is also occasionally encountered in hollow trees or tree crowns (Bonaccorso 1998). The species forms colonies from a few individuals up to thousands, and it has been documented sharing roost sites with other species of *Dobsonia*. It is normally a quiet species, especially when foraging, whereby the only indicator that the species is present is the hollow sound of wing beats produced from membranes that join along the midline of the back. However, when disturbed at the roost, animals can become extremely vocal.

On mainland New Guinea, the great bare-backed fruit bat acts as a keystone frugivore. It is abundant in wet forests and large enough to disperse the fruits of 96% of 400 plant species measured at Crater Mountain, Chimbu Province (Mack and Wright 2005). Important food plants include specics of *Ficus*, *Terminalia*, *Callophylum* and *Eucalyptus* as well as fruit-producing garden plants such as bananas (*Musa* sp.) and papaya (*Carica papaya*). Hide *et al.* (1984) captured this species in bandicoot traps, suggesting it may occasionally venture to the ground for fallen fruits. Females produce one young per year. These are carried short distances within roosts, but are deposited there when females venture out to forage. On Sideia Island, a subadult female was captured on 3 January 1989.

Dobsonia magna is classified as Least Concern under the IUCN Red List; threats include hunting and habitat loss from logging and mining. It is one of the most abundant and widely distributed pteropodids on New Guinea and is encountered in heavily disturbed habitats.

LESSER BARE-BACKED FRUIT BAT

Dobsonia minor (Dobson, 1879)

mie'-nor

In the region under consideration here, the lesser bare-backed fruit bat is recorded from the island of Bagabag via three incomplete skulls in the collections of the American Museum of Natural History, collected by Jared Diamond on 15 October 1970. A series of spirit specimens also exists, collected from nearby Admosin Island on 21 June 1969, and it was reported from Yapen by Bergmans and Sarbini (1985).

Staple foods of the lesser bare-backed fruit bat (*Dobsonia minor*) include figs and the fruits of the exotic species like spiked pepper (*Piper aduncum*) and papaya (*Carica papaya*) (Chien Lee).

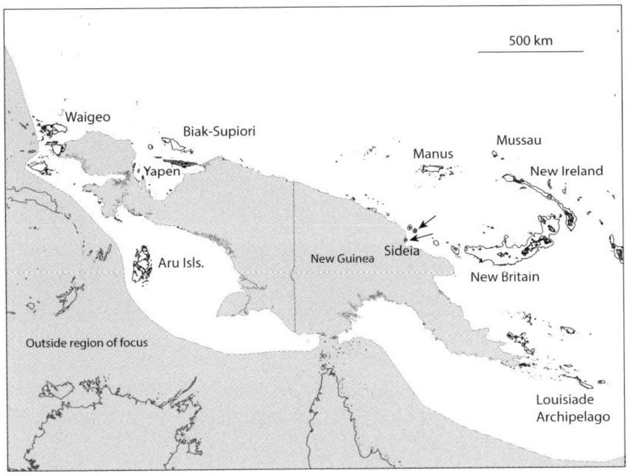

Dobsonia minor			
Size*			
	Mean	**Range**	**Sample**
Head–body	111.2	101.9–135.0	9
Tail	18.4	15.7–20.3	6
Forearm	84.6	78.0–92.0	9
Tibia	31.6	29.0–33.3	9
Hindfoot	20.8	16.0–24.0	4
Ear	18.5	13.0–22.0	9
Weight	77.8	63.0–90.0	9
* From Flannery (1995a) and Bonaccorso (1998)			
Identification This is the smallest species of *Dobsonia*, and its size (forearm length normally below 90 mm) and relatively tiny teeth should be sufficient to differentiate it from other *Dobsonia*. The fur is uniform dark brown.			
Synonyms None			
Subspecies None			
Distribution Yapen (Indonesia); Admosin, Bagabag (Papua New Guinea)			
Extralimital New Guinea			
Conservation status (IUCN) Least Concern			
Etymology Sp.–minor (L.), for its small size relative to other members of the genus.			

Knowledge of species biology is derived from populations on mainland New Guinea. There, it occurs in primary and secondary forests, swamp forest and gardens at lower altitudes (Flannery 1995; Bonaccorso 1998). Radio-tracking studies in Madang Province have shown that contrary to most members of the genus, the lesser bare-backed fruit bat is not an obligate cave roosting species. Instead, animals roost individually among the foliage of trees, palms and shrubs in the understorey or subcanopy (Bonaccorso *et al.* 2002). Home-range sizes averaged 5.1 ha and did not vary between sexes or seasons. Study animals were nocturnal and returned to their roosts shortly before dawn, utilising up to three different sites in close proximity of each other. Staple foods include figs (*Ficus* spp.) and the fruits of the exotic species spiked pepper (*Piper aduncum*) (Bonaccorso 1998), as well as papaya (*Carica papaya*) (Bonaccorso *et al.* 2002).

In the Madang region encompassing Bagabag and Admosin Islands, pregnant females have been encountered in January, May, June and September, and lactating females have been found in February and April (Bonaccorso 1998). Females reproduce once per year, but the timing of births indicates breeding is asynchronous or in biannual peaks. The lesser bare-backed fruit bat is classified as Least Concern on the IUCN Red List because it has a broad distribution, appears relatively common, and is encountered in semi-disturbed habitats.

MOLUCCAN BARE-BACKED FRUIT BAT
Dobsonia moluccensis (Quoy and Gaimard, 1830)

mol'-uk-en'-sis

There remains continued disagreement as to whether *Dobsonia moluccensis* and *D. magna* are conspecific or distinct species. As a precautionary measure to cater for inadequate sampling, Bergmans and Sarbini (1985) suggested *D. moluccensis* be reserved for central Maluku populations, with *D. magna* distributed to the west into New Guinea and northern Australia. Here we have retained *D. moluccensis* and *D. magna* as species pending further sampling and additional taxonomic and genetic research.

The presence of Moluccan bare-backed fruit bat in the Aru Group was identified by Andersen (1912a) on the basis of specimens in the British Museum of Natural History, London and Naturalis Biodiversity Center, Leiden. Bergmans and Sarbini (1985) questioned the origin of these specimens and whether the genus did indeed exist in the Aru Group. However, Aplin and Pasveer (2006) confirmed an extended presence of *Dobsonia* in the Aru Group via an examination of Pleistocene to Holocene deposits on Kobroor. Two species of *Dobsonia* were identified, one of which was tentatively referred to *D. moluccensis*. The second (a smaller species) was referred to *D. viridis* (reported from nearby Kai Islands). Van Strien (1996)

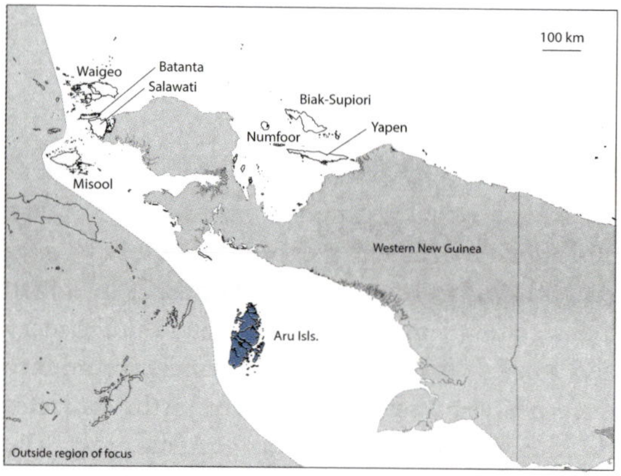

Dobsonia moluccensis			
Size*			
	Mean	**Range**	**Sample**
Head–body	193.9	170.0–210.0	7
Tail	30.3	26.4–34.4	7
Forearm	139.4	132.7–147.3	10
Tibia	64.6	61.9–67.2	7
Hindfoot	–	–	–
Ear	30.9	30.0–31.9	7
Weight	321.2	266.0–360.0	7
* From AM specimens (Buru), Bergmans and Sarbini (1985) (Aru Isls), and Andersen (1912a) (Ambon, Aru Isls)			
Identification A large species of *Dobsonia*, with forearm length slightly shorter than in *D. magna* (however, sample sizes are small and ranges apparently overlap). The fur is uniformly brown to grey-brown on the face and ventrum; the claws are ivory-coloured. The wings join in the midline of the back, giving it a naked appearance.			
Synonyms None			
Subspecies None			
Distribution Aru Isls			
Extralimital Maluku Isls			
Conservation status (IUCN) Least Concern			
Etymology Sp.–from Moluccan Isls (L.), the type locality for the species			

provided the details of *Dobsonia* specimens from Kobroor in several museum collections. The Western Australian Museum holds nine specimens tentatively referred to *D. moluccensis* collected on Wokam by D. J. Kitchener in 1992 and Dr Pat Woolley obtained a single specimen among the 'kitchen midden' and hunter samples collected on Kobroor in 1992 (Aplin and Pasveer 2006).

The Moluccan bare-backed fruit bat has been documented from tropical primary forests, disturbed forests, gardens, and plantations. It is a gregarious and primarily cave-roosting species, but little else is known of its biology. The Moluccan bare-backed fruit bat has a wide distribution, is abundant within its range, and is not listed in a threatened category on the IUCN Red List.

PANAEATI BARE-BACKED FRUIT BAT

Dobsonia pannietensis (De Vis, 1905)

pan'-nee-et-en'-sis

Panaeati bare-backed fruit bat is a very common species on islands such as Woodlark, but little has been documented of its biology. The Australian Museum holds study specimens from most islands in its known distribution.

This species, like many *Dobsonia*, is known to roost in cave colonies and will also occupy tree hollows. The Australian Museum collection includes a series obtained from a cave located at 1800 m

Panaeati bare-backed fruit bats (*Dobsonia pannietensis*) can form large colonies, and will roost in caves, in the crowns of palms and inside the hollows of trees (Pavel German).

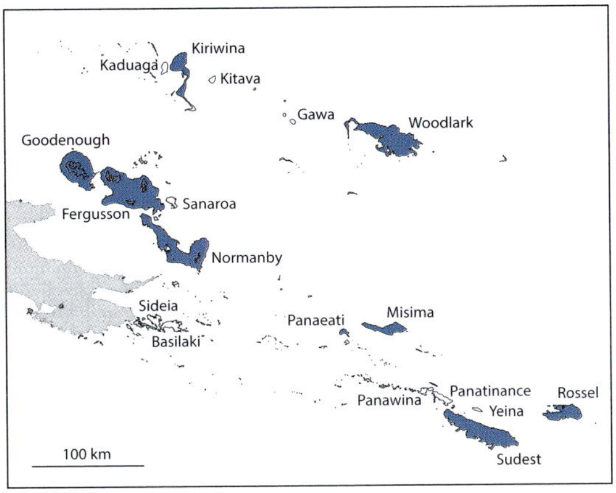

Dobsonia pannietensis			
Size*			
	Mean	**Range**	**Sample**
Head–body	161.1	127.0–203.0	12
Tail	26.1	19.0–36.0	10
Forearm	113.6	104.0–125.0	12
Tibia	53.6	46.0–58.0	6
Hindfoot	33.1	25.0–40.0	8
Ear	24.9	21.0–29.0	8
Weight	196.6	110.0–304.0	8
* From AM specimens (Woodlark)			
Identification The dorsum is grey-brown and the ventrum medium brown with a yellowish wash. Animals from Woodlark are small and often have a pale or reddish patch on the belly. The wings join along the midline of the back giving it a naked appearance. All populations have brownish tan claws and a black crown.			
Synonyms None			
Subspecies *Dobsonia pannietensis pannietensis* De Vis, 1905 (D'Entrecasteaux and Louisiade Isls) *Dobsonia pannietensis remota* Cabrera, 1920 (Trobriand Isls)			
Distribution Fergusson, Goodenough, Kiriwina, Misima, Normanby, Panaeati, Rossel, Sudest, Woodlark			
Extralimital None			
Conservation status (IUCN) Near Threatened			
Etymology Sp.–from Panniet (L.) a former spelling of Pannaeati, the type locality			

elevation on Goodenough on 24 August 1987. Four individuals were encountered in a tree hollow on the eastern slopes of Goodenough in mid-October 1956 (Bonaccorso 1998).

No information is available on diet, but it is probably primarily frugivorous. During the Fifth Archbold Expedition in June 1956, this species (listed by Brass 1959 as *D. moluccensis*) was captured in mist nets near a flowering kapok (*Ceiba pentandra*) planted in a village on Fergusson Island. Several individuals recorded from Rossell Island on 10 October 1956 were hanging in bunches in the tops of coconut palms (*Cocos nucifera*) in secondary forest near sea level. Two caves in high sea cliffs were examined at Waspimat Bay, Woodlark Island on 14 November 1956 and contained only a few individuals. Also on Woodlark, in a large cave approximately one mile north-west of Kulumadau, 'the roof was aglow' with the reflective eyeshine of *Dobsonia* (Brass 1959).

Of five females collected from Goodenough, one had a foetus weighing 9 g (24 August); two were early-term pregnant females; one was non-reproductive (mid-October); and another was a subadult (30 May 1956) (Bonaccorso 1998). A female documented from Woodlark on 9 August 1987 was lactating.

The Panaeati bare-backed fruit bat is classified as Near Threatened on the IUCN Red List because of its small geographical distribution and declining quality of habitat. Large, colonial roosting bats such as *D. pannietensis* are a vulnerable target on islands where they represent a significant source of protein.

BISMARCK BARE-BACKED FRUIT BAT
Dobsonia praedatrix Andersen, 1909a
pree-dah'-trix

This common species is sometimes overlooked because (somewhat unusually for the genus *Dobsonia*) it rarely inhabits caves. For example, it was found to be abundant in the Bismarck Archipelago, but was never encountered in caves despite extensive searches on both New Britain and New Ireland (Smith and Hood 1981). The Bismarck bare-backed fruit bat instead roosts singly or in small groups among foliage or beneath the dead fronds of coconut palms. The greenish fur of the species provides effective camouflage in such situations. However, an Australian Museum team encountered one maternity colony consisting of around 12 individuals that was in a sea cave on Southern New Ireland.

Bonaccorso (1998) reported that this species is found in lowland primary and secondary forests.

The Bismarck bare-backed fruit bat forages for fruits and flowers in sympatry with *D. anderseni* and on New Britain it has been recorded eating the fruits of cashew trees (*Anacardium occidentale*) (Bonaccorso 1998). Smith and Hood (1981) mist-netted this species in approximately equal numbers with *D. anderseni* and noted that blue ticks found on the rump and uropatagium of *D. anderseni* were never present on *D. praedatrix*.

Births occur twice per year, with one peak around June–July and a second in December–January. On New Britain, pregnant females have been recorded on 4 January 1969, 30 July 1993 (late-term) at Kerevat (Bonaccorso 1998). Between December and January 1989 in New Britain's Fulleborn region the Australian Museum recorded 19 females, including two that were clearly pregnant (as examined from external palpation). Three lactating females, a female carrying a newborn young weighing 32 g, and five subadults were also present. The five subadults all weighed 70–90 g and were presumably young born earlier in 1988. On New Ireland and Boang, pregnant females and females with dependent young were recorded on 12 July 1979 and 6 February 1935, and subadults were recorded on New Ireland and Tabar on 24 June 1979 and January 1935 (Bonaccorso 1998).

The Bismarck bare-backed fruit bat is smaller in size than the sympatric *D. anderseni* and does not form large conspicuous colonies. This may mean the species is less exposed to intensive hunting pressure.

Dobsonia praedatrix			
Size*			
	Mean	**Range**	**Sample**
Head–body	152.0	116.0–180.0	8
Tail	29.3	21.0–37.0	7
Forearm	114.0	101.5–125.0	9
Tibia	49.2	43.5–54.0	8
Hindfoot	27.5	22.0–33.0	4
Ear	24.1	21.0–28.0	8
Weight	172.0	120.0–260.0	9
* From AM specimens (New Britain)			
Identification This species can be distinguished from the sympatric *D. anderseni* by its smaller size, brown-based instead of entirely white claws, and greenish colour of the fur and skin. The ears are longer and more slender, and the crown and face are more intensely blackish-brown, than other species of *Dobsonia*.			
Synonyms None			
Subspecies None			
Distribution Boang, Duke of York, Dyual, Lavongai, Mioko, New Britain, New Ireland, Tabar, Ulu, Umboi			
Extralimital None			
Conservation status (IUCN) Least Concern			
Etymology Sp.–predator, feminine (L.) allusion unclear			

Bismarck bare-backed fruit bat (*Dobsonia praedatrix*). This beautiful, greenish coloured fruit bat is restricted to the Bismarck Archipelago (Stephen Richards).

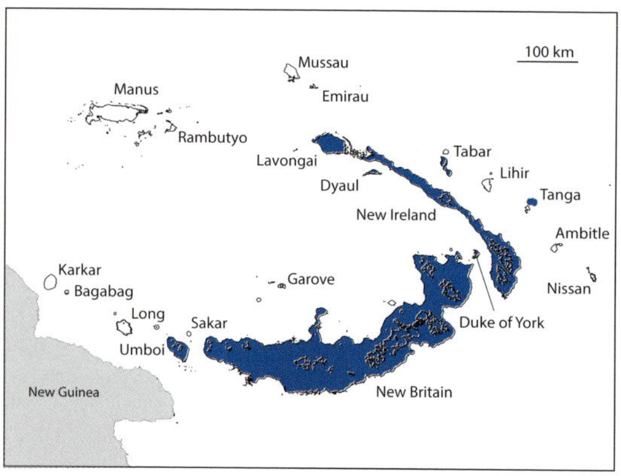

Macroglossus

LESSER LONG-TONGUED BLOSSOM BAT

Macroglossus minimus (Geoffroy, 1810)

mak'-roh-glos'-us min'-ee-moos

The lesser long-tongued blossom bat is an extremely common and widespread species, and one of the most commonly captured bats when mist-netting in the region. It occurs from Thailand and Cambodia, through Wallacea and the Philippines, and into northern Australia and Melanesia as far as Makira in the eastern Solomon Islands. The lesser long-tongued blossom bat inhabits large and small islands, and a range of habitats including mangroves, lowland primary and secondary forests, gardens and village environments. In the South-west Pacific, this species also occurs into high-elevation montane forests. Aplin and Opiang (2011) recorded it as high as 860 m in East New Britain, and McKean (1972) reported it from 1155 m on Bougainville. The Zoologisk Museum, Copenhagen holds specimens from Dyual, Mussau and Nuguria which were collected during the Noona Dan Expedition. Records from the islands off south-eastern New Guinea are inexplicably rare. Koopman (1982) identified a single individual collected on Fergusson by Petersen in 1956, but extensive survey work by the Australian Museum in the D'Entrecasteaux Islands in the 1990s failed to encounter it at all.

The lesser long-tongued blossom bat is primarily nectarivorous, and in the South-west Pacific is known to feed on the flowers of a wide variety of cultivated plants including coconuts (*Cocos nucifera*), Malay apple (*Syzygium malaccense*), and cutnut (*Barringtonia edulis*). It is especially common around the bracts of flowering cultivated and wild bananas (*Musa* spp.). Smith and Hood (1981) reported it was captured in great abundance in the Bismarck Archipelago during their visit in 1979. Many individuals had whitish pollen from papaya (*Carica papaya*) on the face, and they were always found associated with *Rousettus* and *Dobsonia* species. The lesser long-tongued blossom bat is an important pollinator for

Lesser long-tongued blossom bat (*Macroglossus minimus*), Vangunu, Solomon Islands. This is one of the most abundant and widespread pteropodid bats in the region (Tyrone Lavery).

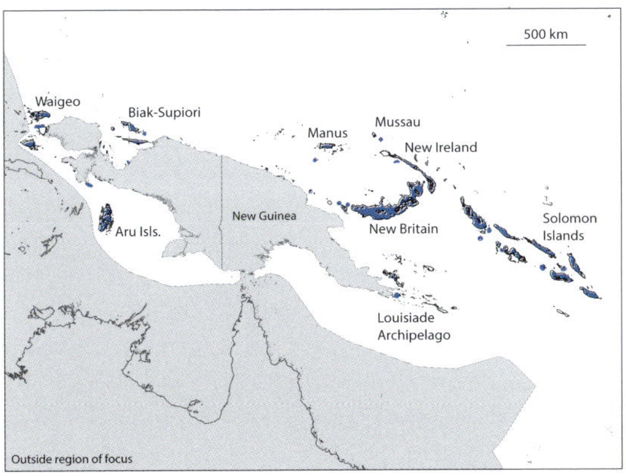

Macroglossus minimus			
Size*			
	Mean	**Range**	**Sample**
Head–body	57.5	38.0–67.5	54
Tail	–	–	–
Forearm	38.7	34.0–50.0	56
Tibia	16.3	13.0–19.2	52
Hindfoot	11.3	10.9–11.6	3
Ear	11.7	8.7–13.8	53
Weight	14.2	10.0–19.0	52
* From Solomon Isls AM and QM specimens, and field measurements			
Identification One of the smallest fruit bats in the region. It is warm brown in colour with short fur, a very narrow snout, greatly reduced incisors, and a very long tongue. On the internal sides of the legs is a very small patagium membrane that is completely absent in *Syconycteris*. Adult males have a raised, red, V-shaped glandular area on the chest.			
Synonyms *Macroglossus novaguinea* Matschie, 1899 *Odontonycteris meyeri* Jentink, 1902			
Subspecies *Macroglossus minimus minimus* (Geoffroy, 1810) (Java, Madura, Bali, Kangean Isl. Lombok) *Macroglossus minimus lagochilus* Matschie, 1899 (Thailand, Vietnam, Malaysian Peninsula, Borneo, Philippines, Sulawesi, Maluku Isls.) *Macroglossus minimus nanus* Matschie, 1899 (Bismarck Archipelago, Admiralty Isls, Kai and Aru Isls, northern Australia) *Macroglossus minimus microtus* Andersen, 1911 (Solomon Isls) *Macroglossus minimus booensis* Kompanje and Moeliker, 2001 (Boo Isl. north-west New Guinea)			
Distribution Adi, Aru Isls, Batanta, Biak-Supiori, Bromsi, Num, Numfoor, Misool, Ratewo, Rumberpon, Waigeo, Yapen (Indonesia); Dyual, Emirau, Fergusson, Kairuru, Karkar, Manus, Mussau, New Britain, New Ireland, Nuguria, Sakar, Sariba, Sideia, Tolokiwa, Umboi (Papua New Guinea); Bougainville, Buka, Choiseul, Fauro, Gatokae, Gizo, Guadalcanal, Kolombangara, Makira, Malaita, Mono, New Georgia, Ngella Pile, Ngella Sule, Ranongga, Rendova, Russell Isls, Santa Isabel, Tetepare, Vangunu, Vella La Vella (Solomon Isls)			
Extralimital Australia, Indonesia, Malaysia, New Guinea, Philippines, Thailand, Timor-Leste, Vietnam			
Conservation status (IUCN) Least Concern			
Etymology Gen.–large tongue (Gr.); Sp.–smallest (L.)			

cultivated plants and native flora; the species regularly excretes large amounts of pollen when captured in mist nets.

In Madang Province, Papua New Guinea, Winkelmann and Bonaccorso (2003) radio-tracked 18 individuals and recorded an average home range of 5.8 ha, and a mean core-use area of 1.5 ha. Hotspots of activity were normally associated with flowering bananas. The species roosted individually among foliage in the subcanopy or canopy with high fidelity to a particular area (Winkelmann and Bonaccorso 2003).

Breeding in the South-west Pacific is asynchronous and pregnant or lactating females can be encountered at most times during the year. Moreover, females in New Guinea are capable of sperm storage (Hood and Smith 1989). Young tend to stay clinging to the mother for the first few days before they are left behind at the roost while mothers forage.

The lesser long-tongued blossom bat is widespread, population size is presumably large, and it is a species accustomed to exploiting disturbed environments. Long-term studies over a 23-year period in Solomon Islands could not detect any decline in abundance of this species associated with widespread commercial logging that has occurred (Lavery *et al.* 2020a).

The black-bellied blossom bat (*Melonycteris melanops*) is a strikingly coloured animal. These bats occupy small stable home ranges and are relatively common (by Joseph Smit in Dobson 1877).

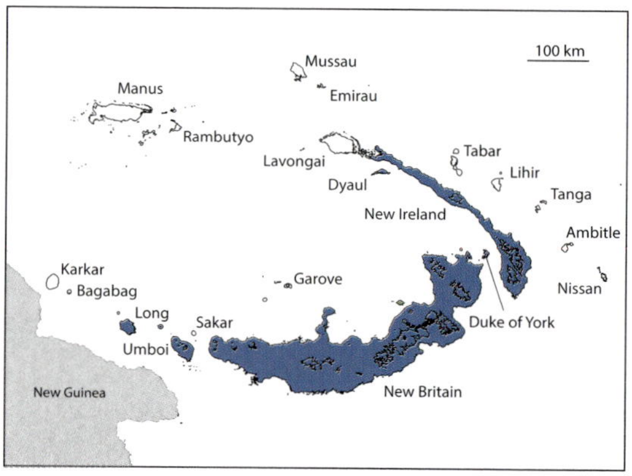

Melonycteris

BLACK-BELLIED BLOSSOM BAT

Melonycteris melanops Dobson, 1877

mel'-oh-nik'-te-ris mel'-ahn-ops

This is a distinctive and beautiful bat with a contrasting bright orange back and blackish belly. Black-bellied blossom bats are widespread in disturbed forests, forest edges, gardens and banana plantations (*Musa* sp.) in the Bismarck Archipelago from sea level to 1600 m elevation. It is most common in banana plantations and relatively rare in primary forests (Smith and Hood 1981; Bonaccorso 1998; Bonaccorso and Winkelmann 2005).

Detailed information on the ecology of this species was collected during a radio-tracking study in West New Britain (Bonaccorso and Winkelmann 2005). Home ranges varied between 0.5 and 9.2 ha (average 2.1 ha), did not vary between the sexes and remained stable over tracking periods of up to 34 nights. Three individuals studied closely for nocturnal activity patterns spent between 24.3% and 35.8% of the sampling period in flight. Mean

Melonycteris melanops			
Size*			
	Mean	**Range**	**Sample**
Head–body	95.8	78.1–110.0	8
Tail	–	–	–
Forearm	60.7	50.0–68.3	8
Tibia	27.2	25.0–28.1	8
Hindfoot	18.5	13.0–24.0	4
Ear	16.3	13.7–19.0	8
Weight	49.1	31.5–63.0	8
* From AM specimens (New Britain and New Ireland)			
Identification This species is very distinctive with a bright orange back and blackish belly. There are distinct white epaulettes on the shoulders. The skin of the feet, head and flight membranes is spotted bright pink.			
Synonyms *Pteropus (Cheiropteruges) alboscapulatus* Ramsay, 1877d			
Subspecies None			
Distribution Dyual, Duke of York, Long, Mioko, New Britain, New Ireland, Tolokiwa, Ulu, Umboi			
Extralimital None			
Conservation status (IUCN) Least Concern			
Etymology Gen.–fruit tree + bat (Gr.), hence fruit-bat; Sp.–black faced (Gr.)			

flight times varied between 20.8 and 30.7 s. During the day, black-bellied blossom bats preferred the undersides of dead and curling banana leaves as roost sites, but were also found in large-leafed trees in closed-canopy forests. Animals used a relatively small number of roosts within core-use areas that typically contained one to six flowering banana plants. Black-bellied blossom bats are sexually dimorphic with males being larger than females (in bodyweight and forearm length). Adults are strongly territorial, excluding other same-sex adults from feeding territories. The larger body size of males is probably indicative of sexual competition.

Lactating and pregnant females have been encountered in two major peaks during the year suggesting two reproductive seasons may occur. Two samples of three females captured on New Ireland in June 1988 and July 1993, comprised two lactating and one pregnant individual on both occasions. Two females examined on 26 December 1988 on New Britain were lactating, and two from Ulu in the Duke of York group were pregnant in July 1993. Black-bellied blossom bat is not considered threatened because the species is most common in disturbed environments and it is widespread on the largest islands of the region.

Mirimiri

FIJIAN MONKEY-FACED BAT

Mirimiri acrodonta (Hill and Beckon, 1978)

mee'-rhee-mee'rhee ak'roh-don'-tah

This is a very cryptic and poorly known species from montane cloud forests, and the only mammal endemic to the Fiji Islands. It has one of the most restricted distributions of all bats, having been recorded by mammalogists only a handful of times at Koroturanga, the summit of Taveuni Island, between elevations of 815 and 1190m (Helgen 2005; Scanlon *et al.* 2013a). It also may occur in the montane forests of the adjacent island of Vanua Levu, but this is yet to be confirmed (Palmeirim *et al.* 2007; Scanlon *et al.* 2013a).

Fijian monkey-faced bat (*Mirimiri acrodonta*), Taveuni, Fiji. One of the world's rarest mammals, it is known to occupy only a few square kilometres of habitat on Taveuni (Pavel German).

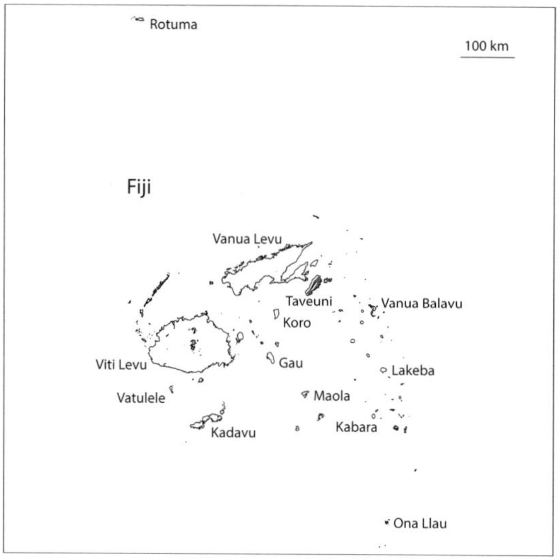

Mirimiri acrodonta			
Size*			
	Mean	**Range**	**Sample**
Head–body	183.3	170–200	3
Tail	–	–	–
Forearm	119.5	111.2–127.2	3
Tibia	52.4	49.4–57.2	3
Hindfoot	–	–	–
Ear	18.4	18.2–18.8	3
Weight	276.5	222–362	4
* From AM specimens (Taveuni) and Watling (2011)			
Identification A small species, dull olive green to yellow all over, with bright orange eyes. Ears are small, short, almost concealed by fur. Upper margin of ear semicircular with no tip. The possession of canines with two cusps reliably separates this species from all Fijian flying foxes.			
Synonyms None			
Subspecies None			
Distribution Taveuni			
Extralimital None			
Conservation status (IUCN) Critically Endangered			
Etymology Gen.–Fijian 'mist', an allusion to the cloud forests this species inhabits; Sp.–pointed tooth (Gr.), an allusion to the many pointed summits of the teeth			

The Fijian monkey-faced bat is rarely encountered, being either naturally scarce or difficult to mist net. Over 1887 mist-net hours carried out in 2009, only one animal was captured. Over a 12-day mist-netting period in May 1990, using 10 mist nets between 880 and 1190 m elevation, three individuals were caught. Two were adult females captured at 1010 m elevation; the third was a male captured at 1190 m. The lower mist nets failed to capture any specimens. There is a distinct change in vegetation at high elevation on Taveuni, from taller forest into montane mossy forest. At least 23 plant species are unique to Taveuni's stunted montane forests at 1200 m (Ash 1988) and conditions are usually misty and wet. It seems *Mirimiri acrodonta* prefers these high-elevation mossy forests.

In 1990 *Pteropus samoensis* was captured in mist nets on the Taveuni summit only on clear nights. The frequent dense mists possibly limit visibility to such a degree that foraging becomes overly difficult. *Mirimiri acrodonta* may be advantaged under these conditions, for the wing membranes meet at the midline of the back giving the species great manoeuvrability, including the ability to fly slowly and avoid obstacles.

The Fijian monkey-faced bat may roost in pairs or small groups among epiphytes. An individual caught at 815 m elevation in 2009 took refuge in a clump of epiphytes in the cloud forest canopy upon release. However, it did not appear to be an established roost site (Scanlon *et al.* 2013a). This species is sexually dimorphic in colour and possibly size (sample sizes are too small to confirm). The fur of the mantle and head in both sexes is long, soft and khaki in colour, but the shorter hair over the back and rump is khaki in females and a golden colour in males. Two females caught in May 1990 and one caught in May 2009 were all lactating, but the nipples were tiny compared with those of *Pteropus* captured at the same time.

Nothing is known of diet, but the teeth are even more sharply cuspidate than in the sister genus of monkey-faced bats (*Pteralopex*) and are among the most cuspidate of all fruit bats. This suggests a peculiar, possibly tough diet. The Fijian monkey-faced bat has an unusual and distinctive call, which unfortunately eludes adequate description.

The limited availability of montane habitat in Fiji is the main threat to *M. acrodonta*, and this is likely to be exacerbated by climate change. Taveuni's cloud forests occur along only about a 10 km stretch of the main range and cover around 6000 ha.

Nesonycteris

FARDOULIS'S BLOSSOM BAT

Nesonycteris fardoulisi (Flannery, 1993)

nees'-oh-nik'-te-ris far'-du-lee'-see

Nesonycteris fardoulisi			
Size*			
	Mean	**Range**	**Sample**
Head–body	84.7	77.2–95.4	19
Tail	–	–	–
Forearm	57.2	52.0–62.3	19
Tibia	23.8	20.8–26.6	19
Hindfoot	17.3	15.3–18.8	12
Ear	13.7	11.6–15.5	19
Weight	38.4	28.0–55.0	16
* KU and QM specimens and field measurements from Makira			
Identification No tail or interfemoral membrane, pelage drab brown, not orange as in *N. woodfordi*. There is abundant pink spotting on the ears, flight membranes and feet. Both males and females are larger than *N. mengermani*.			
Synonyms None			
Subspecies None			
Distribution Makira, Ugi			
Extralimital None			
Conservation status (IUCN) Near Threatened			
Etymology Gen.–island + bat (Gr.); Sp.–for Mr Emmanuel Fardoulis, in recognition of his generous support of the Australian Museum Mammal Section.			

Fardoulis's blossom bat was formerly regarded as a single species spanning Solomon Islands' New Georgia group of islands, Guadalcanal, Malaita and Makira. However, these islands or groups of islands all remained isolated from each other during the Pleistocene, and from the larger landmass of Greater Bukida (combined Buka, Bougainville, Choiseul, Santa Isabel and Ngella). Each of these separate islands or island groups corresponds to a morphologically and genetically distinct taxon described by Flannery (1993). Pulvers and Colgan (2007) showed genetic pairwise distances between major island clades exceeded 6% (mitochondrial gene Cytochrome *b*). More recent RAD-seq genetic data have reaffirmed these deeply divergent clades corresponding to *Nesonycteris fardoulisi*, *N. maccoyi* and *N. mengermani* (T. Lavery, unpublished data). The *N. fardoulisi* clade from Makira was sister to all other clades, a position that probably reflects the extended period over which Makira has remained isolated from other islands in the archipelago.

Nesonycteris have a pugnacious temperament – when captured in a mist net they are quick to bite and are reluctant to release their grip. The holotype (an adult male collected in 1987 near Sesana village) and paratypes are held in the collections of the Australian Museum. Two of these were caught in mist nets set in a new garden by a small river. The garden contained ripening papaya (*Carica papaya*) and was backed by dense but disturbed forest (Flannery 1993). Other bats captured at the same site were *Dobsonia inermis* and a species of *Nyctimene*. Fardoulis's blossom bat was a common capture in mist net surveys in montane forests of East Makira's Bauro Highlands in June 2018. A series of nets positioned along ridgelines from around 800–1000 m elevation captured 12 individuals over four nights. Eight of these were females, of which seven were lactating. Other species caught at the same site were *Macroglossus minimus*, *Nyctimene malaitensis*, *Pteropus cognatus* and *Rousettus amplexicaudatus*.

Like many islands in the Solomon Archipelago, Makira has suffered from unsustainable levels of commercial logging. How this affects Fardoulis's blossom bat is uncertain. Montane forests above 600 m elevation are seldom logged and may offer some refuge for the species. Fordoulis's blossom bat is classified as Near Threatened on the IUCN Red List. This assessment was based on all subspecies formerly included under *Nesonycteris fardoulisi* (*N. maccoyi*, *N. mengermani* and *N. woodfordi schouteni*) and an assessment specific to Makira is needed.

MCCOY'S BLOSSOM BAT

Nesonycteris maccoyi (Flannery, 1993)
mack-koy'-ee

McCoy's blossom bat is a species commonly captured in mist nets on Malaita. The holotype was caught near a fruiting forest tree at Naufe'e Village, East Kwaio at approximately 400 m elevation (Flannery 1993). The paratype, also caught in a mist net, was taken near a fruiting papaya (*Carica papaya*) close to sea level. Other species caught at these sites included *Dobsonia inermis*, *Rousettus amplexicaudatus*, *Macroglossus minimus* and *Nyctimene malaitensis*. Flannery (1993) suggested this species is rarer than other *Nesonycteris*. This could be due in part to hunting pressure on Malaita, as the island supports roughly double the human population density of other similar sized islands in Solomon Islands. Moreover, in East Kwaio, people are highly skilled at catching McCoy's blossom bat with woven fibre nets. These are placed high in the trees to span natural canopy gaps over ridgelines and the stalks of ginger plants (*Alpinia* spp.) are used to create high-pitched whistles that seem to be particularly successful at attracting *Nesonycteris* to the nets.

Nesonycteris maccoyi			
Size*			
	Mean	**Range**	**Sample**
Head–body	86.5	75.7–97.1	19
Tail	–	–	–
Forearm	56.7	50.9–60.4	19
Tibia	24.8	20.6–28.6	19
Hindfoot	–	–	–
Ear	13.3	11.5–15.5	19
Weight	42.3	24.0–56.0	19
* AM and QM specimens and field measurements from Malaita			
Identification No tail or interfemoral membrane, pelage drab brown, not orange as in *N. woodfordi*. The ears, feet and wing membranes lack the pinkish spotting that is found on other species of *Nesonycteris*.			
Synonyms None			
Subspecies None			
Distribution Malaita			
Extralimital None			
Conservation status (IUCN) Near Threatened (assessed as *N. fardoulisi*)			
Etymology Sp.–for Mr Michael McCoy in recognition of his assistance with field work at Naufe'e Village, Malaita, the type locality			

McCoy's blossom bat (*Nesonycteris maccoyi*) is common in more isolated parts of Malaita where human population densities, vegetation clearing, and hunting pressure are lower (Ben Speare).

At high elevation, where human population densities are lower, McCoy's blossom bat is more common. On Mt Tolobusu in late October 2016, this species was found to be extremely abundant. A series of six nets set along a ridgeline were successful in capturing 78 individuals over four nights. The sex ratio of captures was relatively even (52.6% males).

Of seven adult females collected at Su'u Harbour, Malaita on 17–19 March 1991, two were pregnant and two were lactating. On Mt Tolobusu in October 2016, one female was near-term pregnant and two were lactating, but the majority of adult females were post-lactating or nulliparous.

Malaita's high human population density, deforestation for gardens and agriculture, and apparent hunting pressure focused towards bats (including small species) may threaten McCoy's blossom bat. Longer term studies are needed to identify population trends. It was most recently classified on the IUCN Red List as Near Threatened under an assessment that included *N. fardoulisi*, *N. mengermani* and *N. woodfordi schouteni*.

New Georgia Group blossom bat (*Nesonycteris mengermani*) is one of the most common species captured in mist nets throughout its range, but has rapidly declined in the Western Province over the past 25 years (Tyrone Lavery).

NEW GEORGIA GROUP BLOSSOM BAT

Nesonycteris mengermani (Flannery, 1993)

men'-jer-man'-ee

The New Georgia Group blossom bat is found throughout the small and large islands of the Solomon Islands Western Province (excluding Mono and the Shortland islands), and the Russell Islands. Unlike other species in the genus, New Georgia Group blossom bat is sexually dimorphic. Males are more sombre in colour than females, are larger, and have longer canines and more robust heads. Sex ratios can be highly skewed in any one locality.

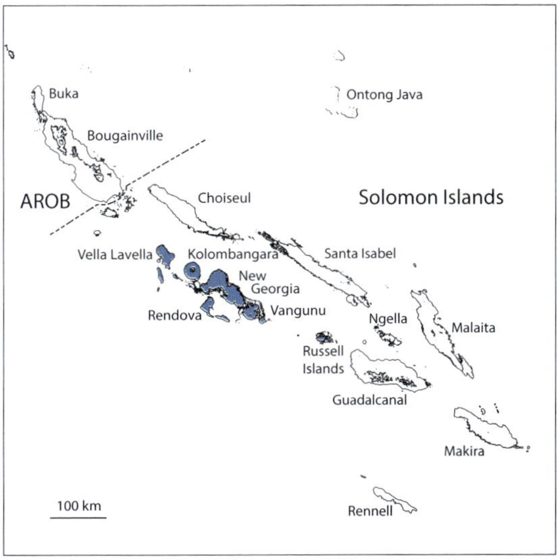

Sexual dimorphism in dentition and crania suggests males and females may exploit different habitats and food sources, or that males fight to defend territories and access to females. The New Georgia Group blossom bat is nectarivorous and a frequent visitor to the flowers of wild bananas (*Musa* sp.), wild ginger (*Alpinia* spp.), heliconias (*Heliconia* spp.) and cutnut (*Barringtonia edulis*)

Nesonycteris mengermani			
Size*			
	Mean	**Range**	**Sample**
Head–body	78.1	53.6–95.0	60
Tail	–	–	–
Forearm	54.3	49.9–58.7	66
Tibia	23.2	19.3–27.9	63
Hindfoot	17.8	14.2–21.3	22
Ear	11.0	8.5–14.4	59
Weight	38.0	29.5–70.0	60
* KU and QM specimens and field measurements from Kolombangara, New Georgia, Ranongga, Vangunu and Vella La Vella			
Identification No tail or interfemoral membrane, pelage drab brown, not orange as in *N. woodfordi*. There is abundant pink spotting on the ears, flight membranes and feet. Male forearm length is smaller than in *N. fardoulisi* and *N. maccoyi*.			
Synonyms None			
Subspecies None			
Distribution Gatokae, Kolombangara, New Georgia, Ranongga, Rendova, Russell Isls, Tetepare, Vangunu, Vella La Vella			
Extralimital None			
Conservation status (IUCN) Near Threatened (assessed as *N. fardoulisi*)			
Etymology Sp.–a euphonious combination of the surnames of Dr Gregory Mengden and Mr Pavel German, who participated in the 1987 Australian Museum expedition to Solomon Isls, and contributed greatly to its success			

(Kress 1985). They are solitary, nocturnal foragers that hover around flowers before perching awkwardly on leaves or hanging from the bracts to feed. This species makes high-pitched calls and full spectrum recordings have shown these have ultrasonic elements (Pennay and Lavery 2017).

Pregnant and lactating females have been recorded from February–April on New Georgia and Vangunu. The young are never carried by their mothers and are left at roost sites while she forages. On Vangunu in November 2011, a juvenile (unable to fly) was found clinging to a cutnut in a village garden that was presumably the roost site.

Commercial logging in the New Georgia group of islands is correlated with a major decline in abundance of this species. Between 1992 (before commercial logging) and 2015 (post widespread commercial logging) mist-net capture rates (number of individuals caught per m^2 mist-net hour) on Kolombangara, New Georgia and Vangunu declined dramatically by approximately 70% (Lavery *et al.* 2020a). Female captures declined disproportionately compared to males, suggesting females could be more dependent on the resources provided by primary forests. In light of the revised distribution of *Nesonycteris mengermani* and this stark decline, closer monitoring of conservation status is urgently needed. The New Georgia Group blossom bat is presently classified as Near Threatened on the IUCN Red List under an assessment that included *N. fardoulisi*, *N. maccoyi* and *N. woodfordi schouteni*.

WOODFORD'S BLOSSOM BAT

Nesonycteris woodfordi Thomas, 1887c

wood'-ford-ee

Woodford's blossom bat is widespread in the north of the Solomons Islands archipelago. The *Nesonycteris* taxon from Guadalcanal was formerly regarded as a subspecies of *Nesonycteris fardoulisi*, but recent genetic data and clear morphological differences reveal it is also better regarded as a subspecies of *N. woodfordi* (Flannery 1993; Lavery unpublished data). It is a species commonly encountered in primary and secondary forests as well as disturbed areas such as village

Woodford's blossom bat (*Nesonycteris woodfordi*), Choiseul, Solomon Islands. This is a common species in most habitats on Mono and among islands that were closely associated during the Last Glacial Maximum (Buka, Bougainville, Choiseul, Santa Isabel, Ngella and Guadalcanal) (Tyrone Lavery).

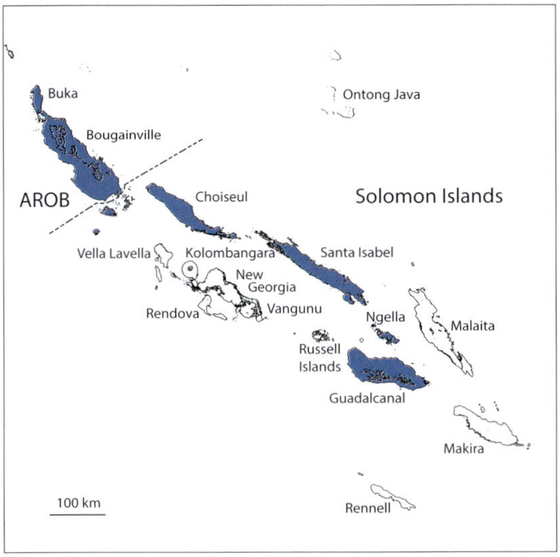

Nesonycteris woodfordi			
Size*			
	Mean	**Range**	**Sample**
Head–body	82.8	59.5–98.4	18
Tail	–	–	–
Forearm	54.9	40.2–63.8	18
Tibia	23.6	20.3–27.4	15
Hindfoot	18.0	15.5–19.9	9
Ear	13.7	10.4–18.0	18
Weight	40.8	25.0–55.0	18
* From AM, KU and QM specimens (Choiseul, Ngella and Santa Isabel)			
Identification *N. w. woodfordi* and *N. w. aurantius* are predominantly orange in colour (duller on the belly) with pinkish spots on the flight membranes, face and feet. In the subspecies *N. w. schouteni* the pelage is drab brown (not orange as in other subspecies), males are smaller in size than *N. maccoyi* and *N. fardoulisi* and there is a greater amount of pinkish spotting on the wings, ears and hindfeet than in *N. maccoyi*.			
Synonyms None			
Subspecies *Nesonycteris woodfordi woodfordi* Thomas, 1887c (Buka, Bougainville, Choiseul, Fauro, Santa Isabel, Shortland) *Nesonycteris woodfordi aurantius* (Phillips, 1966) (Ngella) *Nesonycteris woodfordi schouteni* (Flannery, 1993) (Guadalcanal)			
Distribution Bougainville, Buka, Choiseul, Guadalcanal, Mono, Ngella Pile, Ngella Sule, Santa Isabel			
Extralimital None			
Conservation status (IUCN) Least Concern			
Etymology Sp.–for Charles M. Woodford, District Commissioner of the British Protectorate of Solomon Isls and the collector of the holotype			

gardens and coconut plantations. In the subspecies *Nesonycteris woodfordi aurantius*, females are larger than males, a highly unusual scenario among pteropodids.

The *N. woodfordi schouteni* type series was captured in mist nets in 1987 and held in the collections of the Australian Museum. The nets were set in newly made gardens surrounded by largely undisturbed forest near Kokaleku Village and also caught *Nyctimene bougainville*, *Rousettus amplexicaudatus* and *Dobsonia inermis*. Woodford's blossom bat is not uncommon on Guadalcanal, but nonetheless it is not as frequently captured in mist nets as are species of *Macroglossus*, *Rousettus* and *Nyctimene*. Surveys in the undisturbed primary montane forests of Mt Makarakomburu in 1990 encountered Woodford's blossom bat at elevations between 380 and 1270 m. Similarly, at elevations of 1000–1300 m on the northern side of Mt Popamaneseau in September 2015 this species was detected regularly in mist nets. Of seven individuals captured over four nights, five were female and two were male. Two of the females were post-lactating, two were nulliparous and one was pregnant.

Evidence indicates breeding may be aseasonal. Five females (four lactating and one pregnant) were encountered in the Shortland Islands on 14–15 November; and one lactating and six pregnant females (two of which gave birth after capture) were found from 7–8 May on Ngella. Lactating females have also been found on Choiseul and Santa Isabel in March. As with other species of *Nesonycteris*, females are never captured in mist nets when carrying their young and presumably leave them at the roost sites while they forage.

On Mono Island in July 2019, this species was frequently encountered in early evenings perched on the flat leaves of wild bananas or hovering to feed on the inflorescences. Roost sites are probably chosen in dense understorey foliage. Woodford's blossom bat is common and widespread on the largest islands of the Solomon Islands archipelago and is classified as Least Concern on the IUCN Red List.

Notopteris

FIJIAN LONG-TAILED BAT

***Notopteris macdonaldii* Gray, 1859**

no-top'-te-ris mak-don'-al-dee

The Fijian long-tailed bat occurs in both Fiji and Vanuatu and apparently also occurred on 'Eau, Tonga (as indicated by subfossil deposits) but is now extinct there (Koopman and Steadman 1995). Medway and Marshall (1975) also discussed a 19th century record from Pohnpei, assumed to be a labelling error (Koopman and Steadman 1995). Animals from Fiji are slightly larger than those from Vanuatu (Banks Islands).

Notopteris are among the most unusual of all pteropodid bats, and seem to represent a very ancient, long-isolated lineage. They are the only members of their family to have very long tails that are free from the flight membranes. Their skulls also possess some conserved features, including an infraorbital canal (the opening through which the nerves of the face pass) that opens well forward on the snout as in many other mammals, rather than immediately in front of the eye as in almost all other pteropodids. The wings join over the spine, allowing for a very extensive wing area and thus great manoeuvrability in flight. This gives the lower part of the back a naked, wrinkled appearance. The ear possesses a tiny tragus.

This species was abundant in the lowland secondary forests, gardens and village environments of Vanua Lava and Mota in May 2018. It was mist-netted on Gaua under similar circumstances, particularly in the vicinity of flowering naveli trees

The anatomy of the Fijian long-tailed bat (*Notopteris macdonaldii*) is among the most unusual of all pteropodid bats, possessing several features of early pteropodids, including a very long tail that is free from the flight membranes. It is a species restricted to islands of Fiji and Vanuatu (Pavel German).

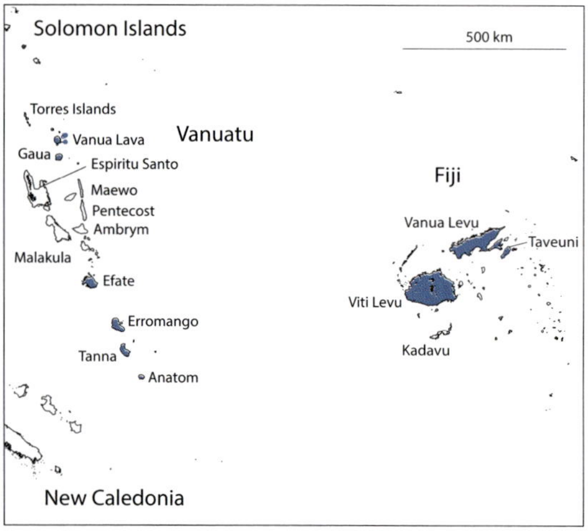

Notopteris macdonaldii			
Size*			
	Mean	**Range**	**Sample**
Head–body	80.3	64.2–110.0	36
Tail	55.3	44.9–63.2	36
Forearm	63.7	57.9–69.0	36
Tibia	34.2	27.9–38.5	36
Hindfoot	–	–	–
Ear	14.9	11.6–19.8	36
Weight	51.5	41.0–73.0	38
* From Mota and Vanua Lava (Vanuatu); Taveuni and Vanua Levu (Fiji)			
Identification Fijian long-tailed bats are very distinctive, with elongate muzzles, large eyes, very long free tails, and wings that join in the midline of the back, giving the back a naked, wrinkled appearance.			
Synonyms None			
Subspecies None			
Distribution Taveuni, Vanua Levu, Viti Levu (Fiji); 'Eua (fossil) (Tonga); Anatom, Efate, Erromango, Gaua, Mota Lava, Mota, Tanna, Vanua Lava (Vanuatu)			
Extralimital None			
Conservation status (IUCN) Vulnerable			
Etymology Gen.–southern + wing (Gr.) Sp.–for Dr John D. Macdonald, collector of the holotype			

(*Barringtonia edulis*). The Fiji long-tailed bat is also common in Fiji's mossy montane forests. The species is a cave obligate and thus it is possible they travel from suitable roosting sites in the lowlands to the montane forests to forage. Four caves containing colonies of this species on Viti Levu are situated at elevations of between 50–150 m above sea level. Population estimates from these roosts in 2008–2009 ranged between 200 and 2000 individuals (Scanlon 2009).

In Fiji, no fruit has been recorded in the diet and the species is presumed to be entirely nectarivorous (Scanlon *et al.* 2014). In fact, Petit *et al.* (2022) revealed a unique relationship with the native tree species *Dillenia biflora*. The flowers of this species contain copious amounts of nectar, but have globose corollas (petals that form a permanently closed cap covering the anthers and stamen). Fijian long-tailed bat is well adapted to remove the corolla to access and pollinate the nectar-rich flowers. Without this removal of the corolla by bats, flowers died and did not produce seeds (Petit et al. 2022). *Barringtonia seaturae* is an important food species on Vanua Levu, where its extended flowering provides a regular food source (Scanlon *et al.* 2013b). *Geissois ternate* also may be an important seasonal species. Dietary overlap is greater with moths than with other bats (Scanlon *et al.* 2013b).

Breeding occurs all year in Fiji (Scanlon *et al.* 2013a). Of 17 females caught on Mota and Vanua Lava in early May 2018, six were nulliparous or immature, seven were lactating, and four were post-lactating. The global population of Fijian long-tailed bat has been estimated to be fewer than 10 000 individuals and is projected to decline from hunting, forest loss and cave disturbance.

NEW CALEDONIAN LONG-TAILED BAT

Notopteris neocaledonicus Trouessart, 1908

nee-oh-kal'-ed-on'-ik-us

The New Caledonian long-tailed bat is very restricted in its distribution and was formerly considered a subspecies of *Notopteris macdonaldii*. It was distinguished from this species primarily by differences in body size (*N. neocaledonicus* being smaller than *N. macdonaldii*). However, measurements of *N. macdonaldii* from the Banks Islands, Vanuatu, were slightly smaller than those from Fiji and closer to the range of *N. neocaldeonicus*. Relationships between populations from these three archipelagos should be further assessed.

The New Caledonian long-tailed bat is dependent on caves as roost sites. These sites are limited and only known to exist in the northern part of New Caledonia. All records of the species are from this part of the island. It may be that the *maquis* (heathland growing on poor soils), which is extensive in the south, or a lack of suitable caves prevents it becoming established there.

Sanborn and Nicholson (1950) recorded this species from caves at Hienghene (where they saw 200 bats) and at Poya where 300 were located. The bats hung onto the walls of the cave in clusters of up to 25 individuals. More recent counts at Poya estimated groups were much larger (150–200 individuals) (M. Oedin pers. comm.). Nelson and Hamilton-Smith (1982) gave a detailed account of a colony in a cave called Naaton in the Valée Seche and recorded a second colony at Grotte de Ninrin Reu, which is located in the next valley to the south. Cave use may vary seasonally; in March–April they apparently inhabit outer chambers and move deeper in for the rest of the year (Nelson

Notopteris neocaledonicus			
Size*			
	Mean	**Range**	**Sample**
Head–body	93.7	85.5–101.0	11
Tail	50.0	43.5–54.0	11
Forearm	58.2	55.0–60.0	10
Tibia	50.0	43.5–54.0	11
Hindfoot	18.3	16.5–20.0	11
Ear	14.5	13.5–15.0	10
Weight	–	–	–
* From Revilliod (1914)			
Identification The only species of bat in New Caledonia with wings that meet in the middle along the spine. It has a long narrow snout, and a long tail free of flight membranes.			
Synonyms None			
Subspecies None			
Distribution Grand Terre (New Caledonia)			
Extralimital None			
Conservation status (IUCN) Endangered			
Etymology Sp.–for New Caledonia (L.)			

and Hamilton-Smith 1982). A deeper chamber inhabited in December was in almost complete darkness, and the bats were roosting in clusters on the ceiling, hanging loosely by their feet. In Grotte de Ninrin Reu, the species was found roosting 300 m into the cave, in a chamber 50 m high.

Like *N. macdonaldii*, the wing membranes join along the spine and this gives the bats great manoeuvrability. They can fly vertically and hover in one location. The New Caledonian long-tailed bat is nectarivorous and this ability is probably well suited to both foraging and avoiding obstacles in dark caves (although Nelson and Hamilton-Smith 1982 found this species unable to avoid obstacles when flying in total darkness). The call is similar to a young *Pteropus*.

Four females caught on 23 December 1965 had young at the teat (average weight of the young was 16.8 g), while another three were lactating, but without young. A further three were non-lactating and pregnant, with the crown–rump length of embryos at 4, 5 and 19 mm. Three of five females caught on 5 January had early embryos (Nelson and Hamilton-Smith 1982). These data suggest that the young are weaned in January.

The extent of occurrence for New Caledonian long-tailed bat is small, as is population size (estimated less that 2500 individuals). Moreover, forest degradation has apparently caused population declines. The greatest single threat to New Caledonian long-tailed bat is disturbance at roosting caves (Mickleburgh *et al.* 1992).

Nyctimene

GREATER TUBE-NOSED BAT

Nyctimene aello Thomas, 1900

nik'-ti-may'-nay ah-ell'-oh

The greater tube-nosed bat is a large and attractively coloured tube-nosed bat. It is widespread on New Guinea, and from the region under consideration

The greater tube-nosed bat (*Nyctimene aello*) is one of the largest tube-nosed bats and is a relatively uncommon species (Michael Pennay).

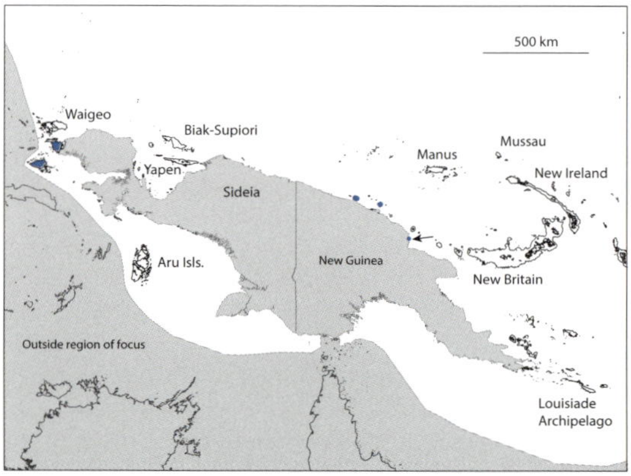

Nyctimene aello			
Size*			
	Mean	**Range**	**Sample**
Head–body	121.8	103.0–144.0	4
Tail	23.5	14.1–31.0	4
Forearm	82.8	77.7–88.1	4
Tibia	31.7	27.4–36.0	4
Hindfoot	19.2	16.0–22.7	4
Ear	18.3	13.4–24.0	4
Weight	81.1	70.3–93.0	4
* From Bonaccorso (1998)			
Identification A large species of *Nyctimene*. The dorsal stripe is much broader than in any other species, extending over a third of the width of the lower back.			
Synonyms *Nyctimene celaeno* Thomas, 1922a			
Subspecies None			
Distribution Salawati, Misool (Indonesia); Admosin, Kadovar, Kairuru (Papua New Guinea)			
Extralimital New Guinea			
Conservation status (IUCN) Least Concern			
Etymology Gen.–moonlight (Gr.), a reference to nocturnal habits; Sp.–a dirge (L.), significance unknown, but possibly for the mournful appearance of the face of many *Nyctimene* species			

here has only been documented on Admosin, Kairuru, Kadovar, Misool and Salawati Islands, which lie in close proximity to the mainland. It may well occur on many more of the satellite islands that lay along the northern coastline of New Guinea.

The greater tube-nosed bat is an uncommon species. It inhabits primary and secondary forests, gardens and swamp forests and roosts solitarily under the leaves of understorey trees, having been encountered in such a roost approximately 5 m from the ground (Bonaccorso 1998). One individual that was radio-tracked used a different roost every day, but always remained within less than 38 m of the original roost site. By night it foraged in an area around 200 m by 200 m (Bonaccorso 1998). The diet includes figs (*Ficus* sp.) and, like many *Nyctimene*, animals will pluck the fruit from trees and carry them in flight to feed nearby. Pregnant females have been encountered in January and February, and a lactating female was recorded on 6 September 1990 (McKean 1972).

COMMON TUBE-NOSED BAT

Nyctimene albiventer (Gray, 1863)

al'-bee-ven'-ter

The species name *Nyctimene albiventer* is currently applied to many morphologically distinct forms of tube-nosed bat and may comprise at least two quite different species complexes (Aplin and Armstrong 2016). Populations of tube-nosed bats from areas under consideration here (Admiralty Islands,

Common tube-nosed bat (*Nyctimene albiventer*). As with all *Nyctimene*, the taxonomy of this species remains uncertain (Pavel German).

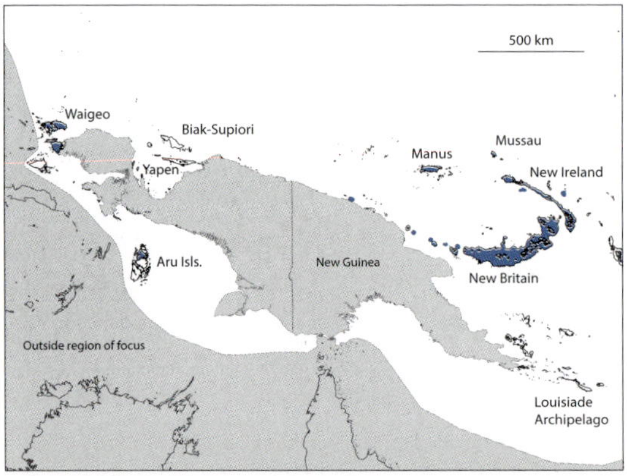

Nyctimene albiventer			
Size*			
	Mean	**Range**	**Sample**
Head–body	82.5	65.5–98.0	4
Tail	21.7	15.4–29.9	4
Forearm	54.3	49.0–58.8	4
Tibia	22.7	18.1–28.5	4
Hindfoot	14.1	11.3–18.4	4
Ear	13.1	8.5–18.6	4
Weight	28.8	20.0–38.0	4
* From Bonaccorso (1998)			
Identification The ears are long and with pointed tips; the width of the dorsal strip is approximately 1.5–2 mm and is sometimes obsolete in the anterior third of back, fading or absent above the shoulder blades.			
Synonyms None			
Subspecies *Nyctimene albiventer albiventer* (Gray, 1863) (North Maluku) *Nyctimene albiventer papuanus* Andersen, 1910 (New Guinea, Bismarck Archipelago)			
Distribution Aru Isls, Salawati, Waigeo (Indonesia); Bagabag, Batanta, Crown, Dyual, Kairuru, Lavongai, Lihir, Long, Manus, Mioko, Mussau, New Britain, New Ireland, Sakar, Tolokiwa, Umboi (Papua New Guinea)			
Extralimital Kai Isls, Maluku Isls, New Guinea			
Conservation status (IUCN) Least Concern			
Etymology Sp.–white belly (L.)			

St Matthias Islands, Solomon Islands, Bismarck Archipelago) have all long needed taxonomic revision. The type locality for *N. albiventer* is in the Maluku Islands, and animals consistent with the description for this species are present on New Ireland, New Britain and other islands of the Bismarck Archipelago.

The common tube-nosed bat occurs in a wide variety of habitats, from primary lowland forests into hill forests, as well as secondary forests and disturbed gardens and plantations. They tend to be more frequently encountered in less disturbed environments. Bonaccorso (1998) reported that this species roosts singly or in pairs (mothers and their young), in dense foliage. The patterned wings help to conceal animals while they are roosting, as when folded around the body they create the appearance of a dried leaf hanging from the vegetation.

On mainland New Guinea, pregnant females can be encountered in almost any month, but seasonal reproduction does seem to exist at some localities (Aplin and Armstrong 2016). Females initially carry the newborn young to forage, but later they are left behind at roost sites. There is no indication populations of common tube-nosed bat are in decline; however, taxonomic revisions may alter the conservation status of South-west Pacific taxa.

SOLOMONS TUBE-NOSED BAT

Nyctimene bougainville Troughton, 1936a

boo'-gan-veel

The Solomons tube-nosed bat has been variously treated as a subspecies or synonym of *Nyctimene vizcaccia* and a distinct taxon. Taxonomy has been further confused by recognition of tube-nosed bats from Malaita and Makira (*N. malaitensis*) as a distinct taxon, a subspecies of *N. bougainville* or a synonym of *N. vizcaccia*. These relationships between these taxa have long been in dire need of taxonomic revision. Here we have retained *N. bougainville*, *N. malaitensi*, and *N. vizcaccia* as species pending such revisions. Natural history data presented here for *N. bougainville* pertain to populations in the Solomon Islands Western Province, where only *Nyctimene major* occurs sympatrically. These two sympatric species can be easily differentiated in the field using forearm length measurements.

The Solomons tube-nosed bat is relatively abundant in most habitats but does seem to be more common in primary forests. It is usually one of the most abundant of all bat species captured in mist nets at lowland sites. This species also occurs into high-elevation montane forests, albeit at lower densities. Over two nights of mist-netting in cloud forest on the summit of Mt Tapalamegutu, Kolombangara, only a single individual was captured. On smaller islands this species is sometimes absent. A week of intensive mist net surveys on Ngella Island in October 2017 failed to detect it at all, whereas *N. major* was abundant. An otherwise diverse pteropodid bat fauna was present, including *Dobsonia inermis*, *Nesonycteris woodfordi*, *Macroglossus minimus*, *Pteropus rayneri* and *Rousettus amplexicaudatus*.

Solomons tube-nosed bats are sexually dimorphic, with males being more of a bluish grey colour with darker wing membranes and females being a paler yellowish brown with lighter coloured wing membranes. They are frugivorous and native figs constitute part of the diet. On spotlighting surveys in lowland forests, the animals are often seen in flight with large fruits in their mouths. These bats roost individually in the understorey among dense foliage and lianas.

Reproduction is apparently synchronous. Of 54 females caught on various islands of the Western Province in April 2014, almost half were reproductively active. One was pregnant, one was carrying a dependent young, 26 were lactating, 18 were regressed or post-lactating, and eight were nulliparous. An additional 19 subadult animals were captured, with forearm lengths ranging from 52.9 to 61.3 mm. Thirteen adult females were caught during surveys on Ranongga, Vella La Vella and Kolombangara in June–July 2017; only one of these was lactating.

The Solomons tube-nosed bat is most common in primary forests but is tolerant of disturbance. A comparison of repeat surveys in the Western Province over a 23-year period following widespread commercial logging did not reveal any decline in abundance of this species (Lavery *et al.* 2020a).

Solomons tube-nosed bat (*Nyctimene bougainville*), Vangunu, Solomon Islands. This is a common species in primary forests of the Solomon Islands archipelago (Tyrone Lavery).

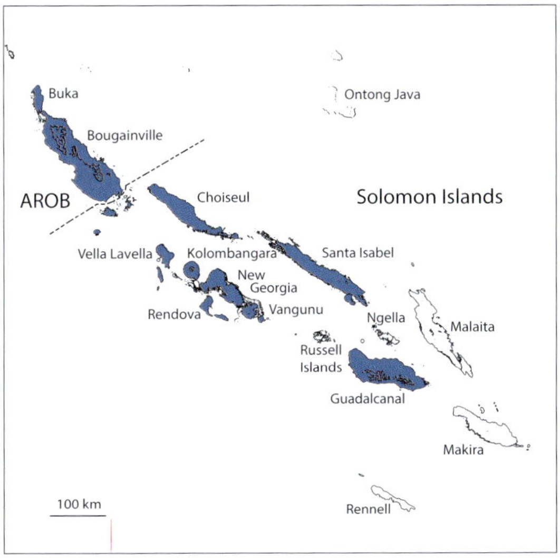

Nyctimene bougainville			
Size*			
	Mean	**Range**	**Sample**
Head–body	73.5	61.0–88.3	109
Tail	20.1	15.8–25.1	10
Forearm	61.0	56.0–66.0	115
Tibia	22.3	19.4–31.5	109
Hindfoot	13.8	12.4–14.5	10
Ear	12.0	8.0–15.2	108
Weight	40.8	30.0–51.0	113
* QM and KU specimens and field measurements from the Western Province, Solomon Isls			
Identification A medium-sized species of *Nyctimene*. It can be differentiated from *N. major* by its smaller size (forearm length less than 70 mm). The dorsal stripe is poorly defined in the region of the shoulders and foreback.			
Synonyms None			
Subspecies None			
Distribution Bougainville, Buka, Choiseul, Fauro, Gatokae, Guadalcanal, Kolombangara, New Georgia, Ranongga, Rendova, Santa Isabel, Tetepare, Vangunu, Vella La Vella			
Extralimital None			
Conservation status (IUCN) Least Concern (*Nyctimene vizcaccia*)			
Etymology Sp.–for Bougainville, the type locality			

MOUNTAIN TUBE-NOSED BAT

Nyctimene certans Andersen, 1912b

ser'-tans

Nyctimene certans			
Size*			
	Mean	**Range**	**Sample**
Head–body	–	–	–
Tail	–	–	–
Forearm	60.0	54.8–67.4	35
Tibia	23.5	18.7–28.8	33
Hindfoot	–	–	–
Ear	11.6	7.1–15.1	30
Weight	–	–	–
* From New Guinea mainland specimens (Irwin 2017)			
Identification Ears short and rounded; on average it is larger and darker dorsally than other medium-sized *Nyctimene* species. The length of the second phalanx of the second metacarpal is greater than 8 mm. The dorsal stripe is narrow, somewhat ill-defined and dark along the posterior portion of the back.			
Synonyms None			
Subspecies None			
Distribution New Britain			
Extralimital New Guinea			
Conservation status (IUCN) Least Concern			
Etymology Sp.–round ear (Gr.)			

Mountain tube-nosed bat has been considered a subspecies of *Nyctimene cyclotis*. However, Petersen (1991) assessed morphological variation in *N. cyclotis* and *N. certans* and recognised each as a distinct species, assigning two specimens from New Britain held in the Bernice P. Bishop Museum, Honolulu, to *N. cyclotis*. They were collected from 1584 m in Warangoi (BBM NG 20914) and Mt Sinewet (BBM NG 20757), East New Britain. Smith and Hood (1981) similarly identified these two incomplete specimens as *N. cyclotis*, but did not encounter the species during their own extensive surveys on New Britain and New Ireland in 1979. Irwin (2017) did not include specimens from New Britain in a re-diagnosis of *N. certans* and *N. cyclotis*; however, all New Guinea specimens east of West Papua were identified as *N. certans*. New Britain specimens are thus here referred to *N. certans*.

Little is known of the biology of this species. It is limited to higher elevation forests, usually above 1600 m elevation (Irwin 2017). Pregnant females have been documented in Marobe and Eastern Highlands Provinces in November and February. Four females captured in March were reproductively inactive.

Irwin (2017) suggested that if museum records reflect natural abundance, this species could be rare. Given the wide distribution, lack of information on population sizes and threats, an IUCN Red List classification of Data Deficient was recommended. However, the most recent IUCN Red List assessment for *N. certans* (2021) classified this species as Least Concern.

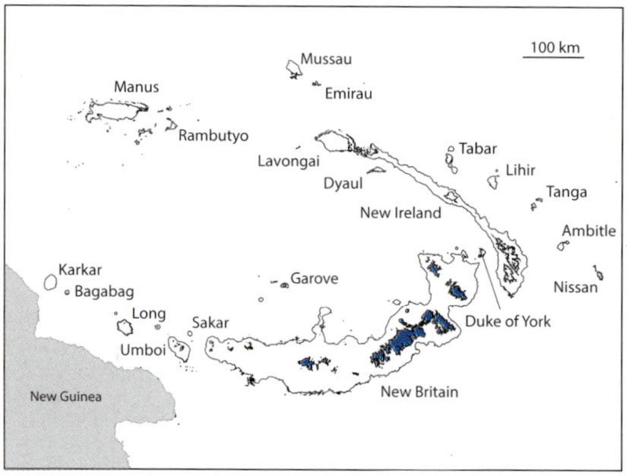

ISLAND TUBE-NOSED BAT

Nyctimene major (Dobson, 1877)

may'-jor

The island tube-nosed bat is broadly distributed with numerous recognised subspecies that may in fact represent distinct species. It can be abundant in the parts of its range where it is the only species of *Nyctimene* present and it is found in gardens, secondary forests, atoll and beach forests, and plantations. The island tube-nosed bat also occurs in lowland primary forests among diverse assemblages of pteropodid bats, usually only on smaller islands, but also in low numbers on larger islands. On Woodlark, for example, it was present in all habitats and one of the species most commonly captured in mist nets. But where it occurs in sympatry with a second *Nyctimene* species, it is less common. In Solomon Islands, island tube-nosed bat is more commonly encountered on small, offshore islands where

Island tube-nosed bat (*Nyctimene major*). This large and broadly distributed tube-nosed bat is variable in its local occurrence, usually rare or absent in primary forests, and more frequently encountered on smaller islands and in areas where other *Nyctimene* species are absent (Tyrone Lavery).

N. bougainville is absent. It is extremely rare or absent from primary forests (where *N. bougainville* is most abundant), unless the island is small (e.g. Ngella Island). It is also usually rare or absent from gardens and disturbed forests on the larger islands.

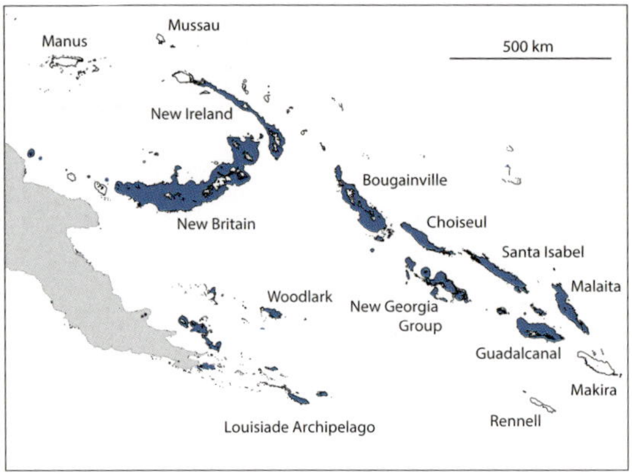

Nyctimene major			
Size*			
	Mean	**Range**	**Sample**
Head–body	99.2	74.5–138.0	40
Tail	25.7	18.0–32.6	12
Forearm	78.5	65.2–86.7	40
Tibia	30.2	23.0–39.3	38
Hindfoot	19.5	15.0–24.5	10
Ear	13.8	11.1–18.0	41
Weight	76.1	48.0–100.0	43
* Field measurements from Western Province, Solomon Isls, AM specimens (Woodlark), and Bonaccorso (1998)			
Identification A large and usually pale-coloured species with a forearm length over 65 mm. The dorsal stripe is narrow and well defined. In subspecies *N. m. lullulae* it is restricted to the lower part of the back.			
Synonyms None			
Subspecies *Nyctimene major major* (Dobson, 1877) (Bismarck Archipelago) *Nyctimene major lullulae* Thomas, 1904b (Woodlark) *Nyctimene major geminus* Andersen, 1910 (D'Entrecasteaux Isls, Louisiade Archipelago) *Nyctimene major scutilus* Andersen, 1910 (Solomon Isls) *Nyctimene major sanctacrucis* Troughton, 1931 (Nendö, Solomon Isls)			
Distribution Alcester, Bagabag, Bougainville, Duke of York, Engineer Isls, Fergusson, Goodenough, Kadovar, Karkar, Kerehikapa, Kiriwina, Marapa, Mioko, Misima, New Britain, New Ireland, Normanby, Rogeia, Rossel, Sakar, Sideia, Sikopo, Sudest, Woodlark (Papua New Guinea); Choiseul, Gizo, Guadalcanal, Kolombangara, Malaita, New Georgia, Nendö, Ngella Pile, Ngella Sule, Nusa Tupe, Pavuvu, Ranongga, Santa Isabel, Shortland, Tetepare, Vangunu (Solomon Isls)			
Extralimital None			
Conservation status (IUCN) Least Concern			
Etymology Sp.–major (L.), for its large size			

The island tube-nosed bat is clearly a capable disperser able to colonise and survive on remote small islands such as Alcester in the Solomon Sea where resources may be limited.

Nyctimene major sanctacrucis, described from a single female specimen presented to the Australian Museum in July 1892 and long recognised as a distinct species, is here recognised as a subspecies of *N. major*. The specimen was probably from Nendö (Santa Cruz), certainly from the Santa Cruz group in Temotu Province, Solomon Islands. Troughton (1931) distinguished *Nyctimene sanctacrucis* from *Nyctimene major* based on the darker pelage in the female holotype (being as dark as males in the latter); by having shorter second phalanges of the third and fourth digits; and by possessing a skull that is relatively short with a proportionally longer rostrum, larger orbital diameter and heavier teeth. Notwithstanding repeated expeditions to the Santa Cruz group of islands, the Nendö tube-nosed bat has not been seen since the holotype was originally collected. Mist-net surveys on Nendö, Reef Islands, Vanikoro and Tinakula between 1990 and 2018 failed to detect it (Lavery *et al.* 2020d). Species of *Nyctimene* are readily captured using this method and thus *N. m. sanctacrucis* must be considered extinct, or the collecting locality erroneous.

Little information is available on the diet of the island tube-nosed bat. On smaller limestone islands it forages among strandline trees such as *Calophylum inophylum*. Bonaccorso (1998) reported that it feeds on fruiting figs (*Ficus* sp.) and seed pods of *Gymnostoma papuanus* by chewing the pod, swallowing the juice and ejecting the wad of fibrous material.

Lactating females have been recorded on Normanby on 17, 22 and 28 December 1988 and on 4, 5 and 6 January 1989. In the Western Province of the Solomon Islands, seven of 15 females caught in April 2014 were lactating. Insufficient data are available from across the species range to understand if it is a seasonal breeder, and how reproductive timing varies regionally. The island tube-nosed bat has a wide extent of occurrence with idiosyncratic local distributions. It is accustomed to more disturbed habitats such as those found on smaller cays.

DEMONIC TUBE-NOSED BAT

Nyctimene masalai Smith and Hood, 1983

mas'-ah-lay

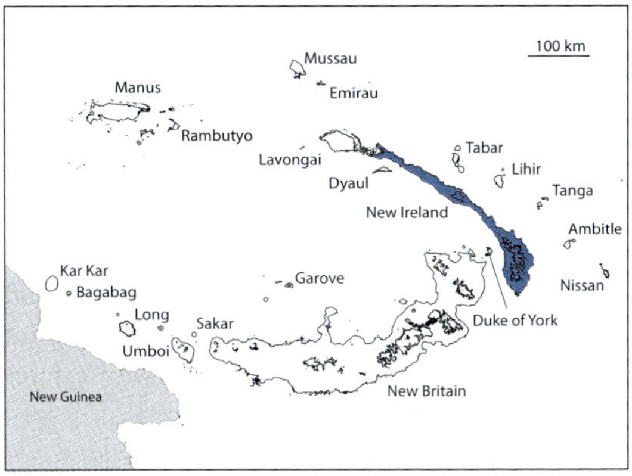

The taxonomic validity of *Nyctimene masalai* is uncertain. It may be a synonym for *N. albiventer*, but has been included here pending the results of a comprehensive revision of the genus *Nyctimene*. Smith and Hood (1983) described this species on the basis of two male specimens from New Ireland. The holotype (LACM 65798) was collected from Ralum at 10 m elevation on 9 July 1979, and the paratype (LACM 65799) from Hilalon at sea level on 19 July 1979. *Nyctimene masalai* was distinguished from other members of the genus by its moderately large body size, possession of a mottled reddish-brown dorsum, a 5 mm-wide mid-dorsal stripe, and a ventrum that is greyish-white with a yellowish-brown wash. The cranium was distinguished as being narrow and elongate; frontal sinuses parallel and inflated; bony palate moderately flat, postdental palate not markedly pandurate; teeth with reduced coronal cusps; and a third upper premolar that is unicuspid and nearly subconical (Flannery and White 1991).

Flannery and White (1991) collected 34 *Nyctimene* specimens on New Ireland and could not

Nyctimene masalai			
Size*			
	Mean	**Range**	**Sample**
Head–body	145.5	94.0–103.0	2
Tail	21.5	21.0–22.0	2
Forearm	65.5	63.5–67.5	2
Tibia	–	–	–
Hindfoot	13.5	13.0–14.0	
Ear	14	14.0	2
Weight	48.8	45.2–52.3	2
* From Smith and Hood (1983)			
Identification It remains uncertain how this species can be conclusively distinguished from other *Nyctimene* that occur in the Bismarck Archipelago. The dorsal stripe is approximately 5 mm wide from the back of the crown to the base of the tail.			
Synonyms None			
Subspecies None			
Distribution New Ireland			
Extralimital None			
Conservation status (IUCN) Data Deficient			
Etymology Sp.–masalai, is taken from the Tolai language and means forest demon or devil			

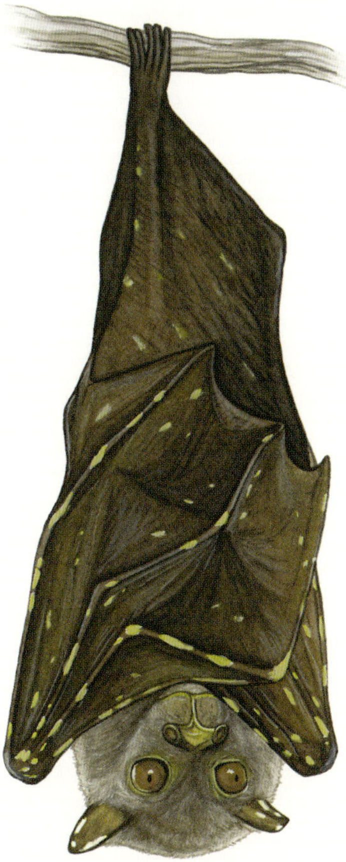

The demonic tube-nosed bat (*Nyctimene masalai*) was described from New Ireland on the basis of two specimens collected in 1979 (Madison Erin Mayfield).

find morphological justification to distinguish *N. masalai* from their sample. They concluded only two species of *Nyctimene* inhabit New Ireland: *N. albiventer* and *N. major*. Emmons and Kinbag (2001) surveyed southern New Ireland and compared the *Nyctimene* specimens they collected with *N. albiventer* from the Maluku Islands and *N. cephalotes* from Sulawesi. They concluded their specimens did not differ from *N. cephalotes* in many significant features but were radically different in many features from *N. albiventer*. Their specimens also shared sexual dimorphism with *N. cephalotes* (males being grey and females paler brown). Nothing is recorded of the ecology of *N. masalai* and it remains classified as Data Deficient on the IUCN Red List.

MALAITA TUBE-NOSED BAT

Nyctimene malaitensis Phillips, 1968

mal-ay-ten'-sis

The Malaita tube-nosed bat belongs to the *albiventer* species group, a group for which species diversity and taxonomic relationships remain confused. It has previously been regarded as a subspecies of *Nyctimene bougainville*, which has been regarded as a subspecies of *N. vizcaccia*. Molecular data from Malaita and Makira (both isolated from other large islands in the Solomon Archipelago by deep sea trenches) are needed to confirm specific status.

This species occurs in a wide variety of habitats, including lowland primary and secondary forests, village gardens and montane forests. It is frugivorous, and individuals have been observed at night carrying the fruits of native figs (*Ficus* sp.). Like *N. bougainville*, the Malaita tube-nosed bat produces a high-pitched 'kissing' call when foraging at night. During surveys on Mt Tolubusu (one of Malaita's highest peaks) in November 2016, the Malaita tube-nosed bat was found to be relatively abundant. At an elevation of 1000 m in the Bauro highlands of Makira in June 2018, five of 29 bats captured over a four night survey were *N. malaitensis*.

Pregnant females have been recorded in May, July and October. On Malaita, three females were mist-netted in July 2016, and two of these were pregnant. Of two females examined in October 2016, one was pregnant and the second was post-lactating. On Makira between 3 and 9 May 2012, four females were captured adjacent to the northern coastline: two were post-lactating, one was pregnant, and the fourth was a subadult. Of five individuals caught at high elevation on Makira in June 2018,

The Malaita tube-nosed bat (*Nyctimene malaitensis*) is found from seas level to the highest peaks of Malaita and Makira, Solomon Islands (Ben Speare).

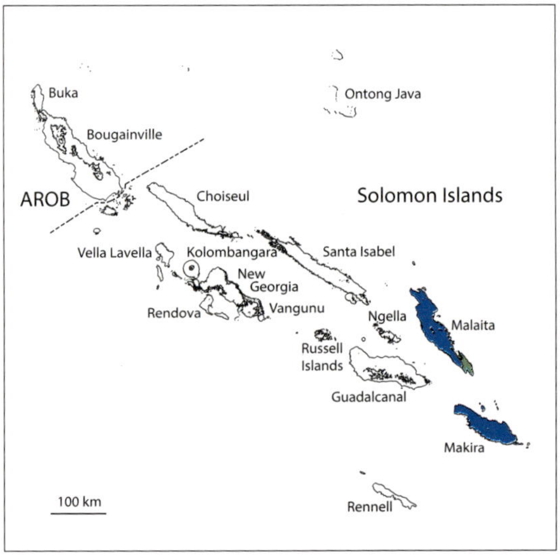

Nyctimene malaitensis			
Size*			
	Mean	**Range**	**Sample**
Head–body	93.5	80.5–118.0	21
Tail	21.2	17.3–23.0	9
Forearm	64.6	60.4–71.0	24
Tibia	24.4	21.4–31.0	23
Hindfoot	14.3	12.4–16.3	16
Ear	13.2	10.6–15.4	24
Weight	47.1	34.0–56.0	20
* KU and QM specimens from Malaita and Makira			
Identification Forearm length (less than 70 mm) is shorter than in *N. major* (greater than 70 mm). Slightly larger than *N. bougainville* (average forearm length is 64.6 mm *v.* 61.0 in *N. bougainville*, but size ranges do overlap). The dorsal stripe is black and well defined from the shoulders to the rump.			
Synonyms None			
Subspecies None			
Distribution Makira, Malaita, Ugi			
Extralimital None			
Conservation status (IUCN) Least Concern			
Etymology Sp.–from Malaita (L.), the type locality			

four were males and one was a female that was lactating.

Unlike many of the larger Solomon Islands, inland villages set in areas of montane forest remain common on Malaita and Makira. Isolation from marine sources of protein increases reliance on forest resources and small bats such as *Nyctimene* are hunted and eaten. Malaitans are skilled at using hand-woven nets to capture bats, but whether this poses a conservation threat to this species needs further assessment. The Malaita tube-nosed bat is presently classified as Least Concern on the IUCN Red List because it is a species commonly captured in a wide range of habitats.

UMBOI TUBE-NOSED BAT

Nyctimene vizcaccia Thomas, 1914a

viz-kak'-kee-ah

The Umboi tube-nosed bat has been variously treated both as a subspecies or synonym of *Nyctimene albiventer*, with *N. bougainville* considered a distinct taxon or a subspecies of *N. vizcaccia*, and *N. malaitensis* treated as a synonym or separate taxon confined to Malaita and Makira in Solomon Islands. During a survey of New Britain's Nakanai mountains, a series of *Nyctimene* individuals examined by Aplin and Opiang (2011) comprised two quite distinct taxa corresponding to a *N. albiventer* type (with a broad U-shaped upper dental arcade), and a *N. vizcaccia* type (with an anteriorly convergent upper dental arcade). No larger-bodied animals (consistent with *N. major*) were captured. Anthony (2001) reported the capture of *N. major* and similarly identified the presence of two distinct populations of smaller-bodied *Nyctimene* (giving a total of three species for New Britain).

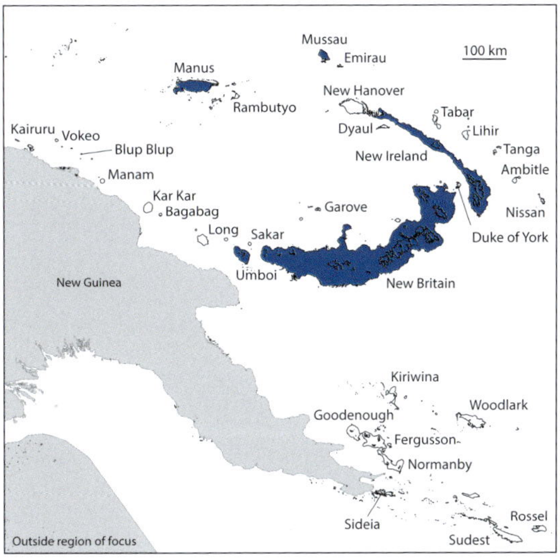

Nyctimene vizcaccia			
Size*			
	Mean	**Range**	**Sample**
Head–body	94.1	74.3–115.0	4
Tail	21.4	18.0–25.5	4
Forearm	59.7	51.6–67.0	4
Tibia	21.7	17.9–24.9	4
Hindfoot	15.9	11.0–22.2	4
Ear	14.1	11.5–16.0	4
Weight	42.0	27.5–55.5	4
* From Bonaccorso (1998)			
Identification External characters available to differentiate this species from other sympatric *Nyctimene* are uncertain. The dorsal stipe begins between the shoulder blades and is rather well marked but not as well marked as in *N. albiventer*.			
Synonyms None.			
Subspecies None			
Distribution Manus, Mussau, New Britain, New Ireland, Umboi			
Extralimital None			
Conservation status (IUCN) Least Concern			
Etymology Sp.–for vizcacha, rodents of the genus *Lagidium*, which share similar varied fur (L.)			

The extremely confusing relationships among *Nyctimene* taxa have long been in need of taxonomic revision. Here we have retained *N. vizcaccia* and *N. bougainville* as distinct species and natural history data are explicitly divided between the Solomon Islands archipelago (*N. bougainville*) and the Bismarck Archipelago (*N. vizcaccia*). We tentatively include Manus and Mussau in the distribution but note that Aplin *et al.* (2015) considered these populations may be undescribed species.

The Umboi tube-nosed bat is relatively abundant in most habitats but does seem to be more common in primary forests. It is one of the most abundant of all bat species captured in mist nets in lowland primary forests. Bonaccorso (1998) reported that it is a dietary specialist that primarily feeds on figs (*Ficus* spp.). Reproductively inactive females have been encountered on Umboi Island February, but two females examined in February on southern New Ireland were pregnant (Bonaccorso 1998, Emmons and Kinbag 2001). Of six individuals captured by Bonaccorso (1998) in June 1995 between 440 and 1250 m elevation on New Britain, three were subadults, one was a non-parous female, one a non-lactating female and one an adult male.

The Umboi tube-nosed bat is listed as Least Concern on the IUCN Red List. Taxonomic revisions of *Nyctimene* will warrant re-examinations of conservation status for most species.

Paranyctimene

UNSTRIPED TUBE-NOSED BAT

Paranyctimene raptor **Tate, 1942a**

pa'-rah-nik-tay-may'-nay rap'-tor

Species of the genus *Paranyctimene* are poorly known. A cryptic taxon (*P. tenax*) was described by Bergmans (2001), including the subspecies *marculus*, for which the type locality was Rumei River, Waigeo. However, no description of the known geographical distribution of *P. tenax* was provided, nor was any delineation against the distribution of *P. raptor*. Thus the distributions of these two species within New Guinea (and a few nearby islands) are confused and any observations of natural history cannot be assigned with any certainty to either species. Here we provide a

Unstriped tube-nosed bat (*Paranyctimene raptor*). The distributions and natural history of *Paranyctimene* species remain uncertain and more than a single species may occur in the region (Pavel German).

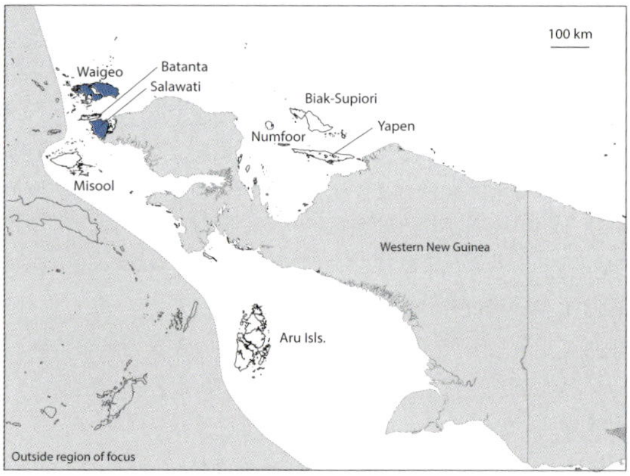

summary of data available for *Paranyctimene* from the region and mainland New Guinea under the first available species name. However, we recognise this information could pertain to either *P. raptor* or *P. tenax*, and that multiple *Paranyctimene* species could in fact be present within the region.

On Salawati in 1992, unstriped tube-nosed bat was uncommon, with only two individuals captured during the survey period. A female caught on

Paranyctimene raptor			
Size			
	Mean	**Range**	**Sample**
Head–body	70.9	70.6–71.2	2
Tail	15.5	12.5–18.5	2
Forearm	49.9	48.8–50.9	2
Tibia	19.6	18.6–20.6	2
Hindfoot	–	–	–
Ear	12.8	12.8	2
Weight	22.0	22.0	2
* From AM specimens (Salawati)			
Identification A small tube-nosed bat with extraordinarily long canines and a distinct greenish colouration to the wing membranes and face. *Paranyctimene* lack the dorsal stripe present in species of *Nyctimene*.			
Synonyms None			
Subspecies None			
Distribution Salawati, Waigeo			
Extralimital New Guinea			
Conservation status (IUCN) Least Concern			
Etymology Gen.–beside Nyctimene (Gr.); Sp.–robber, plunderer (L.), significance unknown			

17 October was mist-netted adjacent to a jackfruit tree (*Artocarpus heterophyllus*) growing in a village near sea level. It was not lactating. A male was taken in a mist net set on the margin of garden and forest at 50 m elevation.

On mainland New Guinea, Bonaccorso (1998) reported *P. raptor* as being common in primary and secondary forests, gardens and swamps from sea level to 1200 m. It roosts individually among forest foliage and has been reported eating native figs (*Ficus* spp.) and introduced piper (*Piper aduncum*). Females give birth to a single young, and the timing of this probably varies geographically (Bonaccorso 1998).

At present both species of *Paranyctimene* are listed as Least Concern on the IUCN Red List. The distributions of both species are presumed to be large and no major threats are known. This may alter with taxonomic reassessments.

Pteralopex

BOUGAINVILLE MONKEY-FACED BAT

Pteralopex anceps **Andersen, 1909b**

te-rah'-loh-pex an'-seps

The Bougainville monkey-faced bat has only been collected from Bougainville Island. Although it has been recorded from sea level to 1900 m, the preferred elevational distribution is apparently higher elevation forests where the sympatric *P. flanneryi* is yet to be documented. Four of seven known museum specimens have been collected above 1100 m (Helgen 2005). The most recent observations of this species also are derived from higher elevation sites in the Rotokas and Kunua sides of Mt Balbi. One of these was captured from a tree hollow and a second was roosting beneath a dense ring of epiphytic ferns that encircled the trunk of a tree (D. Waldien, pers comm.).

No specimens of the Bougainville monkey-faced bat are known from Choiseul or Santa Isabel, but several observations indicate it may occur there. Bowen-Jones *et al.* (1997) reported two observations believed to be this species at an elevation of 850 m on Choiseul. The first group of five individuals was seen in late afternoon in

The Bougainville monkey-faced bat (*Pteralopex anceps*) prefers higher elevation forests. Several recent records of this species (including the individual pictured) have been made from the high-elevation areas of Rotokas and Kunua around Mt Balbi, Bougainville Island (Stephen Richards).

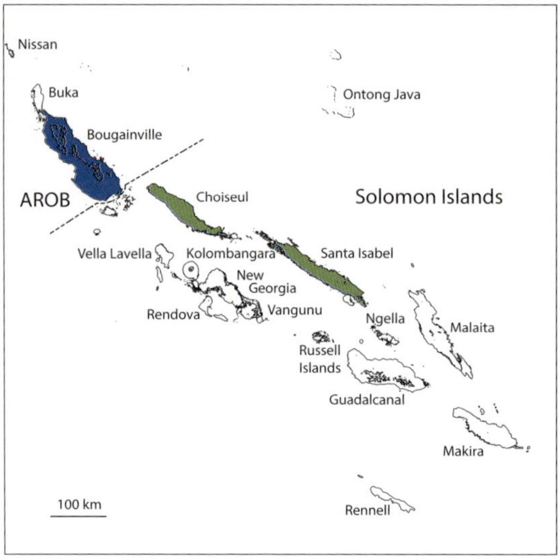

Pteralopex anceps			
Size*			
	Mean	**Range**	**Sample**
Head–body	260.4	252–270	5
Tail	–	–	–
Forearm	154.5	141–160	5
Tibia	–	–	–
Hindfoot	–	–	–
Ear	23.8	22–25	5
Weight	–	–	–
* From Helgen (2005)			
Identification Fur long and thick, not adpressed on the mid-back, and almost completely obscuring the ears. The chest has an expansive yellow or whitish patch.			
Synonyms None			
Subspecies None			
Distribution Bougainville			
Extralimital None			
Conservation status (IUCN) Endangered			
Etymology Gen.–winged + alopex (Gr.) arctic fox; Sp.–two headed (L.), allusion unclear			

primary forest hanging beneath the branches of a large fig tree (*Ficus* sp.), around 12 m above the ground. They were identified as *P. anceps* by the absence of clearly protruding ears, black fur on the head and dorsum, and presence of conspicuous white patches on the ventrum. On the second observation, a single individual was flushed from its roost in a hollow fig tree as a climber scaled the tree. This individual was identified by T. F. Flannery from a photograph. Pikacha (2008) similarly reported an observation of this species in a strangler fig at high elevation on the slopes of Mt Maetembe, central Choiseul.

Helgen (2005) considered ethnobiology recordings by Parnaby (2002b) as evidence that *P. anceps* may be present on Santa Isabel. There, hunters from Hograno district indicated a *Pteralopex* was present, but that it did not feed on coconuts (*Cocos nucifera*) as documented for *P. flanneryi*. Instead, they indicated it chewed bark on the trunks of certain trees, with the chewed pulp ejected onto the forest floor. The large incisors and small molars of *P. anceps* are adaptations that would support such tree-gouging behaviour and exudate feeding (Helgen 2005). Heavy toothwear in museum specimens also suggests hard, abrasive fruits (and possibly nuts) form part of the diet.

Little is known of the reproductive biology of *P. anceps*; however, subadults have been recorded in March and April on Bougainville. Large body size, the tendency to roost in hollow trees and a preference towards primary forests are all traits of *P. anceps* (and monkey-faced bats generally) that predispose them to greater risk of extinction.

GUADALCANAL MONKEY-FACED BAT
Pteralopex atrata Thomas, 1888c
at-rah'-tah

Guadalcanal monkey-faced bat is known from the lowland forests of Guadalcanal to approximately 400 m above sea level. Collections of mammals received by the Australian Museum from northern Guadalcanal in the 1930s included considerable numbers of this species. The Guadalcanal monkey-faced bat is much scarcer in later collections, and today it is rarely encountered in the wild.

Although rarely observed in more recent decades, this is a species that is capable of exploiting more disturbed areas such as gardens and agroforests on the edges of villages and secondary forests. The species was observed visiting mango trees (*Mangifera indica*) in old gardens and village surrounds at Poha River and Kongulai on the outskirts of Honiara, in 1987 and 2014 respectively. Two females were captured at Poha River, but unfortunately reproductive condition was not noted. At Kongulai, fruiting ngali nuts (*Canarium indicum*), wild fig trees (*Ficus* sp.) and coconuts (*Cocos nucifera*) are present on a ridgeline adjacent to secondary forest where the species can reliably be caught in mist nets. Traditional knowledge suggests that the Guadalcanal monkey-faced bat roosts in hollow trees. At dusk one evening in May 1991 in southern Guadalcanal, a single bat (presumed to be this species) was observed leaving its roost in forest to fly high over the canopy. One individual at Kongulai was roosting in a dense mass of vines covering a large tree. The tree was in a valley containing primary forest but surrounded by cleared or heavily disturbed ridgelines.

The Guadalcanal monkey-faced bat (*Pteralopex atrata*) is a large and uniformly black flying fox with mottled black and white wing membranes, and striking red eyes. This subadult was caught in a mist net on the edge of secondary forest adjacent to mango trees and coconut palms (Jonathan 'Tugi P.' Richmond).

Pteralopex atrata			
Size*			
	Mean	**Range**	**Sample**
Head–body	217.1	196.0–237.0	13
Tail	–	–	–
Forearm	137.4	129.0–147.0	13
Tibia	–	–	–
Hindfoot	–	–	–
Ear	17.5	15.0–20.0	13
Weight	472.0	438.0–506.0	2
* From Helgen (2005)			
Identification A large monkey-faced bat, uniformly blackish brown usually with mottled black and white wing membranes. The fur of the mid back is adpressed, eyes are red or orange.			
Synonyms None			
Subspecies None			
Distribution Guadalcanal			
Extralimital None			
Conservation status (IUCN) Endangered			
Etymology Sp.–black (L.)			

On Guadalcanal, *Pteralopex atrata* is probably altitudinally stratified with *Pteralopex pulchra*, with the latter replacing the former at elevations above 1200 m. A single skull from New Georgia held in the collections of Senckenberg Museum, Frankfurt was also identified as *P. atrata* (Helgen 2005). Collected in 1975 by H. L. Bregulla from 'Jericho and Munda Village', it remains the only known specimen of this species from outside Guadalcanal. However, the Guadalcanal monkey-faced bat has not been detected in over 50 000 m^2 hours of mist-net surveys conducted between 2010 and 2019 across the Western Province islands of New Georgia, Vangunu, Kolombangara, Gatokae, Tetepare, Vella La Vella, Ranongga and Rendova. This single record is therefore most unusual and probably erroneous. Alternately, *P. atrata* may have gone extinct on New Georgia since 1975, or if extant must be restricted to the seldom surveyed higher altitude, montane forests. The collecting locality given by H. L. Bregulla could possibly refer to the higher elevation Vinakiki region between Munda (south-west New Georgia) and Jericho (north-west New Georgia). This would be at odds with what is known of the species' distribution on Guadalcanal where it has only been documented from lower elevations.

Little is known of the reproductive biology of this species. Subadults have been captured in October and February. The Guadalcanal monkey-faced bat is undoubtedly threatened by the commercial logging that is widespread throughout the lowland forests of Guadalcanal.

GREATER MONKEY-FACED BAT

Pteralopex flanneryi Helgen, 2005

flan'-er-ee-ee

This is the largest of all *Pteralopex*, and a species rarely encountered across its range. The Australian Museum, Sydney, holds four specimens from Bougainville and Buka; the Bernice P. Bishop Museum, Honolulu, holds two specimens (Bougainville and Choiseul); the United States National Museum of Natural History, Washington, holds three specimens from Bougainville; and the Field Museum of Natural History, Chicago, has one specimen from Santa Isabel. The most recent collection dates for these are from Choiseul in 1964, Bougainville in 1968, and Buka in 1987. A specimen from Santa Isabel (Barora Fa) collected by Leary and Pita in 2000 was also referred to this species by Helgen (2005) on the basis of reported measurements only. Field surveys on Bougainville in the 1990s failed to detect greater monkey-faced bat (S. Hamilton, pers comm.) and expeditions to Santa Isabel (2013 and 2018), Choiseul (2014) and Bougainville (2016 and 2019) similarly failed to record it.

The diet of *Pteralopex flanneryi* includes green coconuts. The animals are able to carry these heavy fruits short distances and open the strong husk with their robust teeth (Tyrone Lavery).

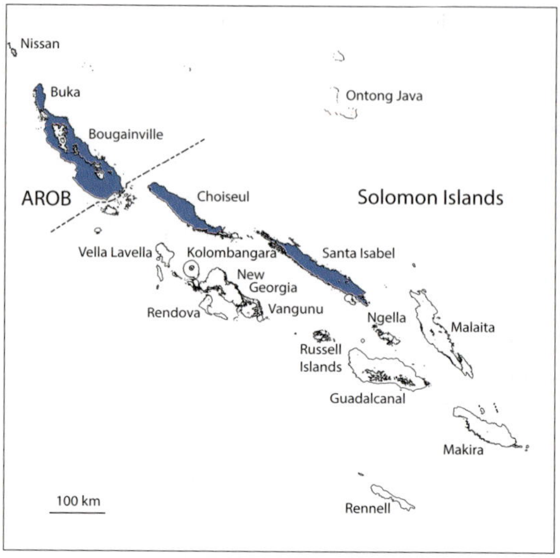

Pteralopex flanneryi			
Size*			
	Mean	**Range**	**Sample**
Head–body	265.1	247–285	11
Tail	–	–	–
Forearm	164.0	159–159	11
Tibia	–	–	–
Hindfoot	–	–	–
Ear	22.5	20–26	11
Weight	–	790	1
* From Helgen (2005)			
Identification Largest of the monkey-faced bats with a forearm length > 158 mm. Fur is overall black, adpressed on the mid-back; tibia and proximal third of the forearm are very sparsely furred.			
Synonyms None			
Subspecies None			
Distribution Barora Fa, Bougainville, Choiseul, Santa Isabel			
Extralimital None			
Conservation status (IUCN) Critically Endangered			
Etymology Sp.–for Professor Tim Flannery in recognition of his studies of Melanesian mammals			

The greater monkey-faced bat has been recorded between sea level and 200 m elevation and occurs syntopically with *Pteralopex anceps* on Bougainville Island. It probably prefers relatively intact lowland forests. In 2016 at Siwai, southern Bougainville, T. H. Lavery was shown a hollow-bearing tree where a pair of this species roosted in an area of lowland forest with an *Areca* palm dominated understorey. They were no longer present at the time the tree was inspected.

The diet includes fruits and green coconuts (*Cocos nucifera*) (Sanborn 1931). Robust teeth enable greater monkey-faced bats to tear into coconut husks to consume the soft meat. They are capable of carrying the heavy fruits short distances from the palm to feed. A lactating female was recorded in July, and subadults were identified in February, April and September (Helgen 2005).

Of the five described species of *Pteralopex*, *P. flanneryi* may well be the species in most urgent need of conservation attention. Records of this species are sparse and it has been documented on few occasions since the 1960s. The likely preference for hollow-bearing trees as roost sites means greater monkey-faced bats are probably greatly impacted by the commercial logging of lowland forests that has rapidly progressed across Choiseul and Santa Isabel.

MONTANE MONKEY-FACED BAT

Pteralopex pulchra Flannery, 1991b

pool'-krah

The montane monkey-faced bat (*Pteralopex pulchra*) is known only from the holotype captured above 1200 m elevation on Mt Makarakomburu, Guadalcanal, Solomon Islands (Madison Erin Mayfield).

The Montane monkey-faced bat remains known only from the holotype, captured at 1230 m elevation on the southern slopes of Makarakomburu, Guadalcanal, in May 1990. This species is apparently an upland specialist restricted to areas of mossy montane forest with emergent *Metrosideros* and palms, and an abundant understorey of ferns and climbing bamboo. *Pteralopex pulchra* is markedly smaller than *P. atrata* and on Guadalcanal replaces this species altitudinally. Intensive mist net surveys at 1000 m and 1300–1500 m elevation on the northern side of Mt Popomanasaeu in 2015 failed to record it. The site at Chupukama (1300–1500 m) was on the transition zone between lowland and montane forests. Although habitat resembled the description given by Flannery (1991b), northern Guadalcanal's rain shadow probably influences the distribution of montane vegetation on the northern side of the island, restricting it to higher altitudes than in the south. The lower altitudinal limit of this species may thus be greater than 1500 m on northern Guadalcanal.

The female specimen was lactating when captured. It was taken in a mist net with *Pteropus*

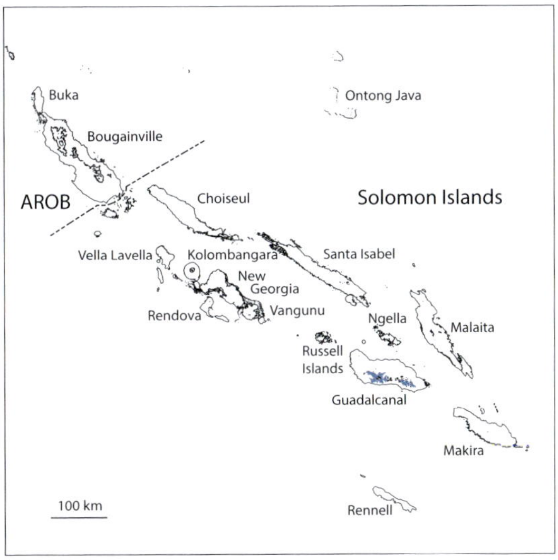

woodfordi, *Nesonycteris woodfordi*, *Nyctimene bougainville* and *Macroglossus minimus*. As is usually the case with members of the genus *Pteralopex*, it was gentle and tractable when handled.

Pteralopex pulchra			
Size*			
	Mean	**Range**	**Sample**
Head–body	–	161.8	1
Tail	–	–	–
Forearm	–	117.9	1
Tibia	–	56.1	1
Hindfoot	–	–	–
Ear	–	16.8	1
Weight	–	280	1
* From the holotype (Guadalcanal)			
Identification A small *Pteralopex*, with heavily mottled black and white wings, long fur, a black head and back contrasting with a yellow belly, and red eyes.			
Synonyms None			
Subspecies None			
Distribution Guadalcanal			
Extralimital None			
Conservation status (IUCN) Critically Endangered			
Etymology Sp.–beautiful (L.)			

Conservation of the montane monkey-faced bat would greatly benefit from increased knowledge of ecology, distribution and population size. However, upper montane Guadalcanal is remote and undisturbed. Most high-elevation villages relocated to the lowlands during the 1970s, and high densities of feral pigs indicate hunting pressure is low above 1000 m elevation. This species is currently classified as Critically Endangered on the IUCN Red List with a cautious designation as Possibly Extinct. It probably survives in Guadalcanal's high-elevation forests, but longer-term survival may be precarious given the restricted area of habitat presently available and predicted impacts of a warming climate.

NEW GEORGIA GROUP MONKEY-FACED BAT
Pteralopex taki Parnaby, 2002b
tah'-kee

This is the most commonly recorded species of the genus *Pteralopex*. The Australian Museum holds numerous records from research in the early 1990s on New Georgia and Vangunu islands and, until recently, this was thought to be the limit of its distribution. Failure to detect it on Kolombangara in the same period led to the conclusion it had been extirpated in the mid-1970s with widespread commercial logging of the island's lowlands. However, surveys in early 2015 recorded six individuals at a garden–forest interface at Vanga on the western side of Kolombangara. More recently, surveys by Piokera Holland recorded another individual at 600 m elevation on Kolombangara and extended the known range to Tetepare and Rendova (P. Holland, pers. obs.). *Pteralopex taki* has not yet been recorded on other large Western Province islands (Gatokae, Ranongga, Vella La Vella) despite extensive surveys there.

The New Georgia Group monkey-faced bat is a quiet and gentle species. On southern Vangunu, T. H. Lavery climbed a strangler fig to examine a known roost for this species. As the tree was scaled, an animal could be heard clambering down the hollow trunk. It rolled out of one of the many gaps in the trunk onto a platform close to his feet, laid

New Georgia Group monkey-faced bat (*Pteralopex taki*), Vangunu, Solomon Islands. This species roosts in small groups inside hollow trees and has been found to be sensitive to commercial logging of primary forests (Tyrone Lavery).

calmly on its back gazing up, and was easily scooped up. Roost sites are usually close to old village sites or village gardens where ngali nut (*Canarium indicum*) and cutnut trees (*Barringtonia edulis*) are common. The roosts consist of small groups of up to 10 individuals within hollow trees, especially strangler figs (*Ficus* spp.) and ngali nuts. These are usually greater than 1.5 m in diameter at breast height and greater than 20 m in height. Preferred cavities are large, with multiple entrances, and typically around 10 m above the ground (Fisher and Tasker 1997). Animals will use the same trees and cavities for months and possibly years.

Like all *Pteralopex*, the New Georgia Group monkey-faced bat has extremely complex and robust teeth relative to other pteropodids. Research has indicated it is largely frugivorous. Fruits are chewed and the hard parts are ejected before the animal stores the pulp in cheek pouches. At least 18 species of fruit were consumed in feeding trials (Fisher and Tasker 1997). Faecal samples contained pollen, bark, rotten wood and moss. No samples contained leaf material, although traditional knowledge indicated they will also take the new leaves of certain trees and vines. The jaws and teeth are powerful – when offered a ngali nut, one animal chewed the exocarp off, broke open the hard coating around the seed and ate the kernel (Fisher and Tasker 1997).

Breeding is probably aseasonal, lactating females having been recorded February–May, and in October. Juveniles have also been captured in mist nets during these months. No female has been observed carrying a young.

Although the New Georgia Group monkey-faced bat has been the most frequently documented of all *Pteralopex*, the species is nonetheless naturally uncommon. Of nine species of pteropodid bat captured in mist nets in the Western Province, *Pteralopex taki* and *Pteropus solomonis* were the least frequently encountered species (Lavery *et al.* 2020a). These bats are not targeted extensively by hunters, but they are easily caught from their hollows by climbing roost trees. The presence of roosting habitat appears to be the most critical factor for their survival. Commercial logging operations target the largest trees that are most likely to bear hollows. However, strangler figs

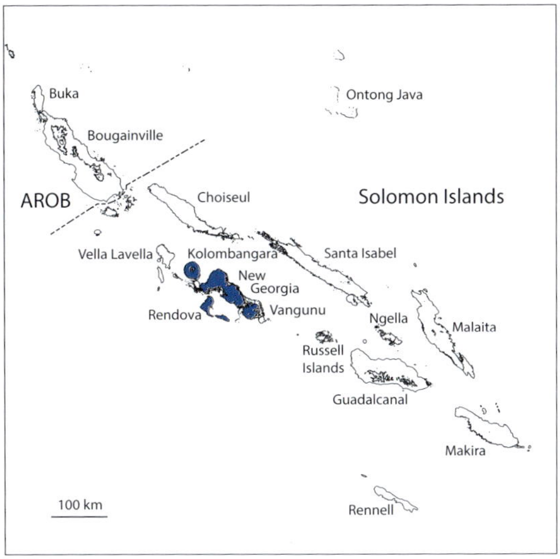

Pteralopex taki			
Size*			
	Mean	**Range**	**Sample**
Head–body	153.5	120.0–202.0	13
Tail	–	–	–
Forearm	115.4	111.2–119.4	12
Tibia	50.5	44.7–55.0	12
Hindfoot	34.3	30.1–40.0	6
Ear	14.3	10.1–19.8	18
Weight	269.6	179.0–354.0	17
* From AM, KU and QM specimens, and field measurements (Kolombangara, New Georgia, Tetepare, Vangunu)			
Identification A small species of *Pteralopex* (forearm < 120 mm), wing membranes mottled black and white, eyes orange, and fur olive brown mixed with pale golden hairs on the venter.			
Synonyms None			
Subspecies None			
Distribution Kolombangara, New Georgia, Rendova, Tetepare, Vangunu,			
Extralimital None			
Conservation status (IUCN) Vulnerable			
Etymology Sp.– for the Vangunu indigenous language name for the species			

often remain undisturbed in the logged forests because they are not commercially viable, and these may continue to provide enough refuge for the species to persist in areas logged as recently as 3 months prior. Nonetheless, capture rates of New Georgia Group monkey-faced bats have declined in the 23-year period between 1992 and 2015 in association with commercial logging (Lavery *et al.* 2020a).

Pteropus

ADMIRALTY FLYING FOX

Pteropus admiralitatum **Thomas, 1894**

te'-roh-poos ad'-mir-ah-li-tah'-tum

The Admiralty flying fox belongs to the *griseus* species group (Almeida *et al.* 2014). It typically inhabits smaller islets across a broad distribution encompassing the Admiralty Islands, Bismarck Archipelago and Solomon Islands. Exceptions to this small island preference are Manus and Malaita. Its body size and external morphology is similar to *Pteropus hypomelanus* and *P. solomonis*, and the three species appear to replace each other ecologically, forming complex checkerboard patterns of distribution across northern Melanesia (Helgen 2007). The true extent of this is probably underappreciated, given many of the smaller islands in the region are yet to be the focus of biological surveys. The islets this species is known to occupy include the Purdy Group and satellite islands of Manus, Tingwon, Tabar, Nuguria, Nissan, Mono, Ontong Java, Dai and Russell Islands. On Malaita and Russell Islands it occurs sympatrically with *P. solomonis*. Several subspecies have been named, identified primarily by differences in colouration of the mantle and chest, and length of the forearm.

On Ontong Java, the Admiralty flying fox forms camps numbering in the hundreds of individuals. On Mono, it was found to be roosting individually

Admiralty flying fox (*Pteropus admiralitatum colonus*), Mono, Solomon Islands. This species has a large geographical range, but its precise island distribution follows a checkerboard pattern whereby it mostly occupies smaller outlying islands to the exclusion of ecologically similar species (Lucas DeCicco).

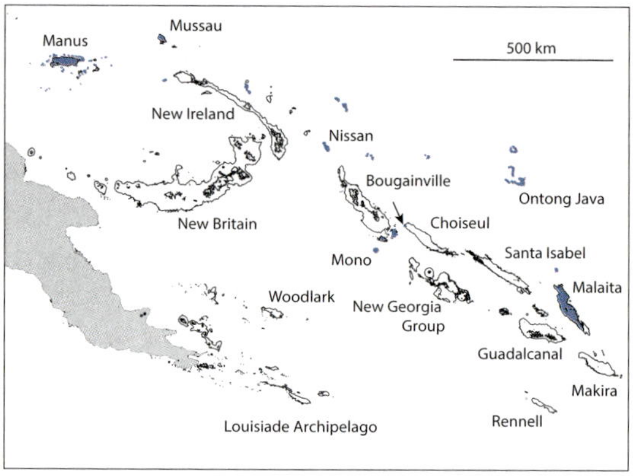

Pteropus admiralitatum			
Size*			
	Mean	**Range**	**Sample**
Head–body	171.6	163.8–190.0	8
Tail	–	–	–
Forearm	118.4	106.1–130.2	9
Tibia	52.4	49.7–54.1	4
Hindfoot	31.5	30.6–33.1	3
Ear	21.2	18.8–28.0	9
Weight	264.9	205.0–334.0	8

* From AM specimens (Manus) and KU specimens (Mono)

Identification A small flying fox that is drab brown in colour with scattered silver hairs over the belly. The mantle is golden yellow, contrasting with the mid-back. It can be separated from *P. solomonis* by possessing a naked tibia rather than being furred to the ankle.

Synonyms None

Subspecies *Pteropus admiralitatum admiralitatum* Thomas, 1894 (Purdy Isls)
Pteropus admiralitatum colonus Andersen, 1908 (Shortland Isls)
Pteropus admiralitatum howensis Troughton, 1931 (?Nukumanu Atoll, Ontong Java)
Pteropus admiralitatum goweri Tate, 1934a (Dai)

Distribution Carteret Islands, Nissan, Nukumanu, Nuugurigia (AROB); Bat, Emirau, Lou, Manus, Mussau, Rambutyo, Tingwon, Tabar, Tong, Pak, Purdy Isls (Papua New Guinea); Dai, Fauro, Malaita, Mbanika, Mono, Ontong Java, Reef (Choiseul Bay), Shortland (Solomon Isls)

Extralimital None

Conservation status (IUCN) Least Concern

Etymology Gen.–wing-foot (Gr.); Sp.–a reference to the Admiralty Isls, the type locality among the dense foliage of trees and palms. Green coconuts and inflorescences (*Cocos nucifera*) may contribute to the diet.

Of four females examined on Manus by a team from the Australian Museum in June 1988, none was lactating. Two out of three females captured on Ontong Java in August were pregnant, carrying embryos weighing 20 and 29 g (Phillips 1968). Two females were caught on Mono in July 2019: one of these was pregnant and the other was immature.

Because the Admiralty flying fox is found on smaller, more exposed islets, individual populations face significant threats from environmental disturbance and disasters such as cyclones. They also are exposed to hunting. However, the species may be accustomed to persisting in disturbed habitats. On Ontong Java, much of the available land was converted to coconut plantation from the 1950s onwards, yet a viable population remains there. Across its distribution, the species is under no apparent threat of extinction.

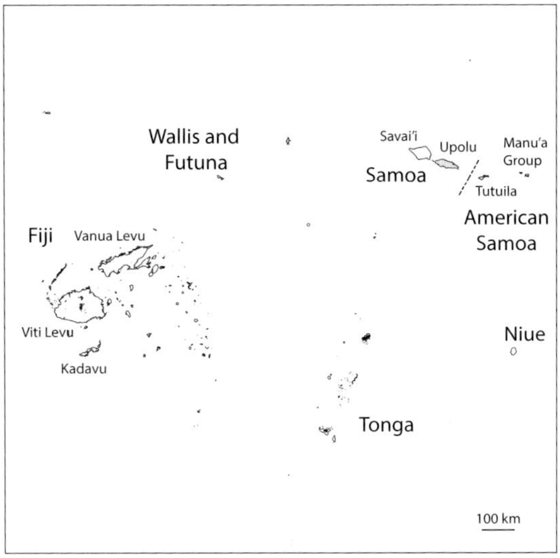

SMALL SAMOAN FLYING FOX

Pteropus allenorum Helgen *et al.*, 2009

al'-en-or'-um

The small Samoan flying fox is known only from the holotype, held in the collection of the Academy of Natural Sciences, Philadelphia, USA. The specimen is a mature subadult, in a poor state of preservation in alcohol, with the skull and mandible removed. It was collected in April 1856 at Apia, Upolu, Samoa, and presented to the museum by a Dr W. S. W. Ruschenberger, but may have been collected by Dr H. C. Caldwell (Helgen *et al.* 2009). It is a small to medium-sized flying fox probably with a brown head, tinged with russet; a golden-brown mantle; dusky brown back; and warm brown limbs and wing membranes (Helgen *et al.* 2009).

Nothing is recorded of biology. Greatly reduced cheek teeth and a gracile mandible suggest the diet is similar to the nectar-feeding species *P. fundatus*, *P. mahaganus* and *P. woodfordi* of Vanuatu and Solomon Islands. *Pteropus allenorum* is presumed extinct, but efforts to locate populations of this species in the more remote parts of Samoa would be worthwhile.

Pteropus allenorum			
Size*			
	Mean	**Range**	**Sample**
Head–body	–	–	–
Tail	–	–	–
Forearm	–	116	1
Tibia	–	–	–
Hindfoot	–	–	–
Ear	–	–	–
Weight	–	–	–
* From Helgen *et al.* (2009) (mature subadult)			
Identification The external appearance of this species is uncertain due to the state of the only known specimen. In relation to *P. samoensis* the teeth are greatly reduced and forearm length is shorter.			
Synonyms None			
Subspecies None			
Distribution Upolu			
Extralimital None			
Conservation status (IUCN) Extinct			
Etymology Sp.–the epithet *allenorum* chosen to honour the name of Allen (plural): zoologist, anthropologist, and physician Harrison Allen (1841–1897); and Allen Drew of Philadelphia			

VANUATU FLYING FOX

Pteropus anetianus (Gray, 1870)

ah'-net-ee-ah'-nus

This brightly coloured flying fox is endemic to Vanuatu and is a member of the *samoensis* species group (Almeida *et al.* 2014). Across its broad distribution, very distinctive subspecies are recognised on different island groups, suggesting

Vanuatu flying fox (*Pteropus anetianus*), Banks Islands, Vanuatu. The Vanuatu flying fox is one of only two extant species of mammal endemic to Vanuatu (Pavel German).

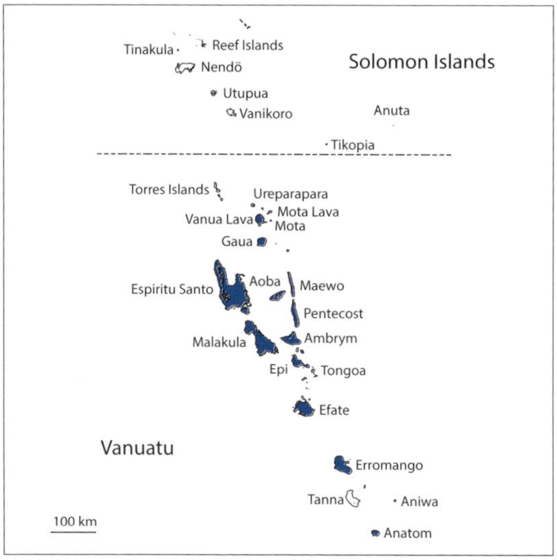

Pteropus anetianus			
Size*			
	Mean	**Range**	**Sample**
Head–body	190.8	155.0–205.0	6
Tail	–	–	–
Forearm	130.8	117.5–135.0	9
Tibia	–	57.3	1
Hindfoot	–	33.0	1
Ear	24.5	20.7–26.0	7
Weight	346.5	200.0–440.0	6
* From AM specimens (Malakula), Andersen (1913) and Sanborn (1930)			
Identification A very variable species, it is medium-sized and usually yellowish-brown in colouration. It is larger than *P. fundatus* (with which it is sympatric in the Banks Group) and has proportionately larger teeth.			
Synonyms None			
Subspecies *Pteropus anetianus anetianus* Gray, 1870 (Aneityum, Erromango) *Pteropus anetianus eotinus* Andersen, 1913 (Malakula, Malo) *Pteropus anetianus bakeri* Thomas, 1925 (Efate, Emao, Nguna) *Pteropus anetianus banksiana* Sanborn, 1930 (Ureparapara, Vanua Lava) *Pteropus anetianus aorensis* Lawrence, 1945 (Aore, Espiritu Santo) *Pteropus anetianus motalavae* Felten and Kock, 1972 (Mota Lava) *Pteropus anetianus pastoris* Felten and Kock, 1972 (Emae, Tongoa)			
Distribution Ambrym, Aneityum, Aoba, Aore, Efate, Emae, Emao, Epi, Erromango, Espiritu Santo, Gaua, Lopevi, Maewo, Malakula, Malo, Mota Lava, Nguna, Pentecost, Shepperd Is. Group, Tongoa, Ureparapara, Vanua Lava			
Extralimital None			
Conservation status (IUCN) Vulnerable			
Etymology Sp.–a reference to Aneityum, the type locality			

the Vanuatu flying fox has limited dispersal ability. Further research into taxonomy and ecology is needed.

The Vanuatu flying fox roosts by day in small quiet groups (Chambers and Esrom 1988), with Baker and Baker (1936) suggesting it is not very social and never forms large colonies. An Australian Museum team recorded it roosting among large numbers of *Pteropus tonganus* on Efate. The same behaviour was reported during 2018 surveys in the Banks Group of islands, but the Vanuatu flying fox was not observed nor captured in mist nets on Mota or Vanua Lava.

The diet includes fruits of *Syzygium* and *Ficus* (Medway and Marshall 1975), and it has been recorded feeding on coconuts (*Cocos nucifera*), breadfruit (*Artocarpus altilis*) and the flowers of *Barringtonia edulis*. In contrast with *P. tonganus*, the Vanuatu flying fox seems to be more reliant on blossoms, and may be partly diurnal or at least feeds prior to dusk. Lawrence (1945) recorded that a Commander Austin saw them flying on the island of Aore in dull weather, and he considered them to be more diurnal that other bats encountered. Peregrine falcons (*Falco peregrinus*) can pose significant threat to Vanuatu flying foxes so these daytime habits do not seem to reflect a release from diurnal predation (Mickelburgh *et al.* 1992). Instead, diurnal behaviour could be an adaptation to reduce competitive interactions with the much larger *P. tonganus*.

Reproduction is apparently seasonal: male testes size peaks from October to January (Baker and Baker 1936). This, in addition to data on pregnant females, suggests births peak from August to September. Females mature at a bodyweight of around 300 g and males at a weight of 325 g (Baker and Baker 1936).

Anecdotal reports have indicated hunting is a major threat to Vanuatu flying fox, especially following cyclone events. Monitoring of population trends across the distribution of this species is needed.

ARU FLYING FOX

Pteropus aruensis Peters, 1867

ahr'-oo-en-sis

This poorly known flying fox has been collected from the Aru Islands on two occasions, at 'Wonambay' (probably Manumbai on Kobroor) and on Wokam (Laurie and Hill 1954). A specimen conferred to *Pteropus melanopogon* (formerly considered conspecific with *P. aruensis*) was among

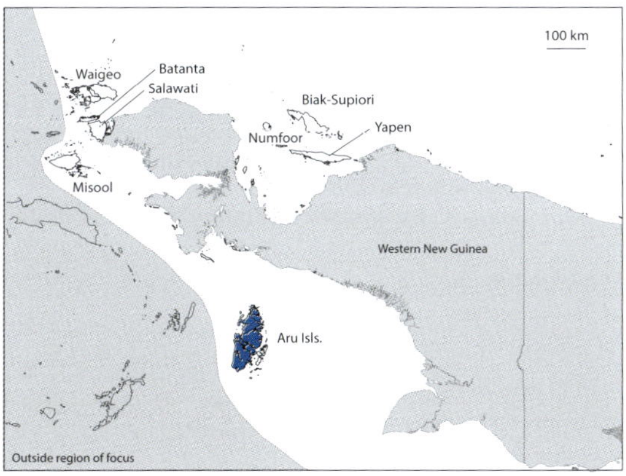

Pteropus aruensis			
Size			
	Mean	**Range**	**Sample**
Head–body	–	–	–
Tail	–	–	–
Forearm	190.5	190.0–191.0	2
Tibia	–	–	–
Hindfoot	–	–	–
Ear	–	–	–
Weight	–	–	–
* From Andersen (1912a)			
Identification The width of the area of dorsal fur on the back is not restricted to a narrow band along the spine (as in *P. melanopogon*), being more of a normal breadth, and is glossy silvery whitish in colour and thinly sprinkled with blackish hairs. *Pteropus keyensis* differs by having fur that is a nearly uniform yellow on the ventrum and dorsum. The forearm length is slightly shorter than that of *P. melanopogon* (196–204 mm) and longer than that of *P. keyensis* (179–187.5 mm).			
Synonyms *Pteropus fumigatus* von Rosenberg, 1867 *Pteropus rubiginosus* von Rosenberg, 1867			
Subspecies None			
Distribution Aru Isls			
Extralimital None			
Conservation status (IUCN) Critically Endangered			
Etymology Sp.–from Aru (L.), the type locality for the species			

the 'kitchen midden' and hunter collections made at Namara on Korboor by Dr Pat Woolley in October–November 1992 (Aplin and Pasveer 2006).

Information on the natural history of the Aru flying fox is severely lacking and it is currently feared extinct because it is a large and presumably conspicuous bat that has not been scientifically documented since the 19th century. Hunting is probably a major threat, and large-scale plantations threaten to transform vast areas of forested habitat for *P. aruensis* in the Aru Islands (Obidzinski *et al.* 2015).

NEW BRITAIN FLYING FOX

Pteropus capistratus Peters, 1876

kap'-ee-strah'-tus

The New Britain flying fox is a member of the *capistratus* species group (Almeida *et al.* 2014) and is one of the most spectacularly coloured of all flying foxes. The face is boldly patterned brown and white and the wings are beautifully venated. *Pteropus capistratus* was formerly considered a subspecies of *P. temminckii*. However, they differ in size, in colour and in the presence of anastomosing brown lines on the wing membranes. They also are geographically separated on either side of New Guinea, in the Bismarck and Maluku Islands respectively.

Flannery and White (1991) discussed known museum records, including two syntypes identified in the description of *P. capistratus* (ZMB 4998 and 49999), which were collected during the S.M.S. Gazelle Expedition. The type locality was given as New Ireland (Nova Hibernia) and specimen 4999 bears the label 'Neu Merkeleg, Gazelle', the former being an abbreviated name for New Ireland. However, Flannery and White (1991) reviewed these records and raised doubt as to whether they were indeed from New Ireland and concluded New Britain or Duke of York was the more likely collecting locality. Additional individuals (NHMUK 77.7.18.1–2, 4) labelled as coming from New Ireland were collected by the Reverend George Brown in the 1870s. Brown was based in the Duke of York Group on Mioko and bought many specimens from local people, who undoubtedly carried them from both nearby New Britain and New Ireland. Specimens in the Museum für Naturkunde, Berlin (ZMB 34068, 6259), labelled as coming from Finschafen (Huon Peninsula) are probably incorrect, as this species has never otherwise been recorded from mainland New Guinea.

On New Britain, *P. capistratus* has been collected from numerous locations up to an elevation of

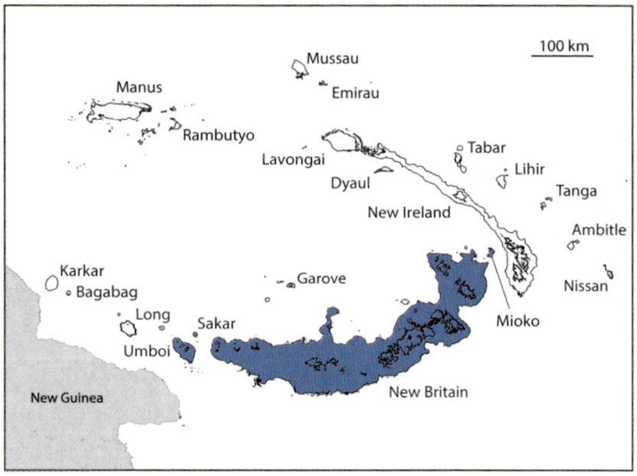

New Britain flying fox (*Pteropus capistratus*) (illustration by Gustav Mützel published by Matschie (1899).

Pteropus capistratus			
Size*			
	Mean	**Range**	**Sample**
Head–body	169.0	145.0–192.0	4
Tail	–	–	–
Forearm	112.3	102.0–122.0	4
Tibia	49.3	43.0–55.0	4
Hindfoot	37.8	31.0–45.0	4
Ear	23.3	20.0–26.0	4
Weight	234.3	165.0–290.0	4
* From Bonaccorso (1998)			
Identification A small flying fox with a face boldly marked with brown and white, and brownish wings with dark brown networks of veins. This species is likely to be confused with only *P. ennisae* of New Ireland. It differs from that species in having the face being much more distinctly marked, with bands on the rostrum and surrounding the eyes being white rather than grey. The mantle is yellow, and the crown and back also have hairs tipped with yellow (absent in *P. ennisae*). *Pteropus personatus* (North Maluku, Indonesia) also possesses a distinct face mask similar to that of *P. capistratus* but does not occur in the region under consideration here.			
Synonyms None			
Subspecies None			
Distribution Duke of York, Mioko, New Britain, Sakar, Umboi			
Extralimital None			
Conservation status (IUCN) Vulnerable			
Etymology Sp.–head + striped (L.)			

1200 m. Smith and Hood (1981) recorded the capture of three individuals sometime between June and August 1979. One individual was located by day while roosting alone under a *Heliconia* leaf, while a male and a female were discovered as they foraged around a Japanese cherry (*Muntingia calabura*) around 2 a.m. The bats were first located because of their bird-like trill. The female was carrying a young and apparently placed it onto the tree upon being disturbed. Japanese cherry is usually planted in more open village areas and gardens, indicating *P. capistratus* may be tolerant of disturbed habitats. *Dobsonia anderseni* and *Rousettus amplexicaudatus* were feeding in the same tree. Surveys on New Britain in 2009 failed to detect this species, probably because efforts were focused on higher elevations where the New Britain flying fox may not be present (Aplin and Opiang 2011).

The New Britain flying fox is one of only two species of mammal (both Old World fruit bats) in which males are known to lactate (Bonaccorso 1998). Broader reproductive biology is poorly known so it is uncertain why this occurs and how it may provide any adaptive advantage.

Few data are available to assess the status of this species. The most recent IUCN Red List assessment classified the New Britain flying fox as Near Threatened. The lowland forests of New Britain have been extensively converted to oil palm plantations and detailed monitoring of the conservation status of New Britain flying fox is needed.

MAKIRA FLYING FOX

Pteropus cognatus Andersen, 1908

kog'-nah-tus

Pteropus cognatus belongs to the *samoensis* species group (Almeida *et al.* 2014), and has been treated as a subspecies of *P. rayneri*. However, genetic data reveal it is certainly best regarded as a distinct species (Lavery, unpublished data). Information available on biology, ecology and conservation status remains limited for the Makira flying fox. The distribution is restricted to Makira, Ugi and Santa Ana. Few specimens exist in museum collections. Andersen (1908) listed one adult skull and two immature skins in the British Museum of Natural History, and Sanborn (1931) reported on a further two specimens, in the American Museum of Natural History and Museum of Comparative Zoology, Harvard. The Australian Museum holds three specimens.

Pteropus cognatus occurs from sea level into montane forests above 900 m elevation. Across this range it is typically an uncommon capture in mist nets. One female examined near sea level south of Kira Kira on 17 November 1987 was nursing a 90 g young. They were found by day hanging in a banana (*Musa* sp.) by a river. The few records suggest that this is a solitary species. This contrasts markedly with the communal roosts seen in *P. rayneri*, but is similar to observations of the closely related *P. rennelli* (which roosts solitarily in mid to upper canopy branches).

The diet includes breadfruit (*Artocarpus altilis*) and other cultivated fruits. Makira flying fox is regularly hunted for food and, along with *P. tonganus*, is targeted with sling shots when foraging in village gardens at night. On Makira, the teeth of flying foxes are an important cultural commodity, used as traditional currency for important occasions such as paying bride price. It is primarily the teeth of *P. tonganus* that are used due

The Makira flying fox (*Pteropus cognatus*) occurs from sea level into montane forests of Makira, Solomon Islands (Tyrone Lavery).

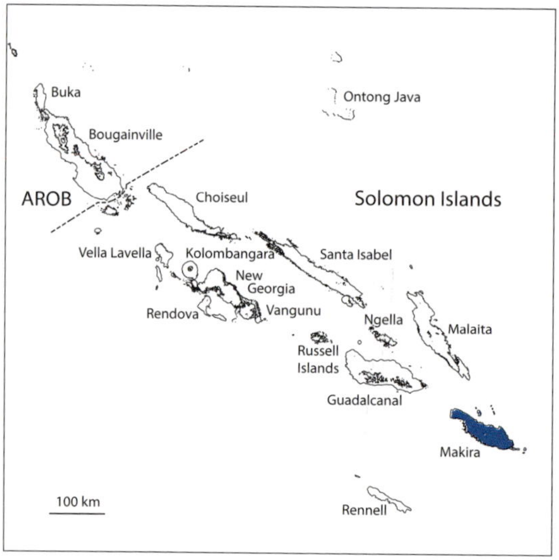

Pteropus cognatus			
Size*			
	Mean	**Range**	**Sample**
Head–body	173.3	166.6–180	2
Tail	–	–	–
Forearm	125.7	121.5–129.2	3
Tibia	61.1	60.8–61.3	2
Hindfoot	–	35.2	1
Ear	22.5	21.9–23.4	3
Weight	288.8	260–305	4
* From AM and KU specimens (Makira)			
Identification A medium-sized flying fox, predominantly brown in colour with a cinnamon mantle. The belly has scattered golden hairs. It can be separated from the sympatric *P. tonganus* by its smaller size, brown rather than black overall colour and lack of a contrasting yellow mantle.			
Synonyms None			
Subspecies None			
Distribution Makira, Santa Ana, Ugi			
Extralimital None			
Conservation status (IUCN) Vulnerable			
Etymology Sp.–related, connected by blood (L.), perhaps a reference to its relatedness to *P. rayneri*			

to their larger size and this apparently directs hunting effort towards that species rather than the smaller *P. cognatus* (Lavery and Fasi 2017). Hunting intensities vary across Makira, being greatest in the Bauro region of the north and lowest in Arosi (west Makira). This pattern corresponds with perceptions that *P. cognatus* is common in Arosi and in decline in Bauro.

Makira, like many islands in the Solomon Archipelago, continues to be heavily logged by commercial companies. This threat is likely to be compounded by subsistence hunting.

SPECTACLED FLYING FOX

Pteropus conspicillatus Gould, 1850

kon'-spis'-ill-ah'-tus

The spectacled flying fox is a large species that belongs to the *griseus* species groups with *Pteropus tonganus* of the South-west Pacific, and *P. alecto* of Australia, New Guinea and Indonesia (Almeida *et al.* 2014). A *Pteropus* molecular phylogeny found three samples of *P. conspicillatus* from New Guinea were paraphyletic, with one sample of the Australian subspecies of *P. alecto* nested among them (Almeida *et al.* 2014). The authors suggested hybridisation was the most likely explanation. Neaves *et al.* (2018) similarly concluded long-term hybridisation is common between *P. conspicillatus* and *P. alecto*.

Spectacled flying fox (*Pteropus conspicillatus*) near Waigeo Island in the Raja Ampat Islands. This species forms large camps comprising thousands of individuals (Chien Lee).

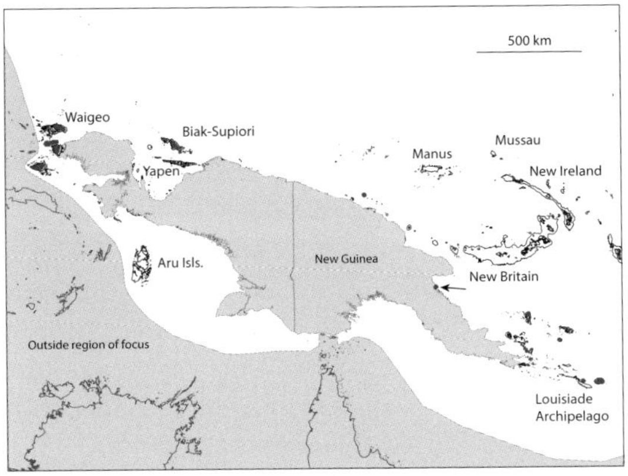

Pteropus conspicillatus			
Size*			
	Mean	**Range**	**Sample**
Head–body	265.5	233.0–295.0	4
Tail	–	–	–
Forearm	179.5	167.5–188.5	4
Tibia	85.0	84.0–86.0	2
Hindfoot	–	–	–
Ear	31.6	30.0–33.0	3
Weight	781.3	510.0–1000.0	4
* From AM specimens (Normanby, Woodlark)			
Identification A large, long-snouted flying fox that is mostly black in colour but with a bright yellow mantle and tallow-green fur surrounding the eyes.			
Synonyms *Pteropus mysolensis* Gray, 1870			
Subspecies *Pteropus conspicillatus conspicillatus* Gould, 1850 (Australia and Papua New Guinea) *Pteropus conspicillatus chrysauchen* Peters, 1862 (north-west New Guinea, Maluku Isls)			
Distribution Batanta, Biak-Supiori, Misool, Owi, Salawati, Waigeo, Yapen (Indonesia); Alcester, Goodenough, Hull, Kiriwina, Lababia, Normanby, Rossel, Sudest, Woodlark, Yeina (Papua New Guinea)			
Extralimital Australia, Maluku Isls, New Guinea			
Conservation status (IUCN) Endangered			
Etymology Sp.–spectacles (N.L.), a reference to the yellow-green fur around the eyes.			

The spectacled flying fox occurs on several islands in the D'Entrecasteaux, Louisiade and Trobriand Groups in Milne Bay, Papua New Guinea. Koopman (1982) identified specimens in the collections of the American Museum of Natural History from Sudest, Rossel, Normanby, Goodenough, Kiriwina and Woodlark islands. Specimens from Hull Island (off Normanby Island), Alcester and Woodlark islands are held by the Australian Museum. It also is known from islands of Cenderawasih Bay, and west of the Vogelkop Peninsula.

This is a gregarious species that forms large camps in stands of *Casuarina*, *Artocarpus* and mangrove forests (e.g. Lae, Madang, and Raboin Island, Papua New Guinea) (Bonaccorso 1998). Camps can number in the thousands of individuals. Australian Museum researchers encountered it in 1989, in a mixed roost with *P. hypomelanus* on Hull Island in the D'Entrecasteaux Group.

The spectacled flying fox forages in primary and secondary forests, coconut plantations and gardens. In the Wet Tropics of Australia it utilises a broad variety of plant resources, with the seed and pulp of figs (*Ficus* spp.) and pollen from the family Myrtaceae being the most common elements identified in faecal samples (Parsons *et al.* 2006). On New Guinea, the diet includes the blossoms of coconut palms and the fruits of plants from the genera *Ficus*, *Eugenia*, *Terminalia* and *Calophyllum* (Bonaccorso 1998).

Spectacled flying foxes breed seasonally, and in Australia females give birth to young between October and December. At the camp on Hull Island four juveniles weighing 240, 260, 380 and 400 g were recorded on 7 January 1989, suggesting the timing of reproduction is similar there. However, Bonaccorso (1998) found that individuals further north at Madang on mainland New Guinea gave birth in July and early August. This is a large, conspicuous species that roosts in exposed areas and is presumably at risk from hunting, especially on islands where sources of dietary protein may be scarce.

LARGE SAMOAN FLYING FOX

Pteropus coxi Helgen *et al.*, 2009

cox'-ee

The large Samoan flying fox is known only by the holotype and paratype collected in 1839–41 and held in the United States National Museum of Natural History, Smithsonian Institution, Washington, DC (USNM 3790 and USNM 3791). The skins of both specimens have been lost, and only crania remain (Helgen *et al.* 2009). They were collected during the USA Exploring Expedition, and the type locality on accompanying labels is identified generally to 'Samoan Is.' or 'Samoan Isle'. Based on

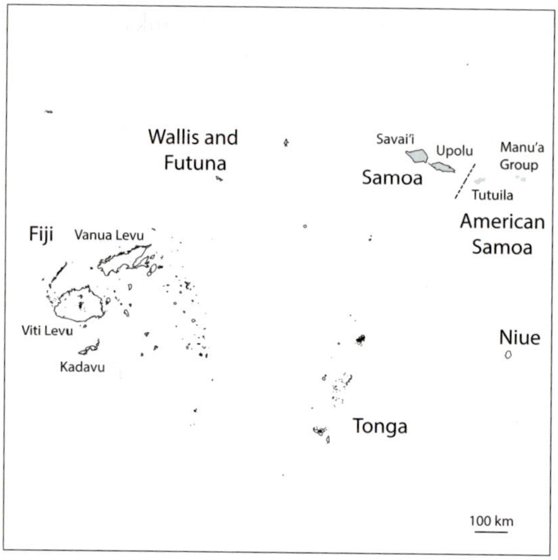

Pteropus coxi			
Size			
	Mean	**Range**	**Sample**
Head–body	–	–	–
Tail	–	–	–
Forearm	–	–	–
Tibia	–	–	–
Hindfoot	–	–	–
Ear	–	–	–
Weight	–	–	–
Identification External appearance is unknown, but it is likely to be larger and with a longer forearm than *P. samoensis* and *P. tonganus*. Fur colouration may be somewhat similar to *P. samoensis*.			
Synonyms None			
Subspecies None			
Distribution Samoan Isls (exact island distribution uncertain)			
Extralimital None			
Conservation status (IUCN) Extinct			
Etymology Sp.–for Dr Paul A. Cox in recognition of his research on flying foxes, tropical ecology, and biodiversity conservation			

morphology of the skull, Helgen *et al.* (2009) suggested this species was likely to be larger, heavier and with a longer forearm than the sympatric species *P. samoensis* and *P. tonganus*. The apparent referral of *P. coxi* specimens to *P. samoensis* by Peale (1848) indicates the external characteristics of the large Samoan flying fox may not have been especially different to those of *P. samoensis*.

Nothing is recorded of biology. *Pteropus coxi* is now presumed to have been extinct for over 100 years, but searches in the remote montane areas of Upolu and Savai'i may be worthwhile.

NEW IRELAND FLYING FOX

Pteropus ennisae Flannery and White, 1991b

en'-is-ee

The New Ireland flying fox belongs to the *capistratus* species group (Almeida *et al.* 2014), formerly considered a subspecies of *Pteropus capistratus* and, prior to that, a subspecies of *P. temminckii* of the Maluku Islands. However, molecular data have shown it is a distinct species (Almeida *et al.* 2014). The holotype in the Australian Museum is an adult female skin, skull, and body in spirit. It was collected on 20 June 1988 in secondary forest near Madina, New Ireland, by T. Flannery, L. Seri, T. Heinsohn and T. Ennis. Also captured on 23 and 26 of June in the same region were a male, and a female paratype. An Australian Museum expedition to southern New Ireland in July 1993 recorded a further two individuals. An adult male was mist-netted along a garden–rainforest boundary in the Suir River area, and a female was mist-netted in logged forest at 300 m elevation in the Weitin Valley. This species was again recorded in January–February 1994 on southern New Ireland during a Conservation International biodiversity assessment. It was found at all four sites surveyed from Weitin River to Hans Meyer Range between elevations of 150 m and 1800 m (Emmons and Kinbag 2001).

The New Ireland flying fox is evidently an uncommon and largely solitary species. The three individuals examined in June 1988 were each caught on separate occasions. One female was discovered in secondary forest feeding on the fruits of a cauliflorous regrowth tree species (possibly a cluster fig *Ficus* sp.),

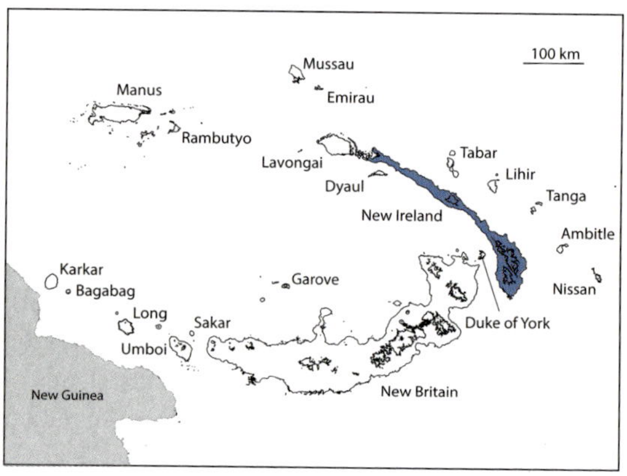

Pteropus ennisae			
Size*			
	Mean	**Range**	**Sample**
Head–body	186.8	155.0–185.0	5
Tail	–	–	–
Forearm	113.5	109.0–118.5	5
Tibia	54.0	49.5–56.5	5
Hindfoot	35.3	32.6–39.0	3
Ear	20.0	18.4–21.4	5
Weight	215.0	170.0–250.0	5
* From AM specimens (New Ireland)			
Identification Differs from *P. capistratus* in having claws darker at tips, wing membranes that are darker with anastomosing dark lines that are less clear, and the face being much less distinctly marked (areas of the face that are white in *P. capistratus* are dirty grey-brown in *P. ennisae*). *Pteropus ennisae* is also browner in overall colour, with a darker venter (fur tips are much darker); it lacks a yellow wash on the crown, the mantle is whitish grey rather than yellow, and the hairs of the back lack yellow tips.			
Synonyms None			
Subspecies None			
Distribution New Ireland			
Extralimital None			
Conservation status (IUCN) Vulnerable			
Etymology Sp.–for Tish Ennis, former research assistant in the mammal section of the Australian Museum, who participated in the expedition during which the type series was collected.			

and a male was similarly feeding on cluster figs. Another female was taken in a mist net. One of the two females examined in June 1988 was lactating, and the other weighed 270 g and was carrying a near-term foetus. The female caught in July 1993 in southern New Ireland was also lactating. *Pteropus ennisae* is classified as Vulnerable on the IUCN Red List due to the small extent of occurrence and threats posed by forest degradation and loss and climate change.

BANKS FLYING FOX

Pteropus fundatus Felten and Kock, 1972

fun-dah'-tus

This small species of flying fox, belonging to the *samoensis* species group (Almeida *et al.* 2014), is endemic to the Banks Group of islands in northern Vanuatu. The distribution as presently understood follows the 1963 collections of H. Bregulla described by Felten and Kock (1972). The holotype (an adult male) and 12 paratypes were collected from Mota, along with an adult male from Vanua Lava, an adult male from Mota Lava and two males from 'Banks Islands'. Extensive surveys in the Banks Group by P. German of the Australian Museum in 1992 detected this species on Mota only. Recent surveys on Vanua Lava and Mota similarly recorded it on Mota only (although the survey effort on Vanua Lava was limited) (Lavery *et al.* 2020d).

Another two *Pteropus* species have been reported from the Banks Group. *Pteropus anetianus* has been collected from Gaua, Mota Lava, Ureparapara and Vanua Lava by the Whitney South Sea Expedition (Sanborn 1930), H. Bregulla in 1963, and P. German in 1992. Specimens of *P. tonganus* have been obtained from Gaua, Mota, Mota Lava and Vanua Lava. The exact distribution of *P. fundatus* and patterns of sympatry with other *Pteropus* need further assessment within this group of islands. The existence of three sympatric *Pteropus* on any of these relatively small islands would be most unusual considering remaining islands of the archipelago support just two species. Although the presence of a far richer pteropodid bat fauna in Vanuatu prior to human settlement is apparent (Hawkins 2015).

Banks flying fox (*Pteropus fundatus*), Mota Island, Vanuatu. This is a small, largely nectarivorous species endemic to the Banks Islands (Tyrone Lavery).

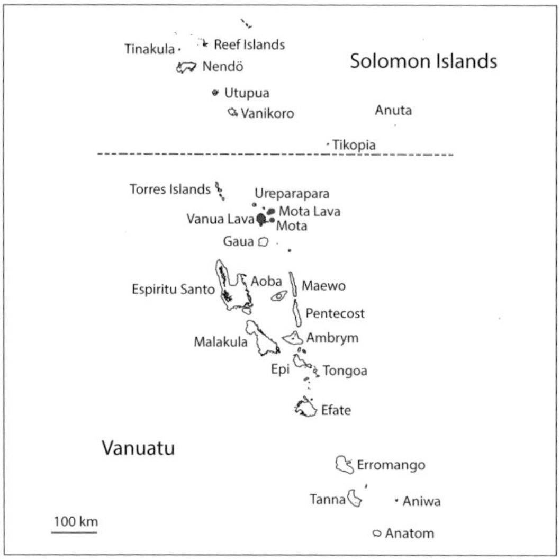

Pteropus fundatus			
Size*			
	Mean	**Range**	**Sample**
Head–body	136.2	124.6–152.0	8
Tail	–	–	–
Forearm	102.7	99.6–107.6	8
Tibia	47.0	42.0–50.1	8
Hindfoot	–	–	–
Ear	15.1	10.9–17.8	8
Weight	155.7	69.5–224.0	8
* From AM specimens and field measurements (Mota)			
Identification The mantle is orange and does not sharply contrast with the greyish brown back. The chin and muzzle are often silver. The Banks flying fox is mostly likely to be confused with *P. anetianus*. It can be separated from this species by its smaller size and much finer cheek teeth.			
Synonyms None			
Subspecies None			
Distribution Mota, ?Mota Lava, ?Vanua Lava			
Extralimital None			
Conservation status (IUCN) Endangered			
Etymology Sp.–to lay a foundation (L.), perhaps a reference to basal morphology			

Pteropus fundatus has highly reduced cheek teeth and probably favours nectar and soft fruits in the diet. Animals recorded on Mota by the Australian Museum in 1992 were in plantations and gardens near Latawora Village, and between Squea and Veverau villages. One was feeding on flowers of a naveli tree (*Barringtonia edulis*), and a group of four were among about two dozen bats that appeared in the area at around 6 p.m. to feed on coconut inflorescences (*Cocos nucifera*) until around 7:30 p.m. They then dispersed gradually and no more were seen after 8:30 p.m. Residents on Mota recognised coconut flowers and the flowers of coastal *Pandanus* species as primary food sources. The Banks flying fox was said to roost in small groups beneath the dense foliage of *Pandanus* growing along Mota's steep coastlines (Lavery *et al.* 2020d).

Two of four females captured in June 1991 were pregnant. Two females caught in early May 2018 were nulliparous. The Banks flying fox is not a major target for hunters on Mota and numerous captures in close proximity to villages are evidence that it is probably tolerant of some level of disturbance to primary forests. Nonetheless, the extremely small distribution leaves it vulnerable to natural disasters (especially cyclones), disease and larger-scale habitat disturbances.

GILLIARD'S FLYING FOX

Pteropus gilliardorum Van Deusen, 1969

gill'-ee-ard-or'-um

Gilliard's flying fox is a member of the *pelagicus* species group (Almeida *et al.* 2014) that has rarely been documented by biologists. The holotype was collected in 1958 in a mist net at 1600 m elevation in the Whiteman Mountains of West New Britain and is held at the American Museum of Natural History (Van Deusen 1969). The United States National Museum of Natural History, Smithsonian Institution, lists three specimens held in the collection, all captured in February 1994 from the Hans Meyer Range on New Ireland. A fifth individual was captured in 1995 by M. Ainie and F. Bonaccorso in East New Britain and is held at the PNG National Museum.

Habitat at the type locality consisted of dissected limestone ridges, supporting closed forest with dense moss and containing *Rhododendron*, scrambling bamboo (*Racemobambos* sp.) and a species of *Pandanus* (Van Deusen 1969). The capture of an adult male at an elevation of 200 m in primary forest near Mu village, East New Britain, in 1995, shows Gilliard's flying fox is not entirely restricted to higher elevations. However, numerous other expeditions to New Britain's lowlands have failed to record this species, and it must certainly prefer montane habitats.

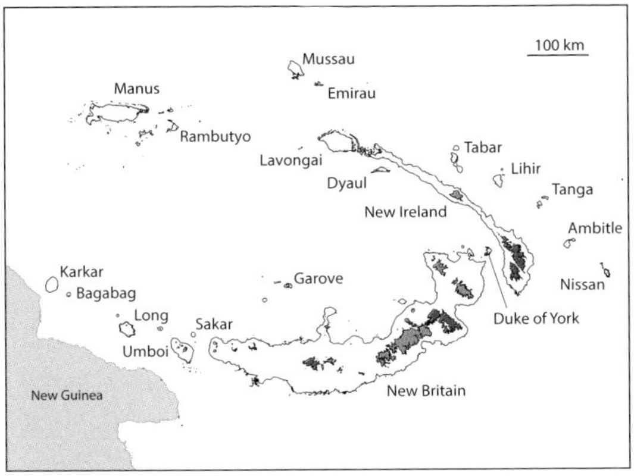

Pteropus gilliardorum			
Size*			
	Mean	**Range**	**Sample**
Head–body	167.3	154.0–178.0	3
Tail	–	–	–
Forearm	117.3	114.0–120.6	2
Tibia	52.3	46.0–56.9	3
Hindfoot	36.7	34.0–40.0	3
Ear	22.3	20.0–25.0	3
Weight	286.0	172.0–400.0	2
* From Bonaccorso (1998) including subadults			
Identification A small, dark brown and thickly furred flying fox. The forearm length is generally < 120 mm, cheek teeth are very reduced, and ears are broad and with rounded tips. The first premolar is minute and the second large on both lower and upper jaws.			
Synonyms None			
Subspecies None			
Distribution New Britain, New Ireland			
Extralimital None			
Conservation status (IUCN) Vulnerable			
Etymology Sp.–named in honour of Dr and Mrs Gilliard, who undertook the expedition during which the holotype was collected. The specific name has been amended in accordance with article 31 of the International Code of Zoological Nomenclature.			

The reduced cheek teeth indicate this species primarily forages on soft fruits and nectar. On New Ireland, L. Emmons was able to attract subadult males into a mist net by using an Audubon bird caller (Bonaccorso 1998). Nothing is known of reproduction; the animals caught in February 1994 were all subadults, as was the holotype captured in December 1958.

The habitat preference for montane habitats means Gilliard's flying fox may be somewhat removed from the impacts of widespread deforestation and conversion to palm oil plantations that has occurred across New Britain's lowlands. However, the distribution is small: it apparently occurs at low densities and thus overall population sizes must be relatively small. Gilliard's flying fox is classified as Vulnerable on the IUCN Red List and further population data are needed.

ISLAND FLYING FOX

Pteropus hypomelanus Temminck, 1853

hie'-poh-mel'-ah-nus

The island flying fox is an extremely widespread and variable member of the *griseus* species group (Almeida *et al.* 2014), encompassing 16 recognised subspecies (Tsang 2019a). The subspecies *Pteropus hypomelanus luteus* (found in the region under consideration here) has a pale brown back and ventrum, with reddish-brown surrounding the front and sides of the neck. The island flying fox prefers smaller islands and is thus more typically recorded from the smaller satellite islands adjacent to New Guinea. It is apparently an infrequent visitor to mainland New Guinea and larger islands, including those of the Bismarck Archipelago.

This is the most common *Pteropus* of the islands of Milne Bay. The species forms small roosts consisting of a few animals, or large colonies, often in coastal locations. One colony on Raboin Island, Milne Bay, was estimated to contain around 5000 individuals (Bonaccorso 1998). On some of the islands of Milne Bay, it is also known to form mixed roosts with *P. conspicillatus*.

This species feeds on a wide variety of fruits, flowers and nectar from both forested and agroforestry habitats. In peninsular Malaysia, plants from at least 14 families were found to contribute to the diet, but figs (*Ficus* sp.) were consistently a dominant food for most of the year (Aziz *et al.* 2017). Few observations of feeding have been reported for the Pacific Islands. A group of around 20 individuals was seen in March 1992 feeding on the blossoms of *Eucalyptus degulpta* around Nagada Harbour, Madang Province (Bonaccorso 1998).

The island flying fox (*Pteropus hypomelanus*) is common on many of New Guinea's satellite islands. It is a highly variable species with 16 subspecies described throughout its range (Pavel German).

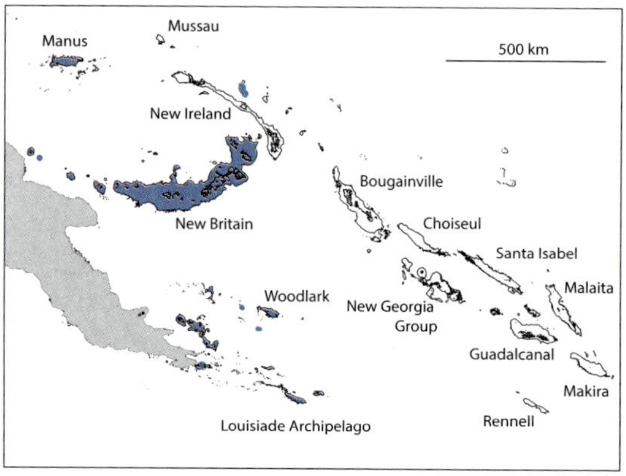

Pteropus hypomelanus			
Size*			
	Mean	**Range**	**Sample**
Head–body	200.8	176.0–228.0	8
Tail	–	–	–
Forearm	131.4	122.0–146.0	11
Tibia	66.1	59.0–72.0	3
Hindfoot	43.3	33.0–50.0	4
Ear	27.2	24.0–33.0	9
Weight	405.7	300.0–550.0	8

* From AM specimens (Normanby and Woodlark, and Milne Bay), and Bonaccorso (1998)

Identification A medium-sized flying fox of variable appearance with broad ears and a yellowish mantle. Subspecies *luteus*, which is found in the region under consideration here, has a paler brown back and ventrum, with reddish-brown surrounding the front and sides of the neck. Among the islands of eastern Papua New Guinea, it is most likely to be confused with *P. admiralitatum*. It differs from *P. admiralitatum* in being a lighter brown, and from *P. solomonis* by the possession of a naked rather than a furred tibia.

Synonyms
Pteropus tricolor Gray, 1870
Pteropus hypomelanus vulcanius Thomas, 1915a

Subspecies *Pteropus hypomelanus hypomelanus* Temminck, 1853 (North Maluku)
Pteropus hypomelanus condorensis Peters, 1869 (satellite islands of the Gulf of Thailand)
Pteropus hypomelanus tomesi Peters, 1869 (satellite islands of north-west Borneo)
Pteropus hypomelanus macassaricus Heude, 1896 (Sulawesi, Talaudm Sangihe Isls, Indonesia)
Pteropus hypomelanus lepidus Miller, 1900 (satellite islands of eastern peninsular Malaysia and south-west Borneo)
Pteropus hypomelanus geminorum Miller, 1903 (satellite islands of southern Myanmar, south-westThailand, western peninsular Malaysia)
Pteropus hypomelanus cayaganus Mearns, 1905 (Philippines)
Pteropus hypomelanus enganus Miller, 1906 (Mentawai Is.)
Pteropus hypomelanus annectans Andersen, 1908 (South Natuna Is.)
Pteropus hypomelanus canus Andersen, 1908 (North Natuna Is.)
Pteropus hypomelanus luteus Andersen, 1908 (north-east New Guinea, Admiralty Isls, Bismarck Archipelago, D'Entrecasteaux, Louisiade and Trobriand Groups)
Pteropus hypomelanus satyrus Andersen, 1908 (Andaman Isls)
Pteropus hypomelanus robinsoni Andersen, in Thomas and Wroughton 1909a (Rumbia Is., Straits of Malacca, Indonesia)
Pteropus hypomelanus fretensis Kloss, 1916 (Jarak Is., Straits of Malacca)
Pteropus hypomelanus simalurus Thomas, 1923c (satellite islands of north-west Sumatra)
Pteropus hypomelanus maris Allen, 1936 (Maldives Isls)

Distribution Alcester, Bagabag, Conflict Group, Fergusson, Goodenough, Hull, Ito, Karkar, Kiriwina, Long, Manam, Manus, Marshall Bennett Isls, Misima, New Britain, Normanby, Rossel, Sideia, Sudest, Tabar, Ulu, Umboi, Woodlark (Papua New Guinea)

Extralimital India, Indonesia, Malaysia, Maldives, Myanmar, New Guinea, Philippines, Thailand, Vietnam

Conservation status (IUCN) Near Threatened

Etymology Sp.–not very black (L.)

The Fifth Archbold Expedition in 1956 recorded mothers with young on Fergusson and Sudest islands on 4 July and 20 August respectively, and a female on Woodlark Island was lactating on 15 August 1987. Females in a large roost on Raboin Island had dependent young on 19 May 1996 (Bonaccorso 1998).

LARGE-EARED FLYING FOX

Pteropus macrotis Peters, 1867

mak-roh'-tis

The large-eared flying fox belongs to the *pelagicus* species group (Almeida *et al.* 2014). The type locality was originally given as Buru, Indonesia (Peters 1867), but it was in fact collected from Wokam in the Aru Islands by von Rosenberg on 7 April 1865 (Andersen 1912a). This is the only record in the Aru Group. Two individuals were recorded on Salawati in October 1992.

The large-eared flying fox prefers drier habitats than its congeners. It is an inhabitant of the drier coastal lowland forests of New Guinea and the Aru Islands, including monsoon forests and savannah woodlands as well as swamp forest, gardens, and primary and secondary rainforests (Ziegler 1982; Bonaccorso 1998). On Salawati it does occur in moist rainforests, although there is a marked dry season there. Three animals recorded by McKean (1972) on the Papuan Peninsula were feeding on the flowers of coconut palms (*Cocos nucifera*) in a region supporting dryish monsoon forest. The two individuals recorded on Salawati 1992 were taken

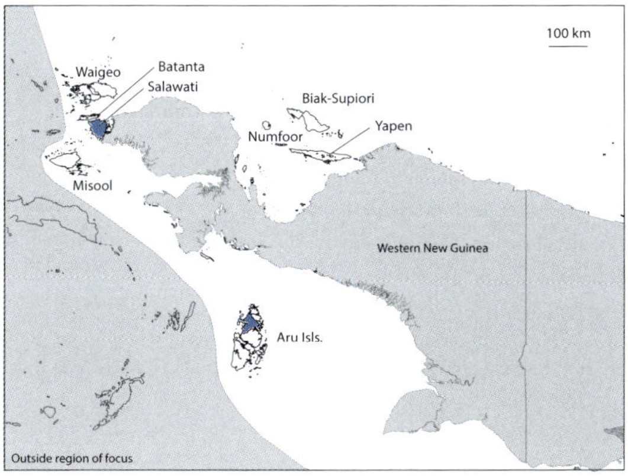

Pteropus macrotis			
Size			
	Mean	**Range**	**Sample**
Head–body	211.3	179.0–244.0	8
Tail	–	–	–
Forearm	135.8	125.0–157.0	10
Tibia	63.9	50.0–70.0	6
Hindfoot	46.5	36.2–55.0	6
Ear	33.6	28.0–37.0	7
Weight	330.4	214.0–480.0	9
* From AM specimens (Salawati), Flannery (1995a) (Sandaun Province, Papua New Guinea), and Bonaccorso (1998)			
Identification A brightly coloured, medium-sized flying fox, with very long ears for its size (the best diagnostic feature), a warm-brown back, and strongly contrasting pale yellow mantle.			
Synonyms None			
Subspecies *Pteropus macrotis macrotis* Peters, 1867 *Pteropus macrotis epularius* Ramsay, 1877e			
Distribution Aru Isls, Salawati			
Extralimital New Guinea			
Conservation status (IUCN) Least Concern			
Etymology Sp.–large ear (Gr.)			

in a mist net set next to a fruiting jackfruit tree (*Artocarpus heterophyllus*).

Limited data are available on reproduction on Salawati or the Aru Islands. Of the two animals examined in October 1992, one of these (a male) was a subadult around 10 months of age. The second was an adult female that was not lactating. In Milne Bay, Papua New Guinea, large-eared flying foxes have been observed copulating in early April (Brass 1956). A pregnant female was caught on the Fly River on 28 October, and a mother with dependent young was caught in Papua New Guinea's Central Province in January (Bonaccorso 1998). The large-eared flying fox has a wide distribution and a large population, and is unlikely to be declining at the rate required to qualify for a threatened category listing.

SANBORN'S FLYING FOX

Pteropus mahaganus Sanborn, 1931

mah-hah'-gan-us

Pteropus mahaganus was named following an Indigenous name for Santa Isabel, Solomon Islands, (Mahaga) by Sanborn (1931) and belongs to the *pelagicus* species group (Almeida *et al.* 2014). Records from the Australian Museum include individuals from Santa Isabel (1995) and Bougainville (2015), and the Queensland Museum holds two specimens collected for the first time on Choiseul in 2014.

The external appearance of *P. mahaganus* is strikingly different to *P. woodfordi*, being larger and rich golden buff in colour. However, these two allopatric species are sister taxa and share similar reduced dental morphology. *Pteropus mahaganus* occupies Bougainville, Choiseul and Santa Isabel islands, where *P. woodfordi* is entirely absent.

Sanborn's flying foxes can be found in primary and secondary forests, the borders between village gardens and forest, and ultramafic forests dominated by ironwood (*Xanthostemon melanoxylon*). It has been encountered from sea level to 400 m and 600 m above sea level on Santa Isabel and Choiseul respectively. This species probably roosts individually or in small groups, in hollow trees in primary forests (Bowen-Jones *et al.* 1997).

Sanborn's flying fox evidently consumes large amounts of nectar, with the diet being somewhat similar to *Pteropus woodfordi*. Sanborn (1931) reported it feeding on young green coconuts (*Cocos nucifera*). However, Bowen-Jones *et al.* (1997) suggested it is in fact the inflorescences rather than young fruit that are preferred. It is an important

Sanborn's flying fox is a sister species to *Pteropus woodfordi* of the southern Solomon Islands (New Georgia group, Guadalcanal and Malaita). It is regularly recorded on surveys using mist nets on Bougainville, Choiseul and Santa Isabel, but is rarely captured in large numbers (Tyrone Lavery).

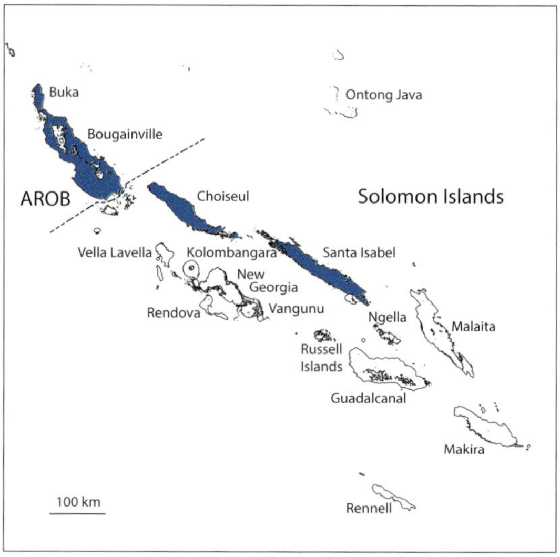

Pteropus mahaganus			
Size*			
	Mean	**Range**	**Sample**
Head-body	211.5	190.0–235.0	8
Tail	–	–	–
Forearm	143.6	126.0–153.3	8
Tibia	68.9	64.0–73.0	8
Hindfoot	45.0	39.9–50.0	4
Ear	23.3	21.3–25.0	8
Weight	430.6	325.0–520.0	8
* From AM specimens (Santa Isabel) and Bonaccorso (1998)			
Identification Cheek teeth reduced, lower first premolar similar in size to second lower molar. Pelage predominantly buff with grey bases, head warm buff, mantle and foreneck light buff.			
Synonyms None			
Subspecies None			
Distribution Bougainville, Buka, Choiseul, Santa Isabel, Tanabuli			
Extralimital None			
Conservation status (IUCN) Least Concern			
Etymology Sp.–from Mahaga (L.), a local name for Santa Isabel			

pollinator of sago palm (*Metroxylon salomonense*) flowers. Little is known about reproduction; a lactating female and a juvenile were identified on Santa Isabel in March 1991.

Bowen-Jones *et al.* (1997) found Sanborn's flying fox to be fairly common on Choiseul: it was caught on numerous occasions across many sites and observed on many other occasions. It remains relatively common on Santa Isabel, Choiseul and Bougainville; however, it is rarely caught in mist nets in large numbers. Sanborn's flying fox may be reliant on primary forests, including old growth hollow trees for roosting. Commercial logging (widespread on Choiseul and Santa Isabel) is thus probably a major threat to the species. Hunting via the use of local snares set on coconut palms is also a threat (Bowen-Jones *et al.* 1997).

MARIANA FLYING FOX

Pteropus mariannus Desmarest, 1822

mar'-ee-an'-us

The Mariana flying fox belongs to the *griseus* species group (Almeida *et al.* 2014). The smaller-bodied *Pteropus tokudae*, once sympatric on Guam, is now extinct and the Mariana flying fox remains the only surviving member of the genus *Pteropus* in the Marianas. It is distributed across almost all Mariana Islands from Guam to Maug, the northernmost island of Farallon de Pajaros being the only island not known to support this species. *Pteropus mariannus* formerly included *P. pelewensis*, *P. yapensis*, *P. ulthiensis*, *P. ualanus*, *P. paganesis* and *P. loochoensis* as subspecies. However, a molecular phylogeny prepared by Almeida *et al.* (2014) classified four of these as different species (*P. pelewensis* Palau, *P. pelewensis yapensis* Yap, *P. ulithiensis* synonymised with *P. pelewensis yapensis*, and *P. ualanus* Kosrae). *Pteropus loochoensis* was retained with *P. mariannus* without rationale. Wiles and Glass (1990) argued oceanic divides in the Marianas were unlikely to facilitate multiple subspecies, and the sample sizes of Yamashima (1932) were too small to justify recognition of the subspecies *P. m. paganensis* (Pagan and Alamagan), a conclusion we have followed here. Moreover, gene flow between various islands of the Marianas Archipelago indicates *P. m. paganensis* is unlikely to be a valid subspecies (Brown *et al.* 2011).

The Mariana flying fox congregates in colonies in areas of native forest on both limestone and volcanic soils, and on coastal strand. Animals roost along cliffs where human disturbance is minimal and there is some protection from winds (Wiles 1992b). Some of the tree species used most commonly for roosting include *Elaeocarpus sphaericus*, *Macaranga thompsonii*, *Guettarda speciosa*, *Ficus prolixa* and *Mammea odorata* (Wiles 1987a; Glass and Taisacan 1988).

The known diet of Mariana flying fox includes fruits, flowers and nectar (Wiles 1987a, 1992a). Over 30 species of food plant have been identified from evidence such as direct observations, discarded fruit, examination of faeces or interviews with island residents. Among favoured foods are the fruits of *Artocarpus mariannensis*, *A. altilis*, *Pandanus tectorius*, *Cycas circinalis*, *Mammea odorata*, *Ficus prolixa* and *Terminalia catappa*, and the flowers of *Ceiba pentandra*, *Erythrina variegata* and *Cocos nucifera*. On Sarigan Island, which supports a relatively impoverished flora, the Mariana flying fox was recorded feeding on just six species of plant (Wiles and Johnson 2004).

Reproduction is aseasonal, and copulation and lactation have been reported throughout the year (Wiles 1987a). The species shows substantial variation in roosting behavior across its range. When in large roosting colonies, the Mariana flying fox will form bachelor groups, harems containing a male and 2–15 females, or remain at the periphery roosting individually. The average sex ratio for a single colony on Guam varied from 37.5 to 72.7 males per 100 females. Young are present year-round with no consistent peak in births. Approximately 7–20% of all harem females possess small to medium-sized young at any one time (Wiles 1992b). Roosting behaviour on Sarigan differs from populations on larger islands. On Sarigan the animals roost individually or form smaller groups typically numbering from 3 to 75 individuals (Wiles and Johnson 2004). Bats roosting solitarily were estimated to make up half of the population. Where animals aggregated into groups, these usually consisted of harems with multiple females. In July

1999 some 30–50% of females possessed young (Wiles and Johnson 2004).

The Mariana flying fox is a delicacy throughout the Marianas and by the 1930s population declines were evident as a result of overhunting with introduced firearms (Coultas 1931). By 1945 the species had become more uncommon and had retreated to Guam's more inaccessible cliff lines (Wiles 1987b). In 1958, species numbers on Guam were considered greatly reduced, with no more than 3000 individuals remaining. By 1983–1984, the extent of declines across the species distribution was stark, fewer than 10 bats remained on Aguiguan, estimates on Guam ranged from 425–500 individuals, 800–1000 remained on Rota, fewer than 50 were found on Saipan, and fewer than 25 were to be found on Tinian (Wiles *et al.* 1989; Wiles and Glass 1990; Stinson *et al.* 1992; Wiles 1992b). More recent estimates indicate populations on individual islands have increased, decreased, and remained relatively stable: for example, Aguiguan (increase from fewer than 10 to 150–200 individuals), Guam (decrease from 425–500 to 119–179 individuals) (USFWS 2009).

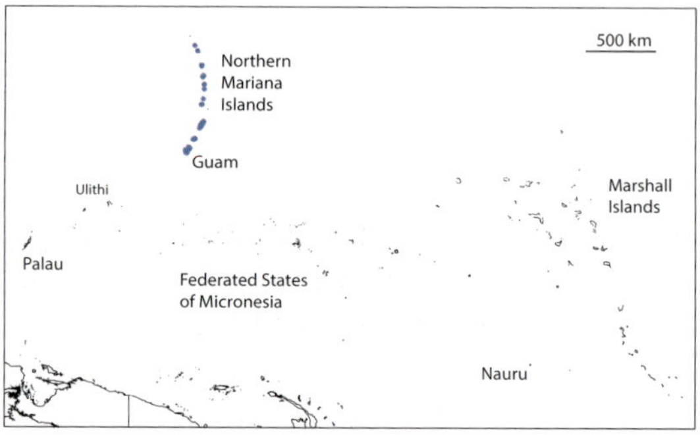

Pteropus mariannus			
Size*			
	Mean	**Range**	**Sample**
Head–body	206.8	140.0–344.0	28
Tail	–	–	–
Forearm	138.2	110.0–165.0	35
Tibia	–	–	–
Hindfoot	45.1	33.0–54.0	20
Ear	24.7	20.0–28.0	26
Weight	381.8	200.0–577.0	13
* From Andersen (1912a); Perez (1972); Brown *et al.* (2011) and AMNH and USNM specimens			
Identification A medium-sized *Pteropus* with the back and rump blackish seal-brown, sprinkled with silver greyish-white hairs. The belly breast and flanks are seal-brown mixed with silvery hairs. The mantle and side of the neck are ochraceous buff.			
Synonyms *Pteropus keraudren* Quoy and Gaimard, 1824 *Pteropus vanikorensis* Quoy and Gaimard, 1830 *Pteropus mariannus paganensis* Yamashima, 1932			
Subspecies *Pteropus mariannus mariannus* Desmarest, 1822 (Marianas) *Pteropus mariannus loochoensis* Gray, 1870 (Okinawa Isls, Japan)			
Distribution Agrihan, Aguiguan, Alamagan, Anatahan, Asuncion, Farallon de Medinilla, Guguan, Maug, Pagan, Rota, Saipan, Sarigan, Tinian (Northern Mariana Isls); Guam			
Extralimital Okinawa Isls, Japan			
Conservation status (IUCN) Endangered			
Etymology Sp.–for the Mariana Isls (L.), the type locality			

Mariana flying fox (*Pteropus mariannus*), Rota, Commonwealth of the Northern Mariana Islands. This species is today the sole surviving member of the genus *Pteropus* that exists in the Marianas (Keiko and Yushi Osawa).

Significant impact on this species was wrought via the introduction of the brown tree snake (*Boiga irregularis*) to Guam after the Second World War. Monitoring from 1984–1988 indicated snake predation was such that almost no juvenile bats survived beyond 1–2 months of age (Wiles 1987a). Illegal hunting, habitat destruction and degradation, major storm events and volcanic activity are all important threats.

POHNPEI FLYING FOX

Pteropus molossinus Temminck, 1853

mol'-oss-een'-us

The Pohnpei flying fox is a member of the *pelagicus* species group and is closely related to *Pteropus pelagicus* and *P. tokudae* of Micronesia and *P. mahaganus*, *P. woodfordi*, and *P. gilliardorum* of the Bismarck Archipelago and Solomon Islands (Almeida *et al.* 2014). The Pohnpei flying fox is recorded only from Pohnpei and nearby Ant and Pakin atolls (Wiles 1992c; Buden 1996a; Buden 1996b). On occasion, the Mortlock Islands have been identified as part of the distribution of this species (e.g. Andersen 1912a). However, this is derived from a single specimen (the type specimen for the synonym *Pteropus breviceps* Thomas 1883) and is now regarded as erroneous (Rainey and Pierson 1992; Buden *et al.* 2013). It was not detected in the Mortlock Islands when surveyed by Buden *et al.* (2013).

Bruner and Pratt (1979) and Jackson (1962) note that this species does not usually form large roosts; instead a large proportion roost as individuals or in pairs outside of colonies. One large colony of 200–300 individuals was observed in forest on top of Sokeh's Rock in 1981 (Wiles 1992c). Females give birth to a single young; unweaned young have been observed on Pohnpei in February, September and November (Jackson 1962).

The Pohnpei flying fox feeds on the fruits of *Pandanus*, endemic palms (*Clinostigma* sp.) and *Exhorrhiza* sp. It also reportedly feeds on the flowers of coconuts (*Cocos nucifera*) and as many as 150–200 individuals have been seen feeding on the flowers of a single kapok tree (*Ceiba pentandra*) (Jackson 1962). Wiles (1992c) commented that many areas at lower elevations on Pohnpei have now been converted to agroforests containing breadfruit (*Artocarpus altilis*), mango (*Mangifera indica*) and bananas (*Musa* spp.). This conversion of native forests to agroforests or secondary forest has probably reduced roosting habitats and influenced the foraging behaviour of the bats.

Prior to the 1980s, the Pohnpei flying fox was described as being common at all elevations (Jackson 1962; Bruner and Pratt 1979; Wiles and

The Pohnpei flying fox (*Pteropus molossinus*) suffered a significant population decline during the 1980s from overhunting for export to Guam (Thibaud Aronson).

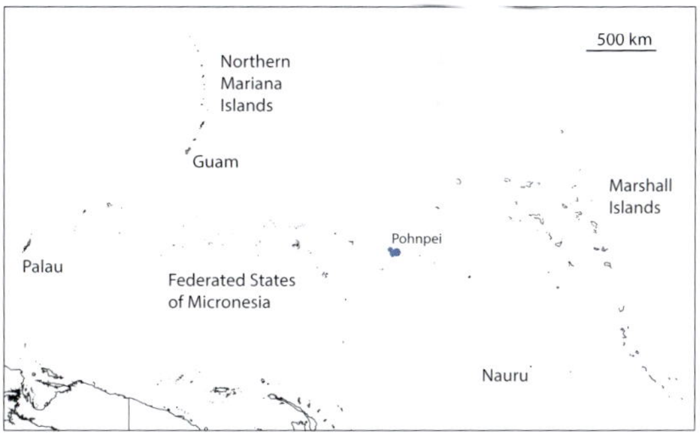

Pteropus molossinus			
Size*			
	Mean	**Range**	**Sample**
Head–body	141.9	105.0–160.0	42
Tail	–	–	–
Forearm	95.7	90.2–100.0	10
Tibia	44.5	43.0–45.7	5
Hindfoot	32.9	27.0–41.0	43
Ear	17.1	14.5–19.5	45
Weight	142.3	69.7–196.2	25
* From Thomas (1883), Andersen (1912a) and AMNH and USNM specimens			
Identification A small flying fox which is brown in overall colour, with reddish-brown on the back and rump.			
Synonyms *Pteropus breviceps* Thomas, 1883			
Subspecies None			
Distribution Ant Atoll, Pakin Atoll, Pohnpei			
Extralimital None			
Conservation status (IUCN) Endangered			
Etymology Sp.–resembling Molossus (L.), allusion unclear as its appearance is quite different from insectivorous bats of the genus *Molossus*.			

Payne 1986). However, by 1990 a significant population decline was clearly evident (driven by the development of a commercial trade in shipments of bats to Guam and Commonwealth of the Northern Mariana Islands) (Rainey 1990). Between 1979 and 1989, 16 210 bats were exported from Pohnpei, comprising a peak of 7164 animals in 1989 (Wiles 1992c,d; Stinson *et al.* 1992). The Pohnpei flying fox was again common on Pohnpei in the late 1990s (Waldien *et al.* 2020), and Buden (1996a, 1996b) reported that the species was again common on Ant Atoll in 1994–95, but uncommon on Pakin Atoll in 1994 (estimated at 100 individuals). Population censuses are needed across the entire species distribution.

Commercial harvest of the Pohnpei flying fox for export ceased in 1990 and no longer threatens the species (Wiles and Brooke 2009). Degradation of lowland forested habitat and conversion for agriculture (including kava *Piper methysticum*) and climate change are now major threats and the Pohnpei flying fox remains classified as Endangered on the IUCN Red List (Waldien *et al.* 2020).

GREAT FLYING FOX

Pteropus neohibernicus Peters, 1876

nay'-oh-hie-ber'-nik-us

The great flying fox is a member of the *grisesus* species group (Almeida *et al.* 2014). It is an extremely large bat (up to 1.6 kg in weight) and is conspicuous wherever it occurs. It was abundant on New Ireland and New Britain in 1979 (Smith and Hood 1981), and in 1988. In 1988 it was extremely rare on Manus Island, and local people reported the population had been destroyed by disease in 1985. Over a few weeks almost all of the flying foxes on the island died, and dead and dying animals were found under well-known roosts. For several years afterwards, none was seen at all, but by 1988 the odd individual was again seen. More recently, individuals of this species were observed on Manus during surveys at Piri Village in 2014 (Aplin *et al.* 2015). However, it was not encountered on Mussau Island (Aplin *et al.* 2015), nor in the Nakanai Mountains of New Britain during surveys in 2009 (Aplin and Opiang 2011). It was similarly not seen on Mussau in 1979 (Smith and Hood 1981) and is therefore presumably absent there. The great flying fox survives on small islands such as Karkar and Sakar; thus its absence from many similar size islands in close proximity to New Guinea (e.g. D'Entrecasteaux Islands) is puzzling.

The great flying fox roosts by day in large colonies in tall, somewhat denuded trees near human habitation or on the seacoast. The largest of these colonies, recorded in the Bismarcks by Smith and Hood (1981), contained 500–600 individuals. Its wing beats are extremely loud when circling overhead in search of food (Flannery 1995a). Two individuals tracked with satellite transmitters on southern New Guinea near the Fly River and Wasi Kussa River travelled total distances of 255 and 960 km over periods of 62 and 120 days respectively

A camp of great flying foxes (*Pteropus neohibernicus*) on New Britain, Papua New Guinea (Keiko and Yushi Osawa).

The great flying fox (*Pteropus neohibernicus*) is the largest flying fox of the region and has a wingspan among the largest of all bats (John Irvine).

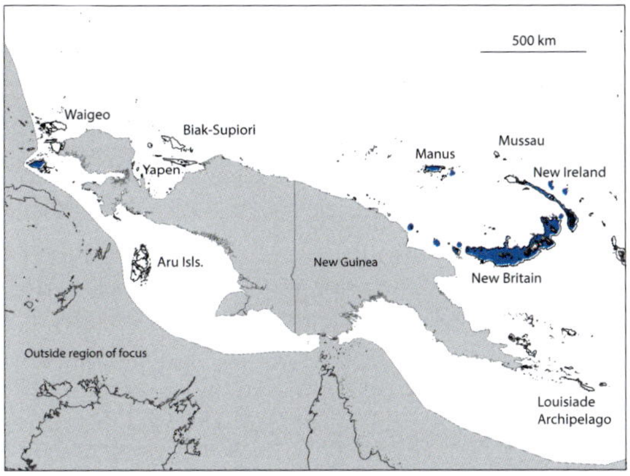

Pteropus neohibernicus			
Size*			
	Mean	**Range**	**Sample**
Head–body	278.0	210.0–425.0	14
Tail	–	–	–
Forearm	186.3	165.0–224.0	16
Tibia	86.9	76.0–100.0	10
Hindfoot	55.4	45.0–78.0	11
Ear	24.7	20.0–31.0	15
Weight	1092.5	630.0–1600.0	12
* New Ireland and Manus specimens, and including measurements from Bonaccorso (1998)			
Identification This is the largest flying fox of the region. The skin is black and the fur on the back is very sparse, appearing almost naked. The colour is usually golden brown overall but is quite variable. The muzzle is very long and narrow compared to other species.			
Synonyms *Pteropus degener* Peters, 1876 *Pteropus melanopogon papuanus* Peters and Doria, 1881 *Pteropus coronatus* Thomas, 1888d *Pteropus rufus* Ramsay, 1891 *Pteropus sepikensis* Sanborn, 1931			
Subspecies *Pteropus neohibernicus neohibernicus* Peters, 1876 (New Guinea, Bismarck Archipelago, Raja Ampat Is.) *Pteropus neohibernicus hilli* Felten, 1961 (Manus)			
Distribution Misool (Indonesia); Duke of York, Karkar, Lihir, Long, Manus, Mioko, New Britain, New Ireland, Rambutyo, Sakar, Tabar, Umboi (Papua New Guinea)			
Extralimital New Guinea			
Conservation status (IUCN) Least Concern			
Etymology Sp.–of New Ireland (L.), the type locality			

(Breed *et al.* 2010). The maximum radial distances travelled from the release site were 31.2 and 73.1 km. One of these animals stayed in mangrove forests and the second remained in savannah and riparian habitats.

Smith and Hood (1981) observed great flying foxes eating kapok (*Ceiba pentandra*) and noted that the bats had stripped the fruit from a *Ficus* sp. within three nights. It also feeds on the fruits of sea almonds (*Terminalia* sp.) and Alexandrian laurels (*Calophyllum* sp.), and the blossoms of *Eucalyptus deglupta*, coconuts (*Cocos nucifera*) and *Trema orientalis* (Bonaccorso 1998).

Females carrying half-grown young have been seen in June in the Bismarck Archipelago (Smith and Hood 1981). An Australian Museum expedition to the Fulleborn area of New Britain found that on 24 December 1988 most females in a colony were carrying young. Four juveniles examined then weighed 180, 190, 210 and 240 g. Births must therefore occur in this area sometime in early December. The great flying fox is widespread and locally common, but could be threatened by localised hunting.

TEMOTU FLYING FOX
Pteropus nitendiensis Sanborn, 1930
nit-end'-ee-en'-sis

The Temotu flying fox is a member of the *samoensis* species group (Almeida *et al.* 2014). It has a highly restricted distribution, occurring only on Nendö, Malo and Tinakula in the remote Temotu Province of Solomon Islands (Lavery *et al.* 2020d). Sanborn (1930) described this species on the basis of two specimens collected from Nendö during the Whitney South Sea Expedition and Troughton (1930) reported a series of seven individuals collected between July and August 1926.

During mist-net surveys on Nendö between 1992 and 2015, the Temotu flying fox was found to be moderately common and the species was encountered with regular frequency in gardens, primary and secondary forests (Lavery *et al.* 2020d). It is primarily a solitary species, roosting individually or in pairs in the mid-canopy where upper branches afford good overhead cover. Roost sites are often on food trees such as coconut (*Cocos nucifera*), mango (*Mangifera indica*) and breadfruit (*Artocarpus altilis*), or on trees adjacent to a food tree. Temotu flying foxes are often seen foraging in mid and late afternoon on coconut and fan palm flowers, and the fruits of roost trees, Malay apple (*Syzygium malaccense*) and cutnut (*Barrintonia edulis*). At least 44 species of plant contribute to the diet in the form of fruits, flowers, nectar or leaves (Lavery *et al.* 2020d).

In April 2018, this species was roosting together in pairs and local residents identified these as male and female mating pairs (C. Posala, pers. comm.). Moreover, when calls of the bats were mimicked, the animals would respond in a territorial manner, flying in from the forest to circle low underneath the canopy and respond to the calls before returning to their roost (C. Posala, pers. comm.). This same behaviour was not observed in September–October of 2014–15 and may be seasonal. Monogamous social organisation is relatively rare in bats but has been reported for *Pteropus samoensis* (McCracken and Wilkinson 2000). Little else is known of breeding biology, but one of two female specimens collected from Nangu village on Nendö in early October 1994 was pregnant.

Temotu flying fox (*Pteropus nitendiensis*), Nendö, Solomon Islands (Ray Pierce).

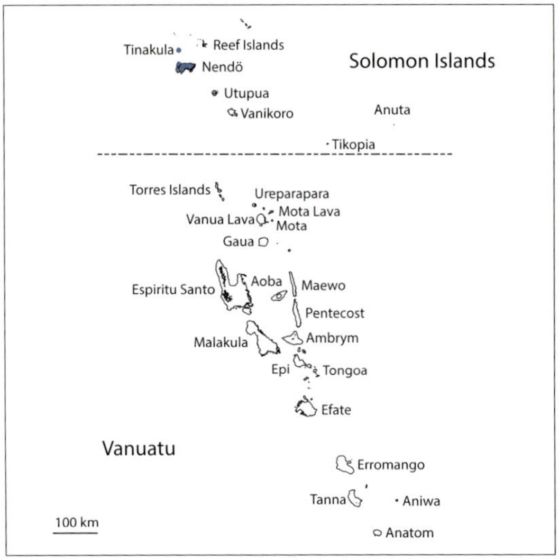

Pteropus nitendiensis			
Size*			
	Mean	**Range**	**Sample**
Head–body	173.0	171.0–175.0	2
Tail	–	–	–
Forearm	118.5	116.0–121.0	5
Tibia	52.3	49.3–54.3	3
Hindfoot	–	–	–
Ear	19.0	18.1–20.5	3
Weight	265.0	210.0–320.0	2
* From AM specimens (Malo, Nendö) and Sanborn (1930)			
Identification A medium-sized flying fox, with shining golden-brown fur and a rich russet mantle. This colouration will separate it from other flying foxes found in Temotu Province.			
Synonyms *Pteropus sanctacrucis* Troughton, 1930			
Subspecies None			
Distribution Malo, Nendö, Tinakula			
Extralimital None			
Conservation status (IUCN) Endangered			
Etymology Sp.– from Nitendi (L.), a former name for the island of Nendö			

Temotu flying foxes are relatively common in disturbed secondary forests and gardens. However, the species has a small extent of occurrence (less than 5000 km^2), and there are significant threats to forest habitats on Nendö stemming from commercial logging and mining interests. Hunting of this species is also relatively common on Nendö (C. Posala, pers. comm.).

ORNATE FLYING FOX

Pteropus ornatus Gray, 1870

or-nah'-tus

The ornate flying fox is the sole member of the *ornatus* species group (Almeida *et al.* 2014). It is one of three *Pteropus* known from New Caledonia and one of two endemic species that occur there. The Ornate flying fox is found throughout New Caledonia, from sea level to at least 1100 m elevation. Revilliod (1914) reported it was the common flying fox of New Caledonia and the Loyalty Islands in the early 1900s. Sanborn and Nicholson (1950) regarded it likewise; the authors inspected multiple camps, two of which were large and contained around 1500 and 4000 individuals each. Other colonies were smaller and made up of around 300 individuals each. In contrast, *Pteropus vetula* was found to be far more common during field surveys in 1990 and 1991. Few ornate flying foxes were seen flying in the early morning and evening. Of eight flying foxes caught in mist nets in the Thio area of central New Caledonia in April and May 1991, only one was *P. ornatus*, the rest being *P. vetula*. However, *P. vetula* is often in small groups and flies lower than *P. ornatus*, and is probably overrepresented in the results of surveys using mist nets (M. Oedin pers. comm.).

The ornate flying fox aggregates by day in large colonies covering several hectares. These are typically situated in rainforests at the upper ends of gullies on slopes (Sanborn and Nicholson 1950). Tree species that commonly support roosts are banyan figs (*Ficus* spp.), *Aleurites molluccana*, *Semecarpus atra*, *Melaleuca quinquenervia*, *Elaocarpus persifolius* and mangroves. Ornate flying foxes will sometimes form mixed roosts with *P. tonganus*. Both species can be found on the same tree, vertically stratified with *P. tonganus* most commonly at the top (M. Oedin pers. comm.).

The distribution of the ornate flying fox across New Caledonia varies seasonally, moving north in the spring months and returning south in the summer and autumn (Sanborn and Nicholson 1950). The species will return to particular roost sites at different times of the year, and movements might be linked to the seasonal flowering periods of forest trees, *Melaleuca quinquenervia* and cultivated fruit trees (Sanborn and Nicholson 1950).

The ornate flying fox forages nocturnally, and the diet includes the fruits of *Elaeocarpus persifolius*, *Semecarpus atra*, *Syzygium* sp., figs (*Ficus* spp.), guava (*Psidium guajava*), mango (*Mangifera indica*), papaya (*Carica papaya*), passionfruit (*Passiflora* sp.) and banana (*Musa* sp) (Sanborn and Nicholson 1950). The flowers or nectar of *Melaleuca viridiflora* and coconut (*Cocos nucifera*) also are eaten. Like many flying foxes, rather than consuming the entire fruit, the ornate flying fox will extract the juices and eject a wad of the fibrous material onto the forest floor (Sanborn and Nicholson 1950).

Sanborn and Nicholson (1950) estimated the gestation period for this species is around 6 months. Breeding is seasonal, with an absence of pregnancies between November and March. Peak pregnancies occur between June and September and by October lactating females with young are present. The ornate flying fox does not reproduce until it is 2 years old and females give birth to just a single young (Sanborn and Nicholson 1950).

Unfortunately, it seems that the ornate flying fox is a species in decline. Disease was reported to have swept through the population in the 1960s and populations may have been extremely slow to recover since (J. P. Révercé, pers comm). Although the hunting of flying foxes is restricted in New

The ornate flying fox (*Pteropus ornatus*) is one of two *Pteropus* species endemic to New Caledonia (Malik Oedin).

Pteropus ornatus			
Size*			
	Mean	**Range**	**Sample**
Head–body	226.2	180.0–265.0	19
Tail	–	–	–
Forearm	148.9	140.0–167.8	19
Tibia	–	72.2	1
Hindfoot	–	–	–
Ear	21.6	19.5–26.5	20
Weight	375.0	325.0–440.0	4
* From AM specimens (New Caledonia) and Revilliod (1914)			
Identification A medium-sized flying fox, the head and belly are a shade of brown, varying from pale to quite dark, with silver or yellow hairs scattered throughout. The mantle is brownish yellow to white in colour, but always lightest in its central section. Long fur all but obscures the ears.			
Synonyms None			
Subspecies *Pteropus ornatus ornatus* Gray, 1870 (Grande Terre, Île de Pins) *Pteropus ornatus auratus* Andersen, 1909c (Lifou, Maré)			
Distribution Grande Terre, Île de Pines, Lifou, Maré			
Extralimital None			
Conservation status (IUCN) Vulnerable			
Etymology Sp.–ornate (L.), perhaps in reference to the beautiful colouring, particularly the silver or yellow frosting on the fur			

Caledonia, illegal poaching and trade are widespread, and flying foxes are valuable (Oedin *et al.* 2019). An ethno-ecological survey of the Northern Province of New Caledonia found an estimated loss of roost sites of 33% over a 40-year period (Oedin *et al.* 2019). However, annual counts of 30 *P. tonganus* and *P. ornatus* mixed camps between 2010 and 2016 found significant inter-annual variation in the numbers of individuals per roost, but no overall decline during the study period. Annual harvesting rates estimated from these surveys were 5–14% of the population.

CHUUK FLYING FOX
Pteropus pelagicus Kittlitz, 1836
pel'-ay-jee'-kus

The Chuuk flying fox is a member of the *pelagicus* species group (Almeida *et al.* 2014), and is endemic to the islands of Chuuk Lagoon, the Mortlock Islands and Namonuito Atoll in the central Caroline Islands of the Federated States of Micronesia. Taxonomic revisions of flying foxes from the Caroline Islands by Buden *et al.* (2013) resurrected the specific name *Pteropus pelagicus* to replace *P. phaeocephalus* from the Mortlock Islands and treated the formerly monotypic species *P. insularis* of Chuuk Lagoon and Namonuito Atoll as a subspecies. The closet relatives are *P. tokudae* and *P. molossinus* (Micronesia), and *P. gilliardorum*, *P. mahaganus and P. woodfordi* (Melanesia) (Buden *et al.* 2013; Almeida *et al.* 2014).

On the atolls of the Mortlock Islands, the Chuuk flying fox usually roosts individually or in groups of 5–10 individuals, but a group of 27 individuals was encountered on Satawan Island in July 2004 (Buden *et al.* 2013). The species favours closed or nearly closed canopy agroforests dominated by coconuts (*Cocos nucifera*) and breadfruit (*Artocarpus mariannensis*), and which commonly include a variety of native tree species (Buden *et al.* 2013). Preferred roosting sites are outside of human settlements, and the crowns of coconuts and breadfruit are commonly used.

This species has been recorded feeding on a variety of fruits. Ripe breadfruit is a common food item and, on all atolls where Chuuk flying fox is present, these trees are invariably a focus for foraging (Buden *et al.* 2013). Traditional knowledge suggests the fruits of banana (*Musa* sp.), papaya

Chuuk flying fox (*Pteropus pelagicus*), Chuuk Lagoon, Federated States of Micronesia. Populations of this species were heavily impacted by hunting for export to Guam in the 1970s and 1980s (Thibaud Aronson).

(*Carica papaya*), *Calophyllum inophyllum*, *Crataeva speciose*, *Pandanus* cf. *tectorius* and *Ficus tinctoria* are eaten. The Chuuk flying fox has also been observed feeding on coconut flowers (Wiles 1992b).

Mating has been observed in December and births have been recorded April–July. On Satawan Atoll in April 2003, two females examined each had a large young clinging to their ventral surface (Buden *et al.* 2013). In July 2004, an estimated 50% of the bats observed on Namoluk Atoll were females with young, and two of five females collected on 21 and 22 July 2004 were pregnant.

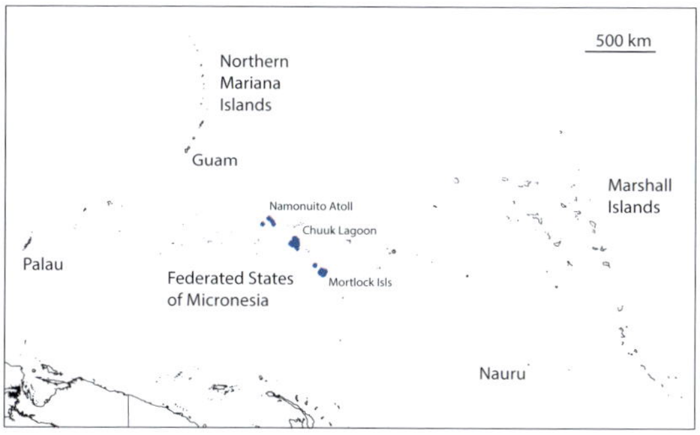

Pteropus pelagicus			
Size*			
	Mean	**Range**	**Sample**
Head–body	158.8	131.0–186.0	18
Tail	–	–	–
Forearm	104.7	90.0–130.0	21
Tibia	47.8	47.0–48.5	2
Hindfoot	32.4	32.0–35.0	9
Ear	21.3	18.0–24.0	17
Weight	176.6	142.0–225.0	11
* From Buden *et al.* (2013), Andersen (1912a) and AMNH specimens			
Identification A small flying fox; the back and rump are dark brown, the centre of the breast bright buff, and the mantle golden-buffy to deep tawny.			
Synonyms *Pteropus phaeocephalus* Thomas, 1883 *Pteropus laniger* Allen, 1890			
Subspecies *Pteropus pelagicus pelagicus* Kittlitz, 1836 (Mortlock Isls) *Pteropus pelagicus insularis* Jacquinot and Pucheran, 1853 (Chuuk Lagoon, Namonuito Atoll)			
Distribution Chuuk Lagoon, Mortlock Isls (?Losap Atoll, ?Nama Is), Namonuito Atoll (?Onoun Is)			
Extralimital None			
Conservation status (IUCN) Endangered			
Etymology Sp.–of the open sea (Gr.), a reference to its habit for travelling between islands.			

In the past, populations of Chuuk flying foxes were heavily impacted by hunting for export to Guam and Commonwealth of the Northern Mariana Islands (Wiles 1992e). On Moen Island in 1976 and 1978, Bruner and Pratt (1979) saw groups of three or four bats flying and noted they were not wary of humans and presumably were not accustomed to hunting. In 1983 colonies estimated to contain around 1000 and 100 individuals were seen at the summits of Tol and Dublon islands respectively (J. Engbring in Wiles 1992b). Wiles (1992a) listed a population estimate of 5628 individuals for Chuuk Lagoon in 1986 (using a count method considered unsuitable) and gave the alarming statistic that, between 1978 and 1989, 5795 individuals were exported to Guam, of which 2507 were exported in 1989 alone (Wiles 1992a). Commercial exports ceased in 1990 after the species was added to the Convention on International Trade in Endangered Species of Wild Fauna and Flora (CITES) (Wiles and Brooke 2009). Field observations in the Mortlock Islands in 2004 estimated the population size to be 925–1200 bats. Most of these occurred on the two largest and southernmost atolls in the chain (Satawan and Lukunor atolls) (Buden *et al.* 2013). The global population is estimated between 1125 and 2250 mature individuals. Threats include sea level rise from climate change and severe cyclones (especially in the atoll portion of the distribution), and habitat degradation in Chuuk Lagoon.

PALAU FLYING FOX

Pteropus pelewensis Andersen, 1908

pel-ue'-en-sis

The Palau flying fox is a member of the *griseus* species group and comprises two subspecies (*Pteropus pelewensis pelewensis*, and *P. p. yapensis*) estimated to have shared a common ancestor around 0.3 million years ago (Almeida *et al.* 2014). Both were subsumed as subspecies of *P. mariannus* (Kuroda, 1938), and treated as such by Koopman (1993), along with *P. mariannus ulithiensis* from Ulithi Atoll. Almeida *et al.* (2014) combined *P. mariannus ulthiensis* with *P. p. yapensis* without rationale, presumably in recognition of morphological similarities (Wiles and Waldien 2020).

Pteropus pelewensis pelewensis is distributed throughout Palau and *P. p. yapensis* occurs on the four main islands of Yap as well as Ulithi. This species has also been recorded in Palau's Southwest Islands, but the subspecies has never been determined (G. Wiles, pers. comm.). Palau flying fox is found in a variety of habitat types, most commonly in upland forest, limestone forest and lowland swamp forest, and less frequently in mangroves, strand forests and savannahs (Wiles *et al.* 1997). On Yap, mangroves are used extensively for roosting, as primary forests on the four main islands have been heavily reduced.

Wiles and Conry (1990) estimated a density of 150–192 bats km^{-2} in the Ngerukewid Islands Wildlife Preserve, with a total population in the reserve of 130–170 bats. On Ulithi Atoll, the population in 1986 was estimated at 1200 bats, corresponding to a density of 280 animals km^{-2} (Wiles *et al.* 1991).

Palau flying fox roosts among Palau's inaccessible vegetation of limestone cliffs, mangroves, and forests of the karst depressions and volcanic substrata (Perez 1968; Wiles *et al.* 1997). It is less gregarious than other closely related flying foxes and infrequently forms large colonies. The animals roost individually, or in small groups of two to four individuals (Wiles and Conry 1990). However, larger colonies of 50–60 or 100–200 bats have been documented, and large numbers of animals were seen dispersing from two roosting areas that were estimated to have held a total of 1500–3000 bats (Wiles *et al.* 1997). During the same study, a regular pattern was observed whereby smaller groups of individuals roosted together for a few hours in the morning and then disbanded, presumably to find individual roosting locations. The authors suggested this variability in camp sizes and behaviour may have been a response to disturbances from hunting or the impacts of severe storms.

The Palau flying fox is nocturnal and peaks in camp departures occur 20–30 min after sunset (Wiles *et al.* 1997). *Pteropus p. yapensis* is known

Yap flying fox (*Pteropus pelewensis yapensis*), Yap, Federated States of Micronesia (Thibaud Aronson).

to feed on the flowers, fruit or leaves of at least 26 species of plant (Wiles *et al.* 1991). At least 54 species of plant are known to contribute to the diet of *P. p. pelewensis*, including the fruits of *Ficus* sp., *Pandanus tectorius*, *Cycas circinalis*, *Eugenia malaccensis*, *Terminalia catappa*, *Neiosperma oppositifolia* and *Mangifera indica*, the flowers of *Gulubia palauensis* and *Intsia bijuga*, and the leaf stems of *Artocarpus mariannensis*. Up to five bats were observed feeding simultaneously in one *Intsia bijuga* tree. Mangrove forests provide seasonally important resources when *Sonneratia alba* is flowering (Wiles and Conry 1990). Food can be defended by individuals by extending the wings forward and clapping them at other bats flying nearby. The resulting sound is loud and audible to humans for at least 15 m (Wiles and Conry 1990).

Data collected from 57 female *P. p. yapensis* revealed births peak from February to April and August to October (Falanruw 1988). The only period during which non-pregnant females were found was May–June. One case of twins has been recorded (Falanruw 1988). Four of 20 bats (*P. p. pelewensis*) seen in a colony on 28 April in Ngetpang, Palau had young attached (Wiles *et al.* 1997).

Palau flying fox was formerly traded in large numbers to Guam and Commonwealth of the Northern Mariana Islands from Yap until 1981, and Palau until 1994 due to a loophole in the Convention on International Trade in Endangered Species of Wild Fauna and Flora (CITES) (Wiles and Brooke 2009). Wiles *et al.* (1997) estimated roughly that up to 2000–5000 *P. p. pelewensis* were killed annually on Palau. Around 112 000 bats from Palau and about 23 410 bats from Yap were exported to Guam from 1975 until exports ceased (Wiles 1992e). By the 1970s there were noticeable declines of both subspecies (Wiles and Payne 1986; Wiles *et al.* 1997). CITES restrictions arrested these trends, and by the mid-1980s populations were beginning to recover. Habitat loss and climate change represent major threats and the Palau flying fox remains classified as Vulnerable on the IUCN Red List (Wiles and Waldien 2020).

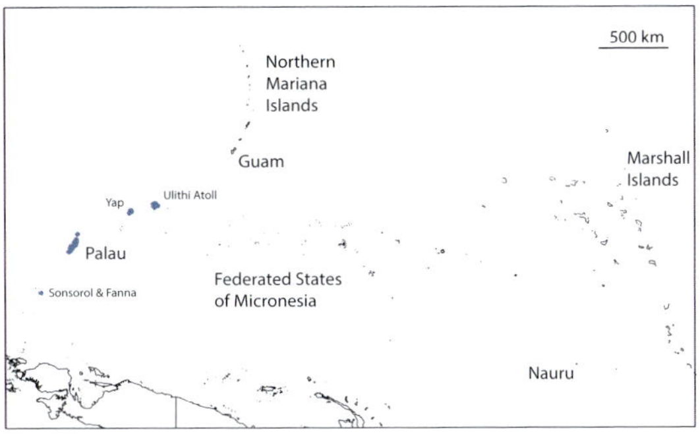

Pteropus pelewensis			
Size*			
	Mean	**Range**	**Sample**
Head–body	186.0	148.0–220.0	22
Tail	–	–	–
Forearm	123.6	104.0–138.0	20
Tibia	53.3	52.0–54.5	2
Hindfoot	37.7	29.0–47.0	26
Ear	23.9	20.0–28.0	27
Weight	279.1	161.9–420.0	19
* From Andersen (1912a), Perez (1968), Falanruw (1988), Wiles *et al.* (1997) and USNM specimens			
Identification A medium-sized flying fox with the back and rump dark brown washed with lighter brown on the back, thinly sprinkled with silver. The breast belly and flanks are also brown, thickly mixed with silvery white hairs. The mantle and neck are yellowish-buff; the tibia is hairless.			
Synonyms *Pteropus mariannus ulthiensis* Yamashima, 1932 (Ulithi)			
Subspecies *Pteropus pelewensis pelewensis* Andersen, 1908 (Palau Archipelago) *Pteropus pelewensis yapensis* Andersen, 1908 (Yap)			
Distribution Palau Archipelago (including Fanna, Kayangel Atoll and Sonsorol), Ulithi Atoll, Yap			
Extralimital None			
Conservation status (IUCN) Vulnerable			
Etymology Sp.–from Pelew (L.), a former name for the Palau Isls			

Yap flying fox (*Pteropus pelewensis pelewensis*), Palau. Subspecies of *Pteropus pelewensis* on Palau and Yap are estimated to have last shared an ancestor around 0.3 million years ago (Margaret Sloan).

LARGE PALAU FLYING FOX

Pteropus pilosus Andersen, 1908

pil-oh'-sus

The large Palau flying fox is a member of the *samoensis* species group (Almeida *et al.* 2014). It is known from only two specimens, both of which were collected before 1874 (Wiles *et al.* 1997). The type specimen was originally lodged with the Godeffroy Museum, and later transferred to the Natural History Museum, London.

Attempts by W. F. Coultas to find this species over a period of 3 months in 1931 were unsuccessful, as were attempts through 1994 that included both field observations and examination of bats being exported to Guam (Wiles and Payne 1986; Wiles 1992e; Wiles *et al.* 1997). The large Palau flying fox is presumed Extinct and has been classified as such on the IUCN Red List since 1988. The causes for its extinction are unknown, but were possibly related to overhunting, forest degradation, or both (Wiles *et al.* 1997).

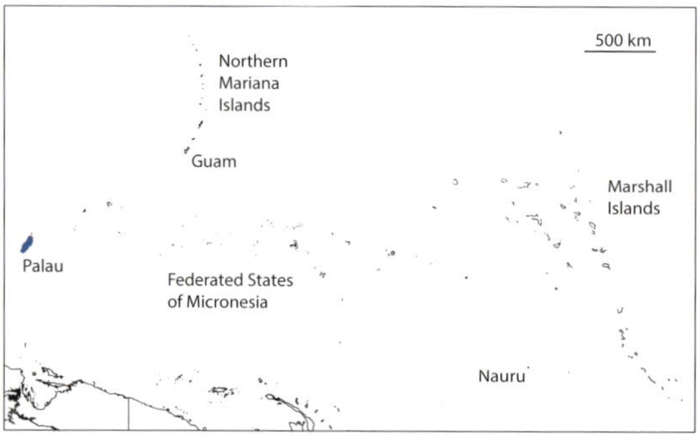

Pteropus pilosus			
Size*			
	Mean	**Range**	**Sample**
Head–body	–	–	–
Tail	–	–	–
Forearm	–	151.5	1
Tibia	–	*ca* 63.0	1
Hindfoot	–	–	–
Ear	–	25.5	1
Weight		–	–
* From Andersen (1908)			
Identification A large flying fox with small ears. The back and rump are dark chocolate and conspicuously sprinkled with long, shiny, white hairs. The breast, belly and flanks are a paler brown than the upperside and the mantle is tawny.			
Synonyms None			
Subspecies None			
Distribution Palau Isls			
Extralimital None			
Conservation status (IUCN) Extinct			
Etymology Sp.–furry (L.)			

GEELVINK BAY FLYING FOX

Pteropus pohlei Stein, 1933

pohl'-ee

The Geelvink Bay flying fox is a member of the *griseus* species group (Almeida *et al.* 2014). It is an attractive flying fox with large eyes and marked colouration. The distribution is extremely limited and biology remains poorly known. The original description was based on 15 specimens collected by Georg Stein between 23 February and 22 March 1931. This series includes a juvenile recorded as having been collected at Servei on Yapen. The remainder are not provenanced more specifically than Yapen.

On Numfoor, Geelvink Bay flying fox was abundant throughout the northern part of the island. There, it was roosting in disturbed forest, either singly or in small, scattered groups. One individual was observed feeding in the early evening in a small (2 m tall) shrub adjacent to the airstrip. It was eating a greenish fruit about 6 cm in length. During a field survey in 1992, the Geelvink Bay flying fox was absent from the environs of Korido Village in southern Supiori. However, it was

The Geelvink Bay flying fox (*Pteropus pohlei*) is a poorly known species with a very small distribution, limited to islands of Cenderawasih (Geelvink) Bay (Christian Langner).

found roosting on the nearby island of Rani, which can be seen on the horizon to the south from Supiori.

One female examined on 29 September was lactating and accompanied by a nearly independent male young that could already fly. Of two other females examined on 30 September, one was lactating and in an early stage of pregnancy (uterus 8 mm in diameter) and the second was in a more advanced stage of pregnancy. Subadults were also caught on Numfoor in late September, all weighing between 150 and 200 g. These data suggest breeding may be seasonal, with births occurring in December–January. The Geelvink Bay flying fox is known from fewer than six locations, the extent of occurrence is small, and it is believed to be declining from major threats posed by hunting and deforestation.

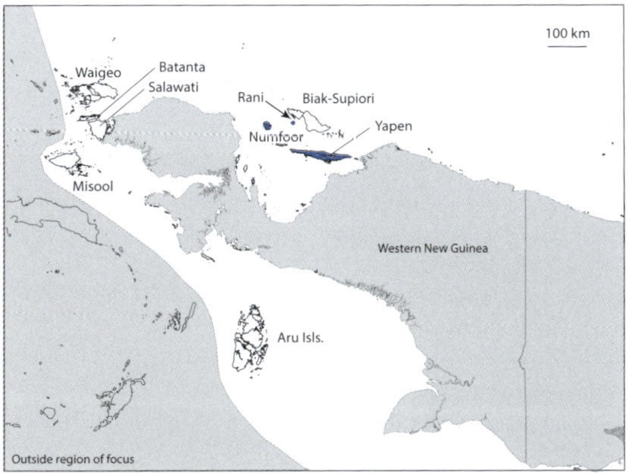

Pteropus pohlei			
Size			
	Mean	**Range**	**Sample**
Head–body	200.2	176.0–215.0	6
Tail	–	–	–
Forearm	130.6	127.2–136.8	5
Tibia	59.3	56.3–62.3	5
Hindfoot	–	30.0	1
Ear	26.1	25.2–28.0	6
Weight	341.8	315.0–360.0	5
* From AM specimens (Numfoor) and AMNH specimen (Yapen)			
Identification A medium-sized flying fox with a golden to orange mantle, a pale brown back with scattered silver hairs, and often a large central blackish patch of fur on the belly. It has very large eyes, which are perhaps the most striking feature of this species in life. The teeth are very unusual – the cheek teeth are small, rounded, high, and widely spaced. They are brownish in colour and lack enamel.			
Synonyms None			
Subspecies None			
Distribution Numfoor, Rani, Yapen			
Extralimital None			
Conservation status (IUCN) Vulnerable			
Etymology Sp.–for Dr Pohle of the Humboldt University Museum, Berlin			

SOLOMONS FLYING FOX

Pteropus rayneri Gray, 1870

ray'-ner-ee

The Solomons flying fox is a member of the *samoensis* species group (Almeida *et al.* 2014). It is a large and spectacularly coloured flying fox with distinct morphological variation across its range and five described subspecies. *Pteropus rennelli* and *P. cognatus* are sister species endemic to Rennell and Makira respectively.

The roosting behaviour of the Solomons flying fox is variable. In many locations it congregates in

On Mono, Solomons flying fox (*Pteropus rayneri*) roosts as a large colony covering an area of forest in the south-west of the island approximately 500 m wide (Lucas DeCicco).

large, exposed colonies numbering in the hundreds to thousands. Roosts at Sasamungga and Chivoko, Choiseul, were estimated to contain 7188 and 5200 individuals respectively (Bowen-Jones *et al.* 1997). However, it can also be found in groups of 3–10 individuals beneath limestone overhangs, at the sunlit entrances to caves, inside hollow strangler figs (*Ficus* spp.) and concealed underneath the dense overhanging foliage of *Pandanus* palms. When in small groups like this, it is much quieter and more easily disturbed.

Large colonies can be found in mangroves and secondary forests, villages and populated areas, especially among Seventh Day Adventist communities where consumption of flying foxes is discouraged. The camps are somewhat mobile, shifting short distances throughout the year following the defoliation of roost trees. One camp situated in sago palms (*Metroxylon salomonense*) at Kena Village (Kolombangara) in late 2014 was reportedly abandoned within 3 months following repeated predation by Sanford's sea eagle (*Haliaeetus sanfordi*). Another large and well-known camp at Viru Harbour, New Georgia, has continued to shift small distances around the village since 2010.

Colonies commence taking to the air 15 to 20 min before dusk and sometimes head offshore to drink saltwater before returning to the forests. One individual fitted with a GPS tracking device on Kolombangara Island was at 260 m above sea level, 5 km inland at 4:04 a.m. and returned to the camp site at 5:50 a.m. (half an hour before sunrise) (C. Posala, pers. comm.). On a late afternoon on Mono in July 2019, one individual feeding in a tree on the northern coastline took flight and headed in the direction of Bougainville, continuing until it was lost from sight.

The Solomons flying fox has been recorded feeding on the flowers of Malay apple (*Syzygium malaccense*) and has been mist-netted in coconut (*Cocos nucifera*) groves on Santa Isabel. Those observed feeding upon Malay apple flowers only did so during a dry period, and on rainy nights the bats did not visit the same trees, presumably because nectar had been washed from the flowers. This species was found to be removing the strong pericarps from large fruits in central Malaita, chewing them to extract juices, and ejecting the remaining fibres onto the forest floor. On Mono in July 2019, Solomons flying foxes were seen picking sago palm fruits (*Metroxylon salomonense*) from the palms and flying to adjacent trees to feed on them. Two of the fruits weighed 203 g and 223 g, probably 40–50% of the animal's bodyweight (Mono average bodyweight = 556 g, $n = 4$).

An Australian Museum expedition to the Western Province in 1990 documented three pregnant females; one female with a newborn juvenile weighing 83 g on New Georgia on 2 July; and three females nursing young weighing 82, 96 and 114 g on Kolombangara on 20 July. Males in a colony at Viru Harbour, New Georgia Island, were defending harems of females and mating in February 2014. Two females roosting beneath a limestone overhang on Gatokae Island in January 2014 were nursing large young. When

Ejecta from Solomons flying fox (*Pteropus rayneri rayneri*) on Malaita Island (Ben Speare).

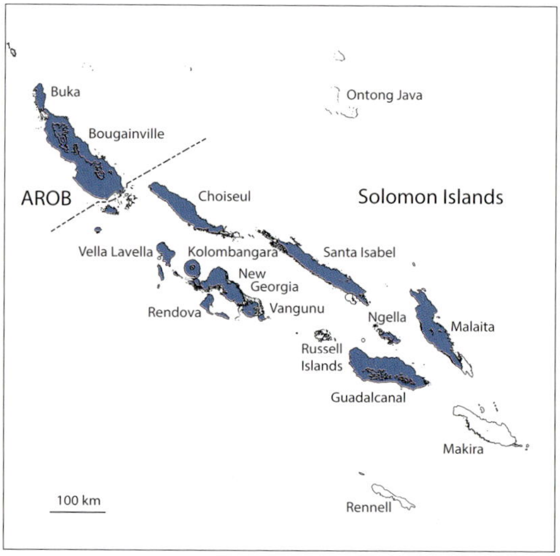

Pteropus rayneri			
Size*			
	Mean	**Range**	**Sample**
Head–body	243.3	185.5–302.0	12
Tail	–	–	–
Forearm	163.7	135.5–187.0	12
Tibia	77.8	65.7–90.5	9
Hindfoot	49.9	37.4–65.0	9
Ear	30.7	25.9–42.0	13
Weight	690.3	435.0–870.0	11
* From AM and KU specimens (Choiseul, Mono, Ngella, Santa Isabel) and Bonaccorso (1998)			
Identification A large flying fox – fur short, pelage highly variable in shades of reddish-brown. Mantle, back and rump often tricoloured; the rump varies from cinnamon-rufous to orange buff. *Pteropus tonganus* is the only other species of similar size in Solomon Isls. It can be differentiated from this by its relatively shorter face and drab pinkish grey, rather than black wing membranes.			
Synonyms None			
Subspecies *Pteropus rayneri rayneri* Gray, 1870 (Guadalcanal, Malaita) *Pteropus rayneri grandis* Thomas, 1887c (Bougainville, Shortland Is., Choiseul, Santa Isabel) *Pteropus rayneri lavallanus* Andersen, 1908 (Vella La Vella) *Pteropus rayneri rubianus* Andersen, 1908 (New Georgia) *Pteropus rayneri monoensis* Lawrence, 1945 (Mono)			
Distribution Bougainville, Buka, Choiseul, Gatokae, Gizo, Guadalcanal, Kolombangara, Malaita, Mono, New Georgia, Ranongga, Rendova, Santa Isabel, Shortland, Simbo, Tetepare, Vangunu, Vella La Vella			
Extralimital None			
Conservation status (IUCN) Near Threatened			
Etymology Sp.–for Dr F. M. Rayner, collector of the holotype			

disturbed, the females left the roost carrying them. McKean (1972) recorded a juvenile captured in September on Bougainville. These data suggest that, at least in the Western Province, breeding may be seasonal with matings occurring at the beginning of the year and births occurring mid-year.

Mass die-offs were reported for this species in 1987 on the islands of Bougainville and Buka and were attributed to disease (Flannery 1990b). However, Bougainville residents often cite such widespread deaths of flying foxes as one of the most noticeable impacts of the Panguna copper mine. *Pteropus rayneri* is a large gregarious flying fox and hunting is presumably a threat. The restriction of firearms in Solomon Islands and Bougainville limits the numbers that can be taken at any one time. Bowen-Jones *et al.* (1997) reported the decline or disappearances of colonies on Choiseul, but could not attribute it to hunting or any other specific threat.

RENNELL FLYING FOX

Pteropus rennelli Troughton, 1929a

ren'-el-ee

The Rennell flying fox is a member of the *samoensis* species group (Almeida *et al.* 2014) endemic to remote Rennell Island in the south-western Solomon Islands. It is a sister species to *P. rayneri* (a more widespread Solomon Islands endemic) and *P. cognatus* (endemic to Makira) (Lavery, unpublished data). Few specimens exist in world museum collections. The holotype, collected by G. A. V. Stanley in 1927, is the only specimen held by the Australian Museum (Troughton 1929a). Two specimens are held in the collections of Universitetets Zoologisk Museum, Copenhagen, collected by Dr T. Monberg at Lavanggu in 1958

The Rennell flying fox (*Pteropus rennelli*) is a poorly known species facing significant threats from logging and mining operations occurring in the western half of Rennell Island. More recently, there have been reports of a mass flying fox mortality event on Rennell, including individuals of both *P. tonganus* and *P. rennelli* (Tyrone Lavery).

(Hill 1962). Single individuals from Nuipani and Hutuna in October 1962 and 1965 respectively are in the collection of the American Museum of Natural History (Hill 1968). In October 2018, two individuals (one a subadult) were caught near Te-Nganno Village at Lake Te-Nganno (Lavery *et al.* 2021). The distribution probably extends to nearby Bellona Island, but this is yet to be confirmed.

Both animals examined in 2018 were caught in an area of mixed primary forest and village gardens. The Rennell flying fox roosts individually or in pairs in primary forests among the lower canopy branches where overhead cover is provided. It is active in late afternoons, when pairs of animals can be seen and heard foraging and squabbling over resources. The adult male caught in 2018 was captured at approximately 5 p.m. in a mist net set 15 m above the ground in a garden area. Nothing is recorded of diet or reproduction.

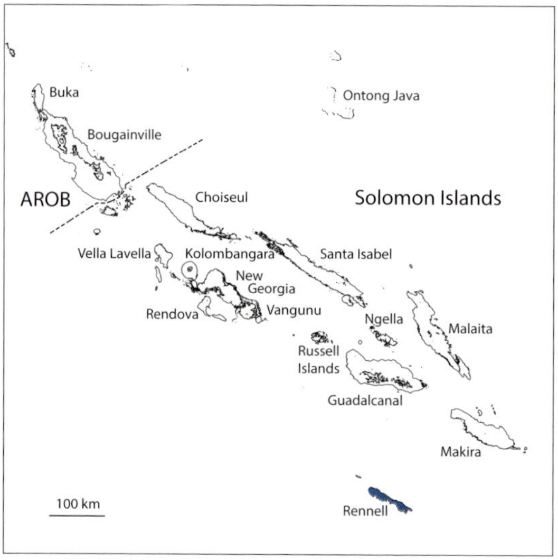

Pteropus rennelli			
Size*			
	Mean	**Range**	**Sample**
Head–body	–	185.7	1
Tail	–	–	–
Forearm	125.1	121.0–128.3	4
Tibia	55.6	53.5–57.7	2
Hindfoot	37.6	36.1–39.0	2
Ear	23.9	23.0–25.2	3
Weight	–	270.0	1
* From KU specimens, Troughton (1929a) and Hill (1968)			
Identification A medium-sized flying fox with blackish central parts on the back and chest. The mantle is chestnut or auburn (in younger animals) and the cheeks and face are greyish.			
Synonyms None			
Subspecies None			
Distribution Rennell			
Extralimital None			
Conservation status (IUCN) Endangered			
Etymology Sp.–for the island of Rennell, the type locality			

West Rennell has faced heavy disturbance from logging and bauxite mining in recent years. Mining involves the complete removal of ore for wholesale export for processing. The depleted substrates that remain will probably have limited ability to support regeneration of native forests and this could have an impact on the conservation status of *P. rennelli*. In January and February 2021, there were reports of a mass flying fox mortality event on Rennell, whereby unexplained deaths of *P. tonganus* and *P. rennelli* occurred. The causes of this remain unknown, but could be related to environmental disturbance and disease outbreak.

SAMOAN FLYING FOX

Pteropus samoensis Peale, 1848

sam'-oh-en'-sis

This beautiful flying fox belongs to the *samoensis* species group, a second radiation of Pacific Island *Pteropus* that includes species from Vanuatu, Solomon Islands, New Caledonia and the Bismarck Archipelago (Almeida *et al.* 2014; Tsang 2019b). Bones recorded in archaeological deposits on 'Eua, Tonga suggest it was present there and may have been extirpated by humans along with species of reptile and bird (Koopman and Steadman 1995). Inter-island dispersal is apparently limited between parts of its known range, and two subspecies are recognised (Russell *et al.* 2016).

On Fiji, the Samoan flying fox occurs from sea level to the highest peaks. However, mist nets set in the summit forests of Taveuni in 1990 were successful at capturing the species only on clear nights. The thick mist that often envelopes Fiji's high island summits may impede foraging or make navigation difficult for this species and prevent it from frequently venturing into higher elevations. Samoan flying fox is found in a variety of habitats including primary and secondary forests, agricultural forest and montane forest, and it appears to be more dependent on forested habitats than *P. tonganus* (Palmeirim *et al.* 2007). White *et al.* (1988) estimated that, in rainforests of Fiji, 56% of all prey remains collected in the roosts of peregrine falcons included bats. In Samoa this species is largely diurnal, whereas in Fiji it is primarily nocturnal, a behavioural difference that probably relates to the presence of peregrine falcons in Fiji and their and absence in Samoa.

Samoan flying fox is unusual among bats in that it forms monogamous breeding pairs. Roosts typically contain between two and eight individuals and often consist of a male, a female and one young (Pierson and Rainey 1992). Unusual instances of larger groups (up to 41 individuals) have been recorded (Pierson *et al.* 1996). Breeding is seasonal and could be cued by the flowering and fruiting seasons of *Syzygium inophylloides* (Banack and Grant 2009). Mating occurs from August to December and births from March to October, peaking in May–June (Banack and Grant 2009). By June–July females can be seen carrying young (Engbring 1989).

The Samoan flying fox (*Pteropus samoensis*) is a medium-sized flying fox found in Fiji and Samoa. It usually roosts in groups of two to eight individuals (Pavel German).

Banack (1998) described the Samoan flying fox as a sequential specialist that targets a limited number of food plants in any one season, but over extended periods has a diverse diet. This strategy was considered a response to the lower variety of food sources available at any one time on small islands. At least 36 plants have been identified from the diet in Samoa (Banack 1998). These include the fruits of *Artocarpus altilis*, *Carica papaya*, *Collospermum samoense*, *Cupaniopsis samoensis*, *Dysoxylum samoense*, *D. moata*, *Fagraea beretiana*, *Ficus graeffi*, *Mangifera indica* and *Planchonella torricellensis*; and flowers of *Barringtonia asiatica*, *Cananga odorata*, *Ceiba pentandra*, *Erythrina* sp., *Freycinetia reineckei* and *Rhus taitensis* (Engbring 1989; Cox 1983; Mickelburgh *et al.* 1992).

The Samoan flying fox was noted as being extremely common on Samoa in the 1840s. By the 1970s, an accelerated decline was evident, driven in large part by hunting for international export to Guam and Commonwealth of the Northern Marianas (Cox 1984). Over a 5-year period in the late 1980s, populations of both *P. samoensis* and *P. tonganus* in American Samoa were estimated to have declined by 80–90% (Craig *et al.* 1994). The population on Tutuila Island (one of the largest) was estimated to contain only 200–400 animals

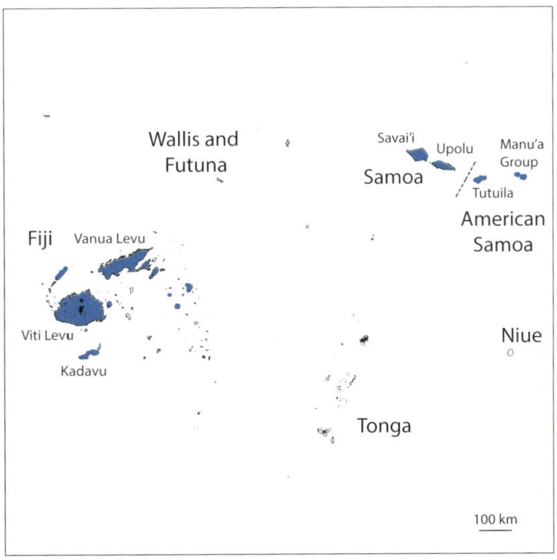

Pteropus samoensis			
Size*			
	Mean	**Range**	**Sample**
Head–body	163.3	120.0–200.0	9
Tail	–	–	–
Forearm	129.8	124.0–133.7	7
Tibia	59.9	53.8–63.4	7
Hindfoot	–	–	–
Ear	21.6	20.1–23.0	5
Weight	337.9	220.0–440.0	7
* From AM specimens (Taveuni, Vanua Levu) and Andersen (1912a)			
Identification A medium-sized flying fox with a light brown back sprinkled with long silver hairs. The belly is brown, often with long yellowish hairs and the mantle and crown of the head are pale yellow. It is readily distinguished from the sympatric *P. tonganus* in its brownish rather than black back, and shorter snout. In flight, *P. samoensis* has a dark triangular shape, a broad appearance to the wings, no deep indentations on the trailing edges of the wings. The sympatric *P. tonganus* has narrower wings, deep indentations on the trailing edges of the wings and a lighter overall colour.			
Synonyms *Pteropus vitiensis* Gray, 1870 *Pteropus whitmeei* Alston, 1874 *Pteropus fiscicollis* Nicholl, 1904 (nomen nudum) *Pteropus ruficollis* Nicholl, 1904 (nomen nudum)			
Subspecies *Pteropus samoensis samoensis* Peale, 1848 (Samoa) *Pteropus samoensis nawaiensis* Gray, 1870 (Fiji)			
Distribution Savaiu, Upolu (Samoa), Tutuila, Ofu, Olosega, Ta'u (American Samoa), Cicia, Kadavu, Nayau, Ovalau, Taveuni, Vanua Balavu, Vanua Levu, Vatu Vara, Viti Levu (Fiji)			
Extralimital None			
Conservation status (IUCN) Near Threatened			
Etymology Sp.–from Samoa (L.), the type locality			

(Craig *et al.* 1994). Inclusion under Appendix II of the Convention on International Trade in Endangered Species (CITES) reduced export hunting pressure from 1989 onwards. However, hunting for local consumption and markets continued as a threat. Hunting can be periodically exacerbated by cyclones, which increase roost exposure by defoliating vegetation. Pierson *et al.* (1996) found roost sites were concentrated around the rims of small volcanic cones, and these sites suffered severe wind damage.

SMALL SOLOMONS FLYING FOX
Pteropus solomonis Thomas, 1904c
sol'-oh-mon'-is

The small Solomons flying fox was originally described in 1904 from a female collected on Gizo Island by Alfred Meek on 10 November 1903. It has long been regarded as a subspecies of *Pteropus admiralitatum* but is certainly a distinct species. In fact, it is more closely allied to *P. rayneri*, with preliminary molecular data showing it belongs to the *samoensis* species group rather than the *griseus* species group of *P. admiralitatum* (Lavery, unpublished data). Male *P. solomonis* possess bristly orange hairs on the shoulder glands, and these are a trait common to males of species in the *samoensis* group. It further differs in possessing a tibia that is furred to the ankle rather than bare (as in *P. admiralitatum*).

The small Solomons flying fox is a common species that occurs on many of the larger Solomon Islands and is replaced by *P. admiralitatum* on small, isolated islands like Mono, Dai and Ontong Java. It can be encountered in both primary and secondary forests, gardens and coastal villages from sea level into montane forests. Ten individuals collected by Hamlin during the Whitney South Seas expedition of the 1920s are now held by the American Museum

Small Solomons flying fox (*Pteropus solomonis*), Vangunu Island, Solomon Islands. This is a Solomon Islands endemic, common in lowland forests of the Western Province (Tyrone Lavery).

of Natural History. They were captured on 30 April on Malaita at an elevation of 900 m.

In the Western Province, this species will roost individually among tree foliage or in small groups of fewer than six animals beneath limestone overhangs. A large hollow tree known as *vasa* (*Vitex cofassus*) at 400 m elevation near Busurata Village, Malaita contained around 30 individuals on 22 May 1991. Three of six females examined were lactating at the time. Lactating females have also been observed in late February on New Georgia and mid-April on Vangunu. Of six females examined in October 2015 in the Solomon Islands Western Province, all were either nulliparous or post-lactating. One female caught on Kolombangara Island on 10 July 2017 was pregnant.

Small Solomons flying foxes seem to be somewhat tolerant of environmental disturbance. A comparison of mist-net survey data from the New Georgia group of islands from 1992 and 2015 revealed capture rates increased over the 23-year period that coincided with widespread commercial logging (Lavery *et al.* 2020a). Presumably secondary forests offer an abundance of favoured food plants (such as fruiting pioneer species) to the benefit of the small Solomons flying fox. It is regularly caught in mist nets set in heavily forested areas, suggesting it is a low-flying species that can forage in vegetation layers beneath the canopy. The small Solomons flying fox has not been assessed under the IUCN Red List. It is widespread and capture rates in the Western Province increased between 1992 and 2015. A classification as Least Concern is probably appropriate.

Pteropus solomonis			
Size*			
	Mean	**Range**	**Sample**
Head–body	135.8	100.2–172.5	18
Tail	–	–	–
Forearm	107.2	97.0–113.1	18
Tibia	49.8	46.4–53.1	17
Hindfoot	31.7	29.2–34.6	9
Ear	17.5	13.0–20.5	18
Weight	196.0	130.0–235.0	17
* From Kolombangara, New Georgia, Ranongga, Vangunu and Vella La Vella			
Identification A medium-sized flying fox with fur of mantle woolly and cinnamon in colour, contrasting with adpressed chocolate brown fur of the mid-back. The tibia is furred to the ankle. Mature males often have a grey or grey-yellow face, and the belly and back also have sparse silver-grey hairs.			
Synonyms None			
Subspecies None			
Distribution Gatokae, Gizo, Guadalcanal, Kolombangara, Malaita, ?Mbanika, New Georgia, Ranongga, Tetepare, Vangunu, Vella La Vella			
Extralimital None			
Conservation status (IUCN) Not Assessed			
Etymology Sp.–relating to the Solomon Isls			

GUAM FLYING FOX
Pteropus tokudae Tate, 1934b
tok'-u-dae

The Guam flying fox is endemic to Guam and belongs to the *pelagicus* species group (Almeida *et al.* 2014). A phylogenetic reconstruction indicated it belongs to a monophyletic Micronesian lineage including *P. molossinus* and *P. pelagicus* (Almeida *et al.* 2014).

The species was first discovered in August 1931 when W. F. Coultas collected two males (at least one of which was an adult) on Guam, 10 August 1931 (Tate 1934b). A third and final specimen, an adult female, was killed by hunters in limestone forest at Tarague Cliff in June 1968 (Perez 1973; Wiles 1987b). At the time it was nursing a young, which escaped capture. Baker (1948) failed to detect the species in 1945, and Perez (1973) interviewed older hunters during the 1960s, all of whom suggested the Guam flying fox was very rare, if not extinct. A possible sighting at Ritidian Point in June 1979 was

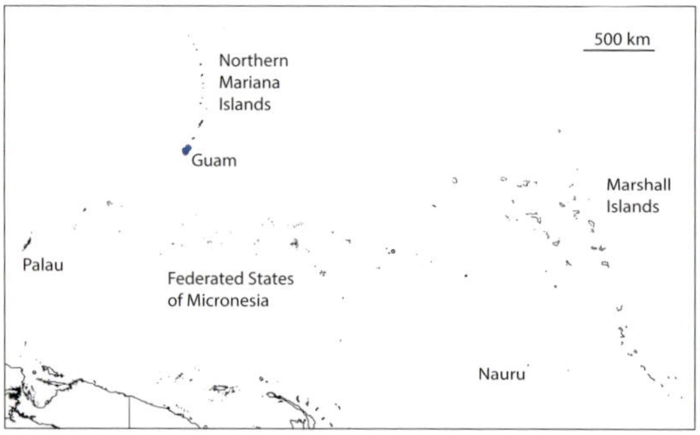

Pteropus tokudae			
Size*			
	Mean	**Range**	**Sample**
Head–body	–	225.0	1
Tail	–	–	–
Forearm	95.0	95.0	2
Tibia	–	70.0	1
Hindfoot	–	–	–
Ear	–	20.0	1
Weight	–	151.8	1
* From Tate (1934b) and Perez (1973)			
Identification A brownish flying fox; it is small and odourless compared to *P. mariannus*, the only other flying fox from Guam.			
Synonyms None			
Subspecies None			
Distribution Guam			
Extralimital None			
Conservation status (IUCN) Extinct			
Etymology Sp.–for Dr M. Tokuda of the Kyoto Imperial University, who worked on the taxonomy of Pacific Island mammals			

never confirmed (Wiles 1987b). Despite considerable field work since the 1970s, focused on Guam's other bats, there have been no further observations of the Guam flying fox. Almost nothing is known of the biology, except that it inhabited native forest. The causes of extinction are unknown, but overhunting and habitat loss are the most likely, with predation by brown tree snakes (*Boiga irregularis*) on any bats that might have occurred in southern Guam also a possible factor (G. Wiles, pers. comm.).

PACIFIC FLYING FOX

Pteropus tonganus Quoy and Gaimard, 1830

ton-gah'-nus

The Pacific flying fox is a member of the *griseus* species group (Almeida *et al.* 2014). It is one of the world's most widespread species of flying fox, found from the more remote eastern Solomon Islands, and extending far west across the Pacific to Tonga, Samoa, Niue and the Cook Islands. There is morphological variation across this range, with three described subspecies. Individuals in the east (subspecies *Pteropus tonganus geddiei*) are larger than those from the west (subspecies *P. t. tonganus*). The Pacific flying fox has been the focus of numerous detailed ecological studies. Given the extremely broad distribution of this species, and because subspecies are well distinguished, the two most broadly distributed subspecies are discussed in detail here.

Pteropus tonganus tonganus

The Pacific flying fox is the common flying fox of Fiji, Samoa, Tonga and the Cook Islands. The population on Rarotonga in the Cook Islands represents the easternmost extension of fruit bats into the Pacific. Tiraa (1992) reported that this population was introduced from Mangaia in the 1870s. The Mangaia population is apparently indigenous, as Kirch *et al.* (1992) identified bones in archaeological deposits there. The remains become less frequent with time, suggesting a population decline following the arrival of humans. Fossil remains have been reported from archaeological deposits on Aitutaki, Cook Islands, where it is now extinct. The farthest eastern limit for Pacific flying fox was identified via five bones (probably from the same individual) recovered from a site on Rurutu, French Polynesia and dated to 1064–1155 CE (Weisler *et al.* 2006). No bones were found in pre-human deposits, so the possibility that *P. tonganus* was transported there from the southern Cook Islands couldn't be ruled out. The absence of flying fox bones in late prehistoric to historic layers further supports this hypothesis of human translocation, or indicates that any natural population was later extirpated (Weisler *et al.* 2006). Pacific flying foxes were apparently translocated by people as pets, having been reported as far afield as Tahiti in the mid-19th century (Martin 1981), so it is possible some populations are the result of introductions.

Periodic population estimates are available for several islands, but long-term monitoring data are lacking. The estimated population size of the Pacific flying fox in Rarotonga in 2002 was 1730 individuals, divided into nine roosts ranging in size from 15 to 491 individuals (Cousins and Compton 2005). On Mangaia the total number of animals was estimated at 78. The island-wide population on Niue was 2040–4080 bats in 1998 (Brooke and Tschapka 2002), and Craig *et al.* (1994) estimated a population of 1500–2500 bats remained on Tutuila, American Samoa, in 1992. On Viti Levu, Wilson and Engbring (1992) recorded a colony of 7000 spread across at least 11 trees on the outskirts of Suva.

Pacific flying foxes remain in camps during the day. The animals usually depart around dusk to forage during the night, although they are sometimes seen in flight in late afternoons (Wilson and Engbring 1992). In the Cook Islands, roost sites were primarily located in the undisturbed forests of steep slopes and ridges in the inner parts of islands that were most inaccessible to humans (Cousins and Compton 2005). The roosting trees were tall, emergent species with sparse leaves (*Cananga ordorata* and *Cerbera manghas*).

On Vanua Levu in the dry season (May–October) of 2009 and 2010, this species formed annual maternity camps containing around 300 females and their young (Scanlon *et al.* 2014). Many of the non-maternity camps were larger, containing 300 to several thousand animals.

Observational data and analysis of bat faecal samples showed that Pacific flying foxes in Fiji obtain sufficient nutrients via the combination of a highly varied plant diet and insectivory (Scanlon *et al.* 2013b). Moth scales were found in 62% of faecal samples examined, and the remains of invertebrates other than moths were even more common, being found in 69% of samples. Studies on Vanua Levu and Taveuni, Fiji, found the floral resources of *Erythrina variegata*, *Freycinetia* spp. and *Syzygium* spp. were shared relatively evenly, making up 61%, 56% and 37% of feeding observations respectively (Scanlon *et al.* 2014). The fruits of breadfruit (*Artocarpus altilis*), papaya (*Carica papaya*), mango (*Mangifera indica*), banana (*Musa* sp.), *Calophyllum inophyllum* and *Syzygium jambos*, and flowers of *Ceiba pentandra*, *Cananga odorata*, *Musa paradisiaca* and *Cocos nucifera*, also are eaten (Cox 1983; Wilson and Engbring 1992).

Pteropus tonganus geddiei

In Vanuatu, both sexes congregate together in large mixed-sex camps from September to January, often in large *Casuarina* trees along the shoreline (Baker and Baker 1936). Many females are pregnant in February and by June they form female-only camps found further inland. Between 1990 and 2019 (Lavery *et al.* 2020d) estimated camp sizes at 23 locations across the islands of Temotu Province, Solomon Islands and Torba Province, Vanuatu. Most contained 100–300 individuals, but two larger camps containing approximately 2780 and 3630 individuals were found on Nendö and Fenualoa (Reef Islands) between October and November 1993. A camp at Mission Harbour, Mota Island, contained around 600 individuals in May 2018. The roost was on a large fig (*Ficus* sp.) in a *tabu* area where the bats were protected from hunting. Two of nine individuals captured on the island were subadults, as was the only individual caught in a mist net on Vanikoro Island in April 2018. Three of four animals examined on Rennell Island in October 2018 were subadults.

Lavery *et al.* (2020d) identified 56 plant species that contribute to the diet of this flying fox species in Temotu Province, Solomon Islands. The Pacific flying fox was observed to target 24 species for their fruit alone (e.g. *Ficus wassa*, *Gnetum gnemon*, *Pandanus* spp.), six species were targeted for nectar (e.g. *Areca catechu*, *Mucuna* sp., *Terminalia catappa*), and both the fruit and nectar or flowers were consumed from 26 species (e.g. *Calophyllum inophyllum*, *Barringtonia edulis*, *Syzygium malaccens*). The leaves of at least one of these plants (papaya *Carica papaya*) were also eaten. Favoured foods reported by Baker and Baker (1936) included the fruit of papaya, banana (*Musa* sp.) and *Ficus copiosa*. Breadfruit was eaten during September–April when it was probably a primary food source. In February, flowers of *Glochidion ramiflorum* and kernels of *T. catappa* were taken, while in March *T. catappa* and guava (*Psidium* sp.) were important. In May, the pulp of *Dracontomelon* sp. was eaten. Signs that this species had been gnawing on coconuts were found on two of the smaller Reef Islands, Makalom Island in 2019, and Matema Island in 1993 (Lavery *et al.* 2020d).

The Pacific flying fox is classified as Least Concern on the IUCN Red List. From an overall perspective the species is very widespread, has a large total population size and does not appear to be declining at a rate that warrants listing under a threatened category. However, a regional approach to the conservation of this species is probably more appropriate. In Fiji, hunting rates do not appear to threaten the species (Palmeirim *et al.* 2007). Yet on Niue, population surveys and consultations with hunters deemed hunting rates to be entirely unsustainable (Brooke and Tschapka 2002). Craig *et al.* (1994) revealed daytime counts of this species in American Samoa had dropped by 90% between 1986 and 1992 as a result of hunting and cyclone disturbance. Similarly, on Makira Island where the species is hunted for its meat and teeth, there is some evidence derived from questionnaire surveys that population numbers are depressed in regions of the island where hunting pressure is high (Lavery and Fasi 2017). Within its broad distribution, Pacific flying foxes should be an important conservation

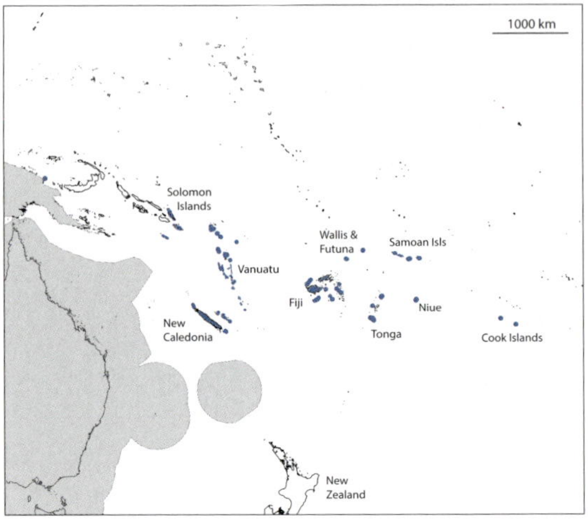

Pacific flying fox (*Pteropus tonganus*), Lotawora Village, Mota Island, Vanuatu. The Pacific flying fox extends further eastwards into the Pacific than any other indigenous mammal (Tyrone Lavery).

Pteropus tonganus			
Size*			
	Mean	**Range**	**Sample**
Head–body	200.1	148.9–263.0	23
Tail	–	–	–
Forearm	146.5	122.0–169.3	25
Tibia	69.7	58.0–83.1	19
Hindfoot	42.0	39.0–45.0	9
Ear	25.2	20.5–30.2	23
Weight	514.5	215.0–765.0	17

* AM and field-measured specimens from Vanuatu (Mota, Vanua Lava, Yesawa), Solomon Isls (Rennell, Vanikoro), and measurements by Revilliod (1914) and Bonaccorso (1998)

Identification A medium-sized to large flying fox with a black back, long muzzle and a bright yellow mantle. Individuals from Vanuatu and further east are larger than those from the west. It is similar to *P. conspicillatus*, but is allopatric with that species.

Synonyms
Pteropus flavicollis Gray, 1870
Pteropus tonganus heffernani Troughton, 1930

Subspecies *Pteropus tonganus tonganus* Quoy and Gaimard, 1830 (Fiji, Wallis and Futuna, Samoa, American Samoa, Tonga, Niue, Cook Isls)
Pteropus tonganus geddiei MacGillivray, 1860 (Solomon Isls, Vanuatu, New Caledonia, Loyalty Isls)
Pteropus tonganus basilicus Thomas, 1915a (Koil and Karkar Is.)

Distribution Manu'a Isls, Ofu, Olosega, Ta'u, Tutuila (American Samoa); Aitutaki†, Mangaia, Rarotonga (Cook Isls); Nairai, Namena, Ovalau, Taveuni, Vanua Levu, Viti Levu, Yasawa (Fiji); Rurutu (fossil) (French Polynesia); Grand Terre, Lifou, Maré, Ouvéa (New Caledonia); Niue; Bam, Blup Blup, Kadovar, Karkar, Koil, Vokeo, Wei (Papua New Guinea); Savai'i, Upolo (Samoa); Duff Isls, Makira, Malaita, Malo, Matema, Nendö, Nupani, Reef Isls, Rennell, Ugi, Vanikoro (Solomon Isls); 'Eua, Tongatapu, Uonukuhahake, Tofua, Vava'u (Tonga); Ambae, Aneityum, Aore, Efate, Emae, Erromango, Espiritu Santo, Hiu, Loh, Maewo, Malakula, Mota, Mota Lava, Pentecost, Tanna, Toga, Tongoa, Ureparapara, Vanua Lava (Vanuatu); Futuna, 'Uvea (Wallis and Futuna)

Extralimital None

Conservation status (IUCN) Least Concern

Etymology Sp.–of Tonga (L.), the type locality

focus on each and every island because of the critical role the species plays in the functioning of island ecosystems (Cox *et al.* 1992). Even where only a moderate population decline has occurred, the corresponding loss of seed dispersal services has been disproportionate and profound (McConkey and Drake 2006).

VANIKORO FLYING FOX

Pteropus tuberculatus Peters, 1869

tue-berk'-ue-lah'-tus

Pteropus tuberculatus belongs to the *samoensis* species group (Almeida *et al.* 2014). It was originally described in 1869 from a skull collected in 1828 during the voyage of the *Austrolabe*. The true origin of this flying fox then remained uncertain for almost a century until Troughton (1927) announced its type locality as Vanikoro Island with the collection of eight specimens (now in the Australian Museum). Sanborn (1931) discussed an additional four specimens held in the American Museum of Natural History.

Following the account by Troughton (1927), the conservation status of the Vanikoro flying fox was not reported for many decades. Scientific surveys in the early 1990s were unsuccessful and it was feared

The Vanikoro flying fox (*Pteropus tuberculatus*) is well known to local residents on Vanikoro. Among the scientific community, its type locality remained confused with Guam for decades and it was subsequently feared extinct until 2014 (Tyrone Lavery).

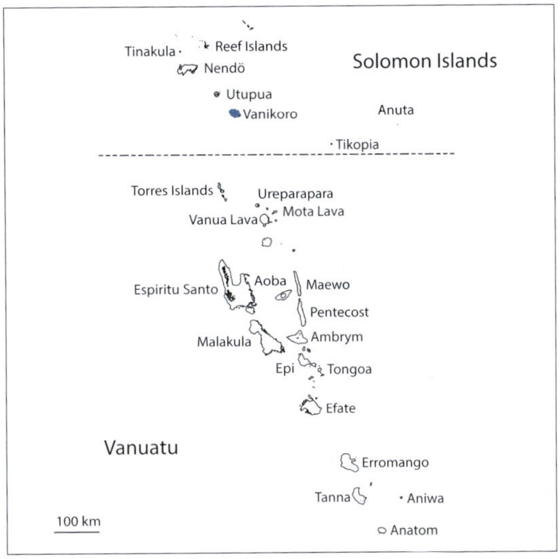

Pteropus tuberculatus			
Size*			
	Mean	**Range**	**Sample**
Head–body	155.3	136.5–170.9	7
Tail	–	–	–
Forearm	120.9	112.0–130.2	12
Tibia	53.4	49.0–61.3	12
Hindfoot	36.5	35.0–38.0	4
Ear	21.1	18.5–26.5	11
Weight	230	168–264	7
* From field measurements (Vanikoro) and Troughton (1927)			
Identification A medium-sized flying fox, dark brown overall, crown brownish tawny-olive. Mantle darker than crown and back. The second lower incisor is very large (four to five times the bulk of the first lower incisor).			
Synonyms *Pteropus vanikorensis* Quoy and Gaimard, 1830 (in part)			
Subspecies None			
Distribution Baine, Tevai (Vanikoro)			
Extralimital None			
Conservation status (IUCN) Endangered			
Etymology Sp.– a reference to possession of tubercles (L.) allusion unclear, possibly a reference to small cusp-like projections on upper canines			

extinct as a result of the extensive logging that occurred on Vanikoro from the 1920s to 1960s (Lees 1990). More recently, however, targeted field surveys found the Vanikoro flying fox to be abundant on the main island of Vanikoro and adjacent Tevai (or Teanu), and it was reported to also exist on Utupua Island (Lavery *et al.* 2020d).

This species roosts individually or occasionally in pairs in primary forests and fruit and nut agroforests close to villages. In 2018 none were encountered in recently logged forests. Roosts are on the lower to mid-canopy branches with good overhead cover. These were often on, or adjacent to, food trees such as mango (*Mangifera indica*), coconuts (*Cocos nucifera*) and breadfruit (*Artocarpus altilis*).

In late September 2015, Vanikoro flying foxes were abundant in coconut plantations, feeding on the inflorescences. In April 2018 they were attracted towards the seasonally abundant breadfruit. On both occasions, animals were most active the early afternoon, between 4 p.m. and 6 p.m. This is presumably to avoid competition with the sympatric *P. tonganus*, which are far larger and tend to be more active closer to dusk and during nocturnal hours.

The American Museum of Natural History collection includes a juvenile collected on 25 August 1926. No juveniles or subadults were caught by Troughton during his visit in July and August 1926. Of seven individuals mist netted in April 2018, four were post-lactating females. One male and one female were subadults.

Logging is widespread on Vanikoro and probably has an impact on the Vanikoro flying fox. However, the species is relatively common in village environments and as such may be able to withstand this type of disturbance. Hunting is not a major concern as the animals are said to have an unpleasant taste.

KOSRAE FLYING FOX

Pteropus ualanus Peters, 1883

ue'-ah-lah'-nus

The Kosrae flying fox was previously treated as a subspecies of *Pteropus mariannus* (Koopman 1993). However, genetic comparisons and examinations of museum specimens reveal it is better regarded as a distinct species that belongs to the *griseus* species group (Almeida *et al.* 2014).

Coultas (1931) observed only a few individuals during a 5-month stay on Kosrae in 1927 and reported that a disease had killed thousands of animals within the previous year. By 1983, a few colonies of 100–400 bats were present, whereas observations in 1989 concluded it remained uncommon (Mickleburgh *et al.* 1992). The most recent data (2013–14) revealed a total population size which is probably still below the carrying capacity for the island (somewhere between 2000 and 3000 individuals spread across at least four colonies) (Hayes and Engbring 2020).

The ecology of Kosrae flying fox remains poorly known. It is a habitat generalist that will roost across the range of environments available on Kosrae from coastal mangroves into the higher elevation forests towards the island's summit at 589 m elevation (Hayes and Engbring 2020). It has been observed feeding on palms, breadfruit (*Artocarpus altilis*) and screw pine (*Pandanus* sp.) (Kittlitz 1836; Hayes and Engbring

The Kosrae flying fox (*Pteropus ualanus*) was formerly considered a subspecies of *P. mariannus*; however, it is a distinct species endemic to Kosrae, Federated States of Micronesia (Thibaud Aronson).

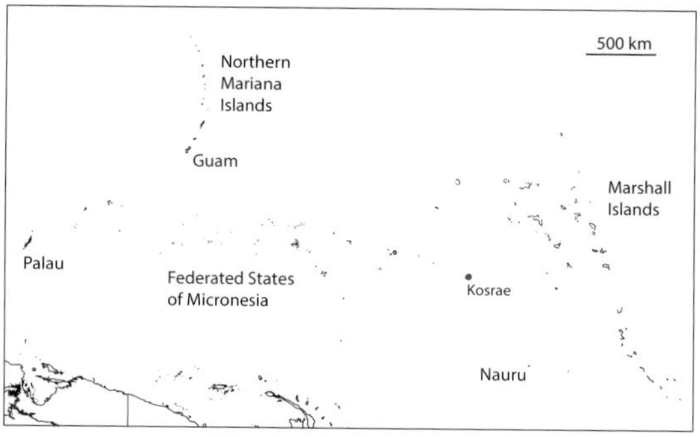

Pteropus ualanus			
Size*			
	Mean	**Range**	**Sample**
Head–body	193.7	162.0–230.0	6
Tail	–	–	–
Forearm	130.3	127.0–133.5	3
Tibia	59.3	59.0–59.5	2
Hindfoot	48.9	42.0–54.0	6
Ear	24.0	21.0–27.0	7
Weight	–	–	–
* From Andersen (1912a)			
Identification A medium-sized flying fox with a blackish brown rump and back, a dark brown belly with scattered silver hairs and a yellow mantle. Overall it is very similar to *Pteropus pelewensis* but has smaller teeth.			
Synonyms None			
Subspecies None			
Distribution Kosrae			
Extralimital None			
Conservation status (IUCN) Endangered			
Etymology Sp.–from Ualan (L.), a former name for the island of Kosrae			

2020). Animals occasionally depart and arrive from colonies during the day, and may sometimes forage during daylight hours (Hayes and Engbring 2020). Overhunting remains a threat to this species, and the small population size and confinement to a single island place it at risk from major disturbance events such as cyclones (Hayes and Wiles 2020).

NEW CALEDONIA FLYING FOX

Pteropus vetula Jouan, 1863

vet'-oo-la

This is a highly unusual species of *Pteropus*. In its external features and cuspidate molar teeth, it resembles a small species of *Pteralopex*, although the teeth are less complex than any member of that genus. Almeida *et al.* (2014) determined it is not particularly closely related to *Pteralopex*. Nonetheless, genetic divergence was sufficient to recognise the New Caledonia flying fox as belonging to its own monotypic *vetula* species group, sister to the *samoensis* species group (Almeida *et al.* 2014).

A single specimen was identified from the Mé Auré Holocene cave deposits (Hand and Grant-Mackie 2011). Few individuals of this species were encountered during surveys in the early and mid-1900s. Revilliod (1914) failed to capture it and Sanborn and Nicholson (1950) obtained only a single individual among numerous *P. ornatus*. Felten (1964a) provided the first indication that the New Caledonia flying fox may be more common, with the capture of 13 individuals, and it was the most common species detected during surveys in May 1991. It is possible it has increased in abundance as larger sympatric *Pteropus* have declined, or that rarity in earlier surveys reflected inadequacies in sampling techniques. Even observational data indicated it may have been the most common *Pteropus* on New Caledonia in 1991. Relative abundances need to be confirmed as *P. ornatus* seems the most commonly observed species today (M. Oedin pers. comm.).

The New Caledonia flying fox has been recorded from both primary and secondary tropical forest. It is entirely nocturnal and roosts at the entrance of caves, or in hollow trees (Brescia 2008). On Mount Koghi, this species was observed flying under the forest canopy (where it may feed), but was not seen soaring above the canopy in early morning and evening as does *P. ornatus*. This behaviour of remaining below the canopy is again analogous to species of *Pteralopex*. Given the dentition, New Caledonia flying fox probably feeds on harder fruits and materials.

Information on reproduction is scarce. Seven individuals captured in the Thio area in early April (now held in collections of the Australian Museum)

The New Caledonia flying fox (*Pteropus vetula*) is a highly unusual species of *Pteropus* with external features, cuspidate molar teeth, and behavioural traits that resemble a small species of *Pteralopex* (Keiko and Yushi Osawa).

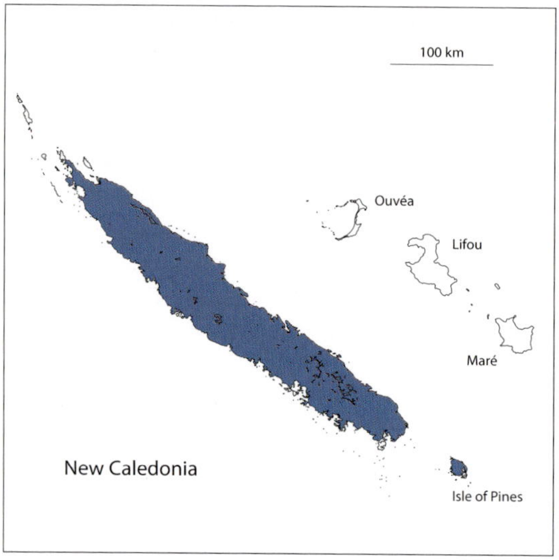

Pteropus vetula			
Size*			
	Mean	**Range**	**Sample**
Head–body	134.0	122.5–142.5	5
Tail	–	–	–
Forearm	99.2	92.3–112.1	5
Tibia	46.4	38.0–54.7	5
Hindfoot			
Ear	14.8	13.0–15.7	5
Weight	152.8	120.0–220.0	5
* From AM specimens (New Caledonia)			
Identification A small dark black to grey flying fox, sometimes with tinge of brown. The fur is dense and luxurious with a subtly demarcated composed of long hairs coloured the same as the body. The ears are short and rounded with a thickened rim and mostly hidden in the fur. The face is short and the molar teeth possess sharp cusps.			
Synonyms *Pteropus germaini* Dobson, 1879 *Pteropus macmillani* Tate, 1942b			
Subspecies None			
Distribution Île de Pines, Grand Terre (New Caledonia)			
Extralimital None			
Conservation status (IUCN) Near Threatened			
Etymology Sp.–grey or of long standing (diminutive) (L.), perhaps an allusion to its colour. Broadly referred to as *P. vetulus*, but the original specific epithet *vetula* is correct.			

were young, probably in their first year of life. The significance of this is uncertain, but it is possible adults avoid mist nets, breeding is seasonal, or that there is heavy mortality in the adult population.

The New Caledonia flying fox is classified as Vulnerable under the IUCN Red List because its extent of occurrence is small, and there is an ongoing decline in extent and quality of habitat.

DWARF FLYING FOX

Pteropus woodfordi Thomas, 1888c

wood'-ford-ee

The Dwarf flying fox is one of the world's smallest *Pteropus*, belonging to the *pelagicus* species group (Almeida *et al.* 2014). It is common throughout its range in the southern and western Solomon Islands and is allopatric with *P. mahaganus*, probably ecologically replaced by this species on Bougainville, Choiseul and Santa Isabel. Although the two species look rather different externally, they have similar cheek teeth and are sister species (Almeida *et al.* 2014).

The dwarf flying fox occurs in a wide variety of habitats from village gardens, secondary forests, and primary lowland and montane forests up to 1230 m elevation. The species has been found roosting in the hollows of large strangler figs (*Ficus* spp.) in lowland forests, and sharing trees with other flying fox species including the New Georgia Group

The dwarf flying fox is one of the world's smallest *Pteropus* and is commonly seen on dusk visiting shoreline coconut palms (Tyrone Lavery).

Pteropus woodfordi			
Size*			
	Mean	**Range**	**Sample**
Head–body	114.5	92.7–145.0	41
Tail	–	–	–
Forearm	90.9	82.5–96.1	41
Tibia	41.6	32.0–46.0	41
Hindfoot	27.4	26.4–28.7	8
Ear	11.7	9.0–14.8	39
Weight	152.2	97.0–206.0	41
* Field measurements from Gatokae, Kolombangara, New Georgia, Tetepare, Vella La Vella and Vangunu			
Identification Smallest flying fox in Solomon Isls (forearm length < 100 mm). Snout and cheek teeth narrow. Colour is overall brown with variable quantities of silver-grey or gold hairs, mantle rufous.			
Synonyms *Pteropus austini* Lawrence, 1945			
Subspecies None			
Distribution Arnarvon Isls (Kerehikapa and Sikopo), Gatokae, Guadalcanal, Kolombangara, Malaita, New Georgia, Ngella Pile, Ngella Sule, Ranongga, Russell Isls (Mbanika and Pavuvu), Tetepare, Vangunu, Vella La Vella			
Extralimital None			
Conservation status (IUCN) Least Concern			
Etymology for C. M. Woodford, District Commissioner of the British Protectorate of Solomon Islands and collector of the holotype			

monkey-faced bat (*Pteralopex taki*) and Solomons flying fox (*Pteropus rayneri*).

The highly reduced cheek teeth are indicative of the mostly nectarivorous diet. Coconut flowers (*Cocos nucifera*) seem to provide an important resource for this species. They are common visitors to the palms along island foreshores and are active and agile climbers among the fronds. Several specimens in the Australian Museum were collected while feeding on coconut flowers and flowering kapok (*Ceiba pentandra*). Lawrence (1945) reported dwarf flying foxes feeding on flowers with lories and lorikeets, among the new fronds in late afternoon. The bats deserted the trees when flowering finished in June. Sanborn (1931) obtained specimens of this species while they were feeding on young green coconuts on Kolombangara Island.

Dwarf flying foxes make a high-pitch chattering call, and adult males produce a pungent, fruity odour and have very large, prominent testes. Lactating females have been captured on Kolombangara Island in May and on New Georgia in early October. Pregnant females have been recorded on New Georgia on 4 May and 9 October. A female carrying a dependent young 43% of the mother's bodyweight was captured in a mist net on Gatokae Island on 31 March 2015.

Within the distribution of dwarf flying fox, the greatest anthropological disturbance is commercial logging. However, comparisons of mist net capture rates in the Western Province over a 23-year period indicated populations have remained stable despite the widespread losses of primary forests (Lavery *et al.* 2020a).

Rousettus

GEOFFROY'S ROUSETTE BAT

***Rousettus amplexicaudatus* (Geoffroy, 1810)**

rue-set'-us am-plex'-ee-kaw-dah'-tus

Geoffroy's rousette bat is a very common and extremely widespread species found from southern China and Myanmar through Wallacea and into Melanesia as far east as Makira in the eastern Solomon Islands. Three subspecies are recognised and the subspecies (*Rousettus amplexicaudatus brachyotis*) that occurs throughout much of the region under consideration here is smaller and more slightly built than the nominal subspecies.

Geoffroy's rousette bat can be found in a wide variety of habitats from lowland and montane primary forests to disturbed villages, gardens and plantations. In fact, in the South-west Pacific it seems to be rarer in primary forests. Smith and Hood (1981) found it to be among the most common bats on New Ireland and New Britain. Similarly, in the Western Province of Solomon Islands it was among the species most commonly captured in mist nets in 1992 and 2015 (Lavery *et al.* 2020a). Koopman (1979) listed it as present on New Ireland, Tabar, Duke of York and Bagabag, and Lidicker and Ziegler (1968) provided a record from Emirau Island. Curiously, extensive survey work has failed to detect it on the large islands off south-east New Guinea (D'Entrecasteaux Islands, Kiriwina, Woodlark and the Louisiade Archipelago). There is no obvious explanation for these absences, but other widespread species show a similar distributional gap in this region (e.g. *Macroglossus minimus*).

Geoffroy's rousette bat is a cave-roosting species, and typically hangs in large clumps on the ceilings. Although it is one of the species most commonly captured in mist nets in the South-west Pacific, roost sites are encountered far less frequently than those of other cave-dwelling bats (e.g. *Dobsonia* species). This may be because Geoffroy's rousette bat populations consolidate into very large colonies among fewer, better suited caves. One roost at Kavolavata, Gatokae, Solomon Islands, was in a vast limestone tunnel system that had been carved out by

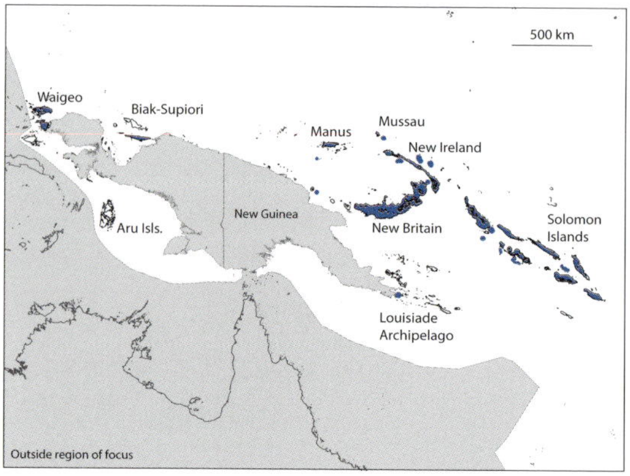

Rousettus amplexicaudatus			
Size*			
	Mean	**Range**	**Sample**
Head–body	84.4	61.9–115.0	48
Tail	17.0	12.3–20.6	11
Forearm	69.2	61.0–78.2	71
Tibia	30.8	25.0–35.5	65
Hindfoot	15.8	9.6–20.0	13
Ear	14.5	10.5–18.0	70
Weight	44.8	30.0–76.0	66
* Field measurements from the Solomon Isls archipelago and Bismarck Archipelago			

Identification A medium-sized cave-dwelling fruit bat with a long snout and sparsely haired back and face. The eyes are large, and there is a claw on the second digit of the wing. Adult males have orange epaulettes of hair on the shoulders.

Synonyms *Eleutherura philippinensis* Gray, 1870
Cynonycteris minor Dobson, 1873
Cynonycteris bocagei Seabra, 1898
Rousettus amplexicaudatus stresemanni Stein, 1933
Rousettus amplexicaudatus hedigeri Pohle, 1952

Subspecies *Rousettus amplexicaudatus amplexicaudatus* (Geoffroy, 1810) (mainland South-east Asia, Philippines, Maluku, New guinea, east Nusa Tenggara)
Rousettus amplexicaudatus infumatus (Gray, 1870) (Sumatra, Krakatau, Java, Bsali, west Nusa Tenggara)
Rousettus amplexicaudatus brachyotis (Dobson, 1877) (Bismarck Archipelago, Solomon Isls)

Distribution Bougainville (AROB); Batanta, Salawati, Waigeo, Yapen (Indonesia); Bagabag, Duke of York, Dyual, Emirau, Lavongai, Lihir, Manus, Mioko, Mussau, New Britain, New Ireland, Sariba, Sideia, Tabar, Ulu (Papua New Guinea); Choiseul, Fauro, Gatokae, Guadalcanal, Kolombangara, Makira, Malaita, New Georgia, Ngella Pile, Ngella Sule, Ranongga, Rendova, San Jorge, Santa Isabel, Tetepare, Vangunu, Vella La Vella (Solomon Isls)

Extralimital Brunei Darussalam, Cambodia, China, Indonesia, Laos, Malaysia, Myanmar, Philippines, Singapore, Thailand, Timor-Leste, Vietnam

Conservation status (IUCN) Least Concern

Etymology Gen.–for the rosettes of orange glandular hairs on each side of the neck in mature males; Sp.–surrounding or encircling tail (L.), significance unknown

Geoffroy's rousette bat (*Rousettus amplexicaudatus*) is a widespread, cave-roosting species. It is one of the species most commonly captured in mist nets and occurs in a range of habitats including primary forests and village gardens (Tyrone Lavery).

a freshwater stream. It was shared with huge colonies of *Hipposideros dinops* and *H. cervinus*. Smith and Hood (1981) found them only in natural caves in the Bismarck Archipelago, being conspicuously absent from excavated tunnels near Rabaul, East New Britain. Geoffroy's rousette bat always roosts just beyond the twilight zone in areas protected from light and when disturbed will fly into the deeper parts of the cave. Species of the genus *Rousettus* can echolocate by tongue-clicking and are thus capable navigators in the darker recesses of caves (Novick 1955).

In the South-west Pacific the diet of Geoffroy's rousette bat includes fruit and flowers. Cultivated trees with small fruits such as Japanese cherry (*Muntingia calabura*) when planted in village environments will attract large numbers of this species in early evenings. The bats pass back and forth over the tree before cautiously perching to feed on the fruits. The fruit of native figs (*Ficus* sp.) must also be a common food item as their seeds are very often passed when Geoffroy's rousette bat is captured in a mist net.

McKean (1972) suggested this species breeds throughout the year on Bougainville. On Buka a newborn was recorded in January 1997, and five of 16 other bats captured were subadults (Bonaccorso 1998). In the Western Province of Solomon Islands, pregnant and lactating females were present in April, May and October, and subadult animals can be captured in most months of the year. Pregnant and lactating females were encountered in December on the south coast of New Britain, and (Bonaccorso 1998) recorded pregnant and lactating females in the Bismarck Archipelago in June, July and August. In the Philippines, breeding is highly synchronised and follows a pattern of seasonal bimodal polyoestry, with births taking place from March–April and August–September (Heideman and Utzurrum 2003). The scant data suggest a similar bimodal pattern could exist in Melanesia.

Geoffroy's rousette bat could be declining in parts of its range as a result of deforestation and persecution. However, in the Western Province of Solomon Islands, capture rates had not declined over a 23-year period during which extensive commercial logging had occurred (Lavery *et al.* 2020a).

Syconycteris

COMMON BLOSSOM BAT

Syconycteris australis **(Peters, 1867)**

sie'-kon-ik'-te-ris os-trah'-lis

The common blossom bat is a widespread and well-studied species found in eastern Australia, mainland New Guinea and nearby islands including Seram and Buru to the west and New Britain, New Ireland and Milne Bay islands to the east. It is a species complex that comprises multiple, coexisting, subtly different forms (Armstrong and Aplin 2011). Aplin and Opiang (2011) considered that the subspecies *Syconycteris australis finschi* of the Bismarck Archipelago almost certainly deserves recognition as a full species.

On certain islands such as those in Milne Bay at the south-eastern end of New Guinea, and New Britain, this is the common nectar-feeding bat. For example, surveys on New Britain in 2014 found the common blossom bat was the most commonly captured bat in mist nets and harp traps at low-elevation sites, and it was captured at a site at 1590 m elevation (Aplin and Opiang 2011). However, on New Ireland it is apparently rare. During several months of field work, Smith and Hood (1981) were able to capture this species on just one occasion. Bonaccorso (1998) included Manus Island in the distribution of common blossom bat; however, Aplin *et al.* (2015) failed to confirm this, as did surveys by Australian Museum researchers in 1988.

The Southern blossom bat occurs sympatrically with the closely related and superficially similar species *Macroglossus minimus*. However, the two can be readily separated by the presence of a very small membrane on the hindlimbs of *M. minimus*, and the complete lack of any such membrane in *S. australis*. These two are ecologically separated on mainland New Guinea, with the diet of *S. australis* incorporating a proportion of fruit, whereas *M. minimus* feeds exclusively on nectar and pollen (Bonaccorso 1998).

Sex ratios of common blossom bat can vary greatly between survey sites and elevations. Bonaccorso (1998) reported that surveys over a 14-month period at Crater Mountain, Papua New Guinea found a male bias of around 2.4:1. At two sites on New Britain, captures at a 200 m elevation site were male-biased (2.9:1), whereas results from a site at 860 m elevation were slightly female-biased (1.6:1) (Aplin and Opiang 2011). Anthony (2001) found a similar pattern on the island, with a male bias at a lowland site (2.5:1), and a site at 800–980 m elevation (3.3:1), and more even ratios above 1000 m (12 males and seven females). The reasons for such biases are unclear, but could reflect genuine demographic variability between sites or sampling biases due to home range differences (e.g. males having larger home ranges at lower elevations and higher probabilities of being intercepted by mist nets).

A radio-tracking study at the Kau Wildlife Area, Madang Province, Papua New Guinea, estimated home-range sizes for 11 individuals of 2.7–13.6 ha (average 5.5 ha) (Winkelmann *et al.* 2000). The bats roosted in riparian or old growth forests among large-leaved understorey trees, and showed high fidelity to a finite area, but not specific trees.

In Melanesia, females give birth to a single young in two breeding seasons per annum. At Crater Mountain, Papua New Guinea, surveys between 12 and 17 March 1992 found three of nine adult females captured were pregnant, indicating births occur there sequentially between November and July (Bonaccorso 1998). Pregnant or lactating females have been encountered in other parts of New Guinea between January and November. The common blossom bat is extremely common, no major threats have been identified, and there are no indications populations are significantly declining.

Syconycteris australis			
Size*			
	Mean	**Range**	**Sample**
Head–body	65.1	59.7–72.0	9
Tail	–	–	–
Forearm	41.6	36.6–46.7	14
Tibia	15.8	12.3–17.7	8
Hindfoot	10.5	7.6–12.5	6
Ear	13.4	9.0–16.0	9
Weight	16.9	11.2–21.0	14
* From AM specimens (Woodlark) and Bonaccorso (1998)			
Identification A small brown, nectar-feeding bat similar to *Macroglossus minimus*, but usually less warm-brown in colour, with larger incisors and entirely lacking any sign of a uropatagium on the internal margins of the legs.			
Synonyms None			
Subspecies *Syconycteris australis australis* (Peters, 1867) (Australia) *Syconycteris australis crassa* (Thomas, 1895) (Trobriand and D'Entrecasteaux Isls) *Syconycteris australis finschi* (Matschie, 1899) (Bismarck Archipelago) *Syconycteris australis papuana* (Matschie, 1899) (New Guinea and nearby islands) *Syconycteris australis keyensis* Andersen, 1911 (Kai Is.) *Syconycteris australis major* Andersen, 1911 (Central Maluku) *Syconycteris australis naias* Andersen, 1911 (Woodlark)			
Distribution Aru Isls, Batanta, Biak-Supiori, Salawati, Yapen (Indonesia); Bagabag, Crown, Fergusson, Goodenough, Kairuru, Kiriwina, Long, ?Manus, Misima, New Britain, New Ireland, Normanby, Rossel, Sakar, Sideia, Sudest, Tolokiwa, Vokeo, Woodlark (Papua New Guinea)			
Extralimital Australia, Maluku Isls, New Guinea			
Conservation status (IUCN) Least Concern			
Etymology Gen.–fig bat (Gr.); Sp.–southern (L.)			

The common blossom bat (*Syconycteris australis*) looks very similar to *Macroglossus minimus*. It can be differentiated from *M. minimus* by its complete absence of a uropatagium membrane on the internal margins of the legs (Angus Fraser).

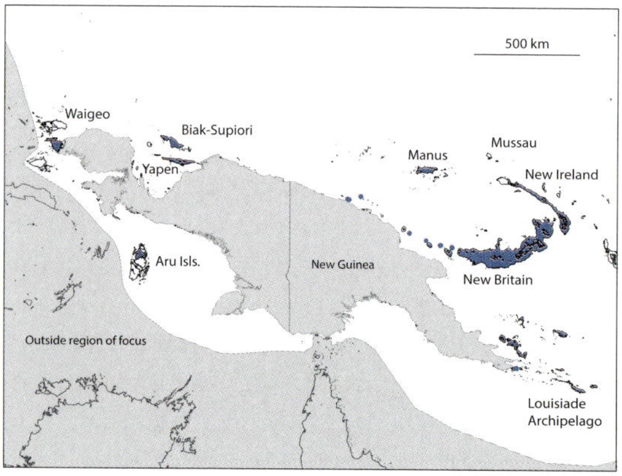

Emballonuridae

The family Emballonuridae consists of 15 genera and over 50 species, and it is widely distributed in tropical and subtropical regions. All possess a tail that protrudes above the uropatagium. Within the region under consideration here, emballonurids are represented primarily by members of the genus *Emballonura* (eight species worldwide, five in the region). Of these, one (*Emballonura serii*) is endemic to the region.

They are widespread, almost ubiquitous bats that are readily seen because they fly early in the evening, and some species roost around villages. *Emballonura semicaudata* is the only member of Yangochiroptera found in Micronesia and islands west and north of Fiji, thus extending further into the Pacific than any other echolocating bat. It is now extinct from many parts of its former distribution and has suffered serious decline where it is still found today. The genera *Mosia* (monotypic) and *Saccolaimus* are each represented in the region by just a single species, both are more widespread into New Guinea, parts of Asia and, in the case of *Saccolaimus saccolaimus*, northern Australia.

The lesser Sheath-tailed Bat (*Mosia nigrescens*) is a widespread and common member of the family Emballonuridae and is one of the most commonly observed bats in early evenings in parts of Melanesia (Michael Pennay).

Emballonura

BECCARI'S SHEATH-TAILED BAT

Emballonura beccarii Peters and Doria, 1881

em-bal'-on-yeu'-rah bek-ah'-ree-ee

Beccari's sheath-tailed bat is a distinctive species with a shortened face and narrow, pointed ears. It occurs from the Bismarck Archipelago and D'Entrecasteaux Islands across to the Maluku Islands. This species seems to be more common east of New Guinea but must be considered uncommon throughout its distribution.

Smith and Hood (1981) encountered Beccari's sheath-tailed bat regularly on New Ireland, but did not find it elsewhere in the Bismarck Archipelago. It was discovered hanging in the twilight zone on the walls of a cave near Lesilaba Mission that was shared with *Emballonura dianae*. A sea cave on Alcester was shared with *Mosia nigrescens* and all 18 Beccari's sheath-tailed bats examined there on 15 August 1987 were male, suggesting the sexes roost separately in particular seasons. Two females collected on New Guinea in October were both lactating, and one was pregnant in August with a 7 mm long embryo (Bonaccorso 1998).

Beccari's sheath-tailed bat forages for insects in riparian zones, dense forests and open areas adjacent to forest (Bonaccorso 1998). The stomach of one animal examined in New Guinea contained traces of a beetle (Vestjens and Hall 1977). Echolocation calls are distinctive but very similar to those of *M. nigrescens*. The pulse shape also resembles

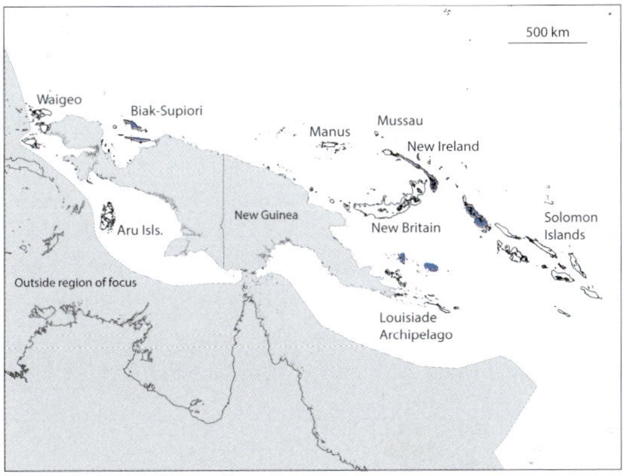

Emballonura beccarii			
Size*			
	Mean	**Range**	**Sample**
Head–body	40.9	34.1–50.0	5
Tail	14.3	10.0–22.0	6
Forearm	39.8	37.5–44.6	8
Tibia	15.3	14.0–17.0	6
Hindfoot	7.3	6.0–8.0	4
Ear	13.9	12.0–18.0	6
Weight	4.5	3.5–5.3	4
* From AM specimens (Alcester), Bonaccorso (1998) and Smith and Hood (1981)			
Identification A very distinctive species, it is small in size being the smallest species of *Emballonura* in the region. The face is 'pushed in', and hair from the crown reaches the tip of the nose when adpressed forward. The eyes are small and largely hidden by the fur and lips. The ears are narrow and pointed and the tragus is rectangular in shape.			
Synonyms *Emballonura meeki locusta* Thomas, 1920			
Subspecies *Emballonura beccarii beccarii* Peters and Doria, 1881 (north-west New Guinea, Biak and Yapen Isls, Bismarck Archipelago and Solomon Isls) *Emballonura beccarii meeki* Thomas, 1896 (south-east New Guinea, Trobriand Isls) *Emballonura beccarii clavium* Thomas, 1915b (Kai Isls)			
Distribution Bougainville (AROB); Biak-Supiori, Yapen (Indonesia); Alcester, Kiriwina, New Ireland, Woodlark (Papua New Guinea)			
Extralimital Kai Isls, New Guinea			
Conservation status (IUCN) Least Concern			
Etymology Sp.–for Dr O. Beccari, collector of the holotype			

Hipposideros with a characteristic frequency of around 60–70 kHz (Armstrong 2019). Beccari's sheath-tailed bat is classified as Least Concern on the IUCN Red List, since it has a wide distribution and presumed large population size.

LARGE-EARED SHEATH-TAILED BAT

Emballonura dianae Hill, 1956

dee-ah'-nee

This is a relatively uncommon bat with a peculiar disjunct distribution. As with many other species in the region, this may be a reflection of incomplete sampling rather than genuine absence. The large-eared sheath-tailed bat is found in the Solomon Islands archipelago (Bougainville, Guadalcanal, Malaita, Rennell and Santa Isabel), New Ireland and New Britain. Echolocation call recordings suggest it also occurs on Mussau (Armstrong *et al.* 2015). Three distinct subspecies have been described (Flannery 1994b). There are significant size differences between these subspecies, with animals from the Bismarck Archipelago and north-west Solomons (Bougainville and Santa Isabel) being larger than those from Malaita, Guadalcanal and Rennell. On mainland New Guinea it has been documented at 1400 m elevation using a mist net placed near a light that was attracting insects (Bonaccorso *et al.* 2008).

Large-eared sheath-tailed bats are commonly found in limestone caves in areas of rainforest where they roost just beyond the twilight zones. They have been encountered sharing caves with several other species. On northern Guadalcanal, Gumburota Cave was shared with *Dobsonia inermis* and *Hipposideros dinops* and in Kovi Cave it was found roosting with these species and *Aselliscus tricuspidatus*, *H. calcaratus*, *H. cervinus*, *H. diadema* and *Miniopterus australis*. An Australian Museum team encountered it with *A. tricuspidatus* and *E. raffrayana* in Plomognake Cave near Ngulahahgi Village on Santa Isabel on 26 November 1987. In

In Kovi Cave, Guadalcanal, large-eared sheath-tailed bat (*Emballonura dianae*) can be found roosting with up to seven other species of bat (Tyrone Lavery).

Madina Cave, New Ireland Smith and Hood (1981) found it with *E. furax* and *E. raffrayana* and near Lesilaba Mission it occurred with *E. beccarii*. *Emballonura dianae* was hanging singly, and widely separated on the walls.

The large-eared sheath-tailed bat has a multiharmonic echolocation call consisting of flat frequency-modulated pulses. The main second harmonic is the characteristic frequency: on Mussau this was approximately 35 kHz (Armstrong *et al.* 2015) and on Guadalcanal it was 34.7 kHz (Pennay and Lavery 2017).

Hill (1971) reported nine specimens (five of which were immature) from Malaita, collected from Riba Cave in October and November. A single female captured at 1400 m on New Britain in June 1995 was lactating, and two of three females examined at Medina, New Ireland, in September 1976 were carrying embryos (Bonaccorso 1998).

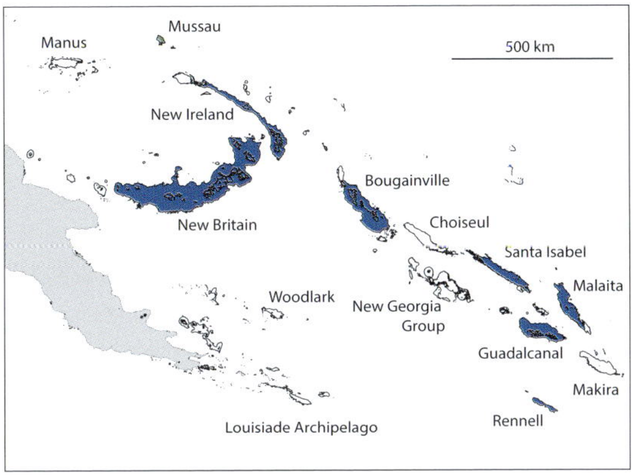

Emballonura dianae			
Size*			
	Mean	**Range**	**Sample**
Head–body	49.8	42.3–53.2	5
Tail	12.6	10.4–14.3	4
Forearm	44.4	41.7–49.4	11
Tibia	17.2	15.7–18.8	11
Hindfoot	8.5	7.9–9.1	4
Ear	15.4	11.8–17.9	10
Weight	7.7	7.0–8.0	3
* From AM and QM specimens (Guadalcanal and Santa Isabel) and Bonaccorso (1998)			
Identification A large emballonurid distinguishable by its very broad ears with rounded tips, pale belly, and tragus with a pointed projection on the front of the upper edge.			
Synonyms None			
Subspecies *Emballonura dianae dianae* Hill, 1956 (Guadalcanal, Malaita, Rennell) *Emballonura dianae fruhstorferi* Flannery, 1994b (New Guinea) *Emballonura dianae rickwoodi* Flannery, 1994b (Bismarck Isls, Bougainville, Santa Isabel)			
Distribution Bougainville (AROB); ?Mussau, New Britain, New Ireland (Papua New Guinea); Guadalcanal, Malaita, Rennell, Santa Isabel (Solomon Isls)			
Extralimital New Guinea			
Conservation status (IUCN) Least Concern			
Etymology Sp.–for the collector of the holotype, Mrs Diana Bradley			

NEW GUINEA SHEATH-TAILED BAT

Emballonura furax Thomas, 1911a

fyue'-rax

The New Guinea sheath-tailed bat is a rarely captured species with a disjunct distribution spanning parts of the central cordillera of Papua New Guinea and Yapen and Biak islands. It is probably more widespread including in West Papua, and acoustic survey data have indicated it can be locally common (Armstrong and Aplin 2017). Its presence on Yapen and Biak islands suggests that taxonomic relationships within *Emballonura furax* and *E. serii* require further assessment (Armstrong and Aplin 2017).

The New Guinea sheath-tailed bat has been recorded in lowland rainforest, hill forest and lower montane forest (Armstrong and Aplin 2017). Acoustic recordings and wing morphology suggest it is a species that forages in open spaces, such as the areas above forest canopies and clearings. Roosts have been recorded in limestone caves and colonies usually contain only a few dozen individuals. The New Guinea sheath-tailed bat has

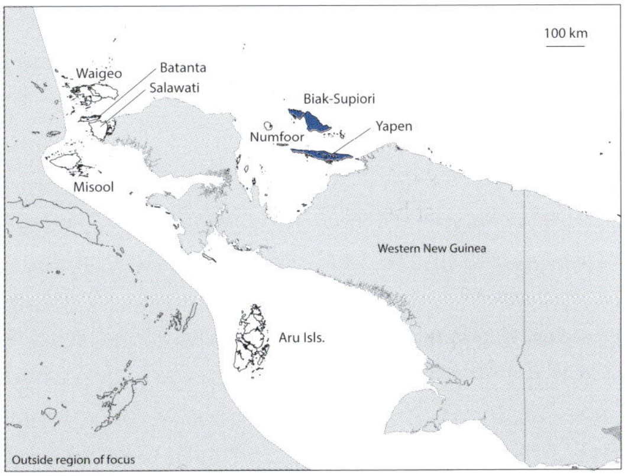

Emballonura furax			
Size*			
	Mean	**Range**	**Sample**
Head–body	56.9	54.5–61.0	4
Tail	14.7	10.0–19.0	4
Forearm	51.1	48.3–53.0	6
Tibia	19.1	18.3–19.8	4
Hindfoot	8.1	6.0–9.5	4
Ear	17.6	16.1–19.4	5
Weight	11.8	9.6–14.3	5
* From Thomas (1911a), McKean (1972), Flannery (1994b) and Bonaccorso (1998)			
Identification A large sheath-tailed bat with a forearm length greater than 48 mm – most other species of the genus in the region have a forearm shorter than this. The tragus is long and narrow compared to *E. dianae*, and the posterior margin is flat, not curling forward as in *E. serii*.			
Synonyms None			
Subspecies None			
Distribution Biak-Supiori, Yapen			
Extralimital New Guinea			
Conservation status (IUCN) Least Concern			
Etymology Sp.–thievish (L.).			

been documented sharing these with larger colonies of *E. dianae* and *E. raffrayana* (Flannery 1995a). Survey data from recent years have indicated this species may be more common than previously assumed and few major threats have been identified.

RAFFRAY'S SHEATH-TAILED BAT

Emballonura raffrayana Dobson, 1879

raf'-ray-ah'-nah

Raffray's sheath-tailed bat is a medium-sized species with a wide distribution, but it is relatively uncommon. It has been recorded on New Ireland (Smith and Hood 1981), Santa Isabel and Malaita (Hill 1971). The Australian Museum holds a large series of specimens from Santa Isabel, New Georgia, Malaita and Guadalcanal.

Raffray's sheath-tailed bat roosts on the vertical walls of caves just beyond the twilight zone. Colonies of 10–30 individuals have been found on mainland New Guinea. One roost in Sandaun Province contained around 10 bats, all of which were males (Flannery 1995a). Smith and Hood (1981) described two specimens of this species from Medina Cave, New Ireland, where it was found co-occurring with *E. dianae* and *E. serii*. However, the forearm lengths recorded (46.6 mm and 47.2 mm) are outside of the normal size range for this species and may instead be measurements taken from *E. dianae*.

In July 2019, a small group of around 10 individuals were encountered roosting among dimly lit large boulders and shallow limestone caves on the coast of Mono Island. It was the only species roosting there. Five animals were examined, of which three were female. Two of these had enlarged teats but were not lactating. On New Guinea in May 1986 at 1000 m elevation in the Thurnwald Range,

Raffray's sheath-tailed bat (*Emballonura raffrayana*) (Madison Erin Mayfield).

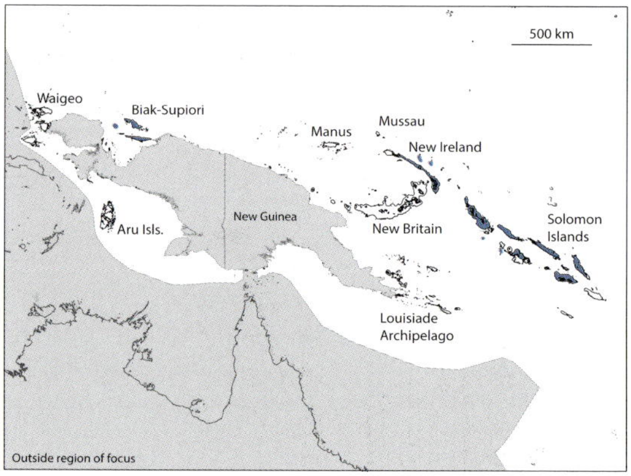

Emballonura raffrayana			
Size*			
	Mean	**Range**	**Sample**
Head–body	47.5	37.5–58.0	16
Tail	12.7	10.5–14.8	11
Forearm	40.1	36.9–42.7	16
Tibia	15.7	12.2–17.9	15
Hindfoot	7.1	5.0–9.0	11
Ear	13.6	11.0–17.0	16
Weight	5.8	3.9–10.2	15
* From AM and KU specimens (Choiseul, Mono), and Bonaccorso (1998)			
Identification Differs from *E. alecto*, *E. serii* and *E. dianae* by its shorter forearm. It has a narrow muzzle and very large eyes, being relatively larger than in any other *Emballonura*.			
Synonyms None			
Subspecies *Emballonura raffrayana raffrayana* Dobson, 1878 (New Guinea and nearby islands) *Emballonura raffrayana stresimanni* Thomas, 1914a (Seram) *Emballonura raffrayana cor* Thomas, 1915b (Solomon Isls, New Ireland and satellites)			
Distribution Bougainville, Buka (AROB); Biak-Supiori, Numfoor, Yapen (Indonesia); Lihir, New Ireland, Tabar (New Guinea); Choiseul, Guadalcanal, New Georgia, Malaita, Mono, Ranongga, Santa Isabel (Solomon Isls)			
Extralimital Maluku Isls, New Guinea			
Conservation status (IUCN) Least Concern			
Etymology Sp.–for the discoverer of the type specimen, M. Raffray.			

Flannery (1995a) reported encountering a roost of this species that contained males, females and dependent juveniles. Bonaccorso (1998) also reported a nursing juvenile encountered on New Ireland in May.

PACIFIC SHEATH-TAILED BAT

Emballonura semicaudata (Peale, 1848)

sem'-ee-kaw-dah'-tah

The Pacific sheath-tailed bat was once a widespread and common species of smaller or more remote islands of the South-west Pacific. It is a medium-sized, brown, cave-dwelling species, and is the only insectivorous bat to have occurred in large parts of the Pacific. The species has four recognised subspecies (Koopman 1997) and mitochondrial DNA sequencing has indicated three of these (*Emballonura semicaudata semicaudata*, *E. s. palauensis* and *E. s. rotensis*) could be sister species (Oyler-McCance *et al.* 2013). The fourth subspecies (*E. s. sulcata*) was not included in the study and requires further evaluation.

Inclusion of the Pacific sheath-tailed bat among the faunas of Vanuatu was derived from a single specimen supposedly collected on Espiritu Santo in 1929. Its original number does not match that in the collector's catalogue, and it is plausible it was actually collected in Fiji. It seems unlikely that the Pacific sheath-tailed bat ever occurred in Vanuatu, as a diverse insectivorous bat fauna exists there and this species is not known to coexist with such a fauna elsewhere.

The Pacific sheath-tailed bat is sexually dimorphic, with females being larger than males. It roosts in the twilight zones of limestone caves and is quick to scatter at signs of light or disturbance. The bats normally hang from the open ceilings rather than crevices, with individuals spaced about 10 cm apart (Wiles and Conry 1990). They emerge from caves on dusk, commencing from around 5 min before sunset and by around 12 min after sunset most individuals have departed (Wiles *et al.* 2011).

This species has a weak and fluttering flight and forages by aerial-hawking insects from above or below native forest canopies, and occasionally from more disturbed areas like roads and coconut plantations (*Cocos nucifera*). Esselstyn *et al.* (2004) found that activity levels (as identified by echolocation call passes) were drastically lower in unforested areas compared to native or non-native forests. Analysis of faecal material from the subspecies *E. s. rotensis* on Aguiguan, Mariana Islands, showed large numbers of wasps, moths and beetles were consumed (Valdez *et al.* 2011). Females give birth to a single young per year. In Aguiguan, this coincides with the rainy season between June and July.

The Pacific sheath-tailed bat has suffered significant declines in many parts of its range and in each case the causes of decline remain unclear. *Emballonura semicaudata rotensis* was reportedly abundant before 1950 in the Mariana Islands, but by the 1950s colonies were dwindling and by 1972 they were apparently gone from all islands except Aguiguan (Lemke 1986). In 2008, the population on Aguiguan was estimated at 359–466 individuals across three known cave roosts, and around 330 of these were roosting in a single cave (Wiles *et al.* 2011). There, Welch *et al.* (2016) suspected that widespread invasion by the introduced weed lantana

(*Lantana camara*) and presence of feral goats (*Capra aegagrus hircus*) might be altering the structure of island vegetation communities to the detriment of the Pacific sheath-tailed bat. Conversely, populations of the subspecies *E. s. sulcata* and *E. s. palauensis* were considered relatively stable in the 1990s (Wiles *et al.* 1997) and up until 2005 in some regions (Waldien and Scanlon 2021). More recent surveys are lacking.

A major decline of *E. s. semicaudata* has occurred in Fiji, exceeding 80% over a period of 10 years (Scanlon *et al.* 2013a). An Australian Museum survey of Viti Levu failed to detect the presence of Pacific sheath-tailed bats in the 1990s. Similarly, Palmeirim *et al.* (2007) concluded it had been extirpated from Viti Levu, but colonies were found on Vanua Levu and a small pair of islets (Aiwa) in the Lau Archipelago. A cave on Yaqeta in the Yasawa Archipelago supported one of the largest colonies, with several hundred individuals present. By the time surveys were completed, between 2008 and 2011, the Pacific sheath-tailed bat was apparently absent from Vanua Levu and was considered probably extinct there (Scanlon *et al.* 2013a). Likewise, it could not be detected on Rotuma in 2018 and was deemed to have been extirpated, with the last reports of the species being from the year 1991 (Cibois *et al.* 2019). Palmeirim *et al.* (2007) considered roost disturbance as playing a part in the decline of Fiji's Pacific sheath-tailed bats, including larger-scale activities such as the mining of guano. However, on Rotuma Cibois *et al.* (2019) suggested cave disturbance was an unlikely cause of decline, as were introduced predators and transformation of primary forests. Use of pesticides or a new pathogen were instead offered as possible explanations.

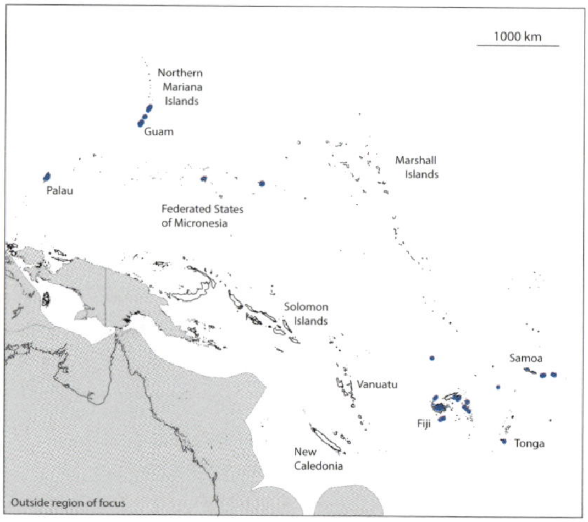

Emballonura semicaudata			
Size*			
	Mean	**Range**	**Sample**
Head–body	47.7	41.0–53.0	7
Tail	17.0	17.0	2
Forearm	45.5	39.0–52.5	15
Tibia	19.7	18.6–21.6	5
Hindfoot	8.0	8.0	2
Ear	11.6	10.0–13.6	7
Weight	6.2	5.1–8.0	9

* From AM specimens (Fiji), Lemke (1986) and Wiles *et al.* (2011)

Identification The only species of bat within its area of distribution that has the tail protruding from the dorsal part of the uropatagium (membrane between the hindlimbs). The dorsal fur is brown and the wing membranes are dark; the eyes are large.

Synonyms *Emballonura fuliginosa* Tomes, 1859

Subspecies *Emballonura semicaudata semicaudata* (Peale, 1848) (Samoa, American Samoa, Fiji, Tonga)
Emballonura semicaudata sulcata Miller, 1911 (Chuuk and Pohnpei Is)
Emballonura semicaudata palauensis Yamashima, 1932 (Palau)
Emballonura semicaudata rotensis Yamashima, 1943 (Northern Mariana Isls and Guam)

Distribution Ta'u† (American Samoa); Chuuk, Pohnpei (Federated States of Micronesia); Aiwa Isls, Kuata, Lakeba, Ovalau, Rotuma†, Taveuni, Vanua Levu†, Viti Levu†, Yaqeta (Fiji); Guam†; Aguiguan, Rota†, Saipan†, Tinian† (Northern Mariana Isls); Anguar, Babeldaob, Koror, Mecherchar, Ngebedangel, Ngeruktabel, Ongael, Peleiu, Ulong (Palau); Savai'i†, Upolu† (Samoa); 'Eau†, Niuafo'ou† (Tonga)

Extralimital None

Conservation status (IUCN) Endangered

Etymology Sp.–part tail (L.), possibly a reference to the fact that only part of the tail is enclosed in the uropatagium

The Pacific sheath-tailed bat (*Emballonura semicaudata*) was a widespread cave-dwelling bat, and the only insectivorous bat to have occurred in large parts of the Pacific. It has undergone drastic declines and is now considered Extinct in many parts of its former range (Bat Conservation International).

Emballonura semicaudata semicaudata has also been extirpated from Tonga, American Samoa and Samoa. Searches of 41 lava tubes across Samoa between 1994 and 1997 found only five contained Pacific sheath-tailed bats and another five caves showed signs the species had been present (Tarburton 2002). It was postulated that a combination of cyclones, agriculture and the associated use of pesticides had been major drivers of decline for the species. Cyclones Ofa (1990) and Val (1991) stripped vegetation and dramatically reduced the abundance of flying insects. A boom in taro (*Colocasia esculenta*) export around the same period also saw increases in the use of insecticides (Tarburton 2002). Few long-term studies of insect abundance are available from Pacific islands. However, if insect declines have occurred in the Pacific to the same extent as on the continents (see Sánchez-Bayo and Wyckhuys 2019), this could be a key reason why the Pacific sheath-tailed bat has declined over much of its range.

Seri's sheath-tailed bat (*Emballonura serii*) is one of the few species of echolocating bat that are endemic to the region under consideration here (Madison Erin Mayfield).

SERI'S SHEATH-TAILED BAT

Emballonura serii Flannery, 1994b

seh'-ree-ee

Seri's sheath-tailed bat is related to but distinct from *Emballonura furax*. Smith and Hood (1981) found it to be abundant on New Ireland but did not encounter it on New Britain. Similarly, Aplin and Opiang (2011) did not record it on New Britain, and Australian Museum expeditions to New Britain and the Duke of York group failed to document it at both locations. Bonaccorso (1998) identified a series of specimens from Los Negros Island in the Admiralty Group as this species, and calls with the typical pulse shape of an emballonurid bat (attributed to *E. serii*) were recorded on both Manus and Mussau islands in 2014 (Armstrong *et al.* 2015). These recordings remain the only indication of its possible presence on Mussau Island. Calls had the typical pulse shape of an emballonurid bat (a flat frequency-modulated call with short-duration frequency sweeps at the beginning and end). The mean characteristic frequency was 45 kHz (Armstrong *et al.* 2015).

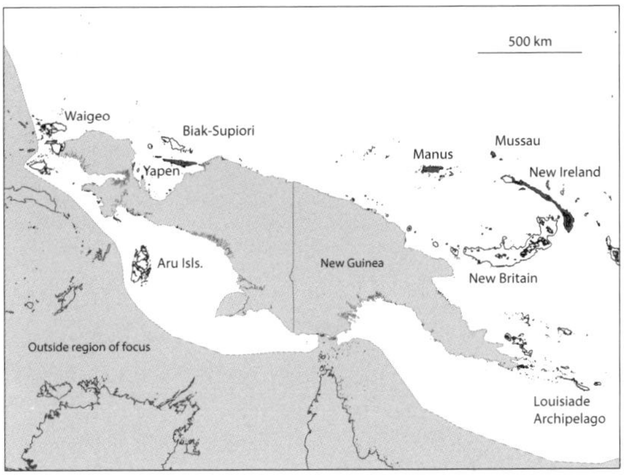

Seri's sheath-tailed bat roosts on walls towards the mouth of caves in the twilight zone. On New Ireland in Matapara Cave near Madina, it was found sharing a cave with *Rousettus amplexicaudatus* and *Pipistrellus angulatus*. This species has been observed foraging over open areas before sunset. Two of four females collected on New Ireland in June 1979 were lactating.

Emballonura serii			
Size*			
	Mean	**Range**	**Sample**
Head–body	45.7	44.6–47.6	3
Tail	12.2	11.3–12.8	3
Forearm	47.6	46.9–48.1	4
Tibia	18.4	18.0–18.9	4
Hindfoot	–	–	–
Ear	15.7	14.4–17.0	4
Weight	6.8	6.0–7.0	4
* From AM specimens (New Ireland)			
Identification Similar to *E. raffrayana* but larger with longer more pointed ears and a larger, more rectangular head. It differs from *E. semicaudata* in its longer, more pointed ears and more robust rostrum, and from *E. dianae* in its narrower, pointed ears and dark belly. It is absolutely smaller than *E. furax* in forearm length. The posterior margin of the tragus is curled anteriorly (not flat).			
Synonyms None			
Subspecies None			
Distribution Yapen (Indonesia); Los Negros, Manus, Mussau, New Ireland (Papua New Guinea)			
Extralimital None			
Conservation status (IUCN) Least Concern			
Etymology Sp.–for Papua New Guinean conservation biologist Lester Seri			

Mosia

LESSER SHEATH-TAILED BAT

Mosia nigrescens **Gray, 1843a**

mos'-ee-ah nig-res'-enz

The lesser sheath-tailed bat is widespread from Sulawesi through New Guinea and into Solomon Islands as far as Makira. It is common among the islands of the Bismarck Archipelago, Bougainville and Solomon Islands, but is conspicuously absent from many east Papuan Islands including most of the D'Entrecasteaux, Trobriand and Louisiade Islands, being known only from Woodlark and Alcester (Koopman 1982). The lesser sheath-tailed bat was formerly placed in the genus *Emballonura*, but was assigned to the monotypic genus *Mosia* by Griffiths *et al.* (1991) based on its hyoid musculature and penile anatomy. Three subspecies are recognised, but the distributions assigned do not correlate with patterns of morphological variation, and a preliminary genetic analysis further indicated diversity does not correlate with the current arrangement (Colgan and Soheili 2008).

The lesser sheath-tailed bat is one of the most commonly observed echolocating bats where it occurs in the region. It is found in a wide variety of habitats, including primary and secondary forests, disturbed village environments and cultivated areas. Small groups of up to six individuals will roost closely together, stacked front to back in a line, sheltered beneath the broad leaves of palms (*Areca* spp.), *Heliconia* and wild bananas (*Musa* sp.). Roost sites are especially common in lowland primary forests adjacent to waterways, where the understorey vegetation is sparser and palms are abundant.

Smith and Hood (1981) found two individuals roosting under a *Heliconia* leaf on New Britain. One roost encountered under a banana leaf on north-west Halmahera, Indonesia, contained a mixed-sex group of six individuals (three adult males, two adult females and one juvenile male) (Hill and Rozendaal 1989). The lesser sheath-tailed bat is less commonly encountered in caves, but will occasionally occupy the twilight zones. Smith and Hood (1981) recorded the capture of four individuals roosting in this manner on New Ireland. One individual was roosting with *E. beccarii* in a sea cave on Alcester Island. A small group occupied northern Guadalcanal's Kovi Cave in 2014 with eight other bats including *Dobsonia inermis*, *E. dianae*, *Aselliscus tricuspidatus*, *Miniopterus australis* and four species of *Hipposideros*. Numerous return visits to the same cave have failed to locate them there again.

Lesser sheath-tailed bats are sexually dimorphic, with females considerably heavier and larger than males. Males are among the smallest of all bats and are the smallest bats in the region. The species is alert with large eyes and good eyesight. Nonetheless, if roost sites are approached cautiously, these bats will remain stationary and only take flight when within just a few metres. Even when disturbed like this, lesser sheath-tailed bats will normally return to roost at the exact same location day after day. Roosts near villages are common and the species takes flight to forage well before dark. Their characteristic flight pattern of tight circles can be seen in the open spaces above village houses, gardens and shorelines as the daylight fades. In the hours following dusk they seem to vacate these areas, presumably heading into forested areas, and by 10 p.m. they can be found resting back in their daytime roosts. Pregnant females have been collected in New Guinea in February, May, June and July (McKean 1972).

Lesser sheath-tailed bat (*Mosia nigrescens*). This species is commonly seen flying above villages in early evenings (Stephen Richards).

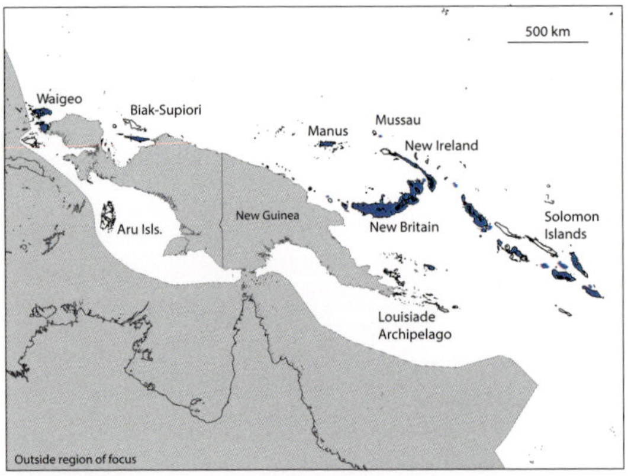

Mosia nigrescens			
Size*			
	Mean	**Range**	**Sample**
Head–body	41.8	30.0–56.0	23
Tail	10.8	7.0–17.5	22
Forearm	34.0	30.1–37.9	13
Tibia	11.9	10.0–13.7	12
Hindfoot	5.9	4.0–7.0	19
Ear	11.7	9.2–15.0	22
Weight	3.7	2.5–5.7	12
* From AM and KU specimens (Gebe, Sanana, Mono), and Bonaccorso (1998)			
Identification A very small species, light brown in colour with a short, blunt, sparsely haired face and large eyes.			
Synonyms None			
Subspecies *Mosia nigriscens nigrescens* Gray, 1843a (Central Maluku) *Mosia nigrescens solomonis* (Thomas, 1904b) (Manus, Woodlark, Bismarck Archipelago, Solomon Isls) *Mosia nigrescens papuana* (Thomas, 1914a) (Sulawesi, north Maluku, New Guinea, Schouten Islands, Kai Isls)			
Distribution Bougainville, Nissan, Sohana (AROB); Biak-Supiori (Indonesia); Alcester, Credner Isls, Duke of York, Emirau, Los Negros, Manus, Mioko, New Britain, New Ireland, Woodlark, Umboi (Papua New Guinea); Choiseul, Fauro, Gatokae, Guadalcanal, Makira, Malaita, Mbanika, Mono, New Georgia, Ngella Sule, Pavuvu, Santa Isabel, Savo, Shortland, Ugi, Vella La Vella (Solomon Isls)			
Extralimital Kai Isls, Maluku Isls, New Guinea			
Conservation status (IUCN) Least Concern			
Etymology Gen.–meaning and derivation unknown; Sp.–becoming black (L.), an inappropriate name because it is light brown in colour with a pale belly			

In the South-west Pacific, echolocation consists of short-duration, frequency-modulated calls, starting with a flatter section around 55–60 kHz followed by a downward sweeping tail. The peak frequency in the second harmonic for Solomon Islands individuals is around 57 kHz and the fundamental harmonic (H1) is around 28–29 kHz. In louder calls and feeding calls, upper harmonics at H3 (85 kHz), and H4 (114 kHz) also may be visible (Pennay and Lavery 2017). The stomach contents of 44 individuals examined in New Guinea contained mostly wingless ants, as well as worker and soldier termites (Vestjens and Hall 1977). The lesser sheath-tailed bat is common in a wide variety of habitat types and there is no reason to suspect it is undergoing major population declines

Saccolaimus

BARE-RUMPED SHEATH-TAILED BAT

Saccolaimus saccolaimus (Temminck, 1838a)

sak'-oh-lay'-mus sak'-oh-lay'-mus

This is a distinctive and very widespread species occurring through parts of India, Sri Lanka, Bangladesh, Cambodia, Thailand and Vietnam, through Wallacea, the Philippines, and into New Guinea, northern Australia and Solomon Islands. Yet it has been detected infrequently in the South-west Pacific. This could be because of its apparent preference for drier habitats that are uncommon; because it is a high-flying species that is difficult to trap; or simply because survey efforts have been lacking. The bare-rumped sheath-tailed bat has recently been added to the inventory of the large island of Malaita (M. Alabai, pers. comm.; Lavery *et al.* 2020b) and probably occurs on many islands in addition to those presented here.

Records from the northern parts of Guadalcanal have been relatively common. Sanborn and Beecher (1947) reported the capture of three individuals from the Teneru River, and another was mist-netted in February 2015 in a disturbed area adjacent to the Mataniko River (S. Galo, pers. comm). Northern Guadalcanal lies in a rain shadow, and it may be that the open grasslands and woodlands of this area provide ideal habitat. On Santa Isabel in July 2014, a high-flying species with audible calls repeatedly circled a cleared helipad in an area of drier ultramafic forest at 400 m elevation. Recorded call sequences, and others passively recorded from Choiseul, were both similar to calls from this species (Pennay and Lavery 2017).

Two individuals were caught in a cave in the La'fea valley of East Kwaio, central Malaita in 2017 (Lavery *et al.* 2020b), and an Australian Museum survey found it to be common in the Siwai district of southern Bougainville in 2016. The La'fea Valley is heavily forested but interspersed with small village gardens and fallow. The site on Bougainville was more disturbed, comprising a mixture of secondary forests and village gardens. The bats were circling above the gardens on dusk and were caught in mist nets set approximately 20 m above the ground.

Bare-rumped sheath-tailed bats are known to roost in groups of a few to hundreds of individuals, in hollow trees, caves and among rocky boulders and crevices. They are alert and can be prone to disturbance, being easily flushed from certain roost locations (Murphy 2002). Breeding in northern Australia is seasonal, with females giving birth to a single young at the beginning of the wet season.

Solomon Islands specimens are small relative to Australian individuals, yet they resemble them externally. Echolocation calls recorded from individuals on Guadalcanal consisted of a very long duration (approximately 18 ms), shallow-frequency sweep commencing at around 30 kHz and dropping to around 25 kHz, with a gentle slope to almost flat appearance (Pennay and Lavery 2017). Calls are multiharmonic, the peak frequency in the second harmonic is around 27 kHz, and the first harmonic at 13 kHz is audible to humans from a long distance. The bare-rumped sheath-tailed bat has a very broad distribution and appears common over much of its range, and few specific threats are known.

Bare-rumped sheath-tailed bat (*Saccolaimus saccolaimus*), Siwai District, Bougainville. This swiftly flying species most often forages in open habitats and has an echolocation call that is audible to the human ear (Tyrone Lavery).

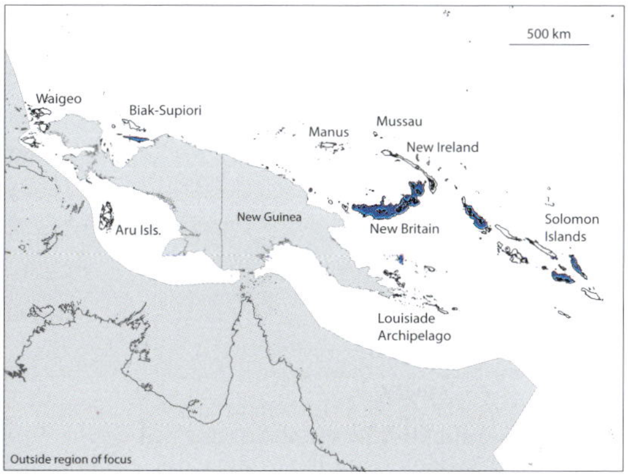

Saccolaimus saccolaimus			
Size*			
	Mean	**Range**	**Sample**
Head–body	91.6	84.2–100.0	6
Tail	27.5	26.0–29.4	10
Forearm	71.5	67.0–78.0	10
Tibia	27.5	26.0–29.4	5
Hindfoot	15.8	13.6–17.0	5
Ear	18.5	16.0–21.0	8
Weight	40.3	33.0–50.0	4
* From AM specimens (Bougainville), Sanborn and Beecher (1947) and Bonaccorso (1998)			
Identification One of the largest sheath-tailed bats in the region. It has a distinctly naked rump, no wing pockets and a distinct throat pouch. The dorsal fur is reddish-brown and usually has small white flecks.			
Synonyms *Taphozous pulcher* Blyth, 1844 *Taphozous granti* Thomas, 1911b			
Subspecies *Saccolaimus saccolaimus saccolaimus* (Temminck, 1838a) (Sumatra and Java) *Saccolaimus saccolaimus crassus* (Blyth, 1844) (India, Sri Lanka, Indochina, peninsular Malaysia) *Saccolaimus saccolaimus affinis* (Dobson, 1875) (Borneo and offshore islands) *Saccolaimus saccolaimus nudicluniatus* (De Vis, 1905) (Australia, New Guinea, Solomon Isls) *Saccolaimus saccolaimus pluto* (Miller, 1910) (Philippines)			
Distribution Bougainville (AROB); Yapen (Indonesia); New Britain (Papua New Guinea); Guadalcanal, Malaita (Solomon Isls)			
Extralimital Australia, Bangladesh, Brunei Darussalam, Cambodia, India, Indonesia, Malaysia, Myanmar, New Guinea, Philippines, Sri Lanka, Thailand, Vietnam			
Conservation status (IUCN) Least Concern			
Etymology Gen. and Sp.–sack + throat (Gr.), a reference to the throat pouch often present in members of the genus			

Hipposideridae

The family Hipposideridae is considerably diverse, consisting of seven genera and around 88 species, widely distributed across Afrotropical, Indo-Malayan and Australasian regions. Members of the family Hipposideridae are very distinctive. The species vary greatly in size (ranking from among the smallest to the largest of echolocating bats in the region), but all possess a broad nose-leaf lacking the spear-like structure found in the nose-leaf of rhinolophids. Hipposiderids are well represented in the region, with 10 species belonging to three genera. The family is an ancient one, with a long Australasian fossil record and considerable Australasian endemism.

In Melanesia, these are the most reliably encountered of the echolocating bats. Most caves found to be inhabited by bats will contain at least one species, often several. Hipposiderids use constant frequency echolocation calls, meaning the call has a flat constant tone at a specific frequency. Some species also have a frequency-modulated component such as a downward sweeping tail that follows the constant frequency. The cochlea of each species is finely tuned towards detecting this very specific constant frequency and there are relationships between the body size of hipposiderid bats, the characteristic frequency of their calls, and the size of insect prey that is most easily detected by that frequency (Francis and Habersetzer 1998). Reference recordings of constant frequency calls can be made when bats are held in the hand and most species call at distinct frequencies. This allows excellent call-identification libraries to be developed, and passive recordings can enable bats to be identified to species with a level of certainty. However, in the region under consideration here, different island populations of the same species do call at very different frequencies and cryptic diversity may be present.

The monotypic genus *Anthops* is endemic to the Solomon Islands archipelago, as are *Hipposideros demissus* and *H. dinops*. *Anthops ornatus* is a very distinctive species and, according to genetic data, is probably sister to all *Hipposideros*. Several species have distributions that extend into Asia (e.g. *A. tricuspidatus*, *H. diadema*), but a significant proportion have distributions that are centred in Melanesia (e.g. *H. calcaratus*, *H. maggietaylorae*).

Anthops

SOLOMONS LEAF-NOSED BAT

***Anthops ornatus* (Thomas, 1888c)**

anth'-ops or-nah-tus

This is a beautiful and unusual leaf-nosed bat. The fur has an overall silver appearance, but this is contrasted by black tipping and black hair bases. The ornate nose-leaf and skin of the face are orange in life and further add to the striking appearance of the species. It is a rarely encountered bat, occurring to an elevation of 411 m on major islands from Buka to Guadalcanal. Lavery *et al.* (2014) estimated the most recent common ancestor shared between *Anthops* and *Hipposideros* occurred in the Miocene some 20 million years ago. Thus, the isolation of this monotypic, endemic genus of insectivorous bat

Diadem leaf-nosed bat (*Hipposideros diadema*), Choiseul, Solomon Islands (Tyrone Lavery).

The Solomons leaf-nosed bat (*Anthops ornatus*) is an unusual and infrequently recorded species endemic to the Solomon Islands archipelago (Evan Bowen-Jones).

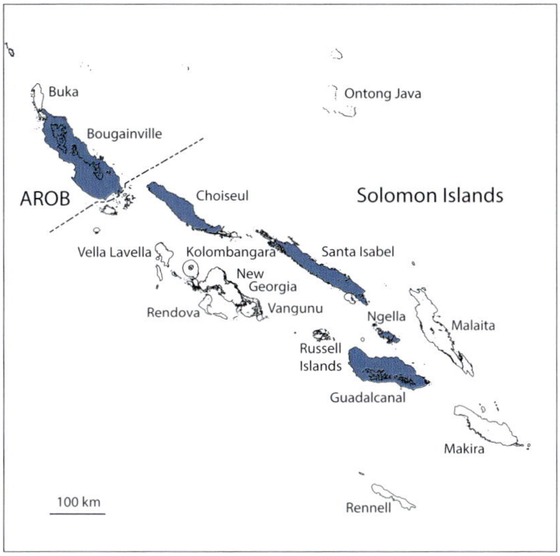

Anthops ornatus			
Size*			
	Mean	**Range**	**Sample**
Head–body	50.3	46.6–53.0	5
Tail	–	6.5	1
Forearm	48.6	37.0–52.5	10
Tibia	22.7	22.0–23.2	5
Hindfoot	9.4	9.3–9.5	2
Ear	18.4	16.5–21.0	5
Weight	–	8.0	1
* From AM specimens (Guadalcanal), Thomas (1888c), Thomas (1888d), Troughton (1936a), and Tate (1941a)			
Identification Nose-leaf unique: the upper margin possesses three projections forming backwards facing spherical cups. Dorsal fur is tricoloured black, silver and black, and the venter silver tipped with a black base.			
Synonyms None			
Subspecies None			
Distribution Bougainville (AROB); Choiseul, Guadalcanal, Santa Isabel, Ngella (Solomon Isls)			
Extralimital None			
Conservation status (IUCN) Data Deficient			
Etymology Gen.–flower-face (Gr.), a reference to the ornate nose-leaf; Sp.–ornate (L.), a further reference to the nose-leaf.			

in the Solomon Islands archipelago is most fascinating.

The Solomons leaf-nosed bat has most commonly been recorded from Guadalcanal, which is the type locality. On 13 May 1990, T. Flannery captured an adult male at an elevation of 40 m at Batani Village, on Guadalcanal's south coast. It flew into a house around 9:00 p.m. and was caught when the owner shut the window. Troughton (1936a) reported a male specimen from Santa Isabel and a female from Bougainville. Sanborn (1931) reported on a specimen taken on Choiseul during the Whitney South Seas Expedition in 1929, and Bowen-Jones *et al.* (1997) recorded the species on the same island. The Bernice P. Bishop Museum, Honolulu, holds a single specimen from Ngella Island.

On 12 November 2011, Patrick Pikacha caught a female carrying young in a mist net set in ultramafic lowland forest (dominated by *Xanthostemon melanoxylon*) on Santa Isabel (Lavery *et al.* 2013a). The animal managed to free itself, but the young had become entangled in the net and was left behind. It was estimated to be less than 2 weeks old, the forearm length was approximately 40% of mean adult length, the eyes were closed, and an umbilical scar was present. Nothing else is known of reproduction.

Bonaccorso (1998) suggested that the Solomons leaf-nosed bat is a foliage-gleaning insectivore, and Lavery *et al.* (2013a) postulated it roosts in hollow trees, given the apparent absence from caves across the known range. If this is the case, it is likely under serious threat from widespread logging of lowland forests.

Aselliscus

TEMMINCK'S TRIDENT BAT

Aselliscus tricuspidatus (Temminck, 1835)

ah-sel-isk'-us trie-kus-pid-ah'-tus

Temminck's trident bat is a common resident of the many limestone cave systems found in the region. It is usually found hanging singly and evenly spaced (30–40 cm apart) in discrete patches of up to 40–50 individuals. It occupies deeper cave recesses and will hang on stalactites, but prefers to occupy the sides of ceilings near cave walls or dome-shaped pockets in the roof.

Smith and Hood (1981) encountered Temminck's trident bat in great numbers in caves and tunnels on New Ireland and New Britain, including one especially large colony of several hundred individuals in a limestone cave near Sohun 2 on New Ireland. In the Solomon Islands archipelago, it is one of the common insectivorous bats that can be seen in early evenings circling over forest clearings and villages. The echolocation call is a very short-duration, constant-frequency tone with a peak frequency of approximately 117 kHz and a frequency-modulated tail that drops around 12 kHz (Pennay and Lavery 2017). Individuals from Mussau Island were found to call at a higher frequency of 125 kHz (Armstrong *et al.* 2015). Harp traps are a relatively unsuccessful survey tool for echolocating bats in Melanesia; however, Temminck's trident bat is one of the few species that can be captured with any regularity. High-frequency calls like those of this species attenuate quickly and this may decrease their ability to avoid collision with a harp trap.

Between June and August 1979, Smith and Hood (1981) found young present at several roosts on New Britain and New Ireland. Juveniles were found on Mussau in October 2014 (Aplin *et al.* 2015). The Australian Museum holds a female specimen from the island of Aore, off Espiritu Santo, Vanuatu, that was lactating when captured on 6 May 1990. The young are left on the cave wall at night while the females forage. However, females will carry even the very large young if disturbed at the roost. Temminck's trident bat is widespread, common on large and small islands, and does not appear to be in decline.

Aselliscus tricuspidatus			
Size*			
	Mean	**Range**	**Sample**
Head–body	39.5	35.9–42.0	16
Tail	23.4	19.2–28.0	8
Forearm	40.7	38.8–44.7	16
Tibia	15.4	13.5–16.9	16
Hindfoot	7.0	6.3–7.8	4
Ear	12.6	9.5–15.0	16
Weight	3.6	2.9–4.8	16
* Field measurements from the Solomon Isls archipelago, Bismarck Archipelago McKean (1972), and AM specimens (Vanuatu)			
Identification A very small insectivorous bat. The nose-leaf is ornamented with three sharp upward projections, a feature that is absent in small species of *Hipposideros*. It is often bright orange and does not have the tricoloured fur of *Anthops ornatus*.			
Synonyms None			
Subspecies *Aselliscus tricuspidatus tricuspidatus* (Temminck, 1835) (Maluku) *Aselliscus tricuspidatus novehebridensis* Sanborn and Nicholson, 1950 (Vanuatu) *Aselliscus tricuspidatus koopmani* Schlitter *et al.*, 1983 (Bismarck Archipelago, Louisiade Archipelago, Woodlark, Trobriand Isls, Solomon Isls) *Aselliscus tricuspidatus novaeguinea* Schlitter *et al.*, 1983 (New Guinea, Misool, Yapen)			
Distribution Biak-Supiori, Misool, Numfoor, Yapen (Indonesia); Duke of York, Fergusson, Goodenough, Kiriwina, Mioko, Misima, New Britain, New Ireland, Normanby, Sudest, Woodlark (Papua New Guinea); Choiseul, Gatokae, Guadalcanal, Makira, Malaita, Mbanika, Nendö, New Georgia, Ngella, Rennell, San Jorge, Santa Isabel, Ugi (Solomon Isls); Aore, Espiritu Santo, Malakula (Vanuatu)			
Extralimital Maluku Isls, New Guinea			
Conservation status (IUCN) Least Concern			
Etymology Gen.–without a saddle (Gr.); Sp.–three cusps (L.), a reference to the projections on the nose-leaf.			

Temminck's trident bat (*Aselliscus tricuspidatus*), Guadalcanal, Solomon Islands. This tiny insectivorous bat is a species commonly encountered in the deeper reaches of Melanesia's limestone caves (Tyrone Lavery).

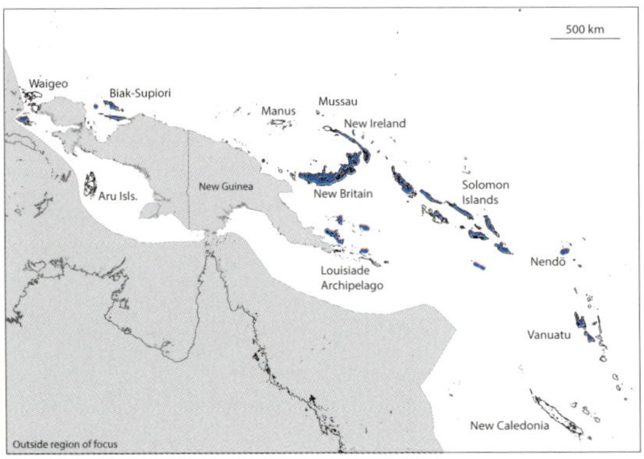

Hipposideros

DUSKY LEAF-NOSED BAT

Hipposideros ater **Templeton, 1848**

hip'-oh-sid'-er-os ah'-ter

The dusky leaf-nosed bat is widespread and comprises seven described subspecies. Some of these probably represent distinct species, including *Hipposideros ater aruensis*, which occurs in the region under consideration here. Phylogenetic reconstructions by Lavery *et al.* (2014) found north Queensland and South-east Asian specimens were polyphyletic and genetic divergence of 23% was estimated between samples from these regions (mitochondrial gene ND2). The type locality for *H. a. aruensis* is the Aru Islands, and further sampling is needed to resolve taxonomic relationships between this and other subspecies, as well as across the distribution of *H. a. aruensis* (Australia, Aru Islands, New Guinea, Bismarck Archipelago, and Woodlark Island).

The dusky leaf-nosed bat is relatively common in the region. Hill (1983) reported two young females captured in a cave on Kairuru Island, off the Sepik north coast of New Guinea. Koopman (1982) recorded four specimens collected by R. Petersen on Woodlark Island in 1956. These are the specimens that Brass (1959) identified as *H. bicolor*, collected at 100 m elevation in an old mine drive (horizontal shaft) on Woodlark's Suloga Peninsula. The entrance to the mine was in a small clearing in primary forest. Smith and Hood (1981) found it to be the sole occupant of a small and shallow cave on New Ireland. On New Britain they found it in several human-made tunnels, where it was always located in small pockets at ends of caved-in passageways, always isolated from other bats.

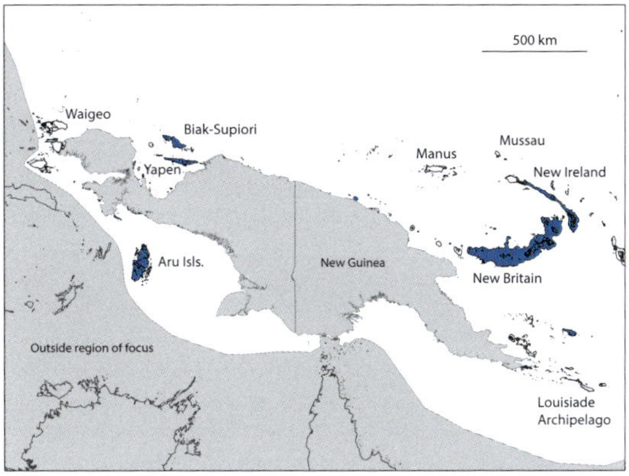

Hipposideros ater			
Size*			
	Mean	**Range**	**Sample**
Head–body	45.3	41.0 51.0	4
Tail	23.8	18.0–30.0	4
Forearm	39.3	36.2–43.0	4
Tibia	17.1	16.0–18.4	3
Hindfoot	7.8	6.0–10.0	4
Ear	20.4	19.0–22.0	4
Weight	6.6	5.5–8.5	4
* From Bonaccorso (1998)			
Identification A small species of *Hipposideros*, black to light brown in colour. It has a simple nose-leaf, lacking accessory leaflets and distinctively broad ears.			
Synonyms *Hipposideros atratus* Kelaart, 1850 *Hipposideros albanensis* Gray, 1866b *Hipposideros wrighti* Taylor, 1934 *Hipposideros gentilis toala* Shamel, 1940			
Subspecies *Hipposideros ater ater* Templeton, 1848 (India and Sri Lanka) *Hipposideros ater aruensis* Gray, 1858 (Eastern Australia, New Guinea, Bismarck Archipelago, Woodlark Isl) *Hipposideros ater gilberti* Johnson, 1959 (Western Australia, Northern Territory) *Hipposideros ater antricola* (Peters, 1861) (Philippines) *Hipposideros ater amboinensis* (Peters, 1871) (Ambon, Maluku Isls) *Hipposideros ater saevus* Andersen, 1918 (Malaysia, Indonesia) *Hipposideros ater nallamalaensis* Srinivasulu and Srinivasulu, 2006 (Eastern Ghats, Andhra Pradesh, India)			
Distribution Aru Isls, Biak-Supiori, Yapen (Indonesia); Kairuru, New Britain, New Ireland, Woodlark (Papua New Guinea)			
Extralimital Australia, India, Indonesia, Malaysia, Myanmar, New Guinea, Philippines, Sri Lanka, Thailand			
Conservation status (IUCN) Least Concern			
Etymology Gen.–horse + iron (Gr.), a horseshoe, an allusion to the shape of the nose-leaf; Sp.–black (L.)			

The dusky leaf-nosed bat (*Hipposideros ater*) is a small insectivorous species of leaf-nosed bat that forages by aerial hawking and foliage gleaning (Mark Sanders).

This species is insectivorous, and forages by aerial hawking and foliage gleaning. Stomach contents of one individual examined in the Northern Territory, Australia contained remains of a cockroach and traces of a moth (Vestjens and Hall 1977). The constant-frequency echolocation calls of New Guinea individuals are at a frequency of 154–164 kHz (Bonaccorso 1998). On mainland New Guinea in the Fly River region, two females were found to be carrying embryos on 15 and 23 October 1936, and a tiny young was documented on 6 December 1936 (Bonaccorso 1998). The dusky leaf-nosed bat is widely distributed and is not likely to be declining at a rate required to qualify for listing in a threatened category.

Spurred leaf-nosed bat (*Hipposideros calcaratus*) Kovi Cave, northern Guadalcanal, Solomon Islands (Tyrone Lavery).

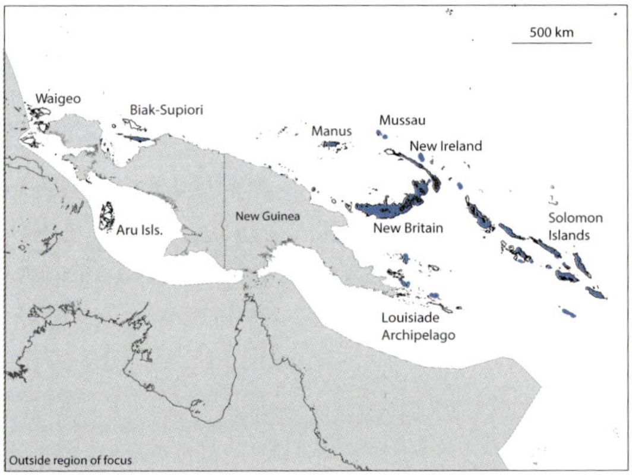

SPURRED LEAF-NOSED BAT

Hipposideros calcaratus (Dobson, 1877)

kal'-kar-ah'-tus

The spurred leaf-nosed bat is relatively common in the Bismarck Archipelago and the Solomon Islands archipelago. Smith and Hood (1983) encountered it in nearly all of the caves visited on New Ireland and New Britain. The bats were roosting in small groups in the mid-sections of caves. It forms large colonies in many of the larger limestone caves set in lowland forests or along coastlines across its distribution. These are often shared with other species of *Hipposideros* (*H. cervinus*, *H. diadema*, *H. dinops*, *H. maggietaylorae*), and *Aselliscus tricuspidatus*, *Dobsonia inermis*, *Miniopterus* spp. and *Myotis moluccarum*. Smaller caves can be occupied solely by this species.

Females carrying dependent young have been observed in October–November and June–July. Of seven females caught on Ngella Island in October 2017, two were lactating and one was carrying a young with a forearm length of 44.3 mm. Approximately half of the females caught from Anapuli cave on Rennell Island in October 2018 were lactating and one female was carrying a young with a forearm length of 31.2 mm. A juvenile (forearm 48.0 mm) in the collections of the Australian Museum was collected on Guadalcanal on 26 November 1987. Females with dependent young have been observed on East New Britain in November 1962, at Madang on mainland Papua New Guinea in July 1992, and on New Ireland in June 1979 (Bonaccorso 1998). Juveniles and subadults are grey in colour.

The characteristics of *H. calcaratus* echolocation calls from Guadalcanal consisted of a short-duration, constant frequency at a peak frequency of around 116 kHz. This is very similar to *Asecllsicus tricuspidatus*, but can usually be distinguished from this species by the slightly longer call duration (more than 4 ms) (Pennay and Lavery 2017). Armstrong *et al.* (2015) cited the characteristic frequency of *H. calcaratus* on mainland New Guinea as being 133 kHz. A species of *Hipposideros* recorded from Mussau Island closely resembled *H. calcaratus* with a proportionately larger nose-leaf and ears. The characteristic frequency of calls was substantially lower (98 kHz) than New Guinea and Solomon Islands animals and it could represent a distinct species (Armstrong *et al.* 2015). The spurred

Hipposideros calcaratus			
Size*			
	Mean	**Range**	**Sample**
Head–body	55.1	47.0–60.0	16
Tail	36.6	27.1–40.5	11
Forearm	52.8	44.3–55.4	23
Tibia	24.5	20.4–26.7	13
Hindfoot	9.0	8.6–13.8	8
Ear	20.0	15.8–22.7	22
Weight	11.0	8.0–13.8	22
* From AM, KU and QM specimens (Solomon Isls)			
Identification The noseleaf is simple, lacking accessory leaflets. Most likely to be confused with *H. maggietaylorae*, but differs from this species by its smaller size, shorter calcar (less than half tibia length), and possession of five instead of four tail vertebrae.			
Synonyms None			
Subspecies *Hipposideros calcaratus calcaratus* (Dobson, 1877) (Bismarck Archipelago, Trobriand Isls, Louisiade Archipelago) *Hipposideros calcaratus cupidus* Andersen, 1918 (New Guinea, Yapen)			
Distribution Bougainville (AROB); Yapen (Indonesia); Duke of York, Emirau, Kiriwina, Manus, Mioko, Misima, Mussau, New Britain, New Ireland, Nissan, Normanby, Tabar, Ulu (Papua New Guinea); Choiseul, Gatokae, Guadalcanal, Makira, Malaita, Mbanika, New Georgia, Rennell, Sandfly, San Jorge, Santa Isabel, Tetepare (Solomon Isls)			
Extralimital New Guinea			
Conservation status (IUCN) Least Concern			
Etymology Sp.–a spur (L.), a reference to the long calcar or heel cartilage of this species			
Extralimital New Guinea			

leaf-nosed bat is common where suitable habitat is present within its wide distribution and it does not appear to be declining at a rate that would qualify it for listing in a threatened category.

FAWN-COLOURED LEAF-NOSED BAT

Hipposideros cervinus (Gould, 1854)

ser-vee'-nus

This is a widely distributed hipposiderid; it is abundant in the region and usually present in any limestone caves occupied by bats. It is frequently the only inhabitant of large caves on smaller islands and in these situations can form enormous colonies with many hundreds of individuals. One colony at the Balek Wildlife Sanctuary, Madang, Papua New Guinea, was estimated to contain some 60 000 individuals (Bonaccorso 1998). The fawn-coloured leaf-nosed bat can be easily recognised by the combination of very small body size and the presence of a pair of lateral accessory leaflets below the main nose-leaf.

On New Ireland and New Britain, the fawn-coloured leaf-nosed bat was present in nearly all of the large caves and human-made tunnels visited by Smith and Hood (1981), but was not encountered by Aplin and Opiang (2011) at high elevation in the Nakanai Mountains of East New Britain. Four of seven caves examined on Mussau Island by Aplin *et al.* (2015) contained groups of 30–300 individuals conferred to this species. Several of the caves were co-occupied by species of *Dobsonia*, *Miniopterus*, *Aselliscus* and a second large-bodied species of *Hipposideros* resembling *H. calcaratus* (Aplin *et al.* 2015).

Aplin *et al.* (2015) examined a Manus Island *Hipposideros* specimen held in the collections of the United States National Museum of Natural History, Smithsonian Institution and tentatively referred it to an undescribed species. This unnamed member of the *bicolor* group is smaller than Emirau Island examples of *H. cervinus* and lacks the accessory leaflets.

The echolocation call of the fawn-coloured leaf-nosed bat is a short-duration, constant-frequency tone and most calls have long frequency-modulated tails dropping approximately 24–25 kHz. There is regional variation in the peak frequency. On Guadalcanal, Solomon Islands, the average peak frequency recorded was 147.2 kHz and on nearby Makira Island it was 138.9 kHz (Pennay and Lavery 2017). Calls recorded from a

Fawn-coloured leaf-nosed bat (*Hipposideros cervinus*), Iron Range, Cape York Peninsula, Australia. This small species of leaf-nosed bat is commonly found in limestone caves set among the coastal lowland forests of Melanesia (Mark Sanders).

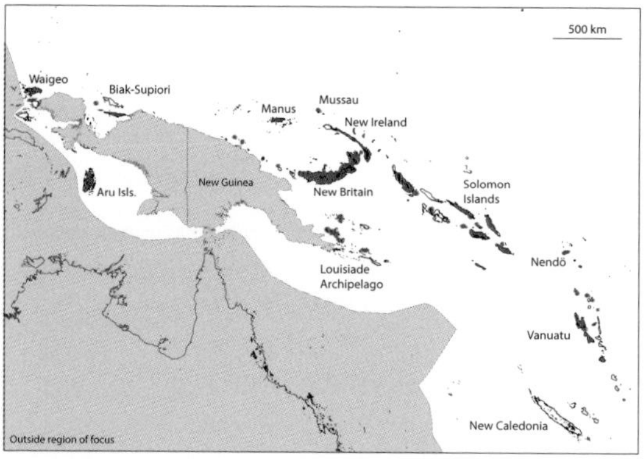

Hipposideros cervinus			
Size*			
	Mean	**Range**	**Sample**
Head–body	47.6	36.6–58.0	35
Tail	28.0	21.9–33.0	19
Forearm	46.4	40.2–50.7	39
Tibia	18.8	13.8–24.5	33
Hindfoot	8.2	7.0–10.5	8
Ear	13.4	10.0–16.8	37
Weight	7.0	4.5–11.0	38
* From Solomon Isls (Rennell, Makira, Guadalcanal, Gatokae), Vanuatu and New Britain			
Identification One of the smallest species of *Hipposideros* in the region, it has two supplementary nose-leafs, a marked antitragal projection on the ear, and usually four tail vertebrae.			
Synonyms *Hipposideros schneidersi* (misprint for *schneideri*) Thomas, 1904d			
Subspecies *Hipposideros cervinus cervinus* (Gould, 1854) (Sulawesi, Talaud, Maluku, Kai and Aru Isls, Cape York Peninsula Australia, New Guinea, Admiralty, Bismarck, Louisiade, and Trobriand Isls, Solomon Isls, Vanuatu) *Hipposideros cervinus labuanensis* (Tomes, 1859) (Malay Peninsula, Sumatra, Borneo, Philippines) *Hipposiderus cervinus batchianus* Matschie, 1901 (Batjan) *Hipposideros cervinus misoriensis* (Peters, 1906) (probably Misoor, Jackson and Groves 2015)			
Distribution Bougainville (AROB); Aru Isls, Batanta, Numfoor, Waigeo, Yapen (Indonesia); Crown, Duke of York, Emirau, Fergusson, Goodenough, Kadovar, Karkar, Kiriwina, Manus, Mioko, Misima, Mussau, New Britain, New Ireland, Normanby, Vokeo (Papua New Guinea); Fauro, Gatokae, Guadalcanal, Kolombangara, Makira, Malaita, Nendö, Rennell, Russell Isls, Sandfly, San Jorge, Santa Isabel (Solomon Isls); Aore, Efate, Espiritu Santo, Loh, Malakula, Malo, Mere Lava, Mota, Mota Lava, Tegua, Toga, Ureparapara (Vanuatu)			
Extralimital Australia, Brunei Darussalam, Indonesia, Malaysia, New Guinea, Philippines			
Conservation status (IUCN) Least Concern			
Etymology Sp.–pertaining to deer (L.), a reference to the colour of the fur			

series of individuals on Mussau Island, tentatively referred to this species, had a characteristic frequency of 145 kHz (Armstrong *et al.* 2015), and on mainland New Guinea the characteristic frequency was between 130 and 136.5 kHz (Leary and Pennay 2011).

A pregnant female from Santa Isabel was examined on 11 March, and a female from Malaita recorded by the Australian Museum was carrying a juvenile when captured on 15 March. Individuals recorded on Mussau in October 2014 were carrying young (Aplin *et al.* 2015). Of two females caught on Sandfly Island, Solomon Islands, on 10 October 2017, one was pregnant and the second was nulliparous. Three females examined on Makira Island on 3 July 2018 had enlarged teats but were post-lactating. Colonies of several thousands of individuals in Madang Province, Papua New Guinea, contained dependent juveniles, independent subadults and mothers in October 1991 (Bonaccorso 1998). The fawn-coloured leaf-nosed bat is one of the most common species of echolocating bat in the region and population declines are not apparent.

MAKIRA LEAF-NOSED BAT

Hipposideros demissus Andersen, 1909b

day-mis'-us

This species was long considered a subspecies of *Hipposideros diadema*, but Kitchener *et al.* (1992) regarded it as a distinct species. Molecular data from Lavery *et al.* (2014) supported recognition of *H. demissus* at the level of subspecies. However, here it is retained as a distinct species due to differences in echolocation call frequency and body size, and the similar levels of genetic divergence between the sympatric species *H. diadema* and *H. dinops*. Further work is needed to resolve the complex relationships between these three taxa.

The Makira leaf-nosed bat is restricted to the island of Makira, nearby Ugi and Pio and presumably Santa Ana at the south-eastern end of the island. The species was abundant in a cave about 1 hour's walk from Sesena Village in the south of Makira in 1987. Hill (1971) recorded 82 individuals from a cave near the Arihito-Gove

The Makira leaf-nosed bat (*Hipposideros demissus*) is endemic to Makira and outlying islands. It is a species that could be vulnerable to overhunting (Tyrone Lavery).

Hipposideros demissus			
Size*			
	Mean	**Range**	**Sample**
Head–body	72.0	53.6–83.0	14
Tail	36.4	30.0–42.6	12
Forearm	67.3	56.0–75.0	14
Tibia	28.6	21.8–33.0	14
Hindfoot	13.2	12.8–13.6	2
Ear	22.3	18.5–24.7	13
Weight	27.0	22.3–33.0	15
* From AM and QM specimens (Makira)			
Identification This is the largest hipposiderid bat found on Makira; forearm length of *H. demissus* (< 70 mm) is smaller than *H. diadema* (> 70 mm).			
Synonyms None			
Subspecies None			
Distribution Makira, Pio, Ugi			
Extralimital None			
Conservation status (IUCN) Vulnerable			
Etymology Sp.–lowly, weak, modest (L.), perhaps a reference to its relatively small size compared with other members of the *H. diadema* complex.			

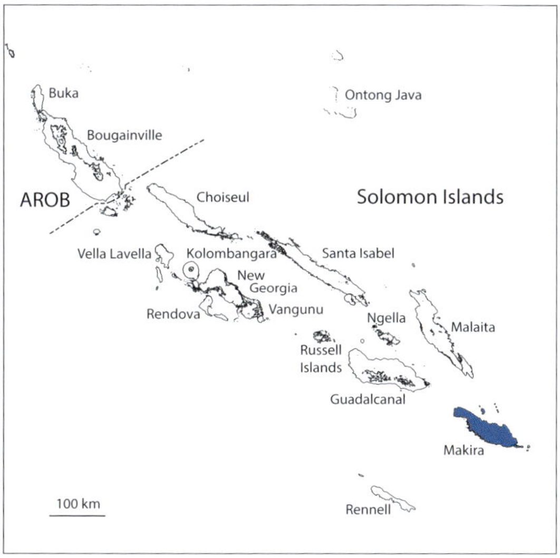

River confluence on October 1953. A large population also inhabits an extensive limestone cave system near Ngora Ngora Village, close to KiraKira.

There can be a striking colour difference between the sexes, with adult males being yellowish-grey and females a bright orange. This does not seem to be the case in all cave systems though and may relate to the sexes roosting separately and being exposed to varying levels of ammonia. The characteristic echolocation call frequency for this species is 59 kHz (Pennay and Lavery 2017).

Of 16 adult females examined from the cave near Sesena Village, 13 were lactating on 18 November 1987, and many juveniles were present. On Ugi Island on 4 December 2012, one male and four females were captured from a cave, one of which was pregnant. This suggests births take place in October–December.

The Makira leaf-nosed bat is not uncommon; however, the species is classified as Vulnerable on the IUCN Red List because its distribution is limited, and it may depend upon relatively few maternity caves. It is threatened by the disturbance of cave roosts, but the full extent of this across Makira is not clear. A cave visited in 2012 near Hauta Village (elevation around 600 m), supported a few dozen individuals of this species roosting alongside *H. cervinus*. When the cave was visited again in 2018, only *H. cervinus* remained, *H. demissus* having apparently been extirpated by hunting.

DIADEM LEAF-NOSED BAT

Hipposideros diadema (Geoffroy, 1813)

die'-ah-dem'-ah

The diadem leaf-nosed bat is a widespread and morphologically variable species found from mainland South-east Asia, throughout the Philippines, Indonesia, New Guinea, Australia and Solomon Islands. Numerous subspecies have been described and require further validation. The diadem leaf-nosed bat shares interesting evolutionary relationships with *Hipposideros dinops* (with which it is sympatric in the Solomon Islands archipelago) and *H. demissus* (an endemic form found on Makira Island). Molecular sequencing has shown *H. diadema*, *H. dinops* and *H. demissus* are all closely related, with genetic divergence in the mitochondrial gene ND2 well below levels often used to infer species (Lavery *et al.* 2014). In fact, genetic divergence between *H. diadema* and *H. dinops* living in the same cave on Guadalcanal was lower than that found between samples of *H. diadema* from Solomon Islands and New Guinea.

This is a large and obvious species that is commonly found roosting in larger caves in the region, often in association with other bats. Smith and Hood (1981) found it in great abundance in almost all caves and tunnels visited on New Britain and New Ireland. The diadem leaf-nosed bat prefers to roost in the deeper recesses of caves; individuals will hang in loosely associated groups separated from other bats, with individuals evenly spaced by 30 cm or more. On New Britain, some individuals returned to the roost 2 h after sunset.

In the South-west Pacific, the diadem leaf-nosed bat is occasionally captured in well-positioned mist nets. A net set on top of a steep, grassland-dominated ridgeline adjacent to a forested gully on Guadalcanal successfully caught a dozen or more individuals (C. Posala, pers. comm.). The net was presumably located on a flight path used by the animals. Smith and Hood (1981) also netted this species on several occasions in early evenings in garden areas or banana groves.

The diadem leaf-nosed bat is insectivorous and uses 'flycatching' foraging tactics. The animals hang from a perch and will make forays out to catch passing prey before returning to the same lookout point to consume them (Smith and Hood 1981; Bonaccorso 1998). An individual observed on Rennell Island demonstrated similar behaviour from a perch on a solar powerline strung between two village houses. Parts of rhinoceros beetles were

The diadem leaf-nosed bat (*Hipposideros diadema*) is a large and extremely widespread species of leaf-nosed bat found from mainland South-east Asia, throughout the Philippines, Indonesia, New Guinea, Australia and Solomon Islands (Angus McNab).

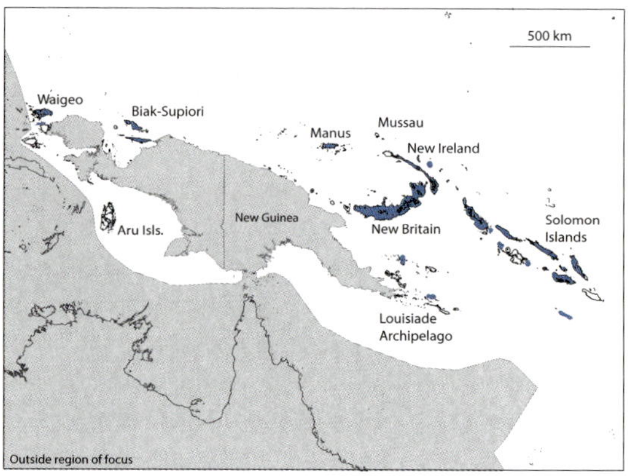

Hipposideros diadema			
Size*			
	Mean	**Range**	**Sample**
Head–body	76.2	57.5–95.0	51
Tail	51.6	41.7–59.8	8
Forearm	75.6	69.8–82.0	53
Tibia	33.0	28.8–36.3	51
Hindfoot	23.3	13.8–18.0	7
Ear	27.1	21.0–37.0	51
Weight	31.9	19.0–48.0	28
* From AM specimens (New Britain), KU and QM specimens (Solomon Isls), and field measurements			

Identification This is the second-largest leaf-nosed bat in the region. Its large size, pale bands on the shoulders, and robust head should help distinguish it. As in *H. dinops*, the nose-leaf has a blunt raised section in the centre, and well-developed lateral supplementary leaflets. Its shorter forearm (usually > 80 mm) should separate it from *H. dinops* (usually < 90 mm).

Synonyms *Hipposideros diadema pullatus* Andersen, 1905a
Hipposideros diadema vicarius Andersen, 1905a
Hipposideros diadema anderseni Taylor, 1934
Hipposideros diadema malaitensis Phillips, 1967

Subspecies *Hipposideros diadema diadema* (Geoffroy, 1813) (east Nusa Tenggara)
Hipposideros diadema nobilis (Horsfield, 1823) (Borneo, Java, Bali, west Nusa Tenggara)
Hipposideros diadema griseus (Meyen, 1833) (New Guinea, Philippines)
Hipposideros diadema nicobarensis (Dobson, 1871) (Nicobar)
Hipposideros diadema masoni (Dobson, 1871) (Indochina)
Hipposideros diadema euotis Andersen, 1905a (Halmahera Group)
Hipposideros diadema oceanitis Andersen, 1905a (Solomon Isls, Bismarck Archipelago)
Hipposideros diadema enganus Andersen, 1907 (Engano)
Hipposideros diadema mirandus Thomas, 1914b (Manus)
Hipposideros diadema custos Andersen, 1918 (Kai)
Hipposideros diadema speculator Andersen, 1918 (Sulawesi)
Hipposideros diadema reginae Troughton, 1937a (Queensland)
Hipposideros diadema trobrius Troughton, 1937a (Trobriand Isls)
Hipposideros diadema natunensis Chasen, 1940 (Bungaran, Natuna)
Hipposideros diadema ceramensis Laurie and Hill, 1954 (Seram, Buru)

Distribution Bougainville, Buka (AROB); Batanta, Biak-Supiori, Waigeo, Yapen (Indonesia); Kiriwina, Lihir, Manus, Misima, New Britain, New Ireland (Papua New Guinea); Choiseul, Fauro, Gatokae, Guadalcanal, Malaita, Ngella Sule, Rennell, Sandfly, San Jorge, Santa Isabel, Vella La Vella (Solomon Isls)

Extralimital Australia, Cambodia, India, Indonesia, Laos, Malaysia, Myanmar, New Guinea, Philippines, Thailand, Timor-Leste, Vietnam

Conservation status (IUCN) Least Concern

Etymology Sp–headband (Gr.).

found under roosts on New Britain (Smith and Hood 1981). Generally, many of the appendages, antennae, elytra and wings of their insect prey are discarded below the feeding perch (Bonaccorso 1998). Diadem leaf-nosed bats will forage up to 10 km away from the day roost (Bonaccorso 1998).

Females give birth to a single young. On New Britain, females were found to be carrying young from June to August (Smith and Hood 1981). One of four females captured in New Britain in late January–early February was in late-term pregnancy (Bonaccorso 1998). Of eight individuals caught on Sandfly Island, Central Province, Solomon Islands, on 10 October 2017, only two were females and one was lactating. Females have pubic teats and the young hang to these with their feet facing towards the head of the mother when being carried in flight.

The echolocation call is a long-duration, constant frequency tone with a regionally variable characteristic frequency. On mainland New Guinea the characteristic frequency was around 56.9 kHz, in Solomon Islands' New Georgia group of islands it was 59.2 kHz, on Manus Island it was 60 kHz, and on Guadalcanal it was almost 6 kHz higher at 65.9 kHz (Leary and Pennay 2011; Armstrong *et al.* 2015; Pennay and Lavery 2017). The diadem leaf-nosed bat has a wide distribution, is common and somewhat tolerant of habitat disturbance, and is not listed under a threatened category on the IUCN Red List.

FIERCE LEAF-NOSED BAT

Hipposideros dinops Andersen, 1905a

dine'-ops

This is an extraordinarily large leaf-nosed bat, with remarkably robust jaws and teeth. Its presence in caves across the region is patchy. It is often absent from many presumably suitable caves, and many caves are inhabited by no more than a few dozen individuals. Large colonies do exist but are rare – one such location was in a deep seaside limestone cave near Kavolavata on Gatokae Island where it numbered in the hundreds of individuals in 2014.

The fierce leaf-nosed bat is regularly encountered coexisting in caves with other species including *Ascelliscus tricuspidatus*, *Dobsonia inermis*, *Emballonura raffrayana*, *Hipposideros diadema*, *H. calcaratus*, *H. cervinus* and *Rousettus amplexicaudatus*. In a cave at Kovi, northern Guadalcanal, it coexists with almost all of these species. One of three female Australian Museum specimens was pregnant when collected from

Gumburota Cave, Kohove River, northern Guadalcanal, on 1 December 1987. The Australian Museum also holds a series of 12 specimens collected from Rahvi Cave, south-east New Georgia, on 12 March 1991. One of the four females collected then was pregnant. One male and one female, also held at the Australian Museum, were collected from Namonisu Cave near the village of Afetafa on the south coast of Malaita. The female was not lactating or pregnant on 19 March 1991.

Tate (1941b) and Hill (1963) asserted that *H. pelengensis* was a subspecies of *H. dinops* despite their distributions being separated by a distance of around 1800 km. However, Lavery *et al.* (2014) demonstrated that these two forms have separate evolutionary relationships with the widespread species *H. diadema*. Moreover, low levels of genetic divergence were identified between *H. dinops* and *H. diadema* in Solomon Islands (mitochondrial gene ND2). In a single cave in northern Guadalcanal, both these species roost together, reproduce at the same time of year, and yet do not seem to interbreed.

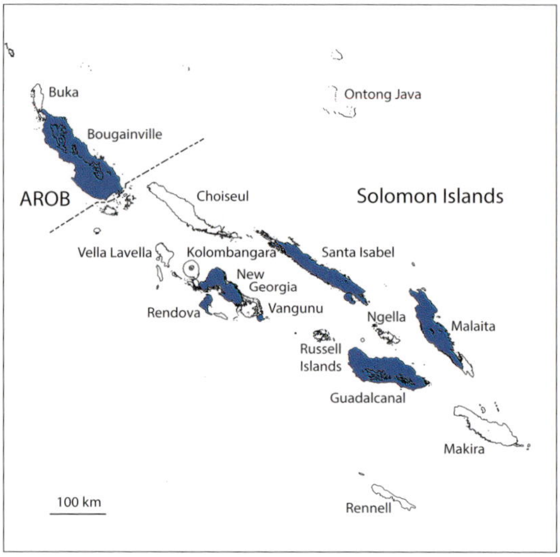

Hipposideros dinops			
Size*			
	Mean	**Range**	**Sample**
Head–body	94.6	88.0–105.0	24
Tail	59.1	52.5–63.0	12
Forearm	92.3	88.0–97.0	24
Tibia	41.1	37.8–44.7	23
Hindfoot	17.6	13.7–19.0	6
Ear	32.4	27.8–35.7	24
Weight	71.8	67.0–78.0	8
* From AM and QM specimens (Bougainville, Gatokae, Guadalcanal, Santa Isabel)			
Identification The largest leaf-nosed bat in the region, with an adult forearm length typically greater than 90 mm. The rostrum and teeth are more robust, and tibia proportionally longer than in *H. diadema*.			
Synonyms None			
Subspecies None			
Distribution Bougainville (AROB); Gatokae, Guadalcanal, Malaita, New Georgia, Rendova, San Jorge, Santa Isabel (Solomon Isls)			
Extralimital None			
Conservation status (IUCN) Vulnerable			
Etymology Sp.–horrific face (Gr.)			

The fierce leaf-nosed bat (*Hipposideros dinops*) (Kovi, Guadalcanal) is an extremely large hipposiderid endemic to the Solomon Islands archipelago. It shares an unusual evolutionary relationship with *H. diadema* and further research is needed to confirm the validity of the species (Tyrone Lavery).

Echolocation frequencies of *Hipposideros* bats scale with body size and these variations in turn determine the size of prey that are optimally detected. The fierce leaf-nosed bat calls at a characteristic frequency of around 55 kHz, and this relatively low frequency corresponds with optimal detection of larger prey. At Viru Harbour on New Georgia Island, this bat was seen at dusk foraging in an open area adjacent to a teak plantation, making broad sweeps from nearby vegetation to catch insects from the air. Given the tremendous body size of this species, it is feasible that terrestrial vertebrates (e.g. frogs) could contribute to the diet (Pavey and Burwell 1997).

The fierce leaf-nosed bat has a smaller distribution than *H. diadema* and is generally less abundant. It is large-bodied and is thus opportunistically hunted when other bats (primarily *D. inermis*) are targeted inside caves. No data are available on population trends, but the species has been listed in a threatened category on the IUCN Red List because very few roosting caves are known and each of these has been found to contain relatively small populations. This leaves fierce leaf-nosed bats susceptible to cave disturbance, hunting and habitat disturbance around roost sites.

MAGGIE TAYLOR'S LEAF-NOSED BAT

Hipposideros maggietaylorae Smith and Hill, 1981

mag'-ee-tay'-lor-ee

Maggie Taylor's leaf-nosed bat is a medium-sized hipposiderid, comprising two described subspecies. On average it is slightly larger than *Hipposideros calcaratus*, but size ranges overlap and the nose-leaf in both species is simple, lacking secondary leaflets. The two species are therefore confused on occasion. However, the head and jaws of *H. maggietaylorae* are long and narrow and contrast with the shorter rostrums of most other species in the genus.

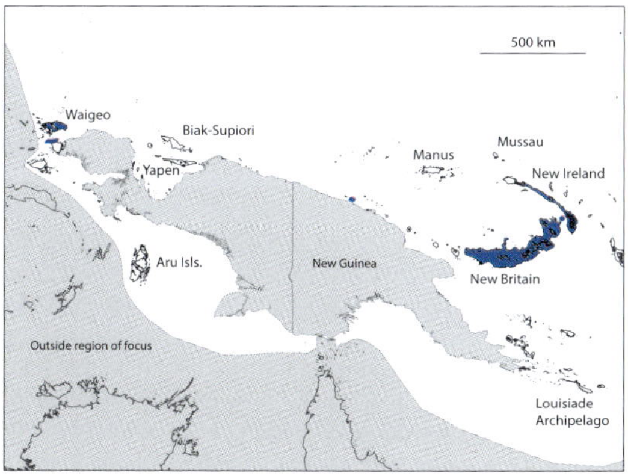

Hipposideros maggietaylorae			
Size*			
	Mean	**Range**	**Sample**
Head–body	65.8	61.2–69.0	4
Tail	39.8	36.4–43.2	4
Forearm	60.0	56.0–67.2	8
Tibia	25.3	24.6–26.6	3
Hindfoot	–	12.6	1
Ear	19.4	18.7–20.0	4
Weight	20.8	16.0–26.0	5
* From AM specimens (New Ireland), Smith and Hood (1981) and Bonaccorso (1998)			
Identification A medium-sized leaf-nosed bat; the nose-leaf lacks secondary leaflets and the tail has four conspicuous vertebrae rather than five found in *H. calcaratus*.			
Synonyms None			
Subspecies *Hipposideros maggietaylorae maggietaylorae* Smith and Hill, 1981 (New Ireland) *Hipposideros maggietaylorae erroris* Smith and Hill, 1981 (New Guinea)			
Distribution Batanta, Waigeo (Indonesia); Duke of York, Kairuru, New Britain, New Ireland (Papua New Guinea)			
Extralimital New Guinea			
Conservation status (IUCN) Least Concern			
Etymology Sp.–for Mrs Reese (Maggie) Taylor, in recognition of her interest in, and most generous sponsorship of, field research in the Bismarck Archipelago			

Smith and Hood (1981) found Maggie Taylor's leaf-nosed bat to be abundant in nearly all of the caves and tunnel systems visited in New Ireland and New Britain. The species was most often found roosting in small groups of three or four individuals with other *Hipposideros* species. At Lakuramau Cave near Madina, New Ireland, it was roosting with *Aselliscus tricuspidatus*, *Dobsonia anderseni*, *H. calcaratus*, *H. cervinus*, *H. diadema*, *Miniopterus* spp. and *Rhinolophus megaphyllus*. Maggie Taylor's leaf-nosed bat was in the main central chamber with *D. anderseni*. Bonaccorso (1998) reported a roost of 50–70 individuals at Yagaum Cave, Marobe Province, mainland New Guinea.

Preferred habitat is tropical lowland rainforest up to 400 m elevation. It is an insectivorous species, emerging from caves after dark to fly low to the ground among the dense vegetation (Bonaccorso 1998). What is known of the reproductive biology of this species has been reported from Yagaum Cave by Bonaccorso (1998). A pregnant female was recorded in August, young and early subadults in November and lactating females in November and January. The author surmised pregnancies must occur from July to October.

A limited number of call sequences recorded for this species on mainland New Guinea showed calls have both constant frequency and frequency-modulated components. The characteristic section was a narrowband constant-frequency tone at around 121 kHz. The long, down-sweeping tail drops 30–50 kHz and the mean pulse duration is 1.38 ms (Leary and Pennay 2011). Maggie Taylor's leaf-nosed bat has a wide distribution, appears to be tolerant of some disturbance, and is not considered to be declining.

BIAK LEAF-NOSED BAT

Hipposideros papua (Thomas and Doria, 1886)

pah'-poo-ah

The Biak leaf-nosed bat occurs on several of the islands of Cenderawasih Bay, the Maluku Islands, and there is a single record from New Guinea. The type locality is Biak Island; however, few data on life history are available from the region under consideration here.

On Numfoor Island, a small colony was occupying an unnamed cave located approximately 1.5 h walk inland from Kampong Sioribo. It shared the cave with swifts and *Dobsonia emersa*. Three adult males, nine adult females and two subadults (weights 9 and 9.5 g) were captured there in early October. Six of the adult females were pregnant – all at approximately the same late stage of pregnancy. Another, captured outside of a roost, was taken by children at around 6:30 p.m. as it flew around the village of Korido, Supiori.

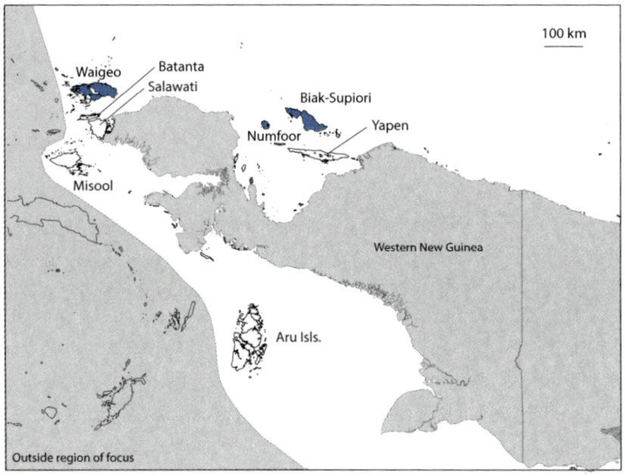

Hipposideros papua			
Size			
	Mean	**Range**	**Sample**
Head–body	53.1	49.9–55.8	4
Tail	27.0	23.6–34.8	5
Forearm	51.2	49.5–53.6	10
Tibia	19.3	18.2–21.7	6
Hindfoot	7.1	6.5–7.7	2
Ear	15.0	10.5–17.5	6
Weight	9.7	8.8–11.0	6
* From AM specimens (Gebe and Halmahera), Hill and Rozendaal (1989) and Bates *et al.* (2007) (West Papua and Maluku Isls).			
Identification A medium-sized leaf-nosed bat superficially resembling a very large *H. cervinus*, but with larger supplementary lateral leaflets, a rudimentary third lateral leaflet, and longer nose-leaf. The claws are reddish-brown.			
Synonyms None			
Subspecies None			
Distribution Biak-Supiori, Numfoor, Waigeo			
Extralimital New Guinea			
Conservation status (IUCN) Least Concern			
Etymology Sp.–for Papua			

In Manok Cave near Tanah Putih, central Halmahera, a single non-lactating female was recorded in November 1991. The cave consisted of a 100 m long arch, open at both ends with a creek flowing through it. It was sharing the site with a large colony of *Rhinolophus euryotis*, and lesser number of *Dobsonia viridis*, *Aselliscus tricuspidatus* and *Emballonura raffrayana*. In Simingit Cave on Gebe, the Biak leaf-nosed bat was fairly common and shared the location with *Myotis stalkeri* and *Miniopterus australis*. It occupied the deepest chamber of the cave. Although the known distribution of Biak leaf-nosed bat is relatively small, it is presumed to be larger (incorporating broader areas of the Vogelkop Peninsula) and it is considered unlikely the species is declining at a rate required for listing in a threatened category.

Miniopteridae

The family Miniopteridae is widespread in Palearctic, Afrotropical, Indo-Malayan and Australasian regions of the world, and relatively diverse, containing around 38 species. However, all belong to a single genus *Miniopterus*. They have a simple rostrum with no nose-leaf, and a tail that is fully enclosed in the uropatagium.

Exactly how many species of *Miniopterus* occur in the region under consideration here is far from settled. The genus is morphologically conservative, and cursory examinations of multiple species in the hand from the same location will often lead the observer to conclude each is nothing more than the slightest size variation on the same morphological design. Body size (judged by forearm measurements) and slight colour variations do offer some direction for identifying species. However, forearm size ranges can be near or slightly overlapping between sympatric species. This becomes even more problematic when considered over larger geographical scales. For example, Ibáñez and Juste (2019) identified six species as occurring in the islands of eastern Papua New Guinea, Solomon, Islands, Vanuatu and New Caledonia, and many of these overlap in the ranges of forearm lengths given (Table 5). Five purportedly overlap in Papua New Guinea and the Solomon Archipelago. Similarly, Bonaccorso (1998) recognised a total of five species (*M. australis*, *M. macrocneme*, *M. medius*, *M. propitristis*, *M. schreibersii*) as occurring on New Britain (4 species) and New Ireland (4 species) in the Bismarck Archipelago.

In Solomon Islands, it is not unusual to find three distinct *Miniopterus* roosting in a single cave. On Makira, this was the case in an extensive limestone tunnel system adjacent to the island's main airstrip at KiraKira. Similarly, on Rennell Island, three taxa were found in a limestone cave that exited onto a steep limestone cliff between Lake Te-Nganno and the open ocean. In both instances, forearm measurements were non-overlapping between the three syntopic taxa (Table 5).

Table 5. Species of *Miniopterus* identified as occurring among Melanesian Islands of the south-west Pacific, forearm length ranges reported in Ibáñez and Juste 2019, and example measurements from sympatric populations on Rennell and Makira Islands, Solomon Islands.

Conferred species	Forearm length (mm)		
	Ibáñez and Juste (2019)	Rennell (*n*)	Makira (*n*)
Miniopterus australis[1,2,3,4]	35.0–41.0	35.1–39.6 (11)	39.5–40.9 (3)
Miniopterus robustior[4]	39.0–42.2	–	–
Miniopterus macrocneme[1,2,3,4]	39.6–45.0	–	–
Miniopterus medius[1,2]	40.0–48.2	–	–
Miniopterus blepotis[1,2]	44.6–51.7	44.3–47.1 (7)	45.2–47.9 (4)
Miniopterus tristis[1,2,3]	51.0–55.0	48.2–49.6 (9)	49.2–49.6 (4)

Distribution: 1. Papua New Guinea; 2. Solomon Archipelago; 3. Vanuatu; and 4. New Caledonia

Great long-fingered bat (*Miniopterus tristis*), Makira, Solomon Islands (Tyrone Lavery).

Detailed molecular and morphological revisions for this complex genus are presently underway and will hopefully clarify the diversity of *Miniopterus* present in this region, better define their distributional limits, and provide external characters that can be used to reliably identify each. Until then, the information presented here should be considered preliminary only.

Miniopterus

LITTLE LONG-FINGERED BAT

Miniopterus australis **Tomes, 1858**

min'-ee-op'-te-rus os-trah'-lis

The little long-fingered bat is a very common and widespread inhabitant of caves in the region, and probably the most encountered species of *Miniopterus*. Prié (2011) found it to be the most common cave-dwelling species on Espiritu Santo, Malo and Aore in Vanuatu during surveys in 2006. Three of 29 caves visited across the three islands supported colonies estimated to comprise more than 50 000 individuals. Sanborn and Nicholson (1950) reported that in New Caledonia it is commonly found roosting in the same cave at the same time with *M. macrocneme*, but no more than

500 individuals were recorded in any one roost. They also found little long-fingered bats roosted in culverts, between the walls of houses, and in a human-made tunnel near Nouméa. In other parts of Melanesia, little long-fingered bats will share cave roosts with a variety of species from the genera *Hipposideros*, *Aselliscus*, *Myotis* and *Dobsonia*.

The little long-fingered bat is an aerial-hawking insectivore that mostly forages in open spaces. In Solomon Islands and Vanuatu, it is one of the species most commonly detected by ultrasonic recorders aimed towards forest clearings and the open air above villages and gardens. However, the species will also forage beneath the canopy (Bonaccorso 1998). Individuals move between roosts seasonally and repeated searches of the same cave at different times of year can reveal varying presence or absence, and fluctuations in population numbers. In a cave in the lowland forests of Guadalcanal (repeatedly surveyed over several years), the little long-fingered bat was not present on 12 May 2012, but on 12 December 2012 small numbers were found. It was again not present on 2 September 2013 and 5 November 2014, but on 18 December 2015 a small group was captured there in a harp trap. Sanborn and Nicholson (1950) similarly found considerable fluctuations in numbers of bats present at roosts that were visited periodically in New Caledonia.

The little long-fingered bat has a very sharply defined breeding season in Vanuatu. Copulations occur from August to September, with the young born in the latter half of December after around 120 days gestation (Baker and Bird 1936). On New Caledonia, births also commence in December, but peak 1 month later than in Vanuatu, with the majority of females found nursing in January (Sanborn and Nicholson 1950). On Rennell, Solomon Islands, one of three females examined in October 2018 was lactating. This species forms large

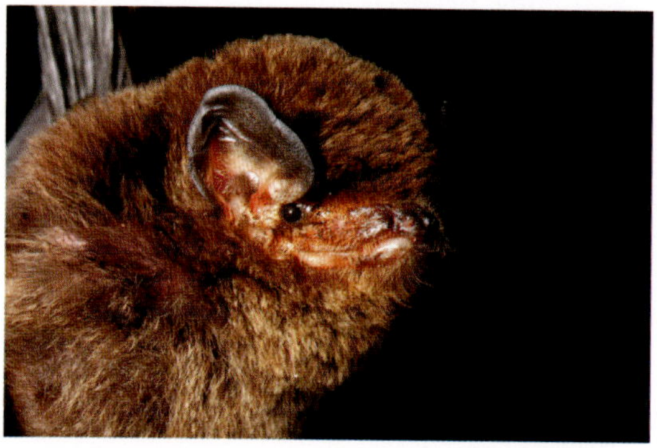

Little long-fingered bat (*Miniopterus australis*), Ngora Ngora, Makira, Solomon Islands. This is a common and widespread species of long-fingered bat that can form large roosting colonies of more than 50 000 individuals (Tyrone Lavery).

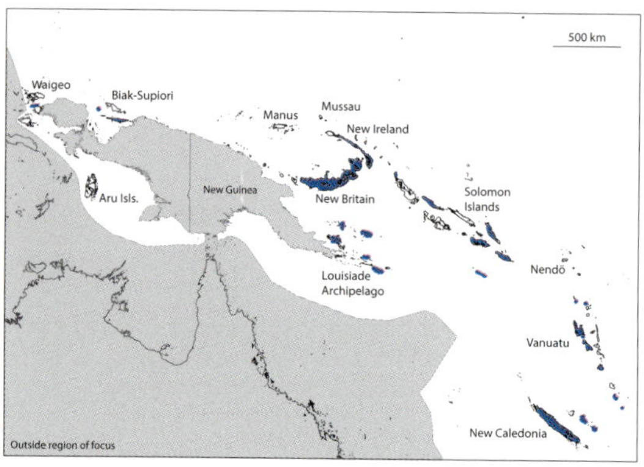

Miniopterus australis			
Size*			
	Mean	**Range**	**Sample**
Head–body	43.0	35.8–50.6	20
Tail	38.8	33.8–44.2	13
Forearm	37.5	35.1–41.0	22
Tibia	15.2	13.5–17.4	20
Hindfoot	7.8	7.2–9.7	9
Ear	9.9	8.1–13.6	19
Weight	4.9	4.0–6.4	18
* AM, QM and KU specimens from Solomon Isls (Rennell, Makira, Guadalcanal) and Vanuatu (Efate)			

Identification A small long-fingered bat which is typical of the genus in having the terminal bone of the third digit long. It can be distinguished from the slightly larger and sympatric *M. macrocneme* by having a tibia that is usually shorter than 18 mm long.

Synonyms None

Subspecies *Miniopterus australis australis* Tomes, 1858 (Loyalty Isls, New Caledonia, Vanuatu)
Miniopterus australis tibialis (Tomes, 1858) (Australia, New guinea, Bismarck Archipelago, Maluku)
Miniopterus australis solomonensis Maeda, 1982 (Solomon Isls)

Distribution Buka (AROB); Batanta, Numfoor, Yapen (Indonesia); Lifou, Maré, Grand Terre, Ouvéa (New Caledonia); Fergusson, Kiriwina, Misima, New Britain, New Ireland, Sudest, Woodlark (Papua New Guinea); Choiseul, Guadalcanal, Makira, Malaita, Rennell (Solomon Isls); Aneityum, Aore, Efate, Espiritu Santo, Malakula, Malo, Tanna, Toga, Ureparapara (Vanuatu)

Extralimital Australia, Brunei, Indonesia (Java, Kalimantan), Malaysia, New Guinea, Philippines, Timor-Leste

Conservation status (IUCN) Least Concern

Etymology Gen.–small wing (Gr.); Sp.–southern (L.)

maternity colonies and the naked young will cluster together in tight groups. A group of 100–150 individuals were found like this by Sanborn and Nicholson (1950) at Bouloupari on 14 January. Although it was 3:30 p.m. and adults had not yet left the cave to forage, the group of young was attended by just one adult female.

Echolocation pulses are very short-duration, frequency-modulated 'hockey stick' shaped calls. The call begins with a rapid frequency sweep that drops from around 82 kHz to 60 kHz and then transitions to a less steep, longer duration characteristic section. The peak frequency of the call is in the first harmonic at the 'knee' after the frequency sweep starts transitioning to a flatter decline. In Solomon Islands, the peak frequency is approximately 55.3 kHz and the characteristic frequency 51.6 kHz (Pennay and Lavery 2017). Calls from individuals on Espiritu Santo had a peak frequency of 52 kHz (Prié 2011). Kirsch *et al.* (2002a) concluded it was not possible to discriminate between the echolocation calls of *M. australis* and *M. macrocneme* on New Caledonia. The little long-fingered bat has a broad distribution, uses a range of habitats (including disturbed areas), and specific major threats have not been identified.

JAVANESE LONG-FINGERED BAT

Miniopterus blepotis (Temminck, 1840)

blep'-oh-tis

The distribution of this species extends from the Malay Peninsula as far east as Makira and Rennell in the Solomon Archipelago. It is relatively uncommon in the region and much work remains to resolve *Miniopterus* taxonomy and confirm the status and distribution of *M. blepotis*. Samples from Solomon Islands are genetically distinct from those from New Guinea (Ibáñez and Juste 2019).

Bonaccorso (1998) reported that the Javanese long-fingered bat occurs in dry woodlands and wet tropical forests in Papua New Guinea, and can be locally abundant. It is an aerial hawking insectivore that mostly forages in open spaces. Surveys on Makira and Rennell in 2018 recorded it in caves on both islands. On Makira it was found roosting with *M. australis*, *M. tristis*, *Hipposideros demissus*, *H. calcaratus* and *H. cervinus*. On Rennell it shared a cave with *Dobsonia inermis*, *Aselliscus tricuspidatus*, *M. australis*, *M. tristis*, *Myotis moluccarum* and *Hipposideros calcaratus*.

Calls recorded from hand-released individuals on Makira Island were short-duration, frequency-modulated 'hockey stick'-shaped pulses starting with a rapid frequency sweep dropping from 70–80 kHz to around 53 kHz and then transitioning to a less steep, longer duration characteristic section around 43–45 kHz. Peak frequency was in the first harmonic at 47.2 kHz where the initial frequency sweep transitions to a flatter decline. Characteristic frequency was 44.1 kHz (Pennay and Lavery 2017).

Of five females examined in a cave on Rennell Island on 24 October 2018, three were pregnant, one was lactating, and one was non-reproductive. This species has not been assessed according to IUCN Red List criteria. It is widespread but little information is available on population sizes or trends within its distribution.

The distribution of Javanese long-fingered bats (*Miniopterus blepotis*) extends from the Malay Peninsula as far east as Makira and Rennell in the Solomon Archipelago (Tyrone Lavery).

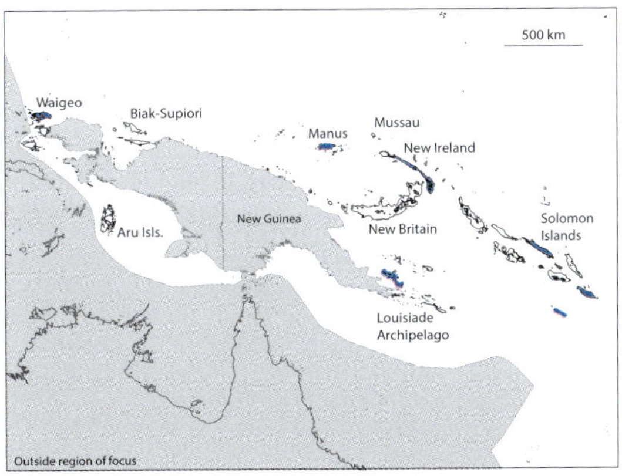

Miniopterus blepotis			
Size*			
	Mean	**Range**	**Sample**
Head–body	54.2	46.0–63.0	16
Tail	53.3	39.0–62.5	12
Forearm	47	44.3–51.7	18
Tibia	19.5	17.1–21.4	15
Hindfoot	8.7	7.0–10.6	15
Ear	11.7	8.2–14.4	16
Weight	9.9	8.5–11.5	14
* From KU and QM specimens (Makira, Rennell)			
Identification A medium-sized species of *Miniopterus*, which is larger than *M. australis* and *M. macrocneme*, and lacks the robust head and rostrum of *M. tristis* and *M. robustior*.			
Synonyms None			
Subspecies None			
Distribution Waigeo (Indonesia); Fergusson, Goodenough, Manus, New Ireland, Normanby (Papua New Guinea); Makira, Rennell, Santa Isabel (Solomon Isls)			
Extralimital Indonesia, Malaysia, New Guinea			
Conservation status (IUCN) Not Assessed			
Etymology Sp.–to see, discern (Gr.), allusion unclear			

SMALL MELANESIAN LONG-FINGERED BAT

Miniopterus macrocneme Revilliod, 1914

mak'-rohk-nay'-may

This is a widespread and common bat distributed from New Guinea to Vanuatu and New Caledonia. However, great uncertainty remains about the distributions and species limits of *Miniopterus* in the Pacific, including small Melanesian long-fingered bat. Ibáñez and Juste (2019) indicated both *M. blepotis* and *M. macrocneme* occur in sympatry on numerous islands, including Makira (along with *M. tristis* and *M. australis*). But surveys of Makira in 2018 found just three forms present: a smaller species consistent with *M. australis*; a large species consistent with *M. tristis*; and a medium-sized species consistent with either *M. blepotis* or *M. macrocneme*. We have tentatively retained overlaps in the distributions of *M. macrocneme* and *M. blepotis*, but it is possible these two species replace each other ecologically in parts of the Pacific.

The small Melanesian long-fingered bat roosts in caves, houses and road culverts. It forms very large colonies numbering in the thousands, and often shares caves with *Notopteris neocaledonicus* (from which it remains separate) and *M. australis*. Out of seven caves surveyed on New Caledonia by Sanborn and Nicholson (1950), four were found to contain both *M. australis* and *M. macrocneme*. The largest colonies reported were at Bouloupari (400–500 individuals), Paita (300–400) and Mount Dore (200). Kirsch *et al.* (2002a) reported this species as rare on New Caledonia and geographically restricted to the extreme north and south-west of Grande Terre, as well as Lifou and Maré islands.

The small Melanesian long-fingered bat is an insectivore that feeds in open space above and around the canopy (Armstrong *et al.* 2019a). Echolocation pulses are very short-duration, frequency-modulated 'hockey stick'-shaped calls with a characteristic frequency of around 50 kHz. On New Caledonia, discriminating between the echolocation calls of *M. australis* and *M. macrocneme* was not possible (Kirsch *et al.* 2002a). Sanborn and

The small Melanesian long-fingered bat (*Miniopterus macrocneme*) forms very large colonies in caves, houses and road culverts. It will share roost sites with *Notopteris neocaledonicus* and *M. australis* (Pavel German).

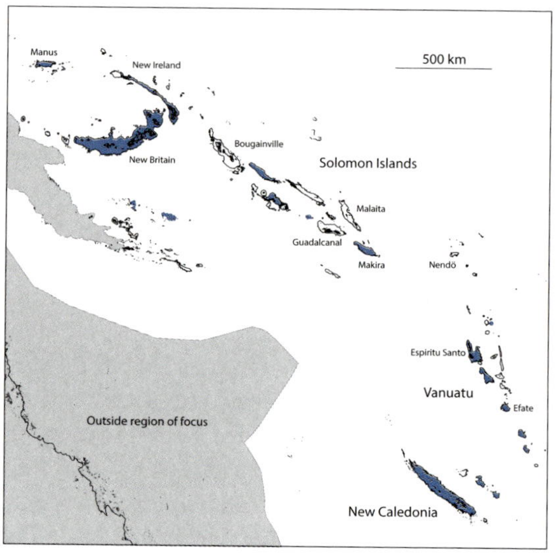

Miniopterus macrocneme			
Size*			
	Mean	**Range**	**Sample**
Head–body	47.1	41.6–63.0	13
Tail	51.5	40.5–64.0	14
Forearm	42.0	39.5–49.0	17
Tibia	18.7	16.7–20.9	17
Hindfoot	8.1	7.0–10.0	10
Ear	10.8	8.6–13.2	14
Weight	9.0	6.9–12.5	8
* From AM specimens (Efate), Revilliod (1914) and Bonaccorso (1998)			
Identification A small species of long-fingered bat, differing from the slightly smaller *M. australis* in having a tibia usually longer than 18.0 mm.			
Synonyms None			
Subspecies None			
Distribution Grand Terre, Lifou, Maré (New Caledonia); Kiriwina, Manus, Misima, New Britain, New Ireland, Woodlark (Papua New Guinea); Choiseul, New Georgia, Makira, Mbanika (Solomon Isls); Aore, Efate, Erromango, Malakula, Malo, Mota, Tanna (Vanuatu)			
Extralimital New Guinea			
Conservation status (IUCN) Least Concern			
Etymology Sp.–large tibia (Gr.)			

Nicholson (1950) indicated copulations probably occur from July to August.

On mainland New Guinea, Flannery (1995a) found the small Melanesian long-fingered bat occurred up to the subalpine zone at 3200 m elevation and it was more common than *M. australis*. At a roost in a shallow limestone crevice at 2600 m elevation, animals were torpid during the day. A roost in a mining adit at 360 m elevation in Papua New Guinea's Western Province was shared with 12 other species of bat, including two species of *Miniopterus* (Flannery 1995a).

The small Melanesian long-fingered bat is threatened by disturbances and destruction of cave roosts, and is eaten in some parts of its range. The species is understood to be widespread and common, but taxonomic and distributional revisions that are underway may alter this and will demand reassessment of conservation status.

INTERMEDIATE LONG-FINGERED BAT
Miniopterus medius Thomas and Wroughton, 1909b
med'-ee-us

The intermediate long-fingered bat is a medium-sized species of *Miniopterus* with a forearm length range that overlaps with four other species presumed to occur in the region (Ibáñez and Juste 2019). It is therefore easily confused with other *Miniopterus*. Taxonomic work underway will clarify the distribution of this species, and until then it is presumed to range from peninsular Malaysia to Papua New Guinea, the Bismarck Archipelago and Buka (*sensu* Armstrong *et al.* 2019b).

The intermediate long-fingered bat is an aerial insectivore that will roost in caves, tunnels and rock crevices (Bonaccorso 1998). It is known from the lowland forests of New Britain and New Ireland and can be abundant in the New Guinea highlands (Bonaccorso 1998). Smith and Hood (1981) found this species to be less abundant on New Ireland than other *Miniopterus*. They encountered it roosting with *M. australis* in several Japanese Second World War tunnels on New Britain. One colony comprising both of these species was estimated to contain tens of thousands of individuals.

One of five adult females caught in January 1997 at Ulatawa Plantation, East New Britain, was in

Intermediate long-fingered bat (*Miniopterus medius*), Thailand. Taxonomic work is underway to clarify the distribution of this species (www.MerlinTuttle.org).

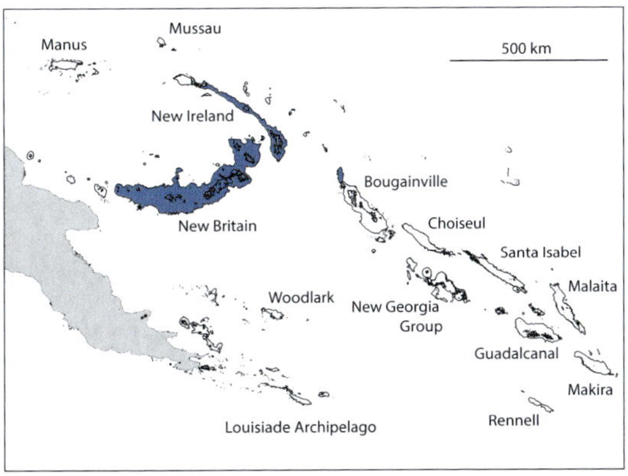

Miniopterus medius			
Size*			
	Mean	**Range**	**Sample**
Head–body	53.6	42.0–60.0	7
Tail	52.5	42.6–59.0	7
Forearm	45.2	40.0–48.4	10
Tibia	19.2	17.2–20.5	6
Hindfoot	9.3	8.0–11.0	5
Ear	11.9	17.2–20.5	6
Weight	10.2	6.0–12.1	7
* From mainland New Guinea Flannery (1995a) and Bonaccorso (1998)			
Identification The fur is paler than other species of *Miniopterus* and the venter more clearly bicoloured. The skull is domed with a high sloping forehead.			
Synonyms None			
Subspecies None			
Distribution Buka, New Britain, New Ireland			
Extralimital Indonesia, Malaysia, New Guinea			
Conservation status (IUCN) Least Concern			
Etymology Sp.–middle (L.), a reference to the intermediate size of this species between *M. australis* and *M. schreibersii* species complexes			

early-term pregnancy, but the other four were reproductively inactive (Bonaccorso 1998). Similarly, females examined in Chimbu Province (New Guinea) from March to May were reproductively inactive, as were two adult females examined on Buka Island, 14–15 February 1997 (Bonaccorso 1998).

No threats have been identified that could affect the intermediate long-fingered bat over its wide distribution. The conservation status of this species will require reassessment following any taxonomic and distributional revisions.

LOYALTY LONG-FINGERED BAT

Miniopterus robustior Revilliod, 1914

roh-bus'-tee-or

The Loyalty long-fingered bat was described from nine individuals (three male, six female) collected at Quépénée (Chépénéhé), a village on the north-west coast of Lifou Island. One of these specimens (a male) is now held in the collections of the Australian Museum. Petersen (1981) identified an additional three specimens from Rawa Ama on the island of Maré; these are held in the American Museum of Natural History.

Revilliod (1914) reported the occurrence of this species (as well as *M. australis* and *M. macrocneme*) at Quépénée, Lifou. Large *Miniopterus* colonies containing average numbers of 1000–1500 individuals exist on Lifou and Maré (Kirsch *et al.* 2002a). But it is quite likely that all three species of *Miniopterus* use these same roosts. Pregnant females were captured in October 2001 (Kirsch *et al.* 2002a). Echolocation calls are frequency-modulated sweeps with a characteristic frequency of 42–45 kHz (Kirsch *et al.* 2002a).

Miniopterus robustior			
Size*			
	Mean	**Range**	**Sample**
Head–body	45.9	43.0–49.0	9
Tail	42.6	37.5–46.0	8
Forearm	40.1	39.0–41.2	8
Tibia	14.9	14.3–15.8	9
Hindfoot	8.3	7.8–9.0	9
Ear	12.2	11.5–13.0	9
Weight	–	–	–
* From the type series Revilliod (1914)			
Identification A small species of long-fingered bat distinguishable from other small members of the genus by its very robust head and rostrum.			
Synonyms None			
Subspecies None			
Distribution Lifou, Maré			
Extralimital None			
Conservation status (IUCN) Endangered			
Etymology Sp.–more robust (L.), a reference to the more robust morphology than its congeners in the Loyalty Isls			

Very little additional information has been recorded of biology. The Loyalty long-fingered bat has the most restricted distribution of any species of *Miniopterus*. It is classified as Endangered on the IUCN Red List because it has a small area of occupancy and there is a continuing decline in the extent and quality of habitat.

GREAT LONG-FINGERED BAT

Miniopterus tristis (Waterhouse, 1845)

tris'-tis

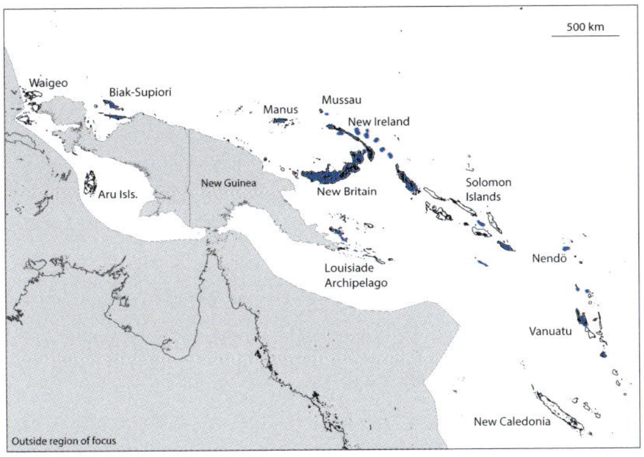

This large-bodied *Miniopterus* has an extremely wide distribution from Sulawesi and the Philippines to Vanuatu, but probably includes multiple distinct species. A multilocus phylogeny by Demos *et al.* (2019) showed *M. australis* and *M. tristis* to be sister taxa to *M. schreibersii*. The great long-fingered bat is less common than the smaller *M. australis* and *M. macrocneme*. Its rather scattered distribution may be an artefact of this rarity, or an indication that many islands in the region lack suitable habitat.

Smith and Hood (1981) reported the great long-fingered bat as having been collected from the St Matthias Group, New Britain and New Ireland. Surveys on Mussau Island in 2014 recorded several colonies of a large *Miniopterus* with adult forearm lengths of 48.4–50.5 mm and bodyweights of 9.7–11.7 g (Aplin *et al.* 2015). Animals from Mussau produced calls that were steeply decreasing broadband frequency sweeps with a characteristic frequency of 40 kHz (Armstrong *et al.* 2015). These were directly comparable to calls passively recorded on Manus Island.

Miniopterus tristis			
Size*			
	Mean	**Range**	**Sample**
Head–body	59.1	56.9–64.6	9
Tail	54.7	48.4–59.6	9
Forearm	48.9	48.2–49.6	12
Tibia	20.2	19.3–21.0	9
Hindfoot	9.4	8.6–10.2	9
Ear	12.5	11.1–13.3	9
Weight	11.6	10.7–13.1	7
* From KU and QM specimens (Makira and Rennell)			

Identification The largest species of *Miniopterus*, with a robust skull and teeth. The sagittal crest is high and in living animals it can be felt when the top of the head is caressed.

Synonyms *Miniopterus bismarckensis* Maeda, 1982
Miniopterus melanesiensis Maeda, 1982

Subspecies *Miniopterus tristis tristis* (Waterhouse, 1845) (Philippines)
Miniopterus tristis celebensis Petersen, 1981 (Sulawesi and Sula Isls)
Miniopterus tristis grandis Petersen, 1981 (New Guinea, including Yapen, Waigeo, Supiori, Biak and Louisiade Archipelago)
Miniopterus tristis insularis Petersen, 1981 (Admiralty Isls, Bismarck Archipelago, Solomon Islands, Vanuatu)
Miniopterus tristis propitristis Petersen, 1981 (Eastern New Guinea)

Distribution Bougainville, Buka (AROB); Biak-Supiori, Yapen (Indonesia); Ambitle, Boang, Duke of York, Dyual, Eloaua, Emirau, Fergusson, Lavongai, Lihir, Manus, Mussau, New Britain, New Ireland, Nissan, Normanby, Tabar Group (Papua New Guinea); Makira, Nendö, Ngella Sule, Rennell (Solomon Isls); Efate, Espiritu Santo (Vanuatu)

Extralimital Indonesia, New Guinea, Philippines

Conservation status (IUCN) Least Concern

Etymology Sp.–sad (L.), a reference perhaps to the dark colour of the fur, or the shape of the face

Great long-fingered bat (*Miniopterus tristis*), Makira, Solomon Islands. This large species is less common than other *Miniopterus* found in the region. It has a large robust head, a pronounced forehead and a high sagittal crest (Tyrone Lavery).

On Makira, echolocation calls of great long-fingered bats released from the hand were short-duration, frequency-modulated 'hockey stick'-shaped pulses. The calls started with a rapid frequency sweep from 70–45 kHz, dropping to around 40 kHz, and then transitioned to a less steep, longer duration section with a characteristic frequency of 36.2 kHz. The peak frequency was in the first harmonic at the 'knee' of the call at 36.5 kHz (Pennay and Lavery 2017).

The great long-fingered bat was recorded on Makira and Rennell in Solomon Islands in 2018. On Makira it was present with *M. australis*, *M. blepotis*, *Hipposideros demissus*, *H. calcaratus* and *H. cervinus*. On Rennell it occupied a cave on a steep cliff overlooking the ocean. Other species present were *Dobsonia inermis*, *Aselliscus tricuspidatus*, *M. australis*, *M. blepotis*, *Myotis moluccarum* and *Hipposideros calcaratus*. Of five females examined on Rennell on 28 October 2018, two were clearly pregnant and one had enlarged teats but was not lactating. Bonaccorso (1998) recorded a lactating female at Javarere Cave, Central Province, Papua New Guinea, on 31 October 1968.

Molossidae

Family Molossidae is one of the most geographically widespread and diverse groups of echolocating bats. It comprises 22 genera and around 126 species, usually rather large in body size. Molossid bats do not possess any elaborations such as a nose-leaf, have ears that are normally conjoined across the top of the head, and have a tail that extends beyond the end of the uropatagium. Most are high-flying, fast-flying species, with echolocation calls that are low-frequency sweeps. The call frequencies of some species are audible to humans.

Molossidae is represented by only four species in the region, two of which are confined to New Britain and smaller islands at the margin of New Guinea. The other two, however, are endemic, one to Solomon Islands and one to Fiji and Vanuatu. Both these endemics are Endangered, primarily from disturbances at the limited roost sites. All species are relatively poorly studied in the region.

Mops

FIJIAN FREE-TAILED BAT

Mops bregullae **(Felten, 1964b)**

mops breg-ull'-ee

The Fijian free-tailed bat was formerly recognised as a subspecies of *Mops jobensis*, but morphological and genetic data support its recognition as a distinct taxon (Ingleby and Colgan 2003). This species has an unusual distribution, being known from only a few islands in the Fiji and Vanuatu island groups. In Vanuatu it occurs at the type locality Malo Island, and Espiritu Santo, and in Fiji it can be found on Taveuni and Vanua Levu. It was once a more widespread species as demonstrated by cave deposits on Viti Levu (Fiji) and 'Eau (Tonga) (Koopman and Steadman 1995). The current disjunct distribution may thus be derived from local extinctions across a formerly broader range.

In Fiji, *M. bregullae* has been recorded in the coastal and lowland forests of Vanua Levu, and into the cloud forests of Des Voeux Peak on Taveuni (Ingleby and Flannery 1991). Echolocation calls are flat and between 20 and 35 kHz (Scanlon *et al.* 2013a). The few roost sites documented can contain many thousands of individuals. An Australian expedition in 1992 recorded the species from Nanumbu Cave (Espiritu Santo), which it shared with a large colony of *Miniopterus australis*. Tan Lensigo near Abounatori Village (Malo) contained a large roost in 1971 (Hill 1983) and 1992. On Vanua Levu there is a major roost near the village of Nakanacagi, now protected as the Nakanacagi Bat

The Fijian free-tailed bat (*Mops bregullae*) has an unusual distribution, being known from only a few islands in the Fiji and Vanuatu island groups (Pavel German).

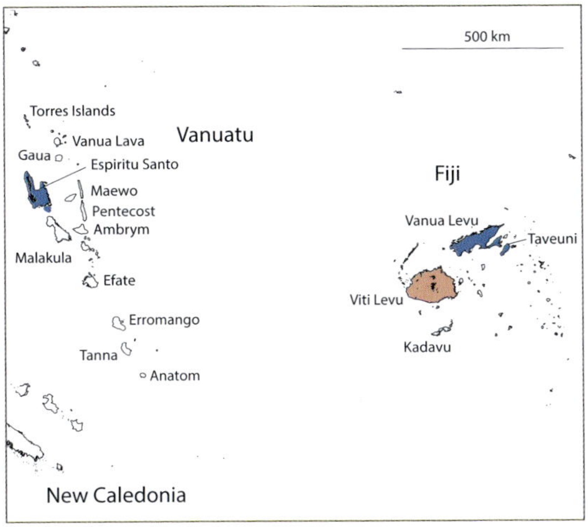

Mops bregullae			
Size*			
	Mean	**Range**	**Sample**
Head–body	65.4	63.0–67.5	4
Tail	43.0	39.7–45.8	4
Forearm	52.3	51.3–53.6	4
Tibia	20.9	20.0–22.3	4
Hindfoot	–	–	–
Ear	18.4	16.6–20.7	4
Weight	20.0	16.0–22.5	4
* From AM specimens (Taveuni, Vanua Levu)			
Identification similar to *M. jobensis* and *M. solomonis* but lacks papillae on ear margins. The forearm is slightly longer than in these two species.			
Synonyms None			
Subspecies None			
Distribution Taveuni, Vanua Levu (Fiji) Espiritu Santo, Malo (Vanuatu)			
Extralimital None			
Conservation status (IUCN) Endangered			
Etymology Gen.–a proper name (Gr.), significance unknown; Sp.–for Herr Bregulla, collector of the holotype			

Sanctuary. The colony there was estimated to contain around 2000 individuals in 2009, 2010 and 2011 (Scanlon *et al.* 2013a). An internal cave assessment estimated the colony size at approximately 4200 in 2014 (NatureFiji-MareqetiViti, unpublished data; Waldien *et al.* 2019).

Fijian free-tailed bats form large maternity colonies. Two of three females captured on Taveuni in November were pregnant (Scanlon *et al.* 2013a). Nakanacagi Cave contained young at various stages of growth, from newborn and naked to furred on 5 December 1990. Many females were pregnant and adult males were also present. In Nunumbu Cave, 14 of 16 females were lactating on 25 April 1992, and adult males were also present. This suggests young were about to be weaned. These data indicate breeding is synchronous and seasonal across Fiji and Vanuatu, with young born late in the year and weaned by April.

The Fijian free-tailed bat is exploited for bushmeat. There are few locations where this species is known to roost and breed, and it is vulnerable to overexploitation and disturbance. Protection of known roost sites is vital for the long-term conservation of this species.

GREATER NORTHERN FREE-TAILED BAT
Mops jobensis (Miller, 1902)

joh-ben'-sis

The greater northern free-tailed bat is a widespread species and the taxonomic relationships between Australia, New Guinea and their satellite island subspecies may need further assessment. Within the region it has been recorded from New Britain and Kairuru Island (Bonaccorso 1998), and Yapen (the type locality). Aplin and Pasveer (2006) identified this species among the archaeological deposits examined on Korbroor, Aru Islands, but it is yet to be recorded among the modern faunas there. *Mops* can be difficult to survey, and this species may be more widespread than currently understood.

The nreater northern free-tailed bat is found in lowland and lower montane habitats, including

Greater northern free-tailed bat (*Mops jobensis*), Cape York Peninsula, Australia (Michael Pennay).

The greater northern free-tailed bat (*Mops jobensis*) will roost in small groups of 10–15 individuals inside buildings, caves, and holes in coconut palms (Angus McNab).

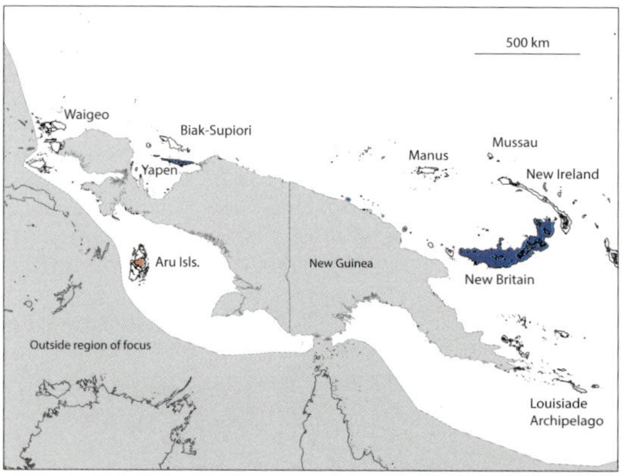

Mops jobensis			
Size*			
	Mean	**Range**	**Sample**
Head–body	62.0	51.0–72.0	9
Tail	35.5	31.3–45.0	9
Forearm	45.6	43.4–50.0	13
Tibia	16.0	14.9–18.5	9
Hindfoot	11.3	10.3–12.6	5
Ear	19.0	16.4–21.3	9
Weight	12.4	10.5–13.5	8
* From New Guinea specimens Hill (1983), Flannery (1995a) and Bonaccorso (1998)			
Identification This is a medium-sized *Mops* that is smaller than *M. bregullae*, but larger than *M. solomonis*, and generally has four minute papillae on each ear.			
Synonyms None			
Subspecies *Mops jobensis jobensis* (Miller, 1902) (Maluku Isls, New Britain, New Guinea, Yapen) *Mops jobensis colonicus* (Thomas, 1906) (northern Australia)			
Distribution Aru Isls (fossil), Yapen (Indonesia); Kairuru, New Britain (Papua New Guinea)			
Extralimital Australia, Maluku Isls, New Guinea			
Conservation status (IUCN) Least Concern			
Etymology Sp.–from Jobi (L.), an older name for Yapen Is.			

primary rainforests and forest clearings. This is a high-flying species with direct and fast flight, usually targeting insects in areas of open space above forest canopies. Roosts are usually small, containing 10–15 individuals inside holes in coconut palms (*Cocos nucifera*) and trees, buildings, and caves in cliffs (Bonaccorso 1998). However, larger colonies have also been found. In Australia, pregnant females have been recorded from October to November; the single young are born in December, and are independent by March–April (Churchill 2008).

Echolocation calls of this species are audible to humans and in Australia have a characteristic frequency of 19.8 kHz (range 16.1–23.6 kHz) (Milne 2002). Greater northern free-tailed bats are usually recorded flying in pairs, resulting in paired call pulses at alternating frequencies. Armstrong and Aplin (2011) recorded calls at around 20 kHz above artificial clearings in the Muller Range, Papua New Guinea that may have been attributable to this species.

SOLOMONS FREE-TAILED BAT

Mops solomonis (Troughton, 1931)

sol'-oh-mon-is

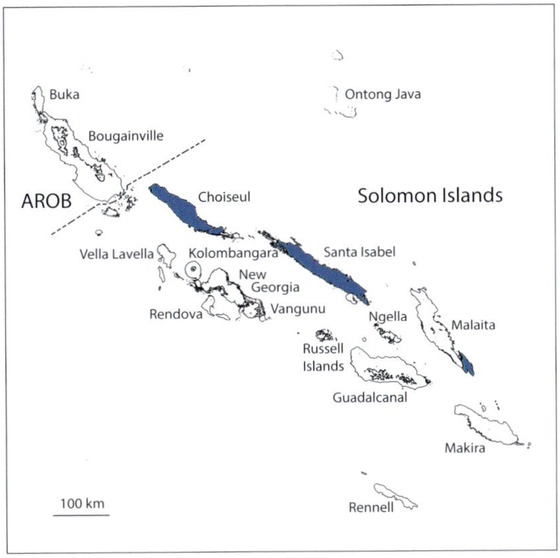

Mops solomonis			
Size*			
	Mean	**Range**	**Sample**
Head–body	63.9	55.6–68.0	8
Tail	34.6	30.0–36.5	8
Forearm	43.0	40.5–45.0	8
Tibia	14.5	13.5–15.5	8
Hindfoot	8.8	8.5–9.0	4
Ear	18.8	16.4–20.5	8
Weight	12.3	11.5–13.0	4
* From AM specimens (Choiseul), and Troughton (1931)			
Identification This is the only species of *Mops* known to occur in Solomon Isls. It can be differentiated from other echolocating bats by possessing a tail that extends well beyond the end of the tail membrane, the distinctive fringe of hairs on the side of each hindfoot, and very wrinkled lips.			
Synonyms None			
Subspecies None			
Distribution Choiseul, Maramasike, Santa Isabel			
Extralimital None			
Conservation status (IUCN) Endangered			
Etymology Sp.–for the Solomon Isls			

This is a little-known species described from a large series of specimens collected from Santa Isabel by N. S. Heffernan Esq. in 1928. Troughton (1931) reported the following from the collector's notes: 'Called at Mufu Point and climbed to the caves and obtained a sackful of bats. All of one kind, and there does not appear to be any other. Proportion of males to females one to six. Attempted photo of the swarm at sunset, estimated to be a mile in length and 50 yards broad before dispersing, but too dark and the film a blank. It is interesting to note that Mufu Point caves have absolutely only one kind bat, whereas a cave only a few miles distant has five varieties and probably more, but the situations are entirely different. Mufu caves open out onto sheer cliffs over the water. The other cave is right on top of a hill in dense forest and about 600 feet [180 m] above sea-level.'

The Solomons free-tailed bat (*Mops solomonis*) has been recorded from very few sites in Solomon Islands and is listed as Endangered on the IUCN Red List (Madison Erin Mayfield).

The Solomons free-tailed bat has been recorded on Choiseul by P. Temple in 1964, and in the north-west of the island at Pavora River by Dr Harry Parnaby and Ian Aujare of the Australian Museum in 1990. The eight adults collected in 1990 were taken via mist nets set over the river rather than from a cave roost. As part of the same work, both researchers visited Mufu Point in 1991 and entered a giant cavern from the sea cliff. No bats were detected inside the cave. However, the cavern ceiling was extremely high and unable to be lit by spotlights, and the exit was not observed on dusk to detect emerging bats (H. Parnaby, pers. comm.). Two females examined on 10 March had reduced teats and were post-lactating.

This species was recently encountered in a large limestone cave at 260 m elevation on Maramasike Island ('small Malaita') (M. Vaghi, pers. comm.). The entrance to the cave was in an area of lowland forest. Recent echolocation call recordings from the Kunua Plains/Mt Balbi region of Bougainville have also indicated it occurs there (J. Novera, pers. comm.). Few other records of the Solomons free-tailed bat exist. Extensive searches of caves across the Solomon Islands archipelago between 2010 and 2019 have failed to find additional populations. Mufu Point was clearly a key site and the lack of recent updates on the status of the colony there is a concern. Additional searches and steps to protect the colony, if it persists, will be an important step for conservation of *M. solomonis*.

Ozimops

BECCARI'S FREE-TAILED BAT

Ozimops beccarii (Peters, 1881)

oz'-ee-mops bek-ar'-ee-ee

Beccari's free-tailed bat was redefined by Reardon *et al.* (2014), and its geographical distribution is now considered to be limited to the Maluku Islands of Indonesia, the New Guinea mainland, and the Bismarck and D'Entrecasteaux Islands. Bonaccorso (1998) recognised two subspecies – *Ozimops beccarii beccarii* from Mimika River, southern West Papua and Maluku, and *O. b. astrolabiensis* from northern Papua New Guinea and the Bismarck Archipelago. Considerable morphological variation suggests cryptic species are likely (Reardon 2017)

Echolocation calls from Akwam and Keboi Kerowa (Papua New Guinea) consisted of a broadband frequency-modulated call with two visible harmonic ranges. The pulse shape was concave or sloped with a down-sweeping initial and up-sweeping tail. The characteristic frequency was between 44 and 50 kHz (mean 47.17 kHz) and the fundamental harmonic 22–27 kHz (mean 24.25) (Leary and Pennay 2011).

Beccari's free-tailed bat has been recorded from several small islands on the northern coastline of New Guinea, on Fergusson Island, and there is a

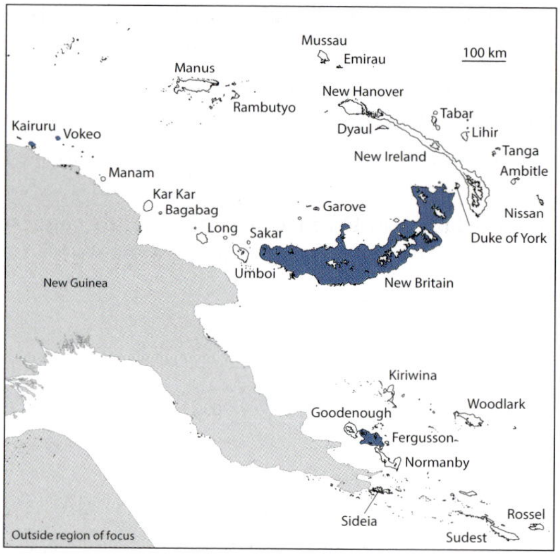

Ozimops beccarii			
Size*			
	Mean	**Range**	**Sample**
Head–body	59.3	54.0–66.0	4
Tail	32.6	27.2–38.0	4
Forearm	34.6	33.0–37.0	9
Tibia	12.4	8.5–15.7	4
Hindfoot	8.4	6.0–10.2	4
Ear	14.5	11.0–18.0	4
Weight	14.8	13.0–17.0	4
* From Bonaccorso (1998), mainland New Guinea and Fergusson Is.			
Identification This is the smallest molossid in the region, with the lips less crinkled than in the species of *Mops* and the rhinarium lacking the beading found along the dorsal edge in those species.			
Synonyms None			
Subspecies *Ozimops beccarii beccarii* (Peters, 1881) (southern West Papua, Maluku) *Ozimops beccarii astrolabiensis* (Meyer, 1899) (northern Papua New Guinea, Bismarck Archipelago)			
Distribution Fergusson, Kairuru, New Britain, Vokeo			
Extralimital Maluku Isls, New Guinea			
Conservation status (IUCN) Least Concern			
Etymology Gen.–ozi, colloquialism for Australia (as the centre of diversity of this group) and mops, a suffix common to many molossid genera; Sp.–for Odoardo Beccari, who collected the holotype			

single record from Kapuluk River, West New Britain Province (Bonaccorso 1998). Armstrong *et al.* (2015) attributed two poor-quality call recordings from Mussau to this species. They were recorded during a single passive session in a coconut grove close to the coast.

Like other molossids, *Ozimops* are high- and fast-flying species. They typically roost in tree cavities rather than caves, and thus can be difficult to survey. One roost on mainland New Guinea contained at least nine individuals of both sexes (Leary and Pennay 2011). The patchy distribution of records in the Bismarck Archipelago and D'Entrecasteaux Islands suggest this species may be more widespread than is currently realised.

Mystacinidae

This family, consisting of two species, is endemic to Aotearoa New Zealand. They are unusual bats in being adapted to a partly terrestrial lifestyle and a very broad diet. Both species of mystacinids have fared poorly since the arrival of humans in Aotearoa New Zealand. *Mystacina robusta* (the larger species) was probably facing extensive range contraction before European colonisation and is now extinct. *Mystacina tuberculata* has also undergone a dramatic range reduction, most of which appears to postdate European arrival.

Mystacina

GREATER NEW ZEALAND SHORT-TAILED BAT

***Mystacina robusta* Dwyer, 1962**

mis'-tah-seen'-ah roh-bus'-tah

Subfossil and fossil remains suggest that the greater New Zealand short-tailed bat was once widely distributed, occurring on the North and South Islands of Aotearoa New Zealand. However, no records of the species exist from the main islands since European colonisation. It was last confirmed from Big South Cape Island (Taukihepa) in 1964 and the last known specimen was mist-netted on Solomon Island (Rerewhakaupoko) by J. A. Mackintosh in April 1965 (King and Lloyd 2021). The species is now presumed Extinct. Comparisons of modern specimens and subfossils revealed there was a clinal decrease in body size from Waitomo and Puketiti Cave in the north to South Canterbury on the South Island. This may have been related to climate and physiology (Worthy *et al.* 1996).

The Museum für Naturkunde, Berlin, holds a subadult female (MZB 2591) in spirit, the skull extracted. The collector is given as 'Gray', but this may have been made in error and instead pertain to the describer of *Mystacina tuberculata* (as this specimen was labelled). Material allegedly obtained from Collingwood on the South Island is held by the Australian Museum, purchased from a Mr James Dall on 17 February 1902. The specimens were well preserved, suggesting they may have been collected relatively recently (Flannery 1987). However, Worthy *et al.* (1996) suggested Dall's specimens were most likely derived from Puai Cave on Big South Cape Island, a hypothesis supported by morphological measurements that were consistent with those of bats from Stewart Island, and smaller than those of fossil specimens from the South Island.

The Bernice P. Bishop Museum, Honululu, holds a badly faded spirit specimen (BM9003), which was collected by H. H. Travers during or before 1902. According to the specimen label it was collected from a cave 'from an island at the extreme end of Stewart Island'. This was most likely either Solomon Island or Big South Cape Island, from which the last records of this species were derived. Both islands are granitic and are covered with forests of *Olearia* spp. and *Metrosideros umbellata*, with *Podocarpus*

Mystacina robusta			
Size*			
	Mean	**Range**	**Sample**
Head–body	64.4	61.0–66.1	3
Tail	17.2	16.4–18.6	3
Forearm	46.6	45.1–48.3	9
Tibia	18.7	18.1–19.7	8
Hindfoot	14.5	12.6–16.0	7
Ear	17.8	16.0–19.1	9
Weight	–	24.0 (estimate)	–
* From AM specimens from the South Is., Hill and Daniel (1985) and Worthy *et al.* (1996)			
Identification In Aotearoa New Zealand, only *Mystacina* have a tail which penetrates the uropatagium. *Mystacina* species can be further distinguished from *Chalinolobus tuberculatus* by the presence of pointed ears that extend past the nose when adpressed forward, and by the presence of tubular nostrils. Of the two *Mystacina* species, this is the larger with a forearm length greater than 45 mm. It is more robust overall, with wide short nostrils, proportionately shorter wing elements, and a more massive rostrum.			
Synonyms None			
Subspecies None			
Distribution Putauhina, Stewart (Rakiura), Solomon (Rerewhakaupoko), Big South Cape Is. (Taukihepa)			
Extralimital None			
Conservation status (IUCN) Critically Endangered (Possibly Extinct)			
Etymology Gen.–moustached (L.); Sp.–robust (L.), for its larger size and more robust proportions relative to *M. tuberculata*			

broadleaf forest at their interiors. Numerous caves exist along their coastlines (King and Lloyd 2021). There were no rats on either Solomon Island or Big South Cape Island until 1963 when *Rattus rattus* was accidentally introduced. The rats reached plague proportions between 1964 and 1967, during which time both *Mystacina* became extinct there (King and Lloyd 2021; Parsons and Toth 2021).

Although little is known of the species' diet, it was probably similar to that of *M. tuberculata*, including arthropods, fruit, nectar and pollen (King and Lloyd 2021). An analysis of stomach contents from two individuals revealed pollen from rata (*Metrosideros umbellata*) and fern spores. There is some indication that the greater New Zealand short-tailed bat was monoestrous, with nursery colonies observed in hollow trees and sooty shearwater (*Ardenna grisea*) burrows on Muttonbird Island from April–May of 1963–1965 (Daniel 1990; King and Lloyd 2021). This species is classified as Critically Endangered on the IUCN Red List, and sadly it is probably already extinct.

LESSER NEW ZEALAND SHORT-TAILED BAT

Mystacina tuberculata Gray, 1843c

tue-ber'-kue-lah'-tah

The Lesser New Zealand short-tailed bat is the only surviving member of the family Mystacinidae. Three subspecies have been described, differing from each other in body size and ear length (Hill and Daniel 1985). This species was once widespread across Aotearoa New Zealand, but has significantly declined in distribution and now exists as numerous fragmented populations (Parsons and Toth 2021).

Lesser New Zealand short-tailed bats are adapted for flight in cluttered habitats with low wing-loading and aspect ratios that facilitate manouverability (Parsons and Toth 2021). They use echolocation to navigate when flying and when hawking flying insects from the air. The bats emerge 1–2 h after dark and are active until an hour before dawn. In flight, they usually remain relatively close to the ground (less than 2 m from the ground in less-cluttered beech-dominated forest) (Parsons and Toth 2021). Their echolocation calls are brief and multiharmonic at low duty cycle during the search phase, and loudest at around 27.7 kHz. The fundamental sweeps from around 37 kHz to approximately 18 kHz (Parsons and Toth 2021). Lesser New Zealand short-tailed bats are primarily insectivorous, but also consume nectar, pollen, and fruit (Parsons and Toth 2021). The diet is dominated by species from the order Lepidoptera, and species

Lesser New Zealand short-tailed bat (*Mystacina tuberculata*) (www.MerlinTuttle.org).

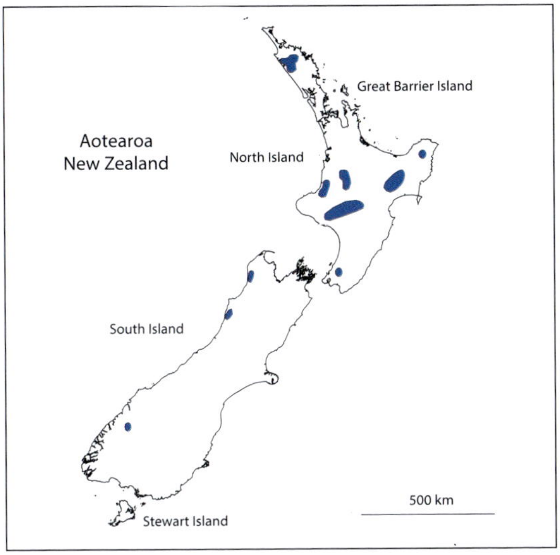

Mystacina tuberculata			
Size*			
	Mean	**Range**	**Sample**
Head–body	–	–	–
Tail	–	–	–
Forearm	42.0	40.2–45.4	15
Tibia	15.5	14.5–16.6	9
Hindfoot	13.3	11.5–16.7	15
Ear	17.1	14.5–19.1	13
Weight	12.4	10.5–14.7	6
* From Hill and Daniel (1985) and (Worthy and Scofield 2004)			
Identification This species can be distinguished from *Chalinolobus tuberculatus* by its pointed ears that extend past the nose when adpressed forward, and by the presence of tubular nostrils. Of the two species of *Mystacina* in Aotearoa New Zealand, this is the smaller, with a forearm length less than 45 mm. It has longer, narrower nostrils, and proportionally shorter wing elements.			
Synonyms *Mystacina velutina* Hutton, 1871			
Subspecies *Mystacina tuberculata tuberculata* Gray, 1843c (extreme south of North Is., South Is. and Codfish Is.) *Mystacina tuberculata aupourica* Hill and Daniel, 1985 (lowland areas of North Is.) *Mystacina tuberculata rhycobica* Hill and Daniel, 1985 (montane areas of North Is.)			
Distribution North and South Isls, Little Barrier Is. (Te Hauturu-o-Toi), Codfish Is. (Whenua Hou)			
Extralimital None			
Conservation status (IUCN) Vulnerable			
Etymology Sp.–a reference to tubercles (L.), but precise allusion unclear			

of Diptera also are commonly consumed (Czenze *et al.* 2018).

Having evolved in the absence of terrestrial mammals, these bats have the remarkable ability to forage on the forest floor. In fact, they use a true quadrupedal walking gait and have skeletal and soft-tissue adaptations for this behaviour (Hand *et al.* 2015). The animals burrow among the leaf-litter and humus in search of invertebrates and can scuttle with rodent- or shrew-like agility. The species can also burrow into rotten wood. When foraging on the ground, the animals switch to emitting echolocation calls at a lower repetition rate, using this for navigation rather than locating prey. In this situation, invertebrates are instead targeted by listening for prey-generated noise, and prominent nostrils aid in the detection of olfactory cues (Jones *et al.* 2003).

Lesser New Zealand short-tailed bats most often roosts in tree hollows but have also been found inside caves, and occasional individuals have been found roosting in houses and huts. Roosts can contain hundreds to thousands of individuals (Parsons and Toth 2021). In the *Nothofagus* forests of Fiordland National Park on the South Island, this species was found to prefer roost sites in more open parts of the forest and also preferred large-diameter, old trees (Sedgeley 2003). The preferred cavities were large, had multiple entrances, and were usually around 10 m from the ground.

This species employs a lek mating system. Females establish communal roosting colonies and males position themselves in nearby 'singing roosts' from which they call during the breeding season. These singing roosts are clustered around the female colonies and are defended by males (Toth *et al.* 2015). Females will visit the singing roosts for mating. The breeding season occurs between February and May.

There is a range of threats to this species. Remaining populations are fragmented and declines in habitat quality and quantity are ongoing. For example, the loss of old-growth roost trees because of forestry practices and clearing for development has had significant negative impacts. Predation by introduced mammals such as stoats (*Mustela erminea*), black rats (*Rattus rattus*) and brushtail possums (*Trichosurus vulpecula*) has also had major impacts. The lesser New Zealand short-tailed bat is thought to have become extinct on Big South Cape Island and Solomon Island due to the introduction of *Rattus rattus* in 1963 (Parsons and Toth 2021).

Rhinolophidae

Rhinolophidae is a diverse family of around 109 species all belonging to the genus *Rhinolophus*. The family is distributed throughout the Old World, with the greatest species diversity found in the Indo-Malayan region. Africa also supports a diverse rhinolophid fauna. All species have an elaborate horseshoe-shaped growth surrounding the nostrils to assist with echolocation, and all lack a tragus.

There are relatively few species in Australia, New Guinea and the South-west Pacific. The region under consideration here has just two species (neither of which is endemic). Both of these have extralimital distributions that include New Guinea, and either the Maluku Islands (*R. euryotis*) or Australia (*R. megaphyllus*).

The eastern horseshoe bat (*Rhinolophus megaphyllus*) is one of two species of rhinolophid found in the South-west Pacific (Angus McNab).

Rhinolophus

NEW GUINEA BROAD-EARED HORSESHOE BAT

Rhinolophus euryotis **Temminck, 1835**

rie'-noh-loh'-fus yue'-ree-oh'-tis

The New Guinea broad-eared horseshoe bat is a very large, distinctive and locally common species. Its distribution spans from the Maluku Islands (where it is apparently most common) to the Bismarck Archipelago and Trobriand Islands where records are scarcer. Smith and Hood (1981) recorded the capture of a single individual in a human-made tunnel on New Britain. This was the first record from the island and intensive searches as part of the same expedition failed to locate more individuals. The Australian Museum holds two specimens from the Fulleborn area of New Britain and Bonaccorso (1998) reported 11 individuals from West New Britain in 1995.

Intensive searches by Smith and Hood (1981), and the Australian Museum in 1987, failed to locate this species on northern New Ireland. It was recorded there by the Australian Museum in 1993, but again was not encountered during surveys on southern New Ireland in 1994 (Emmons and Kinbag 2001).

The New Guinea broad-eared horseshoe bat (*Rhinolophus euryotis*) is a large, distinctive and locally common species of horsehoe bat distributed from the Maluku Islands to the Bismarck Archipelago and Trobriand Islands (Michael Pennay).

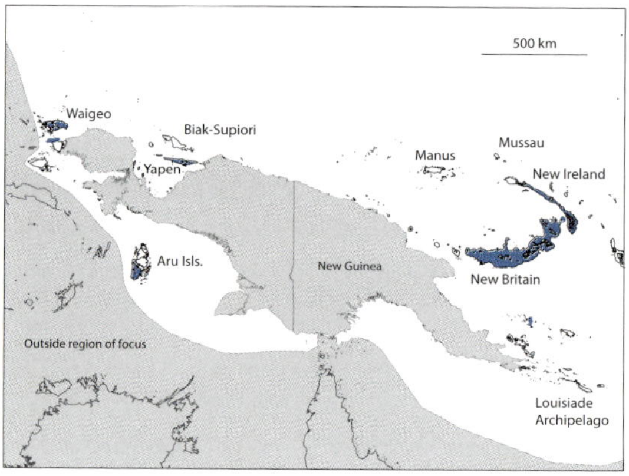

Rhinolophus euryotis			
Size*			
	Mean	**Range**	**Sample**
Head–body	61.4	54.0–73.0	5
Tail	23.1	16.0–28.0	5
Forearm	55.2	50.0–58.5	7
Tibia	26.4	23.0–28.5	7
Hindfoot	13.0	11.0–15.0	7
Ear	23.3	19.0–26.5	7
Weight	16.7	12.8–21.5	7
* From Bonaccorso (1998)			
Identification This is one of the largest *Rhinolophus* species. It has a very broad nose-leaf, which is darkly pigmented except for a white stripe running through the middle of the horseshoe to the upper lip. This white stripe will distinguish it from other *Rhinolophus* of the region.			
Synonyms None			
Subspecies *Rhinolophus euryotis euryotis* Temminck, 1835 (Ambon, Seram, Tanimbar) *Rhinolophus euryotis timidus* Andersen, 1905b (North Maluku Isls, Waigeo, Batanta, Yapen, New Guinea, Bismarck Archipelago, Kiriwina) *Rhinolophus euryotis praestans* Andersen, 1905b (Kai Isls) *Rhinolophus euryotis aruensis* Andersen, 1907 (Aru Isls) *Rhinolophus euryotis burius* Hinton, 1925 (Buru Is.)			
Distribution Aru Isls, Batanta, Waigeo, Yapen (Indonesia); Duke of York, Kiriwina, New Britain, New Ireland (Papua New Guinea)			
Extralimital Kai Isls, Maluku Isls, New Guinea			
Conservation status (IUCN) Least Concern			
Etymology Gen.–nose + crest (Gr.); Sp.– true ear (Gr.), allusion unclear			

Births probably occur in late October. Seven of eight females examined at West New Britain between 23 and 25 August 1995 were pregnant (Bonaccorso 1998). Each had an embryo measuring 17.0–21.0 mm.

The New Guinea broad-eared horseshoe bat is a foliage gleaning insectivore. The echolocation call consists of a narrowband constant frequency with minor frequency-modulated components (Leary and Pennay 2011). On mainland New Guinea the peak frequency of the dominant harmonic was 52–54 kHz.

EASTERN HORSESHOE BAT

Rhinolophus megaphyllus Gray, 1834

meg'-ah-fil'-us

The eastern horseshoe bat is a common species in eastern Australia, eastern New Guinea and the islands off eastern New Guinea. Smith and Hood (1981) reported the eastern horseshoe bat to be in great abundance in caves and human-made tunnels on New Ireland and New Britain. There it was encountered in the mid to deep sections of caves, hanging from the ceiling or upper wall. This is a species that will roost in direct association with species of *Hipposideros* and *Miniopterus*. When disturbed at the roost it displays a characteristic behaviour, moving the head nervously from side to side, with a readily observed twitching movement of the large ears. If further disturbed, it will make short flights to and from the original roost site.

Eastern horseshoe bat is adept at hovering, and forages by gleaning within forest interiors (Bonaccorso 1998). It has a slow, fluttering and manoeuvrable flight, and will venture low to the ground as it hunts for insects. A harp trap placed in a dry gully and mist net set close to the ground on New Britain in 2009 were each successful at catching this species on one occasion; both individuals were males (Aplin and Opiang 2011). It was not recorded during a rapid biodiversity assessment on southern New Ireland in 1994 (Emmons and Kinbag 2001). Stomach contents have been found to contain insects from the orders Hemiptera (family Notonectidae), Coleoptera (families Carabidae, Dystiscidae, Elateridae, Curculionidae), Diptera, Lepidoptera and Hymenoptera (family Formicidae) (Vestjens and Hall 1977).

In Australia, males store sperm for 4 months before copulation in late June. Females will move to maternity sites and around November produce a single young, which is weaned by the end of January

The eastern horseshoe bat (*Rhinolophus megaphyllus*) is a common species in eastern Australia, eastern New Guinea and the islands off eastern New Guinea (Mark Sanders).

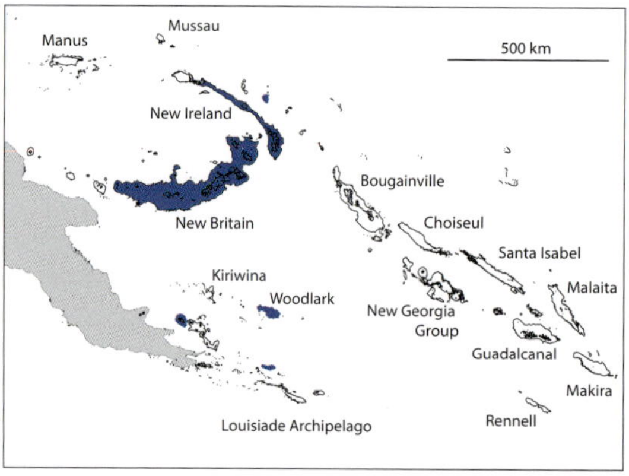

Rhinolophus megaphyllus			
Size*			
	Mean	**Range**	**Sample**
Head–body	49.7	42.7–55.0	6
Tail	22.4	19.0–30.0	6
Forearm	46.8	41.7–59.8	13
Tibia	19.7	17.4–24.0	7
Hindfoot	11.1	9.3–13.0	4
Ear	17.8	16.0–20.0	7
Weight	8.8	6.5–12.0	8
* From AM specimens (New Ireland) and Bonaccorso (1998)			
Identification A smallish *Rhinolophus* with a broad nose-leaf. The sella is narrower than in *R. arcuatus*, *R. philippinensis* and *R. euryotis*, and the ears are relatively short.			
Synonyms None			
Subspecies *Rhinolophus megaphyllus megaphyllus* Gray, 1834 (south-eastern Australia) *Rhinolophus megaphyllus monachus* Andersen, 1905c (Louisiade Archipelago) *Rhinolophus megaphyllus fallax* Andersen, 1906 (New Guinea) *Rhinolophus megaphyllus ignifer* Allen, 1933 (north-eastern Australia) *Rhinolophus megaphyllus vandeuseni* Koopman, 1982 (Bismarck Archipelago)			
Distribution Goodenough, Lihir, Misima, New Britain, New Ireland, Woodlark			
Extralimital Australia, New Guinea			
Conservation status (IUCN) Least Concern			
Etymology Sp.–great leaf (Gr.), a reference to the large nose-leaf			

after a period of around 8 weeks (Pavey and Young 2008). Two females collected on New Ireland 13 June 1979 were not reproductively active (Bonaccorso 1998).

Species of *Rhinolophus* emit constant-frequency echolocation calls with a flat tone at a certain frequency. Reference calls recorded from the Huon Peninsula, New Guinea, had a characteristic frequency of 68 kHz and pulses usually ended with a down-sweeping tail dropping by around 8 kHz in frequency (Robson *et al.* 2012).

Vespertilionidae

Vespertilionidae has the largest number of species of all chiropteran families, and vespertilionids are among the most widespread and abundant of all insectivorous bats. The family comprises some 54 genera and nearly 500 species spanning the wide spectrum of habitats from arid deserts to temperate and tropical forests. All possess a uropatagium that extends to just before the end of the tail.

Fifteen species of vespertilionid are known from the region, distributed among eight genera. New records are common and several species are known from only one or two locations. Species of vespertilionid can be difficult to survey in the South-west Pacific. Many roost in hollows or fissures in trees rather than caves, and these sites can be very difficult to find in the complex tropical forests that occur throughout much of the region. Moreover, their echolocation calls can readily detect mist nets – the dominant survey method used for bats in the Pacific. There is a significant amount of work needed to improve distributional records for most species, and it is almost certain additional species remain to be described.

New Guinea pipistrelle (*Pipistrellus angulatus*) (illustration by Madison Erin Mayfield).

Chalinolobus

NEW CALEDONIA WATTLED BAT

***Chalinolobus neocaledonicus* Revilliod, 1914**

kal'-in-oh-loh'-bus nee-oh-kal'-ed-on'-ik-us

This species was described from a single specimen, an adult male from Canala, Grande Terre. Tidemann

(1986) reported on a further eight specimens (seven from Tiaré) held in the collections of the Natural History Museum, London, and one from Caavatch, held in the Australian National University Zoology Collection. Tidemann (1986) considered *Chalinolobus neocaledonicus* a synonym of *C. gouldii* based on morphological evidence. But this was rejected by Flannery (1995b), who retained it as a full species. Parnaby *et al.* (2020) re-examined key external and cranial criteria from much of the same material used to justify synonymy with *C. gouldii* and found support for the species status of *C. neocaledonicus*. Echolocation recordings indicate it also may be present on Lifou (Kirsch *et al.* 2002a).

Little is known about the natural history of *C. neocaledonicus*. Like other members of the genus in the Australo-Pacific region, it probably favours roost sites in hollow trees. The seven specimens from Tiaré held in the collections of the Natural History Museum, London, were taken from a tree hollow (Parnaby *et al.* 2020). However, roosts within human-made structures have also been reported. The holotype was collected from a house, and large roosts and maternity sites have also been found in houses (Kirsch *et al.* 2002b). In October 2017, 41 adult females were examined from a large colony in a church in Tendo, north-east Grande Terre (Parnaby *et al.* 2020).

New Caledonia wattled bats most likely forage in more open areas of dry forest. Pregnant females have been caught in October–November and births are estimated to occur between late November and December (based on pregnant female bodyweights) (Kirsch *et al.* 2002b).

Echolocation calls from open environments are steep sweeps averaging 10 ms in duration, with a peak frequency around 34.5 kHz, a maximum frequency of 46.3 kHz and minimum frequency of 33 kHz (Kirsch *et al.* 2002a). A series of 38 calls recorded from hand-released individuals were shorter in duration, averaging around 2 ms with a peak frequency around 39.5 kHz. The New Caledonia wattled bat is classified as Endangered on the IUCN Red List because of its small distribution and declining extent and quality of habitat.

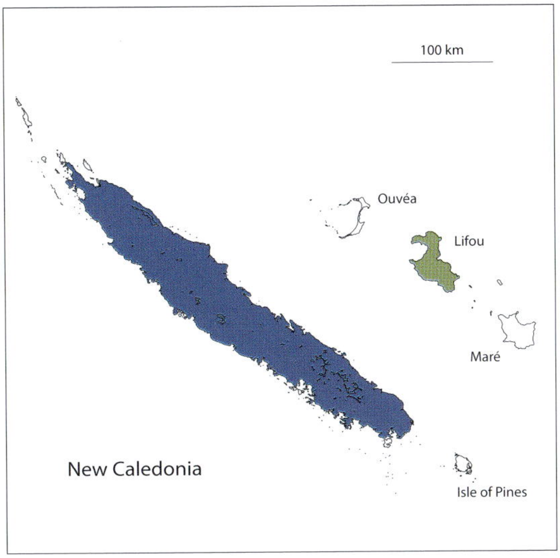

Chalinolobus neocaledonicus			
Size*			
	Mean	**Range**	**Sample**
Head–body	–	50.5	1
Tail	38.1	33.0–43.9	7
Forearm	37.5	35.3–40.6	11
Tibia	16.4	14.3–17.6	7
Hindfoot	7.6	6.1–9.6	6
Ear	–	12.0	1
Weight	9.9	7.1–14.8	8
* From Kirsch *et al.* (2002a) and Parnaby *et al.* (2020)			
Identification The only species of New Caledonian bat with fleshy wattles at the corners of the lips.			
Synonyms None			
Subspecies None			
Distribution Grande Terre, ?Lifou			
Extralimital None			
Conservation status (IUCN) Endangered			
Etymology Gen.–a bridle (hence at the corner of the mouth) + lobe (Gr.) a reference to the fleshy wattles at the corner of the mouth; Sp.–of New Caledonia (L.)			

The New Caledonia wattled bat (*Chalinolobus neocaledonicus*) is endemic to New Caledonia and listed as Endangered on the IUCN Red List (Madison Erin Mayfield).

HOARY WATTLED BAT
Chalinolobus nigrogriseus (Gould, 1856)
nig'-roh-griz-ay'-us

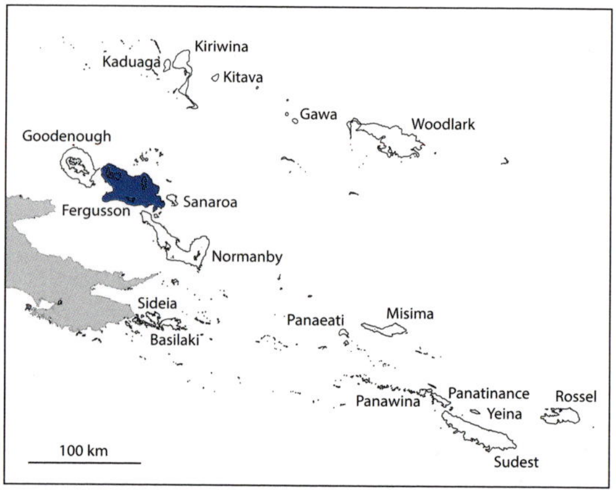

The hoary wattled bat is known from the region by a single specimen originally held by the Queensland Museum and now in the collections of the American Museum of Natural History (Van Deusen and Koopman 1971). It was collected on Fergusson in the D'Entrecasteaux Islands by R. C. English in 1891. Flannery (1995a) noted that on nearby New Guinea this species is seemingly limited to the eastern fifth of the island from Port Moresby to the D'Entrecasteaux Islands. The hoary wattled bat flies fast below the canopy; it is thus more common in vegetation types with sparser understoreys, including riparian woodlands, tall open forests, grasslands and floodplains (Churchill 2008).

Roost sites include rock crevices, buildings and hollow trees. Menzies (1971) documented the presence of a large maternity colony of this species in the roof of a building in Port Moresby. It contained several hundred individuals, including 35 lactating females and 76 juveniles on 6 November. Twin births are common in this species (Menzies 1971; Churchill 2008).

In South-east Queensland the echolocation call is a curved pulse, usually with no tail, and with a characteristic frequency of 37–40 kHz (Reinhold *et al.* 2001). Preferred prey are beetles, ants and moths, but a variety of other insects is consumed (Churchill 2008). In Australia, examination of stomach contents found a variety of food items including arachnids and species from the orders Mantodea, Orthoptera, Dermaptera, Gryllidae, Cicadidae, Coleoptera, Diptera, Lepidoptera and Hymenoptera (Vestjens and Hall 1977).

Chalinolobus nigrogriseus			
Size*			
	Mean	**Range**	**Sample**
Head–body	47.5	42.2–58.2	7
Tail	36.8	26.0–43.0	–
Forearm	36.8	34.0–39.4	7
Tibia	17.9	16.9–20.0	6
Hindfoot	8.1	6.9–9.0	6
Ear	10.9	8.0–13.0	7
Weight	7.2	6.0–9.0	3
* From Flannery (1995a) and Bonaccorso (1998)			
Identification This species has a distinctive black and grey back. It has lobes between the corners of the mouth and the ears, the vertical lobe being much smaller than that of *C. gouldii*.			
Synonyms None			
Subspecies *Chalinolobus nigrogriseus nigrogriseus* (Gould, 1856) (eastern Queensland and north-east New South Wales Australia, New Guinea and D'Entrecasteaux Isls) *Chalinolobus nigrogriseus rogersi* Thomas, 1909b (northern Western Australia and Northern Territory, and north-western Queensland, Australia)			
Distribution Fergusson Is.			
Extralimital Australia, New Guinea			
Conservation status (IUCN) Least Concern			
Etymology Sp.–black and grey (L.), in reference to its grizzled appearance			

Hoary wattled bat (*Chalinolobus nigrogriseus*), Archer River, Australia. This species has been recorded in the South-west Pacific on just one occasion, a specimen that was collected from Fergusson in the D'Entrecasteaux by R. C. English and is held in the American Museum of Natural History (Anders Zimny).

NEW ZEALAND LONG-TAILED BAT

Chalinolobus tuberculatus (Forster, 1844)

tue-ber'-kue-lah'-tus

The New Zealand long-tailed bat is one of only two extant terrestrial mammals native to Aotearoa New Zealand. It is a species with a strong preference for forested habitats and is found in all Aotearoa New Zealand's indigenous forest types (O'Donnell and Borkin 2021). However, it is occasionally recorded from areas of farmland, non-native plantations, shelter belts and the fringes of urban centres (O'Donnell and Borkin 2021). This species roosts communally, predominantly in the high cavities of trees such as southern beech (Nothofagaceae). But it will also roost under exfoliating bark, in the fissures and cracks of dead stags, and in cracks or crevices of bluffs (O'Donnell and Borkin 2021).

The New Zealand long-tailed bat is a moderately fast-flying species that is adapted to foraging within forest gaps and edges, and directly over canopies (O'Donnell and Borkin 2021). Insects taken on the wing make up the bulk of the diet, most commonly from the families Diptera, Coleoptera and Lepidoptera. In the Eglinton Valley, home ranges were among the largest reported for any insect-eating bat (O'Donnell 2001). The bats followed similar movement patterns each night but core areas did not overlap extensively from one night to the next, and home ranges and core areas also did not overlap extensively between individuals.

Chalinolobus tuberculatus			
Size*			
	Mean	**Range**	**Sample**
Head–body	50.4	41.8–63.2	6
Tail	38.5	30.2–45.9	6
Forearm	39.1	36.5–44.4	34
Tibia	16.6	14.0–18.0	4
Hindfoot	–	–	–
Ear	11.0	8.0–14.0	2
Weight	9.1	7.0–12.2	28
* From Daniel (1990), Gillingham (1996), Dekrout (2009)			
Identification This is the only species of bat in Aotearoa New Zealand in which the tail is entirely enclosed within the tail membrane. There are small fleshy wattles at the corners of the mouth and the ears are smaller, broader and more rounded than those of *Mystacina* species.			
Synonyms None			
Subspecies None			
Distribution Aotearoa New Zealand North and South Isls, Great Barrier Is. (Aote), Kapiti Is., Stewart Is. (Rakiura), Little Barrier Is. (Te Hauturu-o-Toi)			
Extralimital None			
Conservation status (IUCN) Vulnerable			
Etymology Sp.–tubercle (L.), possibly a reference to the fleshy tubercles (wattles) present at the corners of the mouth in members of the genus			

The New Zealand long-tailed bat (*Chalinolobus tuberculatus*) is one of only two extant terrestrial mammals native to Aotearoa New Zealand (www.MerlinTuttle.org).

Across the distribution of New Zealand long-tailed bat, reproduction is highly synchronised over a period of less than 3 months (O'Donnell and Borkin 2021). The species reaches sexual maturity at a relatively late age compared to similar-sized vespertilionids (2–3 years of age in females). Females give birth to a single young once a year from November–December and the young are carried by mothers until they reach a large size, up to 80% of the mother's bodyweight (O'Donnell 2002; O'Donnell and Borkin 2021). At this point they are left at the roost while females forage, and after 5–6 weeks of age are able to fly.

Major threats to New Zealand long-tailed bats include habitat loss and fragmentation and predation by introduced black rats (*Rattus rattus*) and stoats (*Mustela erminea*). The problem of predation is greatly exacerbated every 3–4 years during the masting periods of southern beech (*Fuscospora* spp.). Populations of black rats and stoats explode during these periods and population crashes of *C. tuberculatus* follow (Pryde *et al.* 2005). A preliminary viability analysis for a population at Eglington Valley showed an average decline in adult females of 5% per year over a 10-year period.

Kerivoula

LOUISIADE WOOLLY BAT

Kerivoula agnella Thomas, 1908a

ke'-ree-vue-lah ahg-nel'-ah

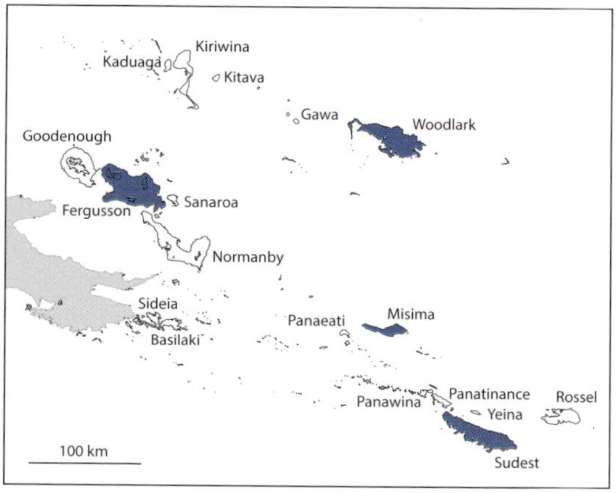

Kerivoula agnella			
Size*			
	Mean	**Range**	**Sample**
Head–body	–	44.0	1
Tail	–	48.0	1
Forearm	–	38.0	1
Tibia	–	–	–
Hindfoot	–	–	1
Ear	–	13.5	1
Weight	–	–	–
* From Thomas (1908a)			
Identification The tragus is spear-shaped and ears funnel-shaped. It is similar in body size to *K. myrella* and slightly larger than *K. muscina* on adjacent mainland New Guinea.			
Synonyms None			
Subspecies None			
Distribution Fergusson, Misima, ?Sudest, Woodlark			
Extralimital None			
Conservation status (IUCN) Endangered			
Etymology Gen.–plantain (banana) bat (Ceylonese); Sp.–derived from the name 'St Aignan Island', a former name for Misima			

This is a very poorly known species, and presumably (like *Kerivoula myrella*) is difficult to detect using mist nets and echolocation recording devices. The type locality is Misima (Thomas 1908a) (NHMUK 1898.4.1.2) and Hill (1965) identified it from Sudest and Misima, but the specimen from Sudest was apparently lost. A single specimen held by the American Museum of Natural History (AMNH 159560) was collected from Iamelele on Fergusson Island on 2 June 1956 by R. F. Peterson. This appears to be the most recent record of the species. Bonaccorso (1998) postulated that the Louisiade woolly bat is a foliage-gleaning insectivore, inhabiting tropical lowland forest and hill forest. Nothing else is known of its biology and it is classified as Data Deficient on the IUCN Red List.

The type locality for the Louisiade woolly bat (*Kerivoula agnella*) is Misima island, Papua New Guinea. This poorly known species is endemic to the islands of Milne Bay, where it inhabits tropical lowland forest and hill forest (Madison Erin Mayfield).

BISMARCK WOOLLY BAT

Kerivoula myrella Thomas, 1914a

mire-el'-ah

Relatively large numbers of Bismarck woolly bats were recorded in the late 19th and early 20th centuries,

Kerivoula myrella			
Size*			
	Mean	**Range**	**Sample**
Head–body	36.8	33.3–40.0	4
Tail	40.3	34.3–46.0	4
Forearm	36.5	31.5–38.5	9
Tibia	16.0	14.2–17.8	4
Hindfoot	8.5	7.0–9.8	4
Ear	12.6	11.4–13.9	4
Weight	5.5	3.5–6.5	3
* From Thomas (1914a) and Bonaccorso (1998).			
Identification A small bat with long, pointed tragus and funnel-shaped ears. Thomas (1914a) notes the canines are larger than in other members of the genus.			
Synonyms None			
Subspecies None			
Distribution Duke of York, Manus, Mioko, New Britain, Umboi			
Extralimital None			
Conservation status (IUCN) Data Deficient			
Etymology Sp.–significance unknown			

but more recent records are rare (e.g. Smith and Hood 1981). Armstrong *et al.* (2015) provided a comprehensive review of the known distribution for this species. Its description was based on one specimen from Manus Island and three from Umboi (Thomas 1914a). Further records were subsequently produced from the Whiteman Range in west New Britain (Koopman 1979) and from Ulatawa Plantation in East New Britain (collected by S. Hamilton in 1997) (Bonaccorso 1998). It has also been recorded from Duke of York Island, and from Mioko Island (Bonaccorso 1998), but has not been recorded from New Ireland despite extensive efforts there.

It seems that the predominant method used for surveying bats in the region (mist-netting) is not effective for Bismarck woolly bat. The most recent record by Armstrong *et al.* (2015) (a female on Manus Island) was taken in a harp trap on 10 October 2014. This was the only animal caught in 23 harp trap-nights across two sites. The trap was set in a vegetation gap beside a shallow stream.

The echolocation call is a steep broadband frequency sweep, low in energy and unlikely to be detected unless the bats pass close to a detector (Armstrong *et al.* 2015). It is thus a species that is also difficult to detect using echolocation recording devices. Calls were passively recorded along narrow tracks running through dense forest understorey on Manus; no detections were made on Mussau Island. The call structure indicated it is likely to be a foliage-gleaning insectivore, adapted to foraging in cluttered environments.

Bismarck woolly bat (*Kerivoula myrella*), Manus Island. This species is known from several islands in the Bismarck Archipelago, but more recent records are rare (Stephen Richards).

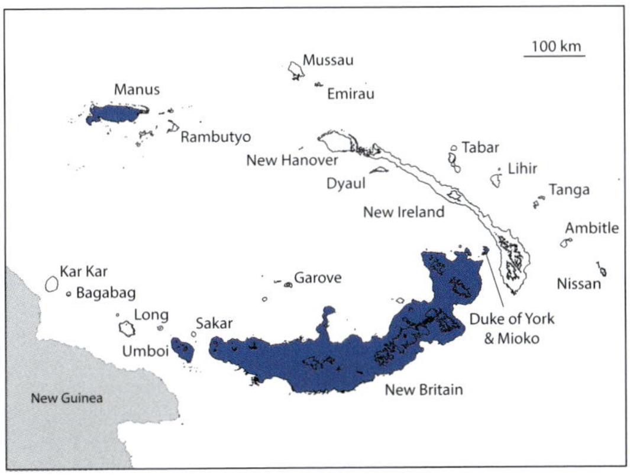

Murina

FLORES FLUTE-NOSED BAT

Murina florium Thomas, 1908a

myue-ree'-nah flor'-ee-um

The Flores flute-nosed bat is a very distinctive insectivorous species because of its tubed nostrils. It is probably more widespread, but is very rarely recorded and thus remains little-known. It is a slow-flying and manoeuvrable species that presumably easily avoids mist nets. In the Iron Range region of Cape York Peninsula, Australia, harp traps have proven effective at catching this species, and more widespread use of this method may lead to

Murina florium being better known in the South-west Pacific. Bonaccorso (1998) identified a specimen of this species in the Papua New Guinea National Museum, collected from New Britain by S. Hamilton, and another in the Natural History Museum, London, from Umboi (1914.4.1.33). Although, this latter specimen is listed in the NHMUK database as having been collected from Manus Island by Albert Meek.

The Flores flute-nosed bat prefers rainforest habitats. It emits low-intensity, wide frequency range, short-duration calls (1.2–2 ms) (Churchill 2008). The species is adapted to foraging in dense forests and capable of hovering. Flores flute-nosed bats mostly forage by gleaning and faecal analyses have found the diet contains a large proportion of spiders and beetles.

Animals roost solitarily among dead leaves and foliage, and abandoned bird nests (Schulz *et al.* 2008). One animal recorded on Mt Otto in the Bismarck Range, New Guinea, was roosting in a hole in the ceiling of a house (Van Deusen 1961). Lactating females have been caught in December,

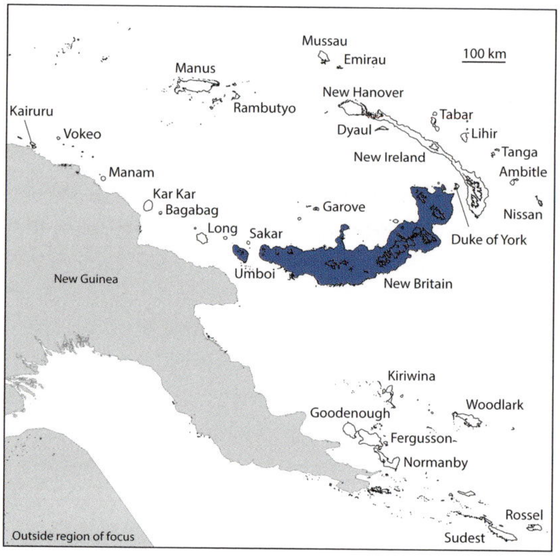

Murina florium			
Size*			
	Mean	**Range**	**Sample**
Head–body	43.9	38.0–50.0	7
Tail	36.3	32.9–39.0	7
Forearm	34.9	33.0–36.9	10
Tibia	17.2	15.5–18.5	6
Hindfoot	8.2	7.6–9.0	4
Ear	12.9	10.6–14.1	7
Weight	6.4	4.3–8.8	8
* From Bonaccorso (1998)			
Identification Outside of Aotearoa New Zealand, this is the only insectivorous bat in the region with tubed nostrils. The fur is reddish and the tibia and forearm are hairy relative to those of most insectivorous bats.			
Synonyms None			
Subspecies *Murina florium florium* Thomas, 1908a (Flores, Sumbawa, Peleng) *Murina florium lanosa* Thomas, 1910a (Ambon, Seram, Gorong) *Murina florium toxopei* Thomas, 1923d (Buru)			
Distribution New Britain, Umboi			
Extralimital Australia, Indonesia, New Guinea			
Conservation status (IUCN) Least Concern			
Etymology Gen.–mouse-like (L.) Sp.–for the island of Flores, Indonesia			

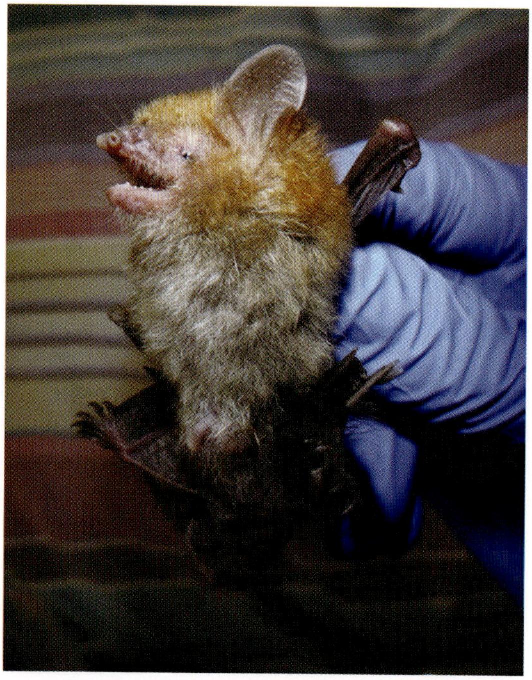

Flores flute-nosed bat (*Murina florium*), Iron Range, Cape York Peninsula, Australia. This species is adapted to foraging in dense forests where it gleans spiders and beetles from foliage (Julie Broken-Brow).

suggesting births occur sometime from October to November. Females captured in April and June have been in non-breeding condition (Schulz *et al.* 2008).

The Flores flute-nosed bat is widespread but occurs at low densities. Although infrequently captured, it is assumed to have a large overall population size and is classified as Least Concern on the IUCN Red List.

Myotis

INSULAR MYOTIS

Myotis insularum (Dobson, 1878)

mie-oh'-tis in-soo-lah'-rum

Both the taxonomy and distribution of this species are very uncertain. The holotype, reportedly collected in the 'Navigator's Islands' (Samoa Islands), is the only known record of *Myotis* in Samoa. It was collected by a 'Mr. Smeltz' before 1878, probably in the 1860s or 1870s, but the exact island locality is not known (Koopman 1984). Findley (1972) considered the insular myotis to resemble the European species *M. mystacinus* most closely. It resembles species of the Palearctic *muricola* or *mystacinus* species groups in possessing relatively small feet. The closest known record of the genus is from Vanuatu (*M. moluccarum*). Although the holotype may represent a mislabelled specimen or a vagrant (Koopman 1984), it has not been conclusively matched to any known species of *Myotis* and possibly represents a distinct, rare or extinct species not collected since the 19th century (Helgen *et al.* 2009). The possibility this species remains extant in the Samoa islands remains worthy of consideration. The insular myotis is listed as Data Deficient on the IUCN Red List.

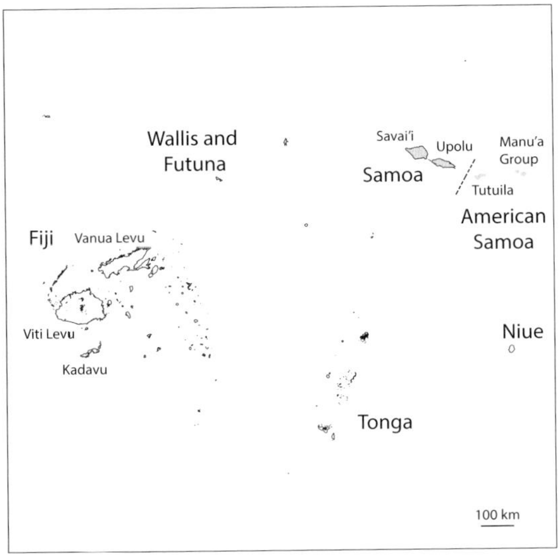

Myotis insularum			
Size*			
	Mean	**Range**	**Sample**
Head–body	–	40.6	1
Tail	–	38.1	1
Forearm	–	35.6	1
Tibia	–	16.5	1
Hindfoot	–	7.6	1
Ear	–	15.2	1
Weight	–	–	–
* From Dobson (1878) (NHMUK 1866.12.27.2)			
Identification *Myotis* possess a tail that is fully enclosed in the tail membrane, rather than a tail that projects through the upper surface of the membrane (as in *Emballonura semicaudata*, the only other insectivorous bat known from remote Oceania). Relatively small feet and short calcars (extending half way to the tail) distinguish it from all other *Myotis* in Oceania.			
Synonyms None			
Subspecies None			
Distribution Samoan Islands (exact locality unknown)			
Extralimital None			
Conservation status (IUCN) Data Deficient			
Etymology Gen.–mouse ear (Gr.); Sp.–of an island (L.), a reference to the insular distribution of the species			

MALUKU MYOTIS

Myotis moluccarum (Thomas, 1915c)

mol'-uk-ah'-rum

This is a widespread species in need of taxonomic revision. The name *Myotis moluccarum* is presently applied to all populations of *Myotis* in New Guinea and islands to the east (Kitchener *et al.* 1995). However, multiple species are probably included under this name. Cooper *et al.* (2001) found marked genetic divergence between a sample from Malakula, Vanuatu, and animals from New Guinea and Australia. Moreover, morphometric analysis identified distinctive characteristics that may warrant recognition of a distinct taxon in Solomon Islands (Kitchener *et al.* 1995).

Koopman (1982) recorded a single specimen from Sudest and another from Normanby. Smith and Hood (1981) reported the capture of 17 individuals from caves on New Ireland and one from a human-made tunnel on New Britain. The Natural History Museum of Los Angeles County holds one specimen from Aore Island off Espiritu Santo, collected in 1993. Acoustic surveys by Armstrong *et al.* (2015) confirmed the presence of a *Myotis* on Manus and Mussau. Recent field surveys have also added it to the fauna inventories of the relatively remote islands of Makira, Mono and Rennell in Solomon Islands, and research is underway to determine taxonomic relationships between these and other islands. In reality, a *Myotis* sp. (whether *M. moluccarum* or a closely related species) probably occurs on most islands throughout the Admiralty, Bismarck, D'Entrecasteaux, Solomon and Vanuatu archipelagos.

On New Britain and New Ireland, this species was found roosting among clusters of long-fingered bats (*Miniopterus* spp.), or in deep cracks and crevices in cave walls (Smith and Hood 1981; Flannery 1995b). None was located during searches of caves on Mussau Island (Armstrong *et al.* 2015). A small number of individuals were caught in a limestone cave located on a steep cliff on Rennell Island in October 2018. The cave was also occupied by *Aselliscus tricuspidatus*, *Dobsonia inermis*, *Hipposideros calcaratus* and three species of *Miniopterus*. A female captured was carrying a hairless young 62% of the mother's bodyweight. Of three females caught on Mono in July 2019, one was lactating and the other two were nulliparous.

The Maluku myotis feeds on aerial insects, aquatic invertebrates and fish. It captures the latter by trailing its enlarged feet over water to scoop them up. On Mono in July 2019 this species was abundant and seen foraging in this manner over a small stream adjacent to the ocean on the northern coast.

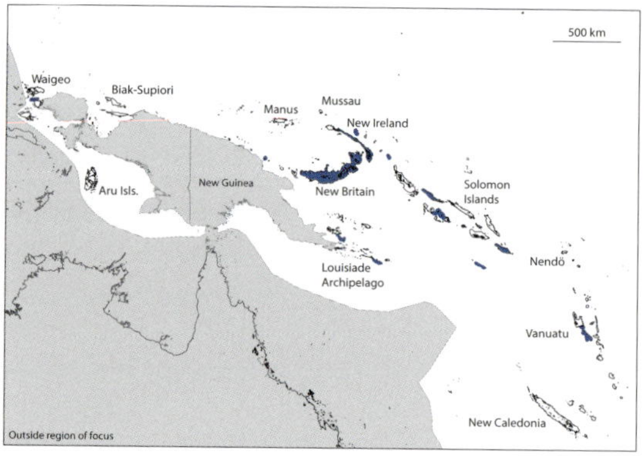

Myotis moluccarum			
Size*			
	Mean	**Range**	**Sample**
Head–body	43.2	39.1–50.6	15
Tail	37.9	34.3–41.4	15
Forearm	40.1	37.9–42.6	14
Tibia	17.4	16.1–18.2	15
Hindfoot	10.6	9.7–11.1	13
Ear	13.9	12.1–15.8	15
Weight	6.9	5.3–9.0	14
* KU and QM specimens from Choiseul, Mono and Rennell			
Identification This species has large hindfeet and long calcars (extending 3/4 of the way to the tail). The region around the eyes is sparsely furred.			
Synonyms None			
Subspecies *Myotis moluccarum moluccarum* (Thomas, 1915c) (Indonesia, Papua New Guinea) *Myotis moluccarum orientis* Hill, 1983 (Vanuatu) *Myotis moluccarum solomonis* (Troughton, 1929b) (Solomon Isls)			
Distribution Nissan (AROB); Batanta (Indonesia), Karkar, New Britain, New Ireland, Normanby, Sudest, Tabar (Papua New Guinea); Choiseul, New Georgia, Makira, Rennell (Solomon Isls); Aore, Malakula (Vanuatu)			
Extralimital Indonesia, New Guinea			
Conservation status (IUCN) Least Concern			
Etymology Sp.–for the Moluccas (Maluku), Indonesia, type locality for the species			

The Maluku myotis (*Myotis moluccarum*) is probably more widespread than currently appreciated, but may comprise additional undescribed species. The individual pictured represents the first record of this species from Makira, Solomon Islands (Tyrone Lavery).

Calls are near vertical frequency sweeps with a very short duration. Those recorded from Makira had a peak frequency of 58.2 kHz. Bandwidth (46 kHz) and slope (16.6 kHz/ms) are far greater than calls from other species of echolocating bat from the region (Pennay and Lavery 2017).

Uncertainties related to taxonomy, distribution and population size have been reflected in conservation assessments for the Maluku myotis. However, no major threats have been identified and it is thus classified as Least Concern on the IUCN Red List.

STALKER'S MYOTIS

Myotis stalkeri Thomas, 1910b

stork'-er-ee

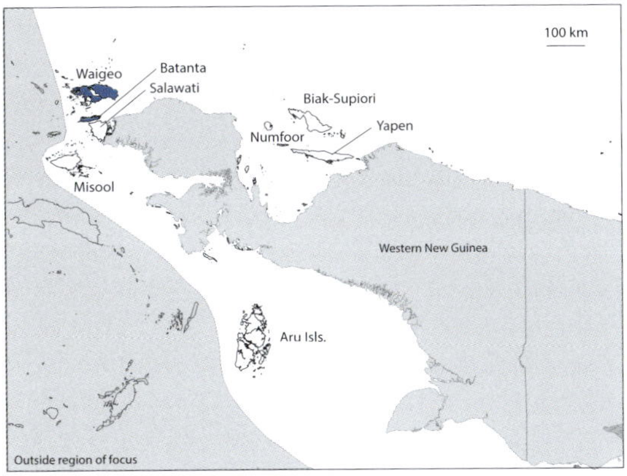

This is a magnificent vespertilionid. When in flight, the overall impression is that of a large, graceful silvery-blue bat, with pale wings and a pale belly. In colour it resembles a small version of the Australian ghost bat (*Macroderma gigas*). The very large feet with sharp, laterally compressed claws, are also striking. The bulk of natural history information for this species is derived from the Kai Islands and Gebe that lie just outside the region of focus.

The holotype was collected at Ara, Kai Islands, in July 1907. Subsequently, in July 1993 the Australian Museum recorded a small colony in the Kai Islands, roosting in a cave 1.5 km north-east of Ohoider Tawun Village on Kai Ketjil. The specimen collected was an adult male found sheltering in a crevice. The colony shared the roost with *Hipposideros cervinus*, *Emballonura alecto*, *Miniopterus australis*, *Taphozous achates* and *Rhinolophus euryotis*. A large colony of around 100 individuals was located in Semingit Cave on Gebe on 30 November 1991. This large cave was located approximately 1 km from the sea and 2 km from a large, tranquil lake. Tall forest was present in the immediate vicinity of the cave, but the surrounding countryside was dominated by heathland and low open forest growing on ultrabasic soils. The cave comprised three main chambers with several large solution holes in the roof of the outermost one. *Myotis stalkeri* were roosting in the apices of these solution holes, and were also present in the inner chambers, although it is uncertain if disturbance had driven them there from their roosts in the solution tubes. They shared the cave with *Miniopterus australis*, *M. blepotis*, *Hipposideros papua* and *Emballonura raffrayana*.

At the time of the visit, four adult *M. stalkeri* (two male, two female) were captured and one of the females was carrying a naked young, weighing 7 g. Another six individuals examined were subadults weighing 12–16 g, and may have been young of the previous year. This suggests seasonal breeding, with births occurring in November–December. If this is

Myotis stalkeri			
Size*			
	Mean	**Range**	**Sample**
Head–body	58.3	49.0–63.6	4
Tail	53.9	49.6–55.5	4
Forearm	49.9	48.6–51.3	4
Tibia	22.1	22.0–22.2	
Hindfoot	–	–	–
Ear	19.6	19.2–20.1	4
Weight	17.0	17.0	2
* From AM specimens (Gebe – extralimital)			
Identification This is a large species of *Myotis*, larger than other species of the region and with extraordinarily large feet, a very pale belly and light grey back.			
Synonyms None			
Subspecies None			
Distribution Batanta, Waigeo			
Extralimital Gebe, Kai Isls			
Conservation status (IUCN) Least Concern			
Etymology Sp.–'For its discoverer Mr W. Stalker, whose sad death by drowning shortly after the landing of the expedition in New Guinea deprives the [British Natural History] Museum of one of its best and most enthusiastic collectors' (Thomas 1910b)			

Stalker's myotis (*Myotis stalkeri*), Kai Islands, Indonesia (Pavel German).

correct, individuals do not reach maturity in the first year of life. An IUCN Red List assessment of population sizes, trends and major threats has indicated Stalker's myotis is not close enough to the criteria thresholds to warrant listing under a threatened category.

Nyctophilus

PAPUAN LONG-EARED BAT

Nyctophilus microtis Thomas, 1888d

nik'-toh-fil'-us mike-roh'-tis

The Papuan long-eared bat was recorded at two sites on New Ireland in 1994 at elevations of 260 and 1200 m (Emmons and Kinbag 2001). A single specimen at the Queensland Museum collected by De Vis from Sudest is also attributable to this species (H. Parnaby, pers. comm.). Armstrong *et al.* (2015) recorded calls of a *Nyctophilus* species on Manus that may be this or a closely related species. The calls were reportedly similar to those recorded for *N. microtus* at various localities on mainland New Guinea, consisting of a steep frequency-modulated sweep with a characteristic frequency of approximately 55 kHz. Robson *et al.* (2012) captured this species on the Huon Peninsula of mainland New Guinea: reference call recordings had a characteristic frequency of 40 kHz.

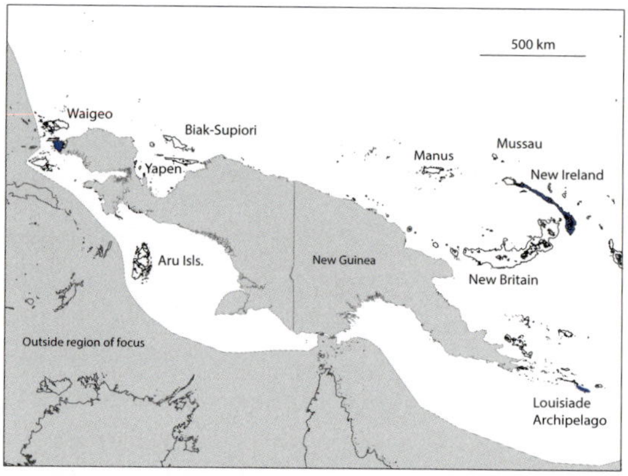

Nyctophilus microtis			
Size*			
	Mean	**Range**	**Sample**
Head–body	50.0	42.5–63.0	5
Tail	41.4	36.0–49.0	5
Forearm	39.6	37.0–41.6	9
Tibia	17.7	16.1–20.3	5
Hindfoot	8.7	7.0–10.0	4
Ear	16.4	14.0–19.4	5
Weight	8.0	5.8–10.5	9
* From Bonaccorso (1998) and Robson *et al.* (2012)			
Identification A small transverse nose-leaf is present on the snout; ears are relatively short and narrow and are not joined by a band across the forehead.			
Synonyms *Nyctophilus microtis bicolor* Thomas, 1915d *Lamingtonia lophorhina* McKean and Calaby, 1968			
Subspecies None			
Distribution Salawati (Indonesia); New Ireland, Sudest (Papua New Guinea)			
Extralimital New Guinea			
Conservation status (IUCN) Least Concern			
Etymology Gen.–night loving (Gr.); Sp.–small ear (Gr.)			

On New Guinea, this species has been found roosting in dried bamboo or high on the interior walls of caves (Bonaccorso 1998). Two of four females trapped in July in Sandaun Province were lactating and two were nulliparous (Flannery 1995a). The Papuan long-eared bat is widely distributed and no major threats to the species are known.

The Papuan long-eared bat (*Nyctophilus microtus*) is one of only two species of *Nyctophilus* known from the region. Echolocation recordings indicate it also may occur on Manus Island (where it has yet to be confirmed) (Madison Erin Mayfield).

NEW CALEDONIAN LONG-EARED BAT

Nyctophilus nebulosus Parnaby, 2002a

neb'-ue-loh'-sus

This poorly known species of long-eared bat has been recorded only four times, on Mont Koghi to the north of Nouméa. Two specimens were captured in a mist net at 430 m elevation on 10 May 1991. The mist net was set across a track in a small clearing in tall closed rainforest. Mt Koghi supports on its upper slopes a remnant of undisturbed rainforest, which is surrounded by *maquis* and grassland. The bats were captured as they were flying along a path around 100 m from the rainforest edge, at a height of approximately 5 m. They were taken at dusk (between 5:45 and 6:15 p.m.) on a dry night, and at least one was flying deeper into the forest. The female was neither lactating nor pregnant when captured.

A *Nyctophilus* (presumably this species) was last detected on Mt Koghi in March 2019 using an echolocation recorder (Bouillard 2018). The calls were typical of *Nyctophilus* species. The pulse shape was linear with a characteristic frequency around 55 kHz. Little extra information is available regarding the biology of the New Caledonian long-eared bat. It is classified as Critically Endangered on the IUCN Red List because the known distribution is small and the area and quality of habitat available for the species are continuing to decline.

The New Caledonian long-eared bat (*Nyctophilus nebulosus*) is a very poorly known species of long-eared bat that has only been recorded on Mont Koghi, north of Nouméa, New Caledonia (Madison Erin Mayfield).

Nyctophilus nebulosus			
Size*			
	Mean	**Range**	**Sample**
Head–body	53.9	52.9–54.9	2
Tail	49.5	48.7–50.2	2
Forearm	42.9	42.4–43.5	4
Tibia	21.0	20.4–21.9	4
Hindfoot	–	–	–
Ear	25.9	24.5–27.3	4
Weight	10.5	9.4–12.0	3
* From Parnaby (2002a)			

Identification This is the only species of *Nyctophilus* known from New Caledonia and can be separated from other insectivorous bats by its long ears joined across the forehead by a flap of skin, and simple nose-leaf. Males are smaller than females and are reddish-brown rather than chocolate brown.

Synonyms None

Subspecies None

Distribution Grande Terre

Extralimital None

Conservation status (IUCN) Critically Endangered

Etymology Sp.–a dual reference to the misty heights of Mt Koghis, the type locality, and the somewhat indistinct (nebulous) nature of diagnostic morphological criteria, which is characteristic of species of *Nyctophilus*

Philetor

ROHU'S BAT

Philetor brachypterus (Temminck, 1840)

fil'-ay-tor brak-ip'-te-rus

This is a widespread and unusual vespertilionid. The genus is monotypic; three subspecies are recognised, but the taxonomy is more complex and cryptic species are likely (Lim *et al.* 2017). Rohu's bat has been recorded very few times from the region under consideration here. On New Britain, Kock (1981) recorded the collection of four males on 14 May 1915 from the Mt Varzin area near Rabaul. In 1994 it was recorded for the first time on New Ireland as part of a rapid biodiversity assessment in the south near the Hans Meyer Range (Emmons and Kinbag 2001). It was caught there at a lowland site (200–260 m elevation) that supported lowland

Rohu's bat (*Philetor brachypterus*), Mindanao, the Philippines. This widespread and unusual species has been recorded very few times from the region under consideration here (Keiko and Yushi Osawa).

Philetor brachypterus			
Size*			
	Mean	**Range**	**Sample**
Head–body	56.9	50.5–65.0	5
Tail	34.6	29.6–40.0	5
Forearm	34.4	30.0–37.8	5
Tibia	13.9	11.5–15.9	4
Hindfoot	8.9	7.8–10.0	5
Ear	13.0	10.0–14.1	5
Weight	11.0	7.1–14.2	5
* From New Guinea and New Britain specimens, Bonaccorso (1998), Flannery (1995a)			
Identification The penis of the male lacks a prepuce and the genital slit of the female is vertically orientated. These are unique features among the bats of the region.			
Synonyms None			
Subspecies *Philetor brachypterus brachypterus* (Temminck 1840) (Java and nearby regions) *Philetor brachypterus rohui* Thomas, 1902b (New Guinea, New Britain) *Philetor brachypterus verecundus* (Chasen, 1940) (Malaya, Borneo)			
Distribution New Ireland, New Britain			
Extralimital Indonesia, New Guinea, Malaysia, Philippines			
Conservation status (IUCN) Least Concern			
Etymology Gen.–lover (Gr.), perhaps a reference to the unusual genitals; Sp.–short winged (Gr.)			

rainforest dominated by the canopy tree *Pometia pinnata* with an open understory.

Rohu's bat roosts in tree hollows from 1.5–4.5 m above the ground (Bonaccorso 1998). It has also been recorded in the hollow trunks of coconut palms and in the disused denning hollow of a lesser tree mouse (*Chiruromys vates*) on mainland New Guinea (Hide *et al.* 1984). Up to 55 individuals have been recorded from a single hollow tree in New Guinea (Koopman 1982). This is a fast-flying species that forages in forest understoreys and subcanopies. The stomach contents of individuals examined by Bonaccorso (1998) contained beetles and Hymenoptera. Limited information on the timing of breeding is available from this region. Rohu's bat is thought to have a stable global population and is classified as Least Concern on the IUCN Red List.

Phoniscus

GOLDEN-TIPPED BAT

Phoniscus papuensis (Dobson, 1878)

fon'-is'-kus pah'-poo-en'-sis

The golden-tipped bat is a distinctive and seldom encountered species of the region. It inhabits rainforests, drier forests, and riparian and coastal *Melaleuca* forests. Golden-tipped bats are specialist feeders (Schulz 2000). The wings have a very low aspect ratio, enabling slow manoeuvrable flight, and the bats have a low intensity echolocation call. These traits are tailored towards snatching spiders from their webs in cluttered environments, but also assist in avoiding mist nets and harp traps and reduce the potential for detecting the species with echolocation recording devices (because the echolocation calls attenuate rapidly).

Roost sites include dead palm fronds, thick foliage, epiphytes, tree hollows, caves, bird nests and houses (Bonaccorso 1998; Law and Chidel 2004). The golden-tipped bat often roosts individually but will also congregate in larger mixed or single-sex groups. A small group was located in the apex of Gua Jepang Cave (Japanese Cave) near Biak Town on 19 September 1992. The roost was in complete darkness approximately 3 m from the cave floor, with a build-up of droppings beneath that indicated the cave had been occupied for an extended period. The individuals were packed tightly together and did not attempt to fly until gently nudged.

The golden-tipped bat (*Phoniscus papuensis*) is a distinctive and seldom-encountered species that will roost in a variety of locations including dead palm fronds, epiphytes, tree hollows and bird nests (George Madani).

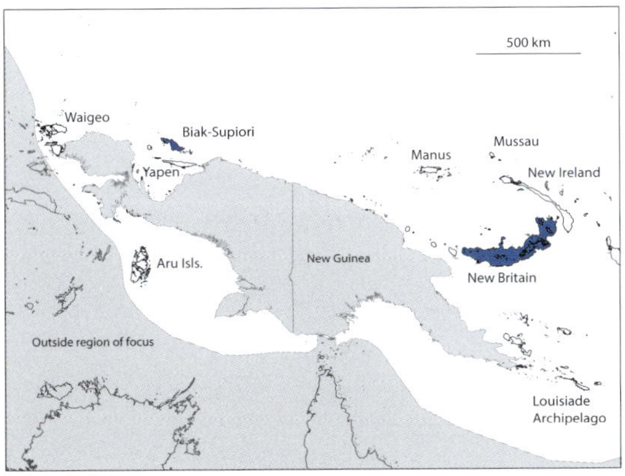

Phoniscus papuensis			
Size*			
	Mean	**Range**	**Sample**
Head–body	47.1	43.6–50.6	5
Tail	46.1	44.5–47.7	5
Forearm	39.4	38.8–40.6	5
Tibia	19.5	19.0–20.0	5
Hindfoot	–	–	–
Ear	15.4	14.9–15.9	5
Weight	9.2	8.8–9.5	5
* AM specimens (Biak-Supiori)			
Identification Immediately recognisable by its golden-tipped fur, very large mouth, spear-like tragus and relatively hairy forearms and legs.			
Synonyms None			
Subspecies None			
Distribution Biak-Supiori, New Britain			
Extralimital Australia, New Guinea			
Conservation status (IUCN) Vulnerable			
Etymology Gen.–little murderer (Gr.); Sp.–from Papua (L.)			

The group on Biak consisted of a single adult male with enlarged testes, two lactating females, another which had just finished lactating, and one pregnant and one nulliparous female. Two juveniles weighing 2.9 and 5.6 g were also present (the smaller was newborn, still retaining a wet umbilical cord). A single, non-reproductive female was captured from a tangle of lawyer cane (*Callumus* sp.) on 16 August 1953, during the Fifth Archbold Expedition in Milne Bay, mainland New Guinea (Brass 1956). Although seldom encountered, the golden-tipped bat probably has a wider distribution than currently appreciated (owing to the difficulties associated with detecting it in surveys).

Pipistrellus

NEW GUINEA PIPISTRELLE

Pipistrellus angulatus (Peters, 1880)

pip'-is-trell'-us ang-yue-lah'-tus

Pipistrellus angulatus has previously been treated as a subspecies of *P. tenius* but is a morphologically distinct species. Koopman (1982) identified pipistrelles on Sudest, Normanby, Fergusson and Goodenough islands that were probably this species. But these were assigned to either *P. tenius collinus* or *P. tenius papuanus*, two former subspecies that have since been elevated to full species (Kitchener *et al.*

1986). Moreover, one of these from high elevation on Goodenough was accessioned as *P. papuanus* but based on body size, cranial morphology and fur characteristics is better ascribed to *P. collinus* (Bonaccorso 1998). The Australian Museum holds specimens from the Bismarck Archipelago, Normanby Island and Solomon Islands. Hill (1971) recorded a single male specimen from a schoolmaster's house at Nuhu on Guadalcanal.

The New Guinea pipistrelle is a very abundant species with no apparent habitat preference. On Guadalcanal it has been captured at elevations of 600 m and 1500 m in primary forest. An animal recorded on New Georgia in 2015 was taken in a harp trap set in an area of disturbed forest that was commercially logged only 3 months prior.

New Guinea pipistrelles are aerial insectivores that emerge in the twilight hours before dusk to feed busily over villages, gardens, plantations, streams and small clearings in primary forest, often alongside *Mosia nigrescens*. They fly in rapid circles with sharp direction changes. Pennay and Lavery (2017) found New Guinea pipistrelles to be one of the most commonly recorded species of echolocating bats in the Solomon Islands archipelago, with free-flight recordings made on most of the major islands visited. The echolocation calls are a short-duration, frequency-modulated 'hockey stick' shape. Mean characteristic frequency in Solomon Islands was 47.5 kHz, and on Manus was 48 kHz (Armstrong *et al.* 2015). Passive call recordings on Choiseul, Tetepare and Vangunu Islands have also indicated the presence of the New Guinea pipistrelle there. It is probably widespread throughout the main Solomon Archipelago.

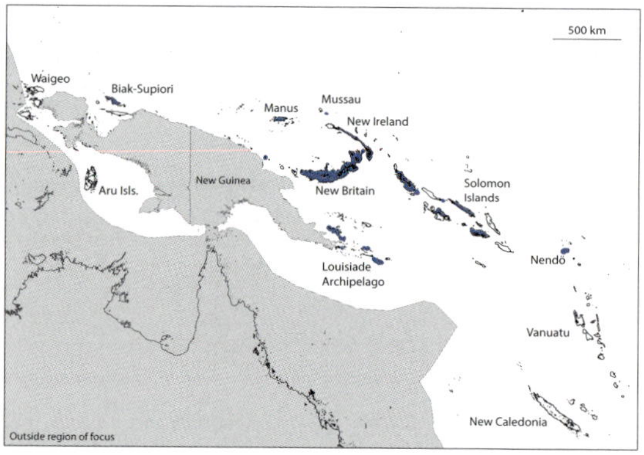

Pipistrellus angulatus			
Size*			
	Mean	**Range**	**Sample**
Head–body	38.6	33.4–48.0	12
Tail	34.4	27.0–41.0	12
Forearm	33.2	31.0–37.4	12
Tibia	14.3	12.8–15.6	12
Hindfoot	7.5	5.3–10.0	10
Ear	11.6	9.3–15.5	12
Weight	4.3	3.2–8.0	7
* From Bonaccorso (1998)			
Identification A small vespertilionid, usually blackish in colour and distinguished from *P. papuanus* by its slightly larger size, longer fur and more furred rostrum.			
Synonyms None			
Subspecies *Pipistrellus angulatus angulatus* (Peters, 1880) (New Guinea region) *Pipistrellus angulatus ponceleti* Troughton, 1936a (Solomon Isls)			
Distribution Bougainville (AROB); Biak-Supiori (Indonesia); Duke of York, Emirau, Fergusson, Goodenough, Karkar, Manus, Mioko, Misima, New Britain, New Ireland, Normanby, Sudest (Papua New Guinea); Fauro, Guadalcanal, Nendö, New Georgia, Santa Isabel (Solomon Isls)			
Extralimital New Guinea			
Conservation status (IUCN) Least Concern			
Etymology Gen.–bat (It.); Sp.–angular, cornered (L.), a reference to the relatively sharply pointed ears			

The New Guinea pipistrelle (*Pipistrellus angulatus*) is one of the most commonly observed insectivorous bats of Melanesia where it emerges shortly before dusk to feed in the airspace above villages, gardens, streams and small clearings in primary forests (Stephen Richards).

Smith and Hood (1981) encountered two colonies on New Ireland in the Bismarck Archipelago. There, the New Guinea Pipistrelle was found only in one of the many caves investigated. The cave, near Laburau was being used as a maternity roost in June, with all 200 bats roosting in a tube-shaped hole in the ceiling around 500 cm wide and 66 cm deep. A second colony was located in a space between the wall and shingles at Madina High School. Two females examined from this colony in June were pregnant, one with an embryo 7 mm long and the other with an embryo 17 mm long. The bats exited the roost on dusk. Flannery (1995a) recorded five individuals roosting in two clumps of dried bamboo at 990 m elevation on New Guinea. Roosting habits thus seem variable. It has never been found in hollow coconut palms (a location favoured by *P. papuanus*) and this has been highlighted as an important point of difference in the ecology of these two sympatric species on New Ireland (Smith and Hood 1981). The New Guinea pipistrelle is widely distributed, appears tolerant of habitat modification, and is not believed to be in decline.

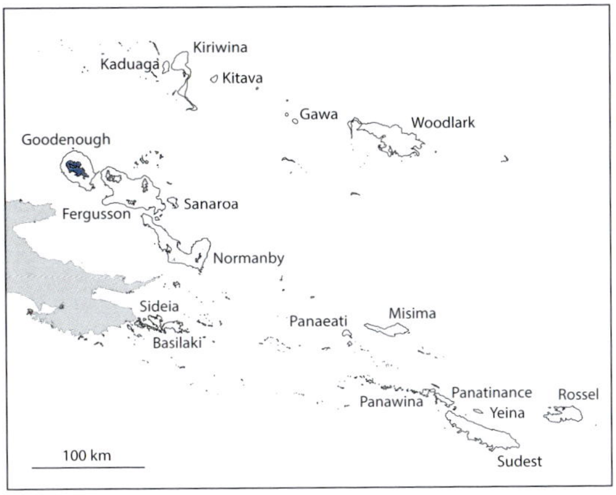

GREATER PAPUAN PIPISTRELLE

Pipistrellus collinus Thomas, 1920

kol'-een-us

The greater Papuan pipistrelle is a high-elevation species that prefers hill forest and mid-montane forest habitats above 700 m. It is known from this region only in the D'Entrecasteaux Islands, where such higher elevation habitats are found in close proximity to mainland New Guinea. Bonaccorso (1998) regarded a single specimen from Goodenough Island (AMNH 157468) as this species. The male animal was captured on 15 October 1953 at 1600 m elevation on Goodenough's eastern slopes. It was accessioned as *Pipistrellus papuanus*, but body size, cranial morphology and fur characteristics are consistent with *P. collinus* (Bonaccorso 1998).

Pipistrellus collinus			
Size*			
	Mean	**Range**	**Sample**
Head–body	47.5	41.3–53.0	4
Tail	37.0	30.0–44.0	4
Forearm	36.4	33.0–39.1	6
Tibia	15.3	13.8–17.2	6
Hindfoot	8.3	6.5–10.0	6
Ear	10.3	7.5–13.0	4
Weight	6.1	4.3–8.5	4
* From Bonaccorso (1998)			
Identification This is the largest *Pipistrellus* species in the region; the dorsal and ventral fur is bicoloured (being black for the bottom two-thirds and brown or reddish at the tips).			
Synonyms None			
Subspecies None			
Distribution Goodenough			
Extralimital New Guinea			
Conservation status (IUCN) Least Concern			
Etymology Sp.–pertaining to hills (L.), a reference to the preferred high-elevation habitats			

The greater Papuan pipistrelle (*Pipistrellus collinus*) prefers high-elevation montane forests above 700 m elevation and, in the region under consideration here, has only been recorded from the D'Entrecasteaux Islands (Madison Erin Mayfield).

Flannery (1995a) reported this species as being sexually dimorphic in colour on mainland New Guinea, with females being reddish or gold-tinged and males more of a drab brown. The greater Papuan pipistrelle is an aerial insectivore that will forage above cleared garden areas (Flannery 1995a). It has been recorded roosting in hollow *Pandanus* trees. Most roosts documented have contained just a single male, but one contained what was probably a harem social system with one male and 11 females. Of six females captured in Milne Bay Province on

15 March 1985, none was reproductively active (Bonaccorso 1998). Two females examined on 30 June 1959 at Pengagl (Eastern Highlands, Papua New Guinea) were lactating.

Echolocation calls recorded for this species from the Huon Peninsula, New Guinea, were frequency modulated with a characteristic frequency around 40 kHz (Robson *et al.* 2012). The call shape consisted of steep linear pulses with upsweeps of 10–30 kHz ending with a slight curve.

No obvious threats to the greater Papuan pipistrelle have been identified and its preferred higher elevation habitats tend to be less prone to large-scale anthropogenic disturbances than in the lowlands. However, climate change is likely to pose significant risk to the small areas of suitable habitat available for this species on high-elevation mountain tops.

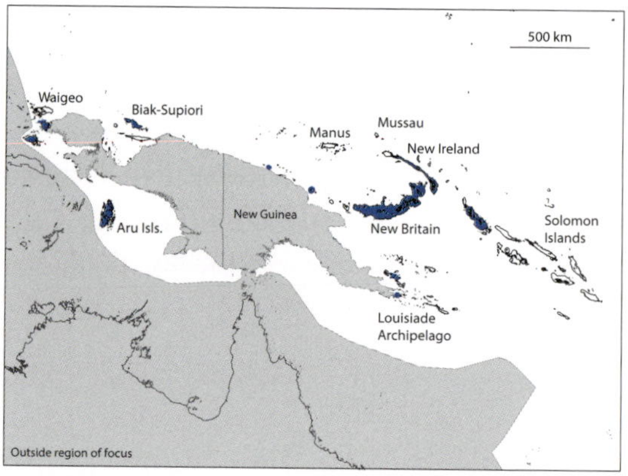

Pipistrellus papuanus			
Size*			
	Mean	**Range**	**Sample**
Head–body	40.2	33.9–50.0	5
Tail	28.6	22.0–35.0	5
Forearm	29.7	27.0–32.2	5
Tibia	12.6	10.9–13.6	5
Hindfoot	6.5	4.5–9.0	5
Ear	10.0	7.5–13.0	5
Weight	4.8	3.9–5.8	3
* From Bonaccorso (1998)			
Identification Slightly smaller than *P. angulatus* with shorter fur and a more extensive, naked rostrum.			
Synonyms *Vesperugo papuanus orientalis* Meyer, 1899			
Subspecies None			
Distribution Bougainville (AROB); Aru Isls, Batanta, Biak-Supiori, Salawati (Indonesia); Fergusson, Kairuru, Karkar, New Britain, New Ireland, Samarai (Papua New Guinea)			
Extralimital Kai Isls, Maluku Isls, New Guinea			
Conservation status (IUCN) Least Concern			
Etymology Sp.–of Papua (L.), a name for the south-eastern part of New Guinea			

LESSER PAPUAN PIPISTRELLE

Pipistrellus papuanus (Peters and Doria, 1881)

pah'-poo-ah'-nus

The lesser Papuan pipistrelle is a common bat adapted to modified environments and encountered as a human commensal species. It has been found up to 1300 m elevation on mainland New Guinea (Bonaccorso 1998).

Documented roost sites have been almost exclusively in small cavities in coconut palms and buildings (Bonaccorso 1998), often in the middle of villages. Brass (1959) reported individuals of this species living in holes in the trunks of coconut palms at Waikaiuna on Normanby Island in April–May 1956. In the absence suitable palms it may roost in tree hollows. Roosts can contain large numbers of individuals and groups of up to 50 have been documented. A roost like this, approximately 2 m from the ground in Bosnik Village, Biak, contained a large number of young when examined on 20 September 1992 (a sample of seven weighed between 2.8 and 3.4 g). All three adults examined were females. Hill (1983) recorded it from the island of Kairuru and Smith and Hood (1981) encountered a large nursery colony inside a coconut palm on New Ireland. The entrance (approximately 10 cm in

The Papuan pipistrelle (*Pipistrellus papuanus*) is a common bat that can exploit disturbed environments and can be encountered as a human commensal species (Stephen Richards).

diameter) was located some 4 m above the ground, with the cavity extending a further 2.5 m upwards.

Across the broad distribution of this species two major breeding peaks are evident (Burgin 2019). However, at Wau in Marobe Province of Papua New Guinea, females breed continuously, possibly due to the more stable climate, rainfall and temperatures (Woolley and Allison 1996). The lesser Papuan pipistrelle has a wide distribution and is tolerant of habitat modification; populations appear to be stable.

WATTS'S PIPISTRELLE

Pipistrellus wattsi Kitchener *et al.*, 1986

wot'-zee

Watt's pipistrelle is known from the region via a single specimen (BMNH 1888.3.16.1) in the Natural

Pipistrellus wattsi			
Size*			
	Mean	**Range**	**Sample**
Head–body	37.3	36.0–40.0	6
Tail	29.5	27.6–31.6	6
Forearm	30.9	29.6–32.2	6
Tibia	12.1	11.7–12.9	6
Hindfoot	5.4	4.8–6.0	6
Ear	10.9	10.0–12.0	6
Weight	–	–	–
* From Kitchener *et al.* (1986)			
Identification A small and delicate species, similar in size to *P. papuanus*. Fur of the ventral surface, chin and throat has fuscous black bases with pinkish buff tips.			
Synonyms None			
Subspecies None			
Distribution Samarai			
Extralimital New Guinea			
Conservation status (IUCN) Least Concern			
Etymology Sp.–for Christopher Henry Stuart Watts of the South Australian Museum			

Watt's pipistrelle (*Pipistrellus wattsi*) is known from the region by just a single specimen from Samarai Island in the Louisiade Archipelago (Madison Erin Mayfield).

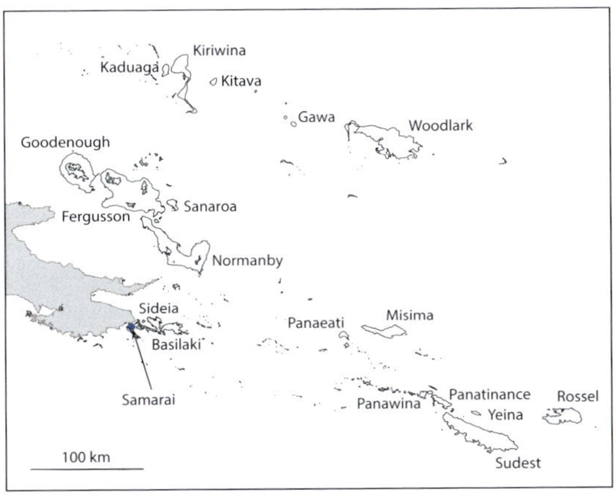

History Museum, London, from Samarai Island in the Louisiade Archipelago (adjacent to the Papuan Peninsula). It was referred to by Kitchener *et al.* (1986), but not listed as a paratype. The distribution of Watt's pipistrelle lies along the southern coastline of the Papuan Peninsula and Gulf of Papua, where it occurs from sea level to 200 m elevation. The species forages in urban areas as an aerial insectivore, and may be more widespread among the islands of Milne Bay (Bonaccorso 1998).

Scotophilus

SODY'S YELLOW BAT

Scotophilus collinus Sody, 1936

scot'-oh-phy'-lus kol'-een-us

Sody's yellow bat belongs to a genus of vespertilionid bats primarily distributed through Asia and Africa. In the region under consideration here, it has been recorded in the Aru Islands only (Sinaga and Maryanto 2008). Almost nothing of the natural history of this species has been reported in the literature. It is presumably found in primary and secondary forests (Sinaga and Maryanto 2008). The species is listed as Least Concern on the IUCN Red List because it is widespread, and is unlikely to be declining at a rate to qualify for listing in a more threatened category.

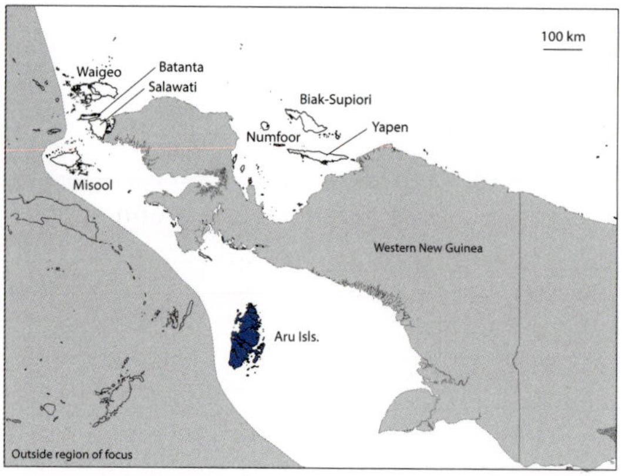

Scotophilus collinus			
Size*			
	Mean	**Range**	**Sample**
Head–body	63.4	55.7–71.0	2
Tail	39.9	27.8–52.0	2
Forearm	48.2	44.5–51.9	2
Tibia	–	–	2
Hindfoot	8.2	6.7–9.7	2
Ear	13.7	12.3–15.1	2
Weight	20.5	18.0–23.0	2
* From Burgin (2019)			
Identification The venter is a pale straw yellow in colour. Ears are small and have transverse ridges, tragus is half of ear length and the tail extends 2–3 mm beyond the uropatagium.			
Synonyms None			
Subspecies None			
Distribution Aru Isls			
Extralimital Indonesia, Malaysia, Timor Leste			
Conservation status (IUCN) Least Concern			
Etymology Gen.–dark lover (Gr.); Sp.–pertaining to hills (L.)			

Introduced species

Almost all islands of the South-west Pacific support at least one species of introduced mammal. In some cases (such as Aotearoa New Zealand) these comprise the overwhelming majority of wild-living species. On other islands just a single species exists (often *Rattus exulans*). In many cases, the history of introduction is well documented. In others it is obscure, having occurred inadvertently or many thousands of years ago. Occasionally, the archaeological record provides clues as to which species have been introduced by humans. In other cases, the actual history may be unknown, but a reasonable argument can be put forward that the species has been introduced.

Species that naturally occur in part of the region and have been translocated to others are dealt with in the main species accounts. The annotated list given here includes only those species that have arrived from outside the region and have established feral populations. Distributions are primarily identified to the level of country or territory only. In the cases of widely introduced species such as the murids, it is likely that many island occurrences have not been reported in the literature as some researchers choose not to record or collect them.

The profiles of introduced species presented here are concise and intended to give an overview of the diversity of non-native mammals in the region. Focus is placed on their history and origins of introduction, and implications for the natural environments to which they have been introduced. Detailed synonymies of species names can be found in the specialist taxonomic literature (e.g. Wilson and Reeder 2005). For comprehensive accounts of the biology, ecology and impacts of the introduced mammals of Aotearoa New Zealand, readers are referred to King and Forsyth (2021).

Artiodactyla

Bovidae

FERAL CATTLE

Bos taurus (Linnaeus, 1758)

All modern cattle are believed to be *Bos taurus*, *B. indicus* or crosses between the two. Cattle are present on many islands of the South-west Pacific, and exist as free-ranging populations in numerous locations. Truly feral populations have been reported from Aotearoa New Zealand and Pagan and Alamagan in Commonwealth of the Northern Mariana Islands. Surveys on Pagan in 2000 and 2010 found the island's forests and grasslands were severely overgrazed due to the abundance of feral cattle, goats (*Capra aegagrus hircus*) and pigs (*Sus scrofa*) (Liske-Clark 2015). Moreover, damage by cattle has contributed to removal of natural vegetation and losses of native flora species (Müeller-Dombois and Fosberg 2013). In Aotearoa New Zealand, feral cattle cause environmental damage by crushing and trampling native vegetation and clearing native forest understoreys, leaving only browse-resistant and unpalatable plants (Parkes 2021a). In subalpine environments this browsing and trampling results in clearings opened up among low-canopied vegetation.

Bos taurus
Native range *Bos taurus* was domesticated in Western Asia from wild aurochs (*Bos primigenius*).
Introduced range (South-west Pacific) Pagan and Alamagan (Commonwealth of the Northern Mariana Isls), Aotearoa New Zealand
Identification The physical characteristics of cattle vary between breeds. Most are stocky with strong necks, prominent dewlaps and long, rectangular bodies. *Bos indicus* breeds typically possess a hump over the shoulders.
Conservation status (IUCN) Not Assessed

WATER BUFFALO

Bubalus bubalis (Linnaeus, 1758)

Water buffalo were introduced to Guam from the Philippines by the Spanish in the 1700s and used widely for farming and transport. A feral population persists today. Surveys on Guam in 2009 estimated the population was 55 individuals, and demographic surveys have estimated yearling calves make up approximately 10% of the population, implying an approximate 10% annual growth rate (DON 2019). Water buffalo can cause significant soil disturbance and damage to wetlands by wallowing, creating erosion and disturbance that promotes invasive weeds.

Bubalus bubalis
Native range *Bubalus bubalis* was domesticated in India and China from wild stock resembling *Bubalus arnee*
Introduced range (South-west Pacific) Guam
Identification A large bovine with large feet and splayed hooves, small ears and a long and narrow face. Both sexes bear large, ribbed horns that are triangular in cross-section with heavy-set bases that are set far apart on the forehead. The hair is sparse, long and ashy gray to black and the long tail is bushy at the tip.
Conservation status (IUCN) Not Assessed

FERAL GOAT

Capra hircus (Erxleben, 1777b)

Domestic goats will readily form feral populations, and globally are regarded as one of the most problematic introduced species for island ecosystems. In the South-west Pacific, feral populations exist in Cook Islands, Fiji, French Polynesia, Guam, New Caledonia, Aotearoa New Zealand, Northern Mariana Islands, Pitcairn Island, Tonga and Vanuatu (Chynoweth *et al.* 2013).

Feral goats will browse and graze an extremely wide variety of plants, and can have serious impacts on the floral diversity and structure of insular vegetation communities. They are agile climbers capable of exploiting steep terrain. By denuding vegetation in these areas in particular, feral goats cause serious land degradation, erosion and losses of topsoil. Browsing by feral goats can significantly

Feral goat (*Capra hircus*) Keauhou, Hawaii (Barry Fackler).

Capra hircus
Native range Feral goats are most likely descended from *Capra aegagrus*, a species native to Armenia, Azerbaijan, Georgia, Iran, Iraq, Pakistan, Russia, Turkey and Turkmenistan
Introduced range (South-west Pacific) Cook Isls, Fiji, French Polynesia, Guam, New Caledonia, Aotearoa New Zealand, Northern Mariana Isls, Pitcairn Is., Tonga, Vanuatu
Identification The coat is short-haired and can be white, black, or brown in colour, or combination of these. Both sexes possess a pair of pointed horns and a beard.
Conservation status (IUCN) Not Assessed

reduce vegetation biomass with consequences for the composition of vegetation understoreys (Chynoweth *et al.* 2013). This process can also increase the potential for invasion by less-palatable weeds, with flow-on effects for island fauna that rely on intact native vegetation.

HIMALAYAN TAHR
Hemitragus jemlahicus (Smith, 1826)

Himalayan tahr were intentionally introduced as a game resource to Aotearoa New Zealand's South Island in 1904 (Forsyth and Tustin 2021). The species has since expanded across the central Southern Alps to occupy some 9600 km^2 and for the period 2016–2018 the total abundance of Himalayan tahr in public conservation lands of Aotearoa New Zealand's Southern Alps was estimated to be 34 292 individuals (Cruz *et al.* 2016; Ramsey and Forsyth 2018). In the absence of hunting, densities of more than 30 animals km^{-2} can be attained, but with regular hunting densities are maintained much lower (Tustin and Challies 1978).

Hemitragus jemlahicus
Native range Himalayan Mountains (China, India, Nepal)
Introduced range (South-west Pacific) Aotearoa New Zealand
Identification Reddish to dark brown in colour, with relatively short legs and a small head. The horns are curved backwards. Males and females are sexually dimorphic. Males possess a mane of thicker hair covering the neck and forequarters.
Conservation status (IUCN) Near Threatened

Himalayan tahr can cause significant, long-term harm to Aotearoa New Zealand's native vegetation communities. Grazing in montane grasslands decreases vegetation cover and tussock height, and trampling probably causes significant impact to herbs (*Celmisia* and *Aciphylla*) and shrubs (*Gaultheria* and *Carmichaelia*) (Cruz *et al.* 2016). Within its native range in the Himalayas, this species is considered to be in decline as a result of declining habitat quality, poaching and competition with livestock.

FERAL SHEEP
Ovis aries (Linnaeus, 1758)

Aotearoa New Zealand's feral sheep are descendants of domestic sheep brought to the country commencing in 1814 (Parkes 2021b). In this early period of Aotearoa New Zealand's colonial history, they were considered widespread, but distinguishing between feral and stray domestic populations was uncertain. Today, feral populations persist on the North, South, Chatham and Pitt islands (Parkes 2021b). The clearest examples of the environmental impacts of this species are derived from smaller islands, where modification to the structure of vegetation communities, prevention of forest regeneration, and disturbance of burrowing petrels and giant snails have been evident (Parkes 2021b).

Ovis aries
Native range Mouflon (*Ovis gmelini*) is thought to be the ancestor of all modern domestic sheep and is native to the Caspian region across eastern Turkey, Armenia, Azerbaijan and Iran.
Introduced range (South-west Pacific) Aotearoa New Zealand
Identification The coat is a woolly fleece of variable length and colour (white, brown and black are common). The horns are grooved and curved backwards, spiraling outwards in males. Both sexes carry horns but these are not always present.
Conservation status (IUCN) Not Assessed

NORTHERN CHAMOIS
Rupicapra rupicapra (Linnaeus, 1758)

Rupicapra rupicapra
Native range Albania, Austria, Azerbaijan, Bosnia and Herzegovina, Bulgaria, Croatia, France, Georgia, Germany, Greece, Italy, Montenegro, North Macedonia, Poland, Romania, Russian Federation, Serbia, Slovakia, Slovenia, Switzerland, Turkey
Introduced range (South-west Pacific) Aotearoa New Zealand (South Is.)
Identification Northern chamois can be identified by their pale-coloured face, broken by dark lateral stripes that pass from nose through the eyes to between the ears and horns. The coat colour varies from light fawn in summer to dark brown in winter. The horns are dark and slender, and curve backwards to form sharp hooks.
Conservation status (IUCN) Least Concern

The northern chamois was intentionally introduced to Aotearoa New Zealand from Austria in 1907 and 1914 for the purposes of establishing a hunting resource (Donne 1924; Forsyth 2021). This species apparently dispersed more rapidly than any other ungulate introduced to Aotearoa New Zealand and now occupies much of the South Island's high country (Caughley 1963; Forsyth and Hickling 1998).

Yockney and Hickling (2000) found that woody plants comprised 83.2% of the dry weight in rumen samples from the South Island's Westland Forests. Two canopy trees, five subcanopy trees, four shrubs, a fern, unidentified grasses and two herbaceous plants were among 16 food items that comprised more than 1% of sample dry weights. Grazing pressure from northern chamois is potentially detrimental to native vegetation communities, and their impacts are probably greatest in alpine grasslands, alpine mire communities and alpine herb fields. However, the actual damage caused by this species is poorly understood. At high densities, and in combination with grazing pressure from sympatric Himalayan tahr (*Hemitragus jemlahicus*) and red deer (*Cervus elaphus*), they probably suppress and threaten more palatable flora species. Northern chamois also cause damage to vegetation and mire wetlands by trampling (Yockney and Hickling 2000).

The natural distribution and population size of northern chamois are both large. Some declines have been evident in parts of Europe, but overall the species is relatively secure.

The northern chamois (*Rupicapra rupicapra*) was intentionally introduced to Aotearoa New Zealand in 1907 for the purposes of establishing a population for hunting (Philippe Pourtallier).

Cervidae

RED DEER
Cervus elaphus (Linnaeus, 1758)

Red deer were introduced to Aotearoa New Zealand from 1851 and by 1923 approximately 1000 animals had been liberated into the wild from direct imports, translocations from previous release sites, and captive-bred populations (Nugent and Forsyth 2021). Initial growth rates of introduced populations were probably high (around 25–30% per year) and the long-term rate of dispersal averaged 1.6 km per year (Caughley 1963). By 1950 red deer were established across much of the central and south North Island, as well as most undeveloped and remote areas of the South Island (Nugent and Forsyth 2021).

Like many of Aotearoa New Zealand's introduced herbivores, red deer can impact native vegetation communities through the selective browsing of preferred or more palatable species. Red deer also kill subcanopy trees by rubbing their antlers and stripping bark. In some areas and communities over certain time scales, these impacts can lead to changes in flora species composition. For example, in northern Fiordland, red deer reduced the density of woody plants in the understorey by as much as 50% in some areas. Preferred food plants also became rarer than those that were less preferred (Stewart *et al.* 1987).

Red deer has a wide natural distribution in Europe, and the population is presumed to be large.

Red deer (*Cervus elaphus*) were liberated in Aotearoa New Zealand on more than 200 occasions between 1851 and 1926 (Philippe Pourtallier).

Cervus elaphus
Native range Algeria, Armenia, Austria, Azerbaijan, Belarus, Belgium, Bosnia and Herzegovina, Bulgaria, Croatia, Czechia, Denmark, Estonia, France, Georgia, Germany, Hungary, Iran, Ireland, Italy, Latvia, Lithuania, Luxembourg, Moldova, Montenegro, Netherlands, North Macedonia, Norway, Poland, Romania, Russian Federation, Serbia, Slovakia, Slovenia, Sweden, Switzerland, Tunisia, Turkey, Ukraine
Introduced range (South-west Pacific) Aotearoa New Zealand (North and South Isls)
Identification A large, red-brown deer; the tail is short and the antlers of males are erect and can have 10 or more points.
Conservation status (IUCN) Least Concern

Although there have been range contractions and population declines in some parts, the species is not considered under major threat and is classified as Least Concern on the IUCN Red List.

WAPITI

Cervus canadensis (Erxleben, 1777a)

In March 1905, 18 wapiti (derived from herds in the eastern United States) were liberated at the head of George Sound on Aotearoa New Zealand's South Island (Latham and Nugent 2021). Aotearoa New Zealand's only wild wapiti herd is now distributed in the South Island's Fiordland National Park and all have hybridised with red deer (*Cervus elaphus*).

Wapiti will browse and graze a diverse array of vegetation including grasses, herbs, shrubs and trees. At high densities, wapiti damage native plants and vegetation communities, and even at moderate

Wapiti (*Cervus canadensis*), Banff National Park, Canada (Elise Short).

Cervus canadensis
Native range Bhutan, Canada, China, Kazakhstan, Kyrgyzstan, Mongolia, Russia, United States
Introduced range (South-west Pacific) Aotearoa New Zealand (South Is.)
Identification A large deer with a light-coloured body contrasting strongly with dark-coloured legs, neck and head. Large creamy patch around rump and tail. Blackish, hairless muzzle. Large ears at an angle from the head. Males carry large, round antlers that are cast off annually.
Conservation status (IUCN) Least Concern

densities can eliminate highly palatable species. These impacts have been documented in habitats from sea level into montane and alpine areas over the approximate 2000 km² of their distribution (Latham and Nugent 2021).

SIKA DEER
Cervus nippon (Temminck, 1838b)

Sika deer were first introduced to Aotearoa New Zealand in 1885 but apparently did not establish (Nugent and Speedy 2021). In 1905 a small group was again introduced and steadily expanded in distribution to now occupy some 9000 km² across the central North Island (Davidson and Fraser 1991; Nugent and Speedy 2021). This species is native to eastern Asia, where it prefers woodlands and forests with dense understoreys (Smith and Xie 2008). It is primarily a crepuscular species that typically forms small single-sex herds. In Aotearoa New Zealand, the rutting season occurs between March and May.

Sika deer are considered to have heavier impacts on Aotearoa New Zealand's native forests than other deer species that have been introduced to the country. They reach high densities and will browse on less palatable plants. This probably inhibits forest regeneration and maintains lower canopy cover in the longer term (Nugent and Speedy 2021).

Cervus nippon
Native range China, Japan, Russia, South Korea, Taiwan, Vietnam
Introduced range (South-west Pacific) Aotearoa New Zealand (North Is.)
Identification A moderately sized deer, chestnut brown in colour with a black dorsal stripe and white rump. The back and flanks often have white spots variable in size and extent.
Conservation status (IUCN) Least Concern

This species is classified as Least Concern on the IUCN Red List because the population in Japan is large and secure. However, in other countries (South Korea, Vietnam) it has undergone decline and possible extinction (Harris 2015).

PHILIPPINE DEER
Rusa marianna (Desmarest, 1822)

The Philippine deer is native to the Philippines where there are four recognised subspecies. However, it was first described from a population introduced to Guam in 1771, and additional introductions followed to the Northern Mariana islands of Rota (around 1880), Saipan (1900), and Pohnpei (1880) (Wiles *et al.* 1999). Extant populations remain on these four islands today. Four deer were released on the island of Pagan in 2020, but it is too early to know if they have established there (G. Wiles, pers. comm.). Philippine

Sika deer (*Cervus nippon*), Dorset, United Kingdom (Barbara Evans).

Philippine deer (*Rusa marianna*) (Shankar S.).

Rusa marianna
Native range Philippines
Introduced range (South-west Pacific) Rota and Saipan (Northern Mariana Isls), Guam and Pohnpei (Federated States of Micronesia)
Identification A medium-sized deer that is nearly uniform medium to dark brown, coarse pelage. The undersides and legs may be somewhat paler, and the insides of the ears and underside of the tail are light brown or whitish.
Conservation status (IUCN) Vulnerable

deer have the potential to establish high population densities when introduced to islands. Wiles *et al.* (1999) estimated densities at Guam's Andersen Air Force Base could be as high as 60–80 animals km^{-2}.

In the Philippines, this is a forest species found in primary and secondary forests as well as open grasslands with the presence of nearby forest patches (MacKinnon *et al.* 2015). In Micronesia, they can be found in most habitats including wetter secondary and primary forests on limestone and volcanic soils, and scrubby secondary growth, grasslands, wetlands, and farmlands (Wiles *et al.* 1999). They tend to avoid urban areas and areas of rugged limestone karst.

Philippine deer can cause substantial damage to native Micronesian vegetation communities when occurring at moderate to high densities. This has been best documented on Guam where extensive damage to ecosystems has occurred since the 1980s (Wiles *et al.* 1999).

Overbrowsing can cause the elimination of herbaceous ground cover and low epiphytic growth, prevent native plant regeneration, and ultimately lead to localised disappearances of certain more vulnerable trees and shrubs (Wiles *et al.* 1999; Wiles 2005). By disturbing natural ground cover and dispersing seeds, Philippine deer can also promote the spread of invasive weeds. Philippine deer are classified as Vulnerable on the IUCN Red List due to population declines in their native range (MacKinnon *et al.* 2015).

JAVAN DEER
Rusa timorensis (Blainville, 1822)

Javan deer were established on New Caledonia in 1870. The species was extremely successful and within 70 years: the original 12 introduced animals had grown to a population of 200 000 or more (Long 2003). The species was introduced by German settlers near Rabaul on New Britain around 1910 and liberated in the Hermit Islands in 1909. From the Hermit Islands, animals were transferred to Arkeb and Luf, with the population on Luf reaching approximately 200 individuals in 1954 before declining (Long 2003). Introduction of this species to Aotearoa New Zealand took place in 1908, with the animals reportedly having been derived from New Caledonia. In the Aru Islands, Healey (1995) suggested Javan deer may have been introduced by the Portuguese. However, Aplin and Pasveer (2006) questioned whether this could have been the case, because they were not reported by Alfred Russel Wallace, and it is doubtful he would have failed to notice this conspicuous species.

The Javan deer is primarily a tropical and subtropical grassland species, but it will inhabit a range of environments. Rouys and Theuerkauf (2003) suggested that in southern New Caledonia it was most abundant in dry open habitats. Droppings were relatively rare in rainforest, common in shrublands and especially abundant where areas of degraded forest bordered pastures. Given the rapid population growth of the species in New Caledonia, complaints about the damage caused to native flora were registered as early as 1882 (Long 2003).

In Aotearoa New Zealand, where the climate is less suitable, this species is less widely distributed than other introduced deer. There, the species

Javan deer (*Rusa timorensis*), Papua New Guinea. This tropical species has been extremely successful in New Caledonia, with smaller populations also present in Aotearoa New Zealand and the Bismarck Archipelago (Angus Fraser).

Rusa timorensis
Native range Indonesia (Bali, Jawa)
Introduced range (South-west Pacific) Aru Isls, New Britain, New Caledonia, Aotearoa New Zealand
Identification Medium-sized deer, with pointed ears and a long narrow tail. Males are a dark reddish-brown in summer and greyish-brown in winter, and females are pale yellowish-red in summer, becoming greyish-red in winter. The chin, throat and belly are cream-coloured in both sexes.
Conservation status (IUCN) Vulnerable

shows preference towards habitats with higher solar radiation, but they do exploit a range of forest types and further contribute to the detrimental grazing pressure of feral herbivores (Allen 2021).

This species is classified as Vulnerable on the IUCN Red List because its total native population on the island of Java is estimated to be fewer than 10 000 individuals and is declining. Habitat loss and degradation, and poaching, are major threats on Java.

SAMBAR
Rusa unicolor (Kerr, 1792)

A single sambar stag and one hind were introduced to Aotearoa New Zealand from Sri Lanka in 1875 (Nugent 2021). The range of this species is restricted to two populations at Manawatu and Bay of Plenty on the North Island. In its native range in South-East Asia, the sambar is primarily a forest species, but can adapt to an extremely wide variety of environmental conditions (Timmins *et al.* 2015). Activity is mostly crepuscular, but this seems to vary geographically and in some parts of the natural distribution there is significant nocturnal behaviour. It is a relatively solitary species with stable groups usually limited to family associations (Timmins *et al.* 2015).

Sambar (*Rusa unicolor*) (Angus McNab).

Rusa unicolor
Native range Bangladesh, Bhutan, Brunei Darussalam, Cambodia, China, India, Indonesia, Laos, Malaysia, Myanmar, Nepal, Sri Lanka, Taiwan, Thailand, Vietnam
Introduced range (South-west Pacific) Aotearoa New Zealand (North Is.)
Identification A very large species with a uniform brown coat. The tail is long and bushy with a black tip, ears large and round, and antlers slightly smaller than those of red deer (*Cervus elaphus*).
Conservation status (IUCN) Vulnerable

Sambar deer are not thought to cause major ecological impacts in Aotearoa New Zealand (Nugent 2021). This is partially due to their restricted range, but also because they tend to occur on forest margins and patches of scrub, and predominantly browse on pastures. They have wider dietary preferences compared to other deer and occur at relatively low density. This species has suffered sustained declines of varying intensity across its natural range. Major threats include loss of habitat and hunting.

FALLOW DEER
Dama dama (Linnaeus, 1758)

Fallow deer were first introduced to Aotearoa New Zealand at a zoological garden in Auckland in 1853 (Nugent and Asher 2021). Subsequently, at least 24 further introductions were made and almost all were successful, establishing the species both on the South and North Islands (Christie and Andrews 1966; Nugent and Asher 2021). Three fallow deer were also imported to Wakaya Island, Fiji in 1880. By 1929 the population there was estimated at 1000 individuals and by 1977 was believed to be around 400 (Chapman and Chapman 1980). Forests on Wakaya were considered to have very little understorey, presumably a result of grazing pressure by fallow deer and feral goats.

This is a gregarious species that under certain circumstances will form herds of 100 animals or more. Breeding in Aotearoa New Zealand occurs April–May, when adult males will establish and defend small mating territories (Nugent and Asher

Fallow deer (derived from Richmond Park, Surrey, UK) were introduced to Aotearoa New Zealand on numerous occasions. Almost all introductions were successful and the species is now present on both the South and North Islands (Marco Messina).

Dama dama
Native range South-west Asia and ancient introduction to the Mediterranean
Introduced range (South-west Pacific) Aotearoa New Zealand (North and South Isls), ?Fiji (Wakaya)
Identification A small species of deer with forelegs that are shorter than the hindlegs, giving a forward-sloping appearance. Fallow deer can be found in four distinctive colour phases of brown, melanistic black, light red-brown with conspicuous white spots, and a leucistic white or cream. The antlers are distinctively broad and shovel-shaped.
Conservation status (IUCN) Least Concern

2021). Fallow deer are successful at exploiting Aotearoa New Zealand's native forests and can cause damage in a similar manner to red deer (*Cervus elaphus*). They primarily feed on the leaves of woody species such as broadleaf (*Griselinia littoralis*), broad-leaved subcanopy trees, and ferns. As selective browsers that target specific plant species, fallow deer can cause changes in the plant composition of forests and at high density can also prevent regeneration by feeding on more palatable plants and seedlings. This species can reach high densities in parts of Aotearoa New Zealand and densities remain high despite managed hunting ballots (Nugent and Asher 2021).

WHITE-TAILED DEER

Odocoileus virginianus (Zimmermann, 1780)

White-tailed deer are native to North America, Central America and northern South America, with over 30 described subspecies. The subspecies found in Aotearoa New Zealand originates from north-eastern U.S.A. and eastern Canada, and was first successfully introduced in 1905 to Stewart Island and Lake Wakatipu for sport hunting (Christie and Andrews 1965; Nugent 2021).

In the Americas, white-tailed deer can be found in a wide range of habitats, including temperate, subtropical and semi-arid forests of North America; and rainforests, deciduous forests and savannahs of Central America and northern South America (Gallina and Lopez Arevalo 2016). They are non-territorial but occupy a well-defined home range.

On Stewart Island births can occur throughout most of the year and the population increased quickly after introduction to become well established by 1917 (Christie and Andrews 1965). The ecological impacts of high population densities are a potential concern for vegetation communities on Stewart Island. White-tailed deer are a browsing species and feed on hardwood trees, preventing regeneration of highly preferred species (Nugent and Challies 1988; Nugent 2021). Woody plants comprised 85.1% (dry weight) of the annual diet, with broadleaf (*Griselinia littoralis*) and supplejack (*Ripogonum scandens*) being the most heavily consumed species. Selection of more palatable species such as broadleaf, *Pseudopanax* spp. and *Coprosma* spp. can lead to these species becoming scarce in the understorey. In 1981 a poisoning program was carried out to reduce densities (Stewart and Burrows 1989). This was followed by increased recruitment of the evergreen shrub *Brachyglottis rotundifolia*, but little evidence is

White-tailed deer (*Odocoileus virginianus*), Calgary, Canada (Elise Short).

Odocoileus virginianus
Native range Belize, Brazil, Canada, Colombia, Costa Rica, Ecuador, El Salvador, French Guiana, Guatemala, Guyana, Honduras, Mexico, Nicaragua, Panama, Peru, Suriname, United States, Venezuela
Introduced Range (South-west Pacific) Aotearoa New Zealand (Stewart Is., South Is.)
Identification The body and sides are reddish- or greyish-brown and the belly and throat white. The muzzle is distinctive, being black and bordered behind by a white band. The tail is brown dorsally and white underneath, and is flicked up when the animals are alarmed, exposing the white underneath.
Conservation status (IUCN) Least Concern

available for broader influences on the composition of Stewart Island's forests.

Suidae

FERAL PIG

Sus scrofa (Linnaeus, 1758)

Feral pigs are a mixture of several strains of domestic swine, Asiatic wild boar and European wild boar and are perhaps the most widespread large introduced vertebrate across the Pacific islands (Wehr *et al.* 2018). According to archaeological and genetic evidence, the pigs found in the South-west Pacific are feral descendants of domestic pigs introduced east of Wallace's Line by early agriculturalists (Larson *et al.* 2007). Genetic evidence indicated there were two human-mediated dispersals into Oceania. One of these was associated with the earlier movements of the Lapita, and the second with more recent Polynesian travel. The more recent pig dispersal linked mainland East Asian pigs with the Philippines, Taiwan and western Micronesia, and the earlier dispersal linked South-east Asia with Melanesia and Polynesia. Although pigs were transported widely around the Pacific by Polynesians, they were never successfully introduced to Aotearoa New Zealand (Tipene 1980; McIlroy and Nugent 2021). The first releases in 1769 by the French explorer de Surville, and in 1773 by Captain James Cook, did not establish (Clarke and Dzieciolowski 1991; McIlroy and Nugent 2021). But more localised releases took place in the following decade, and from 1790 onwards they were introduced with increasing frequency. By 1840 feral pigs were established around most settlements (McIlroy and Nugent 2021).

Pigs now occupy around 93 000 km^2 of Aotearoa New Zealand, having extended their range into Otago, Southland and the West Coast of the South Island (Fraser 2005; Bengsen *et al.* 2017). The prevalence of pigs in other parts of the South-west Pacific varies widely with human population density and hunting intensity. On the heavily populated island of Malaita, Solomon Islands, the only pigs to be found are domesticated, feral pigs having been completely extirpated through hunting. However, on the adjacent island Guadalcanal, pigs can be extremely abundant, especially at high elevation where human villages are absent (Lavery *et al.* 2020c).

Pigs are important for subsistence and recreation and are of immense cultural significance across much of the Pacific, and this can complicate their management. They have a generalist diet and their rooting behaviour causes great disturbance to soils, wetland and waterways both directly and indirectly via erosion and sedimentation. They can also be a serious pest to agricultural subsistence gardens, especially in regions where Christian sects prohibit consumption of pork and densities are not supressed by hunting. Goulding *et al.* (2016) showed the cascading impacts that can occur in island environments when pigs feed on subsistence gardens. In the Louisiade Archipelago, cyclones reduced the availability of coconuts and forest

Pigs (*Sus scrofa*) were prehistorically introduced to the South-west Pacific. Their prevalence today varies widely with human population density and hunting intensity (Angus McNab).

Sus scrofa
Native range Afghanistan, Albania, Algeria, Andorra, Armenia, Austria, Azerbaijan, Bangladesh, Belarus, Belgium, Bhutan, Bosnia and Herzegovina, Bulgaria, Cambodia, China, Croatia, Cyprus, Czechia, Estonia, Finland, France, Georgia, Germany, Greece, Hong Kong, Hungary, India, Indonesia, Iran, Iraq, Israel, Italy, Japan, Jordan, Kazakhstan, North Korea, South Korea, Kyrgyzstan, Laos, Latvia, Lebanon, Liechtenstein, Lithuania, Luxembourg, Malaysia, Moldova, Monaco, Mongolia, Montenegro, Morocco, Myanmar, Nepal, Netherlands, North Macedonia, Pakistan, Palestine, Poland, Portugal, Romania, Russia, San Marino, Serbia, Slovakia, Slovenia, Spain, Sri Lanka, Switzerland, Syria, Taiwan, Tajikistan, Thailand, Tunisia, Turkey, Turkmenistan, Ukraine, Uzbekistan, Vietnam
Introduced range (South-west Pacific) American Samoa, Commonwealth of the Northern Mariana Isls, Fiji, French Polynesia, Guam, Indonesia, Kiribati, Marshall Isls, Nauru, New Caledonia, Aotearoa New Zealand, Niue, Palau, Papua New Guinea, Pitcairn Is., Solomon Isls, Tokelau Isls, Tonga, Tuvulu, Vanuatu, Wallis and Futuna
Identification Compared to their domestic counterparts, feral pigs are smaller and more muscular, with forequarters larger than their hindquarters and a narrowing of the body towards the spine. Feral pigs are predominantly black in colour, but variations of black and white are common.
Conservation status (IUCN) Least Concern

resources used by pigs, resulting in an increase in pressure on gardens. The consequential construction of protection fences around gardens (made from native vegetation) greatly expanded vegetation disturbance and habitat removal with potential impacts for native fauna.

Carnivora

Viveriidae

SMALL INDIAN MONGOOSE

Herpestes auropunctatus (Hodgson, 1836)

The small Indian mongoose was introduced to numerous island nations in the late 1800s to control rodents (including Fiji's Viti Levu in 1883) (Gorman 1975). From Viti Levu it has since spread to occupy at least 11 Fijian islands (Veron *et al.* 2009). An incursion by this species was detected near Satitoa Wharf on Upolu, Samoa, in 2009 (Fisher *et al.* 2011). One individual (a male) was removed and it appeared that no other individuals were present. The small Indian mongoose has also been reported in Nouméa, New Caledonia, where two individuals (a male and female) were trapped (Barun *et al.* 2011), as well as Tongatapu, Tonga.

In its native range, the small Indian mongoose inhabits forests, scrub and open habitats, and can be found in close proximity to areas inhabited by humans (Gilchrist *et al.* 2009). It is mostly diurnal and omnivorous with a diet that varies widely with habitat. On Viti Levu, scat analysis showed littoral crabs were the predominant food in mangrove forests, and 96% of scats from sugarcane fields contained rats. In agricultural lands, birds, cane toads (*Rhinella marina*) and plant matter were found to be important food, and dragonflies and other invertebrates were important in urban areas (Gorman 1975).

The small Indian mongoose (*Herpestes auropunctatus*) was widely introduced to island nations in the late 1800s (including Fiji) to control vertebrate pests. More recently, incidental records have been reported in Samoa, New Caledonia and Tonga (Charles A. Thomas).

Herpestes auropunctatus
Native range Afghanistan, Bangladesh, Bhutan, India, Iran, Iraq, Jordan, Myanmar, Nepal, Oman, Pakistan, Saudi Arabia
Introduced range (South-west Pacific) Fiji (Beqa, Druadrua, Kioa, Macuata-i-wai, Malake, Nananu-i-cake, Nananu-i-ra, Vanua Levu, Viti Levu, Rabi, Yanuca)
Identification The small Indian Mongoose is one of two mongooses found in Fiji, and is the most common and widespread of the two species. It is smaller and more lightly built than *H. fuscus* (the average head–body length of six individuals measured by Morley *et al.* (2007) was 307 mm).
Conservation status (IUCN) Least Concern

The small Indian mongoose is an efficient predator, and Fiji's island faunas (naïve to mammalian predators) have suffered considerable impacts. This species has been implicated in the decline of numerous skinks, ground-dwelling birds, native frogs, and potentially some species of bat (Veron *et al.* 2009).

INDIAN BROWN MONGOOSE

Herpestes fuscus (Waterhouse, 1838b)

The Indian brown mongoose is native to south-west India and Sri Lanka and is one of two mongooses introduced to Fiji (Veron *et al.* 2009). *Herpestes fuscus* and *H. auropunctatus* co-occur on Viti Levu, with *H. fuscus* being the most recent introduction of the two, apparently derived from a pair of animals brought from an unknown source to a private zoo near Suva in the late 1970s.

In its native range, the Indian brown mongoose is most likely crepuscular to nocturnal. It has been recorded in dense evergreen forests and human-modified habitats (including coffee plantations), but is usually associated with native forests (Mudappa 2002). No population estimate is available for this species in Fiji, and no research has been conducted on their activity, diet or impacts. The degree of

Herpestes fuscus
Native range India, Sri Lanka
Introduced range (South-west Pacific) Fiji (Viti Levu)
Identification The Indian brown mongoose is one of two mongoose species found in Fiji. It is much larger and more heavily built than *H. auropunctatus* (head–body length 330–480 mm). Fur is blackish-brown, speckled with yellow or tan, the feet are darker than the body, the tail is bushy and conical.
Conservation status (IUCN) Least Concern

interaction or competition between this species and *H. auropunctatus* on Viti Levu is also unknown. Co-occurrence of two mongoose species could be assumed to impose substantial predation pressure via slightly different dietary preferences and activity patterns, but further research is needed.

STOAT
Mustela erminea (Linnaeus, 1758)

Stoats (along with least weasels *Mustela nivalis*) were deliberately liberated in Aotearoa New Zealand following a determination in New Zealand Parliament that natural predators were needed to control plagues of European rabbits (*Oryctolagus cuniculus*) (King and Veale 2021). Between 1883 and 1892 there were at least 25 shipments that delivered some 7838 stoats and least weasels to Aotearoa New Zealand (King 2017a). Within 6 years following their introduction, there were reports of stoats in locations far from release sites, and where there were no European rabbits to control (King and Veale 2021).

The potential for introduced stoats to cause significant impacts on Aotearoa New Zealand's native birds was forewarned by ornithologists (King and Veale 2021). Indeed, stoats have had serious

Stoat (*Mustela erminea*) (David Brooker).

Mustela erminea
Native range Afghanistan, Albania, Andorra, Austria, Azerbaijan, Belarus, Belgium, Bosnia and Herzegovina, Bulgaria, Canada, China, Croatia, Czechia, Denmark, Estonia, Finland, France, Georgia, Germany, Greece, Hungary, India, Ireland, Italy, Japan, Kazakhstan, Kyrgyzstan, Latvia, Liechtenstein, Lithuania, Luxembourg, Moldova, Mongolia, Montenegro, Netherlands, North Macedonia, Norway, Pakistan, Poland, Portugal, Romania, Russia, Serbia, Slovakia, Slovenia, Spain, Sweden, Switzerland, Tajikistan, Turkey, Ukraine, United Kingdom, United States, Uzbekistan
Introduced range (South-west Pacific) Aotearoa New Zealand (North and South Isls)
Identification the body is long and thin, legs short, eyes black and ears short and round. Fur of the dorsum is chestnut brown and white to yellow on the ventrum. Stoats are larger than least weasels (*Mustela nivalis*) and have a bushy tail tip that is black (a feature not present in weasels). In alpine areas of Aotearoa New Zealand, stoats sometimes undergo a white moult during winter, but the tail tip remains black.
Conservation status (IUCN) Least Concern

negative consequences for Aotearoa New Zealand's predator-naïve faunas, including birds, bats, lizards and invertebrates. They are blamed for the extinction and decline of numerous birds, continue to pose a major threat to Endangered species including takahē (*Porphyrio hochstetteri*) and rock wrens (*Xenicus gilviventris*), and probably impact the population viability of these species (O'Donnell *et al.* 2017). Stoats prey on lizards and can have localised impacts on their populations. They also are known to eat large numbers of native invertebrates such as wētā, but the extent to which this impacts their populations is not well studied. Stoats will also take larger introduced prey such as black rats (*Rattus rattus*), common brushtail possums (*Trichosurus vulpecula*) and European rabbits (Murphy *et al.* 2008). This ability to target common, introduced species may help to sustain larger populations (King 2017b).

FERAL FERRET
Mustela furo (Linnaeus, 1758)

Ferrets were introduced to Aotearoa New Zealand to control the European rabbit (*Oryctolagus cuniculus*). The earliest records of release were 1879 and over the following 50 years more than 75 000 ferrets were introduced to regions infested with European rabbits (Garvey and Byrom 2021). They

Mustela furo
Native range Albania, Algeria, Andorra, Austria, Belarus, Belgium, Bosnia and Herzegovina, Bulgaria, Croatia, Czechia, Denmark, Estonia, Finland, France, Germany, Gibraltar, Greece, Hungary, Italy, Latvia, Liechtenstein, Lithuania, Luxembourg, Moldova, Montenegro, Morocco, Netherlands, North Macedonia, Norway, Poland, Portugal, Romania, Russian Federation, Serbia, Slovakia, Slovenia, Spain, Sweden, Switzerland, Turkey, Ukraine, United Kingdom
Introduced range (South-west Pacific) Aotearoa New Zealand (North and South Isls)
Identification The body is long and narrow with short legs. The fur is woolly, creamy-white to yellow, with a tail that is dark along most of its length (as opposed to only the tip as in *M. erminea*). There is a black face mask covering the eyes.
Conservation status (IUCN) Least Concern

were first liberated in the areas worst affected by European rabbits and from there spread into wider pasture and forested areas until they themselves were considered a pest by the turn of the century (Garvey and Byrom 2021).

Ferrets have had a range of adverse effects on Aotearoa New Zealand's native fauna. They have been found to kill the North Island brown kiwi (*Apteryx mantelli*), affect the breeding of yellow-eyed penguin colonies (*Megadyptes antipodes*), prey on little penguins (*Eudyptula minor*) and hunt underground for burrowing seabirds such as the sooty shearwater (*Ardenna grisea*) (Clapperton 2001). They are also major predators of the black stilt (*Himantopus novaezelandiae*), southern crested grebe (*Podiceps cristatus australis*) and weka (*Gallirallus australis*). Skinks such as McCann's skink (*Oligosoma maccani*), native frogs and rare invertebrates like the robust grasshopper (*Brachaspis robustus*) and kauri snails (*Paryphanta* spp.) are also probably at risk from ferret predation (Clapperton 2001).

LEAST WEASEL
Mustela nivalis (Linnaeus, 1766)

Least weasels were among three species of mustelid (including stoats *Mustela erminea*, and ferrets *M. furo*) deliberately released throughout Aotearoa New Zealand as an attempt to control plagues of European rabbit (*Oryctolagus cuniculus*) that were causing widespread damage to agricultural lands. At least 25 shipments were imported to Aotearoa New Zealand

The least weasel (*Mustela nivalis*) was introduced to Aotearoa New Zealand in an effort to control populations of the introduced European rabbit (*Oryctolagus cuniculus*) (Roger Wasley).

Mustela nivalis
Native Range Afghanistan, Albania, Andorra, Armenia, Austria, Azerbaijan, Belarus, Belgium, Bosnia and Herzegovina, Bulgaria, Canada, China, Croatia, Czechia, Denmark, Estonia, Finland, France, Georgia, Germany, Greece, Hungary, Iran, Islamic Republic of, Italy, Japan, Kazakhstan, North Korea, South Korea, Kyrgyzstan, Latvia, Lebanon, Liechtenstein, Lithuania, Luxembourg, Moldova, Monaco, Mongolia, Montenegro, Netherlands, North Macedonia, Norway, Poland, Portugal, Romania, Russia, San Marino, Serbia, Slovakia, Slovenia, Spain, Sweden, Switzerland, Syrian Arab Republic, Taiwan, Province of China, Tajikistan, Turkey, Turkmenistan, Ukraine, United Kingdom, United States, Uzbekistan
Introduced Range (South-west Pacific) Aotearoa New Zealand (North and South Isls)
Identification The body is long and thin with short legs. Fur colour is brown dorsally and white underneath; the short tail lacks the black tip present in stoats (*Mustela erminea*).
Conservation Status (IUCN) Least Concern

between 1883 and 1892, delivering at least 7838 stoats and least weasels (King 2017a). King (2017b) reported that the proportions of these two species introduced were strongly biased towards least weasels as they were much more abundant among England's lowland farmlands (from which the collections for shipments were derived). Least weasels were thus reportedly the most abundant of the two in the early periods following introductions, but have since been vastly outnumbered by stoats (King 2017b).

Weasels are predators of Aotearoa New Zealand's native birds, their eggs, lizards and invertebrates (King and Murphy 2021). However, given least weasels are generally less abundant than stoats, their impacts are less prominent. The body size of least weasels is also smaller than that of stoats and ferrets, and they are therefore less likely to impact larger native species such as adult kiwi (*Apteryx* spp.).

PALM CIVET

Paradoxurus hermaphroditus (Pallas, 1777)

Aplin and Pasveer (2006) indicated that the common palm civet (*Paradoxurus hermaphroditus*) was present among the modern faunas of Pulau Kobroor, Aru Islands. However, it was completely absent among the examined archaeological deposits and the authors presumed it had been a recent introduction there. Little more has been reported of the distribution, population size or ecology of this species in the region. Gray (1858) listed *P. hermaphroditus* in his report on the species of mammals sent from the Aru Islands by Alfred Russel Wallace, but clearly referred to the habitat as being the Ké (Kai) Islands rather than the Aru Islands. A study of species limits in *Paradoxurus* has since resulted in populations from Mentawai Islands, Borneo and the Philippines being recognised as a distinct species (*P. philippinensis*) (Veron *et al.* 2015). Further study to reaffirm the presence of palm civets in the Aru Islands and confirm species identity is needed.

In its native range, the common palm civet is regarded as an adaptable species known to use a wide variety of habitats from evergreen and deciduous primary forests to monoculture plantations (e.g. oil palm and teak) and village and urban environments (Duckworth *et al.* 2016). It is a partly arboreal and omnivorous species that will catch insects, molluscs and vertebrates such as rats, or forage for fruits (including cultivated species such as mango *Mangifera indica* and banana *Musa* sp.) (Duckworth *et al.* 2016). As an adaptable small carnivore, the common palm civet has potential for significant impacts to the native faunas of the Aru Islands, but such impacts are yet to be studied. Although illegal trade and habitat loss are major threats, it is a widely distributed and adaptable species able to inhabit a broad range of habitats.

Paradoxurus hermaphroditus
Native range Afghanistan, Bangladesh, Bhutan, Brunei, Cambodia, China, India, Indonesia, Laos, Malaysia, Myanmar, Nepal, Pakistan, Philippines, Singapore, Thailand, Vietnam
Introduced range (South-west Pacific) ?Aru Isls
Identification This small carnivore can be distinguished from other species in the region by the presence of a dark mask across the eyes, bordered by a white band across the forehead from above the eyes to below the ears. There is usually a small white patch under each eye and on each side of the nostrils. The body is long and stocky with grey, rather coarse hair, and the ears, lower legs and distal half of the tail are black. Three rows of black markings are present to varying degrees along the length of the body.
Conservation status (IUCN) Least Concern

Felidae

FERAL CAT

Felis catus (Linnaeus, 1758)

Feral cats are widely regarded as one of the greatest threats to global island biodiversity. Based on a review of the literature and IUCN Red List accounts, Medina *et al.* (2011) estimated that feral cats have contributed to more than 14% of modern land vertebrate extinctions on islands and more than 8% of declines to Critically Endangered status. Feral cats have similarly been blamed for a host of extinctions and declines among ground-dwelling and predator-naïve species of the south-west Pacific and have been implicated as contributing to the extinction of iconic species such as the Choiseul pigeon (*Microgoura meeki*) (Tennent 2009).

The true extent to which feral cats impact the faunas of the South-west Pacific is probably yet to be fully appreciated. Melanesia, Polynesia and Micronesia are regional gaps in the study of feral cat impacts and very few studies have investigated feral cat diets on the islands under consideration here (Bonnaud *et al.* 2010). Feral cats have been reported from all countries and territories in the South-west Pacific (Duffy and Capece 2012) and, in reality, are probably present on all islands that have been inhabited by people for substantial periods of time.

From a review of cat dietary studies on 40 of the world's islands, Bonnaud *et al.* (2010) concluded that the main prey items were birds and mammals. Moreover, in a global review Oedin *et al.* (2021) demonstrated that most studies documenting predation of bats by cats were from islands. Medina *et al.* (2011) found the impact of cats on island species was greater if introduced prey such as rats were present. The underlying mechanism may be that abundant rats support denser populations of cats, enabling them to hunt native prey to extinction ('hyperpredation'). On Guadalcanal where two endemic rodents (*Uromys imperator* and *U. porculus*) have not been documented since they were first collected in the late 1800s, feral cats have again been implicated as the primary cause for their demise (Flannery 1991a). Using camera traps, Lavery *et al.* (2020c) showed that feral cats were by far the most commonly detected species across multiple sites and elevations in Solomon Islands. Their abundance was correlated with the frequency with which exotic rodents (*Rattus rattus* and *R. exulans*) were detected on the cameras, suggesting these introduced species may indeed play some part in sustaining feral cat populations.

In New Caledonia, scat analysis by Palmas *et al.* (2017) found that feral cats preyed strongly upon squamates, flying foxes and petrels. Introduced rodents were again an extremely important prey item as indicated by the regularity with which they were found in faecal samples.

Feral cats (*Felis catus*) are the greatest threats to global island biodiversity. They have contributed to more than 14% of modern land vertebrate extinctions on islands and are probably responsible for the decline and extinction of several mammals endemic to the South-west Pacific (Angus McNab).

Felis catus
Native range The ancestor of domesticated cats (*Felis silvestris*) has a very broad distribution, throughout most of Africa, Europe, India, China, Mongolia and further south-west and central Asia.
Introduced range (South-west Pacific) Nearly ubiquitous, especially on islands inhabited by humans
Identification Feral cats are highly variable in their colouration. A significant proportion of most populations are usually tabby with characteristic striped, lined, banded or swirled patterns on the body. Black and tortoiseshell individuals are also common.
Conservation status (IUCN) Not Assessed

The question of how to best manage the threat posed by feral cats to island biodiversity of the South-west Pacific is extremely difficult to answer. Few tools are available for effective control of free-ranging feral cat populations, and this problem is exacerbated by the remoteness of many of the region's islands and the limited resources available for conservation. However, the management of feral cats is certainly one of the most important actions to improve the conservation outlook for a substantial proportion of the South-west Pacific's unique fauna.

Canidae

FERAL DOG

Canis familiaris (Linnaeus, 1758)

The introduction of dogs to the South-west Pacific commenced around 3300 years ago, with the earliest archaeological records being from New Ireland in eastern Melanesia (c. 3000–3300 years before present) (Summerhayes *et al.* 2019). Using an analysis of genetic sequences from archaeological deposits, Greig *et al.* (2018) argued their introduction to the islands of Oceania occurred over three dispersal stages from mainland South-east Asia. The first of these saw dogs reach insular Southeast Asia. A second lineage was associated with Austronesian language speakers of the Lapita Cultural Complex (Greig *et al.* 2018). A third widespread group found across the Pacific (including Polynesia) was linked to post-Lapita movements (Greig *et al.* 2018). Archaeological evidence for the arrival of dogs in the Pacific prior to 2000 years before present is sparse (Matisoo-Smith 2007). Dogs were present on Buka by 1860 years before present (Flannery *et al.* 1988). Kirch and Yen (1982) reported a single dog bone from a deposit on Tikopia dated to 2800 years before present. However, the species is absent in the middle of the sequence and recurs from 1200 CE. People introduced Polynesian dogs (kurī) to Aotearoa New Zealand simultaneously with Polynesian rat (*Rattus exulans*) in the 13th century (Wilmshurst *et al.* 2011; Greig and Rawlence 2021).

The phenotypes of dogs in the South-west Pacific are highly variable, and introductions of

Dogs (*Canis familiaris*) are regularly recorded on camera traps in lowland forests. The morphology of this individual recorded on Guadalcanal is somewhat typical for Melanesia. It was detected (apparently unaccompanied) in a remote area, but was recognised as being an animal belonging to someone in the nearest village (Tyrone Lavery).

Canis familiaris
Native range The original ancestor of domesticated dogs is unknown and may have been distributed across parts of North America, Europe and Asia.
Introduced range (South-west Pacific) Dogs are almost ubiquitous among the South-West Pacific islands that are inhabited by humans.
Identification Dogs are highly variable in their appearance in the region, but are typically small-bodied animals, light brown, dark brown, black or tan in colour. The muzzles are often dark, and tails arched with a ventral brush.
Conservation status (IUCN) Not Assessed

additional breeds subsequent to European colonisation have further increased this diversity in more populated regions. In Melanesia, dogs are typically small, slim-bodied animals, commonly light brown, dark brown, black or tan in colour, with dark muzzles and dorsally or laterally arched tails with a ventral brush. They also are frequently pale to white on the ventrum, with white on the undersides of the chin, chest and paws.

Classifying dogs as being either domestic or feral is problematic for much of the Pacific region. There are certainly animals that are truly domestic (completely reliant on humans for food) or truly feral (with no direct interactions with humans). However, a proportion of animals lie somewhere between these two ends of the spectrum. For example, many animals are closely associated with humans, but must supplement their dietary intake by scavenging in natural environments or hunting native wildlife. In Solomon Islands, hunting dogs taken into the forest to target pigs will catch and consume 'by-catch' such as monitor lizards (*Varanus indicus*). Also in Solomon Islands, reports of feral dogs are common (e.g. on Kolombangara Island). However, widespread camera trap surveys on Choiseul, Guadalcanal, Malaita and Kolombangara recorded few individuals, and the dogs that were detected were identified as hunting animals owned by people in nearby villages (Lavery *et al.* 2020b, 2020c). On Tinian Island, Northern Mariana Islands, both feral and domestic dogs have been present since the mid-1700s (Wiles *et al.* 1990). Feral, semi-feral and domestic dogs can impact native faunas through predation, disturbance and competition. In Micronesia, Polynesia and Melanesia such impacts have been associated with declines in at least 24 vertebrate species (Doherty *et al.* 2017).

Diprotodontia

Macropodidae

HUON TREE KANGAROO

Dendrolagus matschiei (Förster and Rothschild, 1907)

This is the only tree kangaroo known to occur in the region. Its native range is on Papua New Guinea's Huon Peninsula, adjacent to Umboi Island. Its presence on Umboi, noted by Koopman (1979), is almost certainly the result of human introduction, as Umboi is an oceanic island which otherwise has a very depauperate terrestrial mammal fauna. Heinsohn (2010b) similarly suggested it had been introduced through long-term human agency, via animals that were carried as pets, as food, or for trade purposes. He also listed New Britain as a possible location where this species had been introduced, as did Groves (2005) and Martin (2005), citing Mt Agulupella, West New Britain, as the specific locality. This needs validation.

The Huon tree kangaroo has been studied in captivity (to which it adapts better than most species of tree-kangaroo) (Flannery 1995a). A radio-telemetry study of this species in the wild showed 90% of recorded daily locations were in the canopy, at an average height of 18–20 m. The remaining 10% of recorded locations were on the ground (Porolak *et al.* 2014). The average home-range size of 15 individuals was 139.6 ha (in an upper montane forest site). The diet in these upper montane forests included a diverse array of plants from at least 20 families, with a preference towards younger shoots and leaves (Porolak 2008).

Little information is available for this species from Umboi, and additional information on the status of this population is needed. Huon tree kangaroo is classified as Endangered and, as with all tree kangaroos, it is a high conservation priority.

Dendrolagus matschiei
Native range Huon Peninsula, Papua New Guinea
Introduced range (South-west Pacific) Umboi, ?New Britain
Identification The Huon tree kangaroo ranges in size from 6.2 to 7.0 kg. It is generally more sombre in colour than *D. goodfellowi*, and lacks the golden back stripes found in that species.
Conservation status (IUCN) Endangered

BLACK-STRIPED WALLABY

Notamacropus dorsalis (Gray, 1837)

The black-striped wallaby was reported to have been introduced to Kawau Island, Aotearoa New Zealand, in the 1870s. However, Wodzicki and Flux (1967) found people on Kawau did not recognise the description of *Notamacropus dorsalis*, and no skulls of this species were found among 118 collected in September 1966. The only convincing account was of a female with joey poisoned in 1954 and identified by R. Kean (Latham and Warburton 2021). The measurements for this specimen clearly fell outside the range of *N. parma* (the only species also introduced to Kawau with which *N. dorsalis* is likely to be confused) (Wodzicki and Flux 1967). The black-striped wallaby apparently never established or is now extinct in Aotearoa New Zealand.

Notamacropus dorsalis
Native range Eastern Australia
Introduced range (South-west Pacific) Aotearoa New Zealand (Extinct)
Identification A medium-sized wallaby with a dark mid-dorsal strip that runs from the neck to the rump. There is a white cheek-stripe and also a horizontal white stripe on the thigh.
Conservation status (IUCN) Least Concern

Black-striped wallaby (*Notamacropus dorsalis*), Durikai State Forest, Queensland, Australia (Mark Sanders).

TAMMAR WALLABY
Notamacropus eugenii (Desmarest, 1817)

Notamacropus eugenii
Native range South Australia and South-western Australia
Introduced range (South-west Pacific) Aotearoa New Zealand (Kawau and North Is.)
Identification The fur is dark and grizzled, the limbs and sides are rufous and the ventrum pale grey-buff.
Conservation Status (IUCN) Least Concern

Tammar wallabies were taken to Aotearoa New Zealand by Sir George Grey who was Governor of South Australia 1841–45 and Governor of New Zealand 1845–53 and 1861–68. The animals were introduced to Kawau Island and later to Rotorua on the North Island sometime around 1870 (Latham and Warburton 2021). The species has continued to spread since its introduction and, assuming even low rates of expansion, Latham *et al.* (2018) predicted the distribution of tammar wallaby could span 50 km^2 across Aotearoa New Zealand's North Island within 50 years.

In Australia, tammar wallaby populations are considered stable, but much rarer than indicated in historical records (Burbidge and Woinarski 2016). They occupy habitats where open grazing grasslands are mixed with denser vegetation that provides shelter. They are herbivores that primarily eat grasses but also graze on shrubs. In Aotearoa New Zealand, the tammar wallaby has become a pest and, although preferring edge habitats, will live solely in native podocarp and mixed hardwood forests, feeding on a variety of native plants (Latham *et al.* 2018).

Using DNA analysis, Taylor and Cooper (1999) confirmed the Kawau Island tammar wallabies were probably derived from a South Australian mainland population that had subsequently become extinct. A group of the Kawau animals were repatriated to South Australia in 2003 and slowly reintroduced to Innes National Park on Yorke Peninsula between 2004 and 2008 (Kemp 2010).

PARMA WALLABY
Notamacropus parma (Waterhouse, 1846)

The Parma wallaby was feared extinct in Australia until 1965 when an introduced population of macropods on Kawau Island, Aotearoa New Zealand, was identified as this species (Wodzicki and Flux 1967). A remnant population was also subsequently rediscovered in Australia in 1967 near Gosford, New South Wales. Parkes and Murphy (2003) indicated the population on Kawau was stable and comprised fewer than 10 000 individuals over an area of 23 km^2.

In Australia, the parma wallaby occurs in wet forests with a thick understorey, dry forests and rainforests. It is mainly solitary and nocturnal, emerging after dusk to feed on a variety of grasses and herbs (Maynes 2008). Kawau Island was once

Tammar wallabies were taken to Aotearoa New Zealand by Sir George Grey and conservative estimates of population expansion suggest the distribution could span across 50 km^2 of Aotearoa New Zealand's North Island within 50 years (Mark Sanders).

The Parma wallaby (*Notamacropus parma*) is listed as Near Threatened in Australia; an introduced population survives on Kawau Island, Aotearoa New Zealand (Mark Sanders).

Notamacropus parma
Native range Eastern Australia
Introduced range (South-west Pacific) Aotearoa New Zealand (Kawau Is.)
Identification A small species of wallaby with brown to reddish-brown dorsal fur and pale grey to cream-coloured throat, chest and belly. There is a white stripe on the muzzle, a white margin at the base of the ear and an indistinct dark mid-dorsal stripe from the top of the head to the mid-back.
Conservation status (IUCN) Near Threatened

Notamacropus rufogriseus
Native range Eastern Australia
Introduced range (South-west Pacific) Aotearoa New Zealand (South Is.)
Identification This is the largest of the introduced macropods found in Aotearoa New Zealand. The fur of the shoulders and neck is tinged rufous, the hindfeet, paws, tail and tips of ears are black-tipped and there is a paler cheek-stripe and sometimes a dark stipe between the ears.
Conservation status (IUCN) Least Concern

covered with diverse forest that was extensively cleared. The Parma wallaby occurs there with other species of wallaby and the brushtail possum (*Trichosurus vulpecula*), so little is reported of the direct ecological impacts of this species (Latham and Warburton 2021).

RED-NECKED WALLABY

Notamacropus rufogriseus (Desmarest, 1817)

The red-necked wallabies found in Aotearoa New Zealand originate from Tasmania and Bass Strait islands and were first liberated at Christchurch in 1870 (Wodzicki and Flux 1967; Latham and Warburton 2021). In 1874, one male and two females from Christchurch were liberated on Te Waimate Station in the eastern Hunter Hills and by 1916 these had increased to number in the thousands. By 1947 red-necked wallabies had increased to such an alarming extent that between 1947 to 1956 control was implemented and over 68 000 individuals were removed (Wodzicki and Flux 1967). However, this reportedly did not even negate the natural rate of reproduction. Today, the known breeding populations of red-necked wallabies occupy some 5322 km^2 of the South Island and, assuming low rates of spread, the distribution is predicted to expand by 550 km^2 by 2065 (Latham *et al.* 2018).

Red-necked wallaby (*Notamacropus rufogriseus*), Barrington Tops, New South Wales, Australia. Individuals of this species from Tasmania and Bass Strait were released at Christchurch in 1870 and the species distribution expanded quickly across Aotearoa New Zealand's South Island (Mark Sanders).

The red-necked wallaby is primarily a grazer but will also browse palatable shrubs and trees. This can contribute to the depletion of forest understoreys in Aotearoa New Zealand and prevent the regeneration of palatable species (Warburton 2005). The conservation status of the red-necked wallaby is secure in its natural distribution in Australia.

BRUSH-TAILED ROCK-WALLABY

Petrogale penicillata (Gray, 1827)

The brush-tailed rock-wallaby was introduced to several of Aotearoa New Zealand's islands in the

Brush-tailed rock-wallabies (*Petrogale penicillata*) captured near Sydney in the late 19th century were introduced to several locations in Aotearoa New Zealand (Angus McNab).

Petrogale penicillata
Native range Eastern Australia
Introduced range (South-west Pacific) Aotearoa New Zealand (Kawau Is.)
Identification A dark-brown wallaby with a long-tail tipped with a prominent brush of hair. The feet, paws and tips of ears are black, and there is a pale cheek stripe and black dorsal stripe from the forehead to behind the ears.
Conservation status (IUCN) Vulnerable

early 1870s. Animals released on Motutapu spread across a narrow neck of land to reach adjoining Rangitoto Island, and by 1912 were reported as being numerous (Wodzicki and Flux 1967). The species was also released onto Kawau Island. Using genetic data, (Eldridge *et al.* 2001) established that these Aotearoa New Zealand populations were probably founded by animals captured near Sydney in the late 19th century. Brush-tailed rock-wallaby was also liberated on Oahu, Hawaii in 1912.

Brush-tailed rock-wallaby is a species that inhabits rugged rocky outcrops in eastern Australia. The species has undergone widespread population declines and extinctions, and the naturalised Aotearoa New Zealand population remaining on Kawau thus represents a potentially valuable conservation resource for the species.

The swamp wallaby (*Wallabia bicolor*) was presumably introduced to Aotearoa New Zealand's Kawau Island by Sir George Grey sometime in 1870 (Angus McNab).

SWAMP WALLABY
Wallabia bicolor (Desmarest, 1804)

The swamp wallaby was presumably introduced to Kawau Island by Sir George Grey sometime in 1870 (Wodzicki and Flux 1967). In Australia, the swamp wallaby usually occurs in forest types with denser canopy layers and understoreys. Similarly, early reports of habitat preference on Kawau suggested this species preferred scrubby vegetation and was lower in abundance than *Notamacropus eugenii* and *N. parma*, which were also introduced to the island (Wodzicki and Flux 1967). Little information is available on the ecology on current population status of this species on Kawau. In Australia, the swamp wallaby is relatively common and widespread, and it is classified as Least Concern on the IUCN Red List.

Wallabia bicolor
Native range Eastern Australia
Introduced range (South-west Pacific) Aotearoa New Zealand (Kawau Is.)
Identification A large and dark-coloured wallaby, with a paler rufous-orange to yellow chest and belly.
Conservation Status (IUCN) Least Concern

Phalangeridae

COMMON BRUSHTAIL POSSUM
Trichosurus vulpecula (Kerr, 1792)

Common brushtail possums were introduced to Aotearoa New Zealand from Australia in 1858 to establish a fur industry (Pracy 1974). Two subspecies with different phenotypic forms were introduced, *Trichosurus vulpecula vulpecula* (from south-eastern Australia) and *T. v. fuliginosus* (from Tasmania) (Pattabiraman et al. 2022). Despite their presence in Aotearoa New Zealand for over 100 years and

Common brushtail possum (*Trichosurus vulpecula*), Cradle Mountain, Tasmania, Australia. This species was introduced to Aotearoa New Zealand to establish a fur industry and has since caused widespread damage to native flora and fauna (Mark Sanders).

Trichosurus vulpecula
Native range Australia
Introduced range (South-west Pacific) Aotearoa New Zealand (North and South Isls)
Identification A thickly furred body with a black, bushy, prehensile tail. The underside of the tail is lightly furred. The eyes are large, brown and protruding; the ears are pointed and hairless on the inside. The rhinarium is pink, and the hindfeet have a pair of fused toes.
Conservation status (IUCN) Least Concern

interbreeding between populations, two main colour phenotypes corresponding to each subspecies remain today (Pattabiraman et al. 2022).

Common brushtail possum populations expanded rapidly, and the species is now ubiquitous throughout Aotearoa New Zealand's main islands. Common brushtail possums reach high densities and can cause dieback in forests dominated by preferred food trees (e.g. *Metrosideros* or *Weinmannia*) (Cowan and Glen 2021). Dieback typically occurs 15–25 years after an area is colonised, when populations reach an unsustainable peak before dropping back to lower levels (Batcheler 1983). More broadly, the ability of forests to regenerate can be reduced by common brushtail possums because fewer flowers, fruit and seeds are produced (Cowan and Glen 2021).

Common brushtail possums compete with, and directly prey on, Aotearoa New Zealand's native fauna. They compete with native birds for the same resources, including food, and nest sites required by hollow-nesting species (Cowan and Glen 2021). They also eat the eggs or kill the nestlings and adults of a range of native birds, including the iconic brown kiwi (*Apteryx australis mantelli*). Cowan and Moeed (1987) found half of faecal samples examined in the Orongorongo Valley contained the remains of invertebrates. This predation pressure represents a serious threat to easily detected, slow-moving species such as Aotearoa New Zealand's large native land snails (many of which are rare, e.g. *Wainuia*, *Powelliphanta* and *Placostylus*) (Cowan and Glen 2021).

Erinaceomorpha

Erinaceidae

EUROPEAN HEDGEHOG

Erinaceus europaeus (Linnaeus, 1758)

European hedgehogs were deliberately introduced to Aotearoa New Zealand in the late 19th century (Jones 2021). The earliest importations were prompted by little more than sentiment; however, Brockie (2010) revealed later introductions were justified as a control measure for slugs and snails that also had been inadvertently introduced and become pests. By the 1940s and 1950s European hedgehogs had colonised most lowland areas of the South and North Islands (Brockie 2010).

European hedgehog (*Erinaceus europaeus*), southern Sweden (Anders Zimny).

Erinaceus europaeus
Native range Austria, Belgium, Croatia, Czechia, Denmark, Estonia, Finland, France, Germany, Ireland, Italy, Latvia, Luxembourg, Netherlands, Norway, Poland, Portugal, Russia, Slovenia, Spain, Sweden, Switzerland, United Kingdom
Introduced range (South-west Pacific) Aotearoa New Zealand (North and South Isls)
Identification A small round body shape, with short legs, and reduced neck and tail. The dorsal coat that consists of short brown and white spines makes this species readily identifiable.
Conservation status (IUCN) Least Concern

European hedgehogs are omnivorous, and primarily consume invertebrates. A study of gut contents from animals captured in the upper Waitaki Basin found the most commonly eaten prey were Coleoptera, Lepidoptera, Dermaptera, Hymenoptera and Orthoptera (Jones *et al.* 2005). They probably have major impacts on rarer invertebrates. Wētā (*Hemiandrus* spp.) remains occurred in 22% of samples, and the gut contents of one adult male alone contained 283 wētā legs. As predators, they severely affect the recruitment and population recovery of native snails (*Wainuia urnula*). Eggshells and lizard remains have also commonly been found in gut contents (Jones *et al.* 2005). Predation of eggs and chicks is a significant problem for ground-nesting birds such as the double-banded plover (*Charadrius bicinctus*).

Lagomorpha

Leporidae

EUROPEAN HARE

Lepus europaeus (Pallas, 1778)

The first successful release of European hares to Aotearoa New Zealand was derived from England via Phillip Island in Victoria, Australia (Norbury and Flux 2021). By 1876 this species had been liberated to at least six sites and had begun to cause widespread damage (Norbury and Flux 2021). European hares initially spread rapidly in Aotearoa New Zealand, reaching a peak in abundance around 1903, before subsequently declining for unknown reasons (Norbury and Flux 2021).

European hares frequently co-occur with European rabbits and domestic sheep, so their specific impacts on Aotearoa New Zealand's ecology are difficult to pinpoint. They generally occur at lower densities than European rabbits and do not construct burrows (Norbury and Flux 2021). In alpine vegetation communities, ponds and wetlands, they can damage sensitive native vegetation and prevent the regeneration of native plants. However, in certain scenarios, European hares may help maintain the biodiversity of indigenous plants and associated invertebrates by suppressing introduced weeds and exotic grasses.

The European hare (*Lepus europaeus*) was introduced to Aotearoa New Zealand and quickly increased in number, reaching peak abundance in the early 1900s before subsequently declining (Caroline Legg).

Lepus europaeus
Native range Albania, Austria, Belarus, Belgium, Bosnia and Herzegovina, Bulgaria, Croatia, Czechia, Denmark, Estonia, Finland, France, Germany, Greece, Hungary, Iran, Iraq, Israel, Italy, Latvia, Liechtenstein, Lithuania, Luxembourg, Moldova, Montenegro, Netherlands, North Macedonia, Poland, Romania, Russia, Serbia, Slovakia, Slovenia, Spain, Switzerland, Syrian Arab Republic, Turkey, Ukraine
Introduced range (South-west Pacific) Aotearoa New Zealand (North and South Isls)
Identification A large yellow-brown lagomorph with long, black-tipped ears and a tail that is black above and white below. The eyes have a golden iris, rather than a brown iris as found in European rabbit (*Oryctolagus cuniculus*).
Conservation status (IUCN) Least Concern

EUROPEAN RABBIT

Oryctolagus cuniculus (Linnaeus, 1758)

European rabbits were successfully introduced to Aotearoa New Zealand and French Polynesia (Manui Island). The species has now been successfully eradicated from Manui in response to overgrazing, soil erosion and the threat posed to native biotas (Thibault and Bretagnolle 1999; Griffiths *et al.* 2019). They were also introduced to Rawaki, Kiribati, in the 19th century, causing losses of vegetation, erosion, and the loss of eggs and chicks of seabirds such as petrels and shearwaters. This population has also been successfully exterminated (Pierce et al. 2011). The first European rabbits introduced to Aotearoa New Zealand were a domestic breed released onto Motuara Island in Queen Charlotte Sound by Captain James Cook in 1777 (Norbury and Duckworth 2021). Domestic and wild individuals were subsequently introduced to numerous locations throughout the country in an effort to establish the species, but the specific time at which feral populations took hold is uncertain (Norbury and Duckworth 2021). European rabbits quickly reached plague proportions, bringing major environmental and economic impacts for Aotearoa New Zealand.

European rabbits can have significant negative impacts on native flora and their grazing can lead to declines in more palatable and favoured plant species. Their high-intensity, selective grazing will eventually lead to changes in plant species composition, whereby vegetation communities are dominated by less palatable or non-preferred plants, and preferred plants are rare (Norbury and Duckworth 2021). Heavy grazing also increases the opportunity for invasive weeds to become established. The extensive warren systems dug by European rabbits cause and exacerbate soil erosion, leading to further degradation of vegetation communities and deteriorations in water quality. European rabbits probably also cause secondary impacts on Aotearoa New Zealand's native fauna via 'hyperpredation' by supporting higher densities of introduced predators (such as feral cats *Felis catus*, and ferrets *Mustela furo*) (Norbury and McGlinchy 1996).

Oryctolagus cuniculus
Native range France, Portugal, Spain
Introduced range (South-west Pacific) Aotearoa New Zealand (North and South Isls)
Identification Grey-brown dorsally with a white belly and short white fluffy tail. Differs from the European hare in being smaller, possessing ears that are edged with black rather than tipped with black, and in having a brown rather than a yellow iris.
Conservation status (IUCN) Endangered

European rabbits (*Oryctolagus cuniculus*) introduced to Aotearoa New Zealand quickly reached plague proportions and caused immense environmental and economic impact (Angus McNab).

The European rabbit has become a major pest in many parts of the world where it has been introduced. However, in its native range in Spain, Portugal and north-western Africa, the species is in fact threatened by habitat loss, disease and hunting.

Primates

Cercopithecidae

COMMON LONG-TAILED MACAQUE

***Macaca fascicularis* (Raffles, 1821)**

The common long-tailed macaque has been introduced to numerous parts of the world, including Angaur, Palau, in the early 1900s. It is a semi-terrestrial and omnivorous species that tolerates an extremely wide range of habitats, including agricultural areas and a variety of forest types.

The population on Anguar is reportedly derived from a pair of animals brought to the island by German phosphate miners (Poirier and Smith 1974). The species has become a serious pest. The monkeys prey on native wildlife including crabs and birds, and destroy farm crops and gardens in Angaur, such as taro (*Colocasia esculenta*) (McGregor and Bishop 2011). In fact, Angaur has Palau's lowest recorded bird diversity, a status that has been blamed on the presence of the common long-tailed macaques, as well as introduced rodents (Ketebengang and Gupta 2011). In 1972 a national law was passed in Palau prohibiting transport of macaques outside of Angaur. However, enforcement is lacking and captive common long-tailed macaques are regularly transported as pets to other islands, including Koror where the majority of Palau's residents live. There is significant risk of the establishment of populations on other islands.

The common long-tailed macaque is listed as Vulnerable on the IUCN Red List and is threatened by excessive hunting, trafficking and persecution. However, it is tolerant of a wide variety of habitats and habitat change, and has thus proven successful at establishing introduced populations under the right conditions.

Macaca fascicularis
Native range Burma, Thailand, Cambodia, Laos, Malaysia, Philippines, Indonesia (Sumatra, Borneo, Java, Bali, and eastern Lesser Sunda Isls), Vietnam
Introduced range (South-west Pacific) Angaur, Palau
Identification The upperparts are dark brown, underparts light grey and the tail is dark grey-brown. Skin of the feet and ears is black, and the muzzle is light greyish-pink. Eyelids often have prominent white markings.
Conservation status (IUCN) Vulnerable

Common long-tailed macaque (*Macaca fascicularis*), Ubud, Bali, Indonesia. This species has become a serious pest in Angaur, Palau, where it preys on native wildlife and destroys farm crops and gardens (Shankar S.).

Rodentia

Muridae

HOUSE MOUSE

Mus musculus (Linnaeus, 1758)

Mus musculus
Native range *Mus musculus* is a Palearctic species. It has been so widely introduced that it now occurs across the globe on all continents except for Antarctica, and many islands.
Introduced range (South-west Pacific) American Samoa, Caroline Isls, Cook Isls, Fiji, French Polynesia, Guam, Loyalty Isls, Nauru, New Caledonia, Aotearoa New Zealand (North and South Isls), Northern Mariana Isls, Palau, Papua New Guinea, Pitcairn, Solomon Isls, Tonga, Tuvalu, Vanuatu
Identification A small furry rodent, with a long thin tail that is about the same length as the body. Long whiskers. Grey-brown above, with white, grey or brown belly. Female has 10–12 nipples on the belly.
Conservation status (IUCN) Least Concern

The house mouse is a widespread and successful invasive mammal. It is typically a commensal species found in a range of human dwellings and is widespread in towns throughout the South-west Pacific. It will sometimes expand to establish feral populations such as in the temperate forests and subalpine tussocks of Aotearoa New Zealand (Murphy and Nathan 2021). However, the house mouse is rarely found among the tropical native forests of the region.

The house mouse was first recorded in Aotearoa New Zealand when the *Elizabeth Henrietta* ran ashore on Ruapuke Island off Stewart Island in 1824 (Murphy and Nathan 2021). Tate and Archbold (1935b) reported it as present in the Caroline Islands, New Caledonia, Loyalty Islands, Solomon Islands and Vanuatu. Barbehenn (1974) reported a decline on Guam between 1958 and 1969 following introduction of Asian house shrew (*Suncus murinus*).

The effects of house mice on island ecosystems are less well understood compared to those of introduced rats (*Rattus* spp.). It is an omnivorous species and has a flexible diet to take advantage of variable food sources. This broad diet means impacts can be equally broad, especially when occurring at high densities. In Aotearoa New Zealand selective predation of seeds can prevent forest regeneration and alter species composition (Murphy and Nathan 2021). House mice are also a key food source for stoats and feral cats and, during times of food abundance, population booms can support high densities of predators that in turn threaten native fauna (Murphy and Nathan 2021).

The house mouse (*Mus musculus*) is a common commensal species that has now been introduced to most inhabited Pacific islands (M. Berroneau).

POLYNESIAN RAT

Rattus exulans (Peale, 1848)

The Polynesian rat is a widespread, natural element of the mainland and insular faunas of South-east Asia. It was prehistorically introduced across much of the South-west Pacific, being absent from near Oceania before the Holocene. Analyses of genetic data have shown that at least two populations of Polynesian rat were introduced into Oceania (Matisoo-Smith and Robins 2004). By 1860 years before present, it had been introduced to Buka Island in the Solomon Archipelago (Flannery and Wickler 1990), and is estimated to have arrived in Aotearoa New Zealand in the 13th century along with humans (Wilmshurst *et al.* 2011). The Polynesian rat occurs on a vast number of islands in the South-west Pacific (probably more than any other introduced rat species) (Matisoo-Smith and Robins 2004). For example, in Marshall Islands, Polynesian rats were found on 30 of 35 atolls, while highly invasive black rats (*Rattus rattus*) were found on only nine (Spennemann 1997). The diet is omnivorous and includes the mycorrhizal fungi *Rhizophagus clarus* and *R. fasciculatum* (American Samoa, Adler *et al.* 2018), insect exoskeletons, and fungal spores possibly from *Hysterangium* sp. and a member of Cortinariaceae (Goodenough Island, Elliott and Vernes 2021). Studies from Aotearoa New Zealand have shown the Polynesian rat to be a generalist forager from the subsoil into forest canopies, and including insects, spiders, earthworms, landsnails, lizards, flowers, leaves, buds, bark, fruit, and shoot apices in its diet (Wilmshurst and Ruscoe 2021).

In pre-European times, the Polynesian rat was apparently abundant throughout Aotearoa New Zealand. But following European colonisation, further mammal introductions (including *R. rattus*, *R. norvegicus* and *Mus musculus*) and land use changes led to declines (Wilmshurst and Ruscoe 2021). Today, the Polynesian rat is more or less restricted to Aotearoa New Zealand's smaller offshore islands (Taylor 2010).

The morphology of this species varies with elevation on larger tropical islands. In lowland forests they are larger, lighter in colour, and with coarse, spiny dorsal fur. In higher elevation cloud forests, they are smaller, often melanistic, and with much finer, softer fur. Similar patterns are evident in the Philippines (L. Heaney, pers. comm.). The exact reasons for this variation are unknown but may be related to Gloger's rule, which predicts darker coloured animals in more humid/vegetated environments (Delhey 2017).

The introduction of Polynesian rats has caused a range of detrimental impacts for Pacific Island faunas. They are known to affect tuatara (*Sphenodon punctatus*) via competition and predation of eggs and juveniles, and they predate the eggs and hatchlings of green turtle (*Chelonia mydas*) and hawksbill turtle (*Eretmochelys imbricata*) (Meier and Varnham 2004). Robinet *et al.* (1998) suggested that on Wallis and Futuna Polynesian rats have had some limited impact on threatened Ouvea parakeets (*Eunymphicus uvaeensis*) via egg predation. Plants make up a large part of the diet and impacts on a range of species via predation of flowers, fruits and seeds have been reported (Harper and Bunbury 2015).

The Polynesian rat is present in a wide variety of habitats, including disturbed or agricultural lands,

Polynesian rat (*Rattus exulans*) Kolombangara Island, Solomon Islands. This species was prehistorically introduced widely throughout the Pacific and is very common in primary forest interiors of some islands, especially where native rodents are rare or absent (Tyrone Lavery).

Rattus exulans
Native range Bangladesh, Cambodia, Indonesia, Laos, Malaysia, Myanmar, Thailand, Vietnam
Introduced range (South-west Pacific) American Samoa, Cook Isls, Fiji, French Polynesia, Guam, Kiribati, Marshall Isls, Federated States of Micronesia, Nauru, New Caledonia, Aotearoa New Zealand, Niue, Northern Mariana Islands, Palau, Papua New Guinea, Samoa, Solomon Isls, Tokelau, Tonga, Tuvalu, Vanuatu, Wallis and Futuna
Identification The dorsal fur is brown, the belly is grey with white tips, and the feet are pale with distinctive dark bands on the outer edges of the hindfeet. The ears are long and cover the eyes when pulled forward; the tail is about the same length as body. Females have eight nipples.
Conservation status (IUCN) Least Concern

secondary or disturbed forests, and primary lowland and montane forests. Its prevalence in native forests varies and is probably related to the presence or absence of native rodent faunas.

HIMALAYAN FIELD RAT

Rattus nitidus (Hodgson, 1845)

Barbehenn (1974) noted that 11 specimens of the Himalayan field rat, held in the collections of the United States National Museum of Natural History, Smithsonian Institution, were collected on Peleliu in 1945. A further two specimens were collected by Enders at Ngetbang, Babeldoab, in 1949.

The Himalayan field rat is primarily a terrestrial rodent that is active by day and night. In South-east Asia it is variously recorded in forested habitats, croplands, and as a house rat in upland villages (Burgin 2017). In southern China, it can be a major agricultural pest, and the species is known to consume crops such as wheat, corn, rice and potatoes. Little has been reported of the ecology or impacts of this species on Palau. In fact, these is no information available relating to the species on Palau since Barbehenn (1974), and it is possible it is no longer extant there.

Rattus nitidus
Native range Bhutan, China, India, Myanmar, Nepal, Thailand, Vietnam
Introduced range (South-west Pacific) Palau
Identification The dorsal fur is short, soft and dark greyish brown; the ventrum is dull grey and not sharply demarcated from the dorsum. Tail length is approximately equal to, or shorter than head–body length and uniformly coloured, and the dorsal surfaces of the hands and feet are white. Females have six pairs of mammae.
Conservation status (IUCN) Least Concern

BROWN RAT

Rattus norvegicus (Berkenhout, 1769)

The brown rat is geographically widespread in the South-west Pacific, but it is a commensal species that normally does not inhabit native forests. It is thus patchily distributed, and normally found only in the densest of urban centres. For example, in Marshall Islands, Polynesian rats (*Rattus exulans*) were recorded on 30 of 35 atolls, black rats (*R. rattus*) on nine, and brown rats (*R. norvegicus*) on only two atolls (Spennemann 1997). In a review of the global insular distribution of *Rattus* spp., atolls or cays made up 19% of the 42 recorded island locations for the brown rat (Harper and Bunbury 2015). In Solomon Islands the brown rat is common along Honiara's Mataniko River, capitals of the Western Province (Gizo marketplace), and Malaita (Auki), but has only once been recorded outside urban centres (in East Kwaio, Malaita – the circumstances of which are uncertain).

Brown rats live in New Caledonia's coastal capital Nouméa and it is possible the distribution extends to coastal reserves where mangroves or coconut plantations may offer a suitable habitat (Rouys and Theuerkauf 2003). Ekstrom *et al.* (2000) reported brown rats from nature reserves of New Caledonia, but Rouys and Theuerkauf (2003) presumed the authors had misidentified *R. rattus* and suggested the occurrence of brown rats in nature reserves is unlikely.

There are very few documented instances of brown rats impacting native island flora or fauna (Harper and Bunbury 2015). This may be because black rats were often introduced before brown rats through much of the region, preventing the

Brown rat (*Rattus norvegicus*) (Craig Lindsay).

Rattus norvegicus
Native range China, Japan, Russia
Introduced range (South-west Pacific) Marshall Isls, New Caledonia, Aotearoa New Zealand, Solomon Isls
Identification A large-bodied *Rattus* with a tail that is clearly shorter than head–body length and short ears that do not fold over the eye when pressed forward.
Conservation status (IUCN) Least Concern

establishment of brown rats in native habitats and restricting them to disturbed areas where native biotas are already reduced or absent (Harper and Bunbury 2015).

BLACK RAT

Rattus rattus (Linnaeus, 1758)

Rattus rattus
Native range India, Pakistan
Introduced range (South-west Pacific) Nearly ubiquitous
Identification Ears are large, covering the eyes when folded forward, tail is long (approximately 1.5 times head–body length, and the tail scales form overlapping rings. Distinguishing *R. tanezumi* from *R. rattus* using morphological features is almost impossible.
Conservation status (IUCN) Least Concern

The taxonomy of *Rattus rattus* is complex. Six main lineages in the *R. rattus* clade have been recognised as four distinct species (*R. rattus*, *R. sakeratensis*, *R. tanezumi* and *R. tiomanicus*) (Aplin *et al.* 2011). The precise distributions of *R. rattus* and *R. tanezumi* are poorly resolved due to a lack of comprehensive sampling and genetic sequencing across the South-west Pacific. However, globally, lineage I (*R. rattus*) is most widespread and is presumed to be almost ubiquitous among the Pacific Islands.

Few invasive species have proven more successful or problematic than black rats. It is a species that is extremely adaptable, and will readily invade inhabited or uninhabited islands alike, occupying native forests, villages and towns, and agricultural areas. In the Pacific, black rats predominantly feed on fruits, seeds and other vegetative material, as well as insects (Shiels *et al.* 2014). However, they also represent a major threat to a variety of island vertebrates. Through predation of eggs, chicks and adults, black rats are a primary threat to numerous Pacific Island birds, particularly seabirds and narrow-range endemics naïve to mammalian predators (Shiels *et al.* 2014). They have also been reported feeding on green sea turtle hatchlings (*Chelonia mydas*), and other reptiles. On Big South Cape Taukihepa Island, Aotearoa New Zealand, the greater New Zealand short-tailed bat (*Mystacina robusta*) and the lesser New Zealand short-tailed bat (*M. tuberculata*) both went extinct following the introduction of *R. rattus* (Bell *et al.* 2016). However, direct observation of bats in the diets of Pacific Island *R. rattus* are lacking. The impacts of black rats on native rodents of the South-west Pacific (e.g. through competition and introduction of disease) also are unknown.

The black rat (*Rattus rattus*) is one of the species most commonly recorded on camera traps baited with coconut in Melanesia (Tyrone Lavery).

ORIENTAL HOUSE RAT

Rattus tanezumi (Temminck, 1844)

Rattus tanezumi is a member of the *R. rattus* complex, comprising six lineages recognised by four distinct species (*R. rattus*, *R. sakeratensis*, *R. tanezumi* and *R. tiomanicus*) (Aplin *et al.* 2011). Within this clade, lineages II, III and IV correspond to *R. tanezumi*. Without comprehensive sampling and genetic sequencing of *Rattus* from across the region, the precise distribution of *R. tanezumi* in the South-west Pacific is difficult to determine. Lineage II is most widespread and has been detected in Micronesia, Fiji and New Guinea (Aplin *et al.* 2011).

Rattus tanezumi
Native range Afghanistan, Bangladesh, Bhutan, Cambodia, China, India, Japan, North Korea, South Korea, Laos, Malaysia, Nepal, Taiwan, Thailand, Vietnam
Introduced range (South-west Pacific) Federated States of Micronesia, Fiji, Guam, Marshall Isls, Northern Mariana Isls, Palau
Identification Ears are large, covering the eyes when folded forward, tail is long (approximately 1.5 times head–body length, and the tail scales form overlapping rings. Distinguishing *R. tanezumi* from *R. rattus* using morphological features is almost impossible.
Conservation status (IUCN) Least Concern

The oriental house rat is a highly adaptable, commensal species that is most commonly found in villages and agricultural areas, but also occurs in primary and secondary native forests. It is highly arboreal and omnivorous. *Rattus rattus* is widely regarded as one of the most problematic of introduced species for island ecosystems. Nonetheless, on islands where *R. tanezumi* has been introduced, it most likely causes similar issues for native faunas as does *R. rattus*. *Rattus tanezumi* is apparently less widely distributed in the Pacific and for this reason alone is presumed unlikely to present the same scale of threat as *R. rattus*.

Soricomorpha

Soricidae

JAVANESE SHREW

Crocidura maxi (Sody, 1936)

The identification of the Javanese shrew on Wokam, Aru Islands, by Kitchener *et al.* (1994) represented the first occasion a species of shrew had been reported from an island on the Australian continental shelf. A male and two females were collected by hand in *Imperata cylindrica* grassland behind dense mangroves and adjoining a garden with sweet potatoes (*Ipomoea batatas*). The females were both recently pregnant, and one was found within a small nest made from grass. The presence of this species in the Aru Islands is in sympatry with a diverse native mammal fauna including dasyurids such as *Sminthopsis virginiae*.

In its native range, the Javanese shrew occurs in lowland and montane forest, but it is a poorly known species and little else has been reported of its ecology. The life history, population size and potential impacts of this species in the Aru Islands is unknown. Javanese shrew is listed as Least Concern on the IUCN Red List as it has a relatively wide distribution and a presumed large population.

Crocidura maxi
Native range Indonesia (Java to Timor), Timor-Leste
Introduced range (South-west Pacific) Aru Isls
Identification The snout is pointed and short, and the eyes small. The tail is relatively thick with and has moderately dense bristles along the proximal one-third of its length. The feet, tails and pelage are dark brown to black.
Conservation status (IUCN) Least Concern

ASIAN HOUSE SHREW

Suncus murinus (Linnaeus, 1766)

The Asian house shrew is a human commensal species that occupies a very wide range of habitats, including forests, grasslands, plantations, pasture, cultivated fields, and suburban and urban areas (Hutterer *et al.* 2016). It was first reported from Guam in 1953 and was probably introduced there from the Philippines (Peterson 1956a). Asian house shrew spread rapidly and by the end of the second year was found over most of the northern three-quarters of the island. By 1958 its distribution spanned the entire island including remote forested areas (Fritts and Rodda 1998). This species has spread to the Northern Mariana Islands, Palau, Chuuk and Marshall Islands, and has become locally abundant in some locations (e.g. Wiles *et al.* 1990).

The diet is omnivorous, including plant material, invertebrates, human food waste and vertebrates. Fritts and Rodda (1998) reported two of Guam's terrestrial lizards (pelagic gecko *Nactus pelagicus* and Mariana skink *Emoia slevini*) disappeared following introduction of the Asian house shrew.

Suncus murinus
Native range Afghanistan, Bangladesh, Bhutan, Brunei Darussalam, Cambodia, China, India, Indonesia, Laos, Malaysia, Myanmar, Nepal, Pakistan, Singapore, Sri Lanka, Taiwan, Thailand, Vietnam
Introduced range (South-west Pacific) Aru Isls, Federated States of Miconesia, Guam, Marshall Isls, Northern Mariana Isls, Palau
Identification The fur is short and velvety, ranging in colour from light grey-brown to black. The snout is pointed and short, and the tail thick and almost hairless, with few sparse hairs. The ears are rounded and forward facing, and the eyes small.
Conservation status (IUCN) Least Concern

Glossary

Accession The formal act of legally accepting an object or objects to a museum's or institution's permanent collection.

Allopatric Refers to species that have geographical distributions which do not overlap (see also **Sympatric**).

Anastomosing Cross-connecting, as in grooves in the enamel of teeth or veins in the wings of some bats.

Antitragus/Antritragal A flap-like structure on the ears of some bats, which lies in the position occupied by the earlobe in humans.

Apical At or near the tip of a structure.

Arboreal Living primarily in the trees (see also **Terrestrial**).

Arthropods Invertebrate animals with an exoskeleton, a segmented body, and paired jointed legs (e.g. insects, spiders, centipedes, millipedes and crustaceans).

Bract A small leaf with a relatively undeveloped blade, from the axil of which arises a flower or branch of an **Inflorescence.**

Browser An animal that feeds on the leaves and twigs of shrubs rather than grasses. See also **Graze**r.

Caecum A blindly ending branch of the first part of the large intestine.

Calcar A cartilaginous spur arising from the ankle in some bats.

Carnivorous Flesh-eating, feeding on other animals (see also **Insectivore**)

Cauliflorous Producing flowers on older, thickened branches, trunk or base of roots.

cf. Abbreviation for confer (L.), consult; usually prefixed to a species name tentatively applied to a specimen or form of uncertain identification; for example, *Pogonomys* cf. *macrourus* refers to a specimen or population of *Pogonomys* comparable to *macrourus* yet sufficiently different to suggest it might be an unnamed species requiring taxonomic clarification.

CITES Abbreviation for the Convention on International Trade in Endangered Species. This convention regulates the trade in wildlife species (and their products) between signatories, especially those that are or could be significantly affected by trade.

Climax In reference to vegetation, the final stage of development of a forest. See also **Primary forest** and **Secondary forest**.

Cloaca In vertebrates, the single posterior opening for the digestive tract and urinary and female reproductive ducts.

Crepuscular Active during twilight hours at dawn and dusk. See also **Diurnal** and **Nocturnal**.

Cryptic Hidden or difficult to find. A cryptic species can also be one that is not obviously different from other closely related species.

Cuspidate Refers to teeth with many sharp cusps or points.

Distal Towards or at the free end of a structure such as a limb. Opposite of **Proximal**.

Distribution The overall geographical area known to encompass all individuals of a given species.

Disturbed forest Forest which is regrowing after disturbance, and is yet to reach climax.

Diurnal Active by day. See also **Crepuscular** and **Nocturnal**.

DNA Deoxyribonucleic acid. The material of inheritance used in one method to estimate evolutionary relationships between species.

Dorsal On or near the back.

Dorsum The back or upperparts.

Echolocation The method of navigation and food location using reflected sound or ultrasound to determine the location of objects.

Elliott trap A custom-made, rectangular aluminium box trap with a spring door that shuts after an animal has entered. Used widely in Australia to catch small mammals such as murids and dasyurids, but other animals such as frogs, lizards and crabs are also often caught. They can be set on the ground or in trees.

Endemic Unique to the place being discussed. Compare with **Indigenous**.

Epaulette A decoration worn on the shoulders of certain uniforms, used to refer to the tufts of hair that surround scent glands on the shoulders of some bat species.

Epiphyte A plant attached to another plant and usually without roots growing into the soil.

Exoskeleton The tough external covering of an invertebrate animal such as an insect.

Extralimital Outside the limits of the region under consideration. In the context of the mammal fauna of the South-west Pacific, we refer to the extralimital distributions of species that occur outside the geographical region of Melanesia, Micronesia and Polynesia.

Exudate A substance that has been exuded. Usage here is usually in reference to the gums that can seep from certain trees, especially where parts of the tree have been damaged.

Flight membrane An expanse of skin that in bats is stretched between the limbs, hands and body and used to facilitate flight.

Folivore A species that feeds on the leaves of trees.

Frugivore A species that feeds on fruits.

Gardens Areas set aside for the cultivation of subsistence plant foods. Gardens can be within the village environs or as much as a day's walking distance away.

Gracile Of light build, small, slender.

Grazer An animal that eats grasses or herbs.

Guard hair A type of hair that is longer and thicker than other hairs that form the body fur of mammals.

Habitat A place where an animal makes its home, incorporating the components required to find shelter, water, food, and successfully reproduce. Habitat requirements are specific to each species and include biotic (plants and animals) and abiotic (soils, rainfall, temperature) conditions.

Harp trap A custom-made composite trap, comprising a rectangular frame between which numerous parallel fine fishing lines are tautly strung (giving it a harp-like appearance), and a collecting bag suspended beneath. Used for trapping smaller species of bats.

Herbivore A species that feeds on plants. A herbivore can be more specifically identified as a **Folivore**, **Frugivore**, **Browser** or **Grazer**.

Holotype One specimen designated in the original species or subspecies description, and which acts as the physical standard for reference to augment the written description. See **Type specimen.**

Home range The area habitually traversed by an individual animal.

Indigenous Native to the area under consideration. See also **Endemic**.

Inflorescence A group or cluster of flowers arranged on a branch or system of branches.

Insectivore A species that feeds on insects and other arthropods.

Interfemoral membrane A **flight membrane** that stretches between the legs in some bats. It may incorporate the tail.

Lactating Producing milk from the mammae (milk-secreting organs).

Lek mating system A mating system whereby males congregate in order to attract and display to females.

Mantle The area on the back around the shoulders, used here in reference mainly to bats.

Melanistic An unusual individual with skin or fur darker in colour than that which is typical for a species.

Mist net A long, tall net made of very fine thread, usually strung taut between trees or two poles set into the ground and used to catch birds and bats.

Morphology Form, or the study of form of organisms.

Mounted specimen (or exhibition specimen) A specimen, usually in a museum or other institution's collection, which has been professionally prepared in a naturalistic posture, usually for the purposes of display. In the 19th century many mounted specimens were used for species descriptions, but this is now uncommon. The skull and other bones may or may not be removed. Compare with **Study skin**.

Nectarivore A species that feeds on the nectar produced by flowering plants.

Nocturnal Active by night. See also **Crepuscular**, **Diurnal**.

Nominotypical Used here for a subspecies which bears the species name, for example *Hipposideros diadema diadema*.

Non-volant Land-based mammals that are unable to fly. Opposite of **Volant**.

Nose-leaf An outgrowth of skin on the face of certain bats that aids **Echolocation**.

Nulliparous Not yet having given birth. See also **Parous/ Non-parous**.

Olfactory Pertaining to the sense of smell.

Opportunistic Used here in reference to diet to denote an ability to feed on a wide variety of foods depending on their availability and abundance.

Paratype One of the specimnes used to prepare the original description of a new species or subspecies. A paratype is a **Type specimen** that helps define what the scientific name of a species or subspecies represents.

Parous/Non-parous Bearing (or not bearing) offspring.

Patagium An expanse of skin between the forelimb and hindlimb used in gliding or flying.

Pathogens Microbes causing disease.

Pelage Fur or hair covering the body.

Polyoestry Having more than one oestrus cycle annually or during a breeding season.

Prehensile Able to grip. Used here to refer to the tails of possums and murids.

Primary forest Forest which has reached a climax stage.

Proximal Towards the near or fixed end of a structure. Opposite of **Distal**.

Range A term with the same meaning as distribution, the area known to be occupied by a species. For measurements, refers to the smallest and largest measurements.

Regrowth Vegetation communities which are regrowing and progressing towards climax communities following disturbance.

Relict Surviving from the past. A relict species is one or one of a few surviving from a group

Riparian Pertaining to the banks of a river or waterway.

Rostrum The face or snout in front of the eyes, that part of the skull anterior to the eye sockets.

Sagittal crest A narrow ridge of bone running down the dorsal midline of the skull.

Scat An animal faecal dropping.

Secondary forest Vegetation communities which are regrowing and progressing towards climax forest following disturbance.

Sella A slender fleshy projection above the nostrils that forms part of the nose-leaf of horseshoe bats, the shape of which can be used for species identification.

Sexual dimorphism Differences between the sexes of a species.

Snap trap A commercially made, simple trap of the typical mouse-trap kind, in which a metal bar springs back onto the animal when the bait is taken. Used for small mammals. Differs from **Elliott trap**.

Squamate A member of the order Squamata, which includes scaled reptiles: lizards, snakes and amphisbaenians (worm lizards).

Study skin A specimen, usually in a museum or other institution's collection, in which the skin has been professionally prepared for study in a standard posture suitable for research showing the external taxonomic features of the animal. The skull is usually – although not always – removed. Compare with **Mounted specimen**.

Subspecies An interbreeding population within a species, differing measurably from one or more other populations and usually geographically separated from these.

Sympatric A term used to describe populations, varieties or species that occur in the same place at the same time. Usually used to refer to species of the same genus living at least partly in the same area.

Synonymy Recognition that two scientific names refer to the same animal population. A synonym is a scientific name that applies to a taxon that is now recognised under another name.

Syntype One of several specimens used to prepare the original description of a new species or subspecies, when a single specimen was not indicated as the holotype in the original description. Modern species descriptions now require designation of a single specimen, the **Holotype**. In past centuries, several specimens were used as the standard for the species, all of equal status as standards for the species or subspecies.

Taxon (plural taxa) The scientific name of a category of animals recognised by taxonomists, usually refers to species or subspecies.

Taxonomic revision A scientific work determining the **Taxonomic status** of related organisms.

Taxonomic status The precise status – subspecies, species or other category – under which an animal lineage has been classified. Also refers to whether a published name is currently considered to be a real biological entity or is a synonym.

Taxonomy The classification and naming of animals into subspecies, species, genera and other categories.

Terrestrial Living on the ground. Compare with **Arboreal**.

Territory The area defended by an individual animal.

Tragus A flap situated on the front of the outer aperture of the ear.

Trap-night A measure of the intensity of trapping obtained by multiplying the number of traps used by the number of nights on which they were used.

Type locality The locality at which a **Type specimen** was collected, or determined by a later researcher to have been collected at.

Type specimen The specimen which forms the basis of a scientific description of a species under the rules of binomial nomenclature. Type specimens are usually lodged in a museum. The holotype is the name-bearer for the species; other specimens of the species, lodged at the same time as the holotype, are referred to as **Paratypes**. Many other kinds of types which fulfil technical purposes under the International Code of Zoological Nomenclature exist.

Uropatagium See **Interfemoral membrane.**

Venated Having the arrangement of the veins clearly visible, here used to refer to the wings of bats.

Venter or ventrum The underside of the body.

Ventral On or near the undersurface, usually of the body.

Vocalisation The sound made by an animal in contexts such as territory-marking, mating, aggression or distress.

Volant Able to fly. Opposite of **Non-volant**.

Zygomatic plate A flattened bony plate formed at the front of a zygomatic arch (the equivalent of the cheekbone in humans), typical of rodent skulls.

Appendix A: Annotated faunal list for the nations and territories of the region

American Samoa

Terrestrial

Emballonura semicaudata (locally extinct), *Pteropus samoensis*, *Pteropus tonganus*

Introduced

Canis familiaris, *Felis catus*, *Mus musculus*, *Rattus exulans*, *Rattus norvegicus*, *Rattus rattus*, *Sus scrofa*

Marine

Balaenoptera edeni, *Globicephala macrorhynchus*, *Kogia breviceps*, *Kogia sima*, *Megaptera novaeangliae*, *Mesoplodon ginkgodens*, *Orcinus orca*, *Peponocephala electra*, *Pseudorca crassidens*, *Stenella attenuate*, *Stenella coeruleoalba*, *Stenella longirostris*, *Steno bredanensis*, *Tursiops truncates*, *Ziphius cavirostris*

Autonomous Region of Bougainville (AROB)

Terrestrial

Anthops ornatus, *Aselliscus tricuspidatus*, *Dobsonia inermis*, *Emballonura beccarii*, *Emballonura dianae*, *Emballonura raffrayana*, *Hipposideros calcaratus*, *Hipposideros cervinus*, *Hipposideros diadema*, *Hipposideros dinops*, *Macroglossus minimus*, *Melomys bougainville*, *Miniopterus australis*, *Miniopterus medius*, *Miniopterus tristis*, *Mosia nigrescens*, *Myotis moluccarum*, *Nesonycteris woodfordi*, *Nyctimene bougainville*, *Nyctimene major*, *Pipistrellus angulatus*, *Pteralopex anceps*, *Pteralopex flanneryi*, *Pteropus admiralitatum*, *Pteropus mahaganus*, *Pteropus rayneri*, *Rousettus amplexicaudatus*, *Saccolaimus saccolaimus*, *Solomys ponceleti*, *Solomys salebrosus*

Introduced

Canis familiaris, *Felis catus*, *Phalanger breviceps*, *Rattus exulans*, *Rattus norvegicus*, *Rattus praetor*, *Rattus rattus*, *Sus scrofa*

Marine

Balaenoptera edeni, *Balaenoptera musculus*, *Balaenoptera omurai*, *Dugong dugon*, *Feresa attenuata*, *Globicephala macrorhynchus*, *Grampus griseus*, *Kogia breviceps*, *Kogia sima*, *Lagenodelphis hosei*, *Megaptera novaeangliae*, *Mesoplodon densirostris*, *Orcinus orca*, *Peponocephala electra*, *Physeter macrocephalus*, *Pseudorca crassidens*, *Stenella attenuata*, *Stenella coeruleoalba*, *Stenella longirostris*, *Steno bredanensis*, *Tursiops aduncus*, *Tursiops truncatus*, *Ziphius cavirostris*

Commonwealth of the Northern Mariana Islands

Terrestrial

Emballonura semicaudata, *Pteropus mariannus*

Introduced

Bos taurus, *Canis familiaris*, *Capra hircus*, *Felis catus*, *Mus musculus*, *Rattus exulans*, *Rattus norvegicus*, *Rattus tanezumi*, *Rusa marianna*, *Suncus murinus*, *Sus scrofa*

Marine

Balaenoptera borealis, *Balaenoptera edeni*, *Feresa attenuata*, *Globicephala macrorhynchus*, *Kogia breviceps*, *Kogia sima*, *Megaptera novaeangliae*, *Mesoplodon densirostris*, *Mesoplodon ginkgodens*, *Orcinus orca*, *Peponocephala electra*, *Pseudorca crassidens*, *Stenella attenuata*, *Stenella coeruleoalba*, *Stenella longirostris*, *Steno bredanensis*, *Tursiops trruncatus*, *Ziphius cavirostris*

Cook Islands

Terrestrial

Pteropus tonganus

Introduced

Canis familiaris, *Capra hircus*, *Felis catus*, *Mus musculus*, *Rattus exulans*, *Rattus norvegicus*, *Rattus rattus*, *Sus scrofa*

Marine

Balaenoptera edeni, *Balaenoptera musculus*, *Feresa attenuate*, *Globicephala macrorhynchus*, *Grampus griseus*, *Hydrurga leptonyx*, *Kogia breviceps*, *Kogia sima, Lagenodelphis hosei, Megaptera novaeangliae, Mesoplodon ginkgodens, Orcinus orca, Peponocephala electra, Pseudorca crassidens, Stenella attenuate, Stenella coeruleoalba, Stenella longirostris, Steno bredanensis, Tursiops truncates, Ziphius cavirostris*

Federated States of Micronesia

Terrestrial

Emballonura semicaudata, Pteropus mariannus, Pteropus molossinus, Pteropus pelagicus, Pteropus pelewensis, Pteropus ualanus

Introduced

Canis familiaris, Capra hircus, Felis catus, Mus musculus, Rattus exulans, Rattus norvegicus, Rattus tanezumi, Rusa marianna, Sus scrofa

Marine

Balaenoptera edeni, Feresa attenuata, Globicephala macrorhynchus, Kogia breviceps, Kogia sima, Lagenodelphis hosei, Megaptera novaeangliae, Mesoplodon densirostris, Mesoplodon ginkgodens, Orcinus orca, Peponocephala electra, Physeter macrocephalus, Pseudorca crassidens, Stenella attenuata, Stenella coeruleoalba, Stenella longirostris, Steno bredanensis, Tursiops truncatus, Ziphius cavirostris

Fiji

Terrestrial

Emballonura semicaudata, Mirimiri acrodonta, Mops bregullae, Notopteris macdonaldi, Pteropus samoensis, Pteropus tonganus

Introduced

Canis familiaris, Capra hircus, Dama dama, Felis catus, Herpestes auropunctatus, Herpestes fuscus, Mus musculus, Rattus exulans, Rattus norvegicus, Rattus rattus, Rattus tanezumi, Sus scrofa

Marine

Balaenoptera edeni, Balaenoptera musculus, Balaenoptera physalus, Feresa attenuata, Globicephala macrorhynchus, Herpestes auropunctatus, Herpestes fuscus, Indopacetus pacificus, Kogia breviceps, Kogia sima, Megaptera novaeangliae, Mesoplodon densirostris, Mesoplodon ginkgodens, Orcinus orca, Peponocephala electra, Physeter macrocephalus, Pseudorca crassidens, Stenella attenuata, Stenella coeruleoalba, Stenella longirostris, Steno bredanensis, Tursiops truncatus, Ziphius cavirostris

French Polynesia

Terrestrial

No known extant indigenous terrestrial mammals.

Introduced

Canis familiaris, Capra hircus, Felis catus, Mus musculus, Rattus exulans, Rattus norvegicus, Rattus rattus, Sus scrofa

Marine

Balaenoptera edeni, Feresa attenuata, Globicephala macrorhynchus, Grampus griseus, Kogia breviceps, Kogia sima, Lagenodelphis hosei, Megaptera novaeangliae, Mesoplodon ginkgodens, Orcinus orca, Peponocephala electra, Pseudorca crassidens, Stenella attenuata, Stenella coeruleoalba, Stenella longirostris, Steno bredanensis, Tursiops truncatus, Ziphius cavirostris

Guam

Terrestrial

Emballonura semicaudata, Pteropus mariannus, Pteropus tokudae†

Introduced

Bubalus bubalis, Canis familiaris, Capra hircus, Felis catus, Mus musculus, Rattus exulans, Rattus norvegicus, Rattus tanezumi, Rusa marianna, Suncus murinus, Sus scrofa

Marine

Balaenoptera borealis, Balaenoptera edeni, Feresa attenuata, Globicephala macrorhynchus, Grampus griseus, Kogia breviceps, Kogia sima, Megaptera novaeangliae, Mesoplodon densirostris, Orcinus orca, Peponocephala electra, Pseudorca crassidens, Stenella attenuata, Stenella coeruleoalba, Stenella longirostris, Steno bredanensis, Tursiops truncatus, Ziphius cavirostris

Indonesia (east of Lydekker's Line)

Terrestrial

Aselliscus tricuspidatus, Dactylopsila trivirgata, Dasyurus albopunctatus, Dendrolagus inustus,

Distoechurus pennatus, Dobsonia beauforti, Dobsonia emersa, Dobsonia magna, Dobsonia minor, Dobsonia moluccensis, Dorcopsis muelleri, Echymipera clara, Echymipera kalubu, Echymipera rufescens, Emballonura beccarii, Emballonura raffrayana, Emballonura serii, Hipposideros ater, Hipposideros calcaratus, Hipposideros cervinus, Hipposideros diadema, Hipposideros maggietaylorae, Hipposideros papua, Hydromys chrysogaster, Macroglossus minimus, Mammelomys rattoides, Melomys rufescens, Miniopterus australis, Miniopterus blepotis, Miniopterus tristis, Mops jobensis, Mosia nigrescens, Murexia longicaudata, Myoictis melas, Myoictis wallacei, Myotis moluccarum, Nyctimene aello, Nyctimene albiventer, Nyctophilus microtis, Paramelomys moncktoni, Paramelomys naso, Paramelomys platyops, Paranyctimene raptor, Peroryctes raffrayana, Petaurus biacensis, Petaurus papuanus, Phalanger gymnotis, Phalanger mimicus, Phalanger orientalis, Phoniscus papuensis, Pipistrellus angulatus, Pipistrellus papuanus, Pogonomelomys bruijnii, Pseudochirops albertisii, Pseudochirulus canescens, Pteropus aruensis, Pteropus conspicillatus, Pteropus macrotis, Pteropus neohibernicus, Pteropus pohlei, Rattus jobiensis, Rattus leucopus, Rattus praetor, Rhinolophus euryotis, Rousettus amplexicaudatus, Saccolaimus saccolaimus, Sminthopsis virginiae, Spilocuscus maculatus, Spilocuscus papuensis, Spilocuscus wilsoni, Syconycteris australis, Thylogale browni, Thylogale brunii, Uromys boeadii, Uromys caudimaculatus, Uromys emmae, Zaglossus bruijnii

Introduced

Crocidura maxi, Felis catus, Mus musculus, Paradoxurus hermaphroditus, Rattus exulans, Rattus norvegicus, Rattus rattus, Rattus tanezumi, Rusa timorensis, Suncus murinus

Marine

Balaenoptera acutorostrata, Balaenoptera edeni, Balaenoptera musculus, Dugong dugon, Feresa attenuata, Globicephala macrorhynchus, Grampus griseus, Kogia breviceps, Kogia sima, Lagenodelphis hosei, Macaca fascicularis, Megaptera novaeangliae, Mesoplodon densirostris, Mesoplodon ginkgodens, Orcinus orca, Peponocephala electra, Physeter macrocephalus, Sousa sahulensis, Stenella attenuata, Stenella coeruleoalba, Stenella longirostris, Steno bredanensis, Tursiops aduncus, Tursiops truncatus, Ziphius cavirostris

Kiribati

Terrestrial

No known extant indigenous terrestrial mammals.

Introduced

Canis familiaris, Felis catus, Mus musculus, Rattus exulans, Rattus rattus, Sus scrofa

Marine

Balaenoptera edeni, Balaenoptera musculus, Feresa attenuata, Globicephala macrorhynchus, Kogia breviceps, Kogia sima, Lagenodelphis hosei, Megaptera novaeangliae, Mesoplodon densirostris, Mesoplodon ginkgodens, Mesoplodon hotaula, Orcinus orca, Peponocephala electra, Physeter macrocephalus, Pseudorca crassidens, Stenella attenuata, Stenella coeruleoalba, Stenella longirostris, Steno bredanensis, Tursiops truncatus, Ziphius cavirostris

Marshall Islands

Terrestrial

No known extant indigenous terrestrial mammals.

Introduced

Canis familiaris, Felis catus, Mus musculus, Rattus exulans, Rattus norvegicus, Rattus rattus

Marine

Balaenoptera edeni, Feresa attenuata, Globicephala macrorhynchus, Kogia breviceps, Kogia sima, Megaptera novaeangliae, Mesoplodon densirostris, Mesoplodon ginkgodens, Orcinus orca, Peponocephala electra, Physeter macrocephalus, Pseudorca crassidens, Stenella attenuata, Stenella coeruleoalba, Stenella longirostris, Steno bredanensis, Tursiops truncatus, Ziphius cavirostris

Nauru

Terrestrial

No known extant indigenous terrestrial mammals.

Introduced

Canis familiaris, Felis catus, Mus musculus, Rattus exulans, Rattus rattus, Sus scrofa

Marine

Balaenoptera edeni, Balaenoptera musculus, Feresa attenuata, Globicephala macrorhynchus, Kogia breviceps, Kogia sima, Lagenodelphis hosei, Megaptera novaeangliae, Megaptera novaeangliae Oceania subpopulation, Mesoplodon densirostris, Mesoplodon ginkgodens, Orcinus orca, Peponocephala electra, Physeter macrocephalus, Steno bredanensis, Tursiops truncatus, Ziphius cavirostris

New Caledonia

Terrestrial

Chalinolobus neocaledonicus, Miniopterus australis, Miniopterus macrocneme, Miniopterus robustior, Notopteris neocaledonica, Nyctophilus nebulosus, Pteropus ornatus, Pteropus tonganus, Pteropus vetula

Introduced

Canis familiaris, Capra hircus, Felis catus, Mus musculus, Oryctolagus cuniculus, Rattus exulans, Rattus norvegicus, Rattus rattus, Rusa timorensis, Sus scrofa

Marine

Arctocephalus forsteri, Balaenoptera acutorostrata, Balaenoptera edeni, Balaenoptera musculus, Balaenoptera omurai, Balaenoptera physalus, Delphinus delphis, Dugong dugon, Feresa attenuata, Globicephala macrorhynchus, Grampus griseus, Indopacetus pacificus, Kogia breviceps, Kogia sima, Megaptera novaeangliae, Mesoplodon densirostris, Mesoplodon ginkgodens, Orcinus orca, Peponocephala electra, Pseudorca crassidens, Stenella attenuata, Stenella coeruleoalba, Stenella longirostris, Steno bredanensis, Tursiops truncatus, Ziphius cavirostris

Aotearoa New Zealand

Terrestrial

Chalinolobus tuberculatus, Mystacina robusta†*, Mystacina tuberculata*

Introduced

Capra hircus, Cervus canadensis, Cervus elaphus, Cervus nippon, Dama dama, Erinaceus europaeus, Felis catus, Hemitragus jemlahicus, Lepus europaeus, Mus musculus, Mustela erminea, Mustela furo, Mustela nivalis, Notamacropus dorsalis, Notamacropus eugenii, Notamacropus parma, Notamacropus rufogriseus, Odocoileus virginianus, Oryctolagus cuniculus, Petrogale penicillata, Rattus exulans, Rattus norvegicus, Rattus rattus, Rupicapra rupicapra, Rusa timorensis, Rusa unicolor, Sus scrofa, Trichosurus vulpecula, Wallabia bicolor

Marine

Arctocephalus forsteri, Arctocephalus tropicalis, Balaenoptera acutorostrata, Balaenoptera bonaerensis, Balaenoptera borealis, Balaenoptera edeni, Balaenoptera musculus, Balaenoptera physalus, Berardius arnuxii, Caperea marginata, Cephalorhynchus hectori, Cephalorhynchus hectori ssp. Maui, Delphinus capensis, Delphinus delphis, Eubalaena australis, Globicephala macrorhynchus, Globicephala melas, Grampus griseus, Hemitragus jemlahicus, Hydrurga leptonyx, Hyperoodon planifrons, Kogia breviceps, Kogia sima, Lagenorhynchus cruciger, Lagenorhynchus obscurus, Leptonychotes weddellii, Lissodelphis peronii, Lobodon carcinophaga, Megaptera novaeangliae, Mesoplodon bowdoini, Mesoplodon densirostris, Mesoplodon ginkgodens, Mesoplodon grayi, Mesoplodon hectori, Mesoplodon layardii, Mesoplodon peruvianus, Mesoplodon traversii, Mirounga leonina, Orcinus orca, Phocarctos hookeri, Phocoena dioptrica, Physeter macrocephalus, Pseudorca crassidens, Stenella attenuata, Stenella coeruleoalba, Stenella longirostris, Tasmacetus shepherdi, Tursiops truncatus, Ziphius cavirostris

Niue

Terrestrial

Pteropus tonganus

Introduced

Canis familiaris, Capra hircus, Felis catus, Rattus exulans, Rattus norvegicus, Rattus rattus

Marine

Balaenoptera edeni, Feresa attenuata, Globicephala macrorhynchus, Kogia breviceps, Kogia sima, Megaptera novaeangliae, Mesoplodon ginkgodens, Orcinus orca, Peponocephala electra, Physeter macrocephalus, Pseudorca crassidens, Stenella coeruleoalba, Stenella longirostris, Steno bredanensis, Tursiops truncatus, Ziphius cavirostris

Palau

Terrestrial

Emballonura semicaudata, Pteropus pelewensis, Pteropus pilosus†

Introduced

Felis catus, Macaca fascicularis, Mus musculus, Rattus exulans, Rattus nitidus, Rattus norvegicus, Rattus tanezumi, Suncus murinus

Marine

Balaenoptera edeni, Balaenoptera musculus, Dugong dugon, Feresa attenuata, Globicephala macrorhynchus, Kogia breviceps, Kogia sima, Lagenodelphis hosei, Macaca fascicularis, Megaptera novaeangliae, Mesoplodon densirostris, Mesoplodon ginkgodens, Orcinus orca, Peponocephala electra, Physeter macrocephalus, Pseudorca crassidens, Stenella attenuata, Stenella coeruleoalba, Stenella longirostris, Steno bredanensis, Tursiops truncatus, Ziphius cavirostris

Papua New Guinea (excluding mainland and Autonomous Region of Bougainville)

Terrestrial

Aselliscus tricuspidatus, Chalinolobus nigrogriseus, Chiruromys forbesi, Dactylopsila tatei, Dobsonia anderseni, Dobsonia magna, Dobsonia minor, Dobsonia pannietensis, Dobsonia praedatrix, Dorcopsis atrata, Echymipera davidi, Echymipera kalubu, Echymipera rufescens, Emballonura beccarii, Emballonura dianae, Emballonura raffrayana, Emballonura serii, Hipposideros ater, Hipposideros calcaratus, Hipposideros cervinus, Hipposideros diadema, Hipposideros maggietaylorae, Hydromys chrysogaster, Hydromys neobritannicus, Kerivoula agnella, Kerivoula myrella, Macroglossus minimus, Melomys arcium, Melomys lutillus, Melomys matambuai, Melomys rufescens, Melonycteris melanops, Miniopterus australis, Miniopterus blepotis, Miniopterus macrocneme, Miniopterus medius, Miniopterus tristis, Mops jobensis, Mosia nigrescens, Murexia murex, Murina florium, Myotis moluccarum, Notamacropus agilis, Nyctimene aello, Nyctimene albiventer, Nyctimene certans, Nyctimene major, Nyctimene masalai, Nyctimene vizcaccia, Nyctophilus microtis, Ozimops beccarii, Paramelomys moncktoni, Paramelomys platyops, Petaurus papuanus, Phalanger breviceps, Phalanger intercastellanus, Phalanger lullulae, Phalanger orientalis, Philetor brachypterus, Phoniscus papuensis, Pipistrellus angulatus, Pipistrellus collinus, Pipistrellus papuanus, Pipistrellus wattsi, Pogonomys fergussoniensis, Pogonomys cf. *macrourus, Pteropus admiralitatum, Pteropus capistratus, Pteropus conspicillatus, Pteropus ennisae, Pteropus gilliardorum, Pteropus hypomelanus, Pteropus neohibernicus, Pteropus tonganus, Rattus detentus, Rattus mordax, Rattus praetor, Rhinolophus euryotis, Rhinolophus megaphyllus, Rousettus amplexicaudatus, Saccolaimus saccolaimus, Spilocuscus kraemeri, Spilocuscus maculatus, Syconycteris australis, Thylogale browni, Uromys caudimaculatus, Uromys neobritannicus*

Introduced

Axis axis, Canis familiaris, Felis catus, Mus musculus, Rattus exulans, Rattus norvegicus, Rattus rattus, Rattus tanezumi, Rusa timorensis, Sus scrofa

Marine

Balaenoptera acutorostrata, Balaenoptera edeni, Balaenoptera musculus, Dugong dugon, Feresa attenuata, Globicephala macrorhynchus, Grampus griseus, Kogia breviceps, Kogia sima, Lagenodelphis hosei, Macaca fascicularis, Megaptera novaeangliae, Mesoplodon densirostris, Mesoplodon ginkgodens, Orcinus orca, Peponocephala electra, Physeter macrocephalus, Sousa sahulensis, Stenella attenuata, Stenella coeruleoalba, Stenella longirostris, Steno bredanensis, Tursiops aduncus, Tursiops truncatus, Ziphius cavirostris

Pitcairn

Terrestrial

No known extant indigenous terrestrial mammals.

Introduced

Capra hircus, Felis catus, Mus musculus, Rattus exulans

Marine

Balaenoptera edeni, Feresa attenuata, Globicephala macrorhynchus, Hydrurga leptonyx, Kogia breviceps,

Kogia sima, Megaptera novaeangliae, Mesoplodon ginkgodens, Orcinus orca, Peponocephala electra, Pseudorca crassidens, Stenella coeruleoalba, Steno bredanensis, Tursiops truncatus, Ziphius cavirostris

Samoa

Terrestrial

Emballonura semicaudata (locally extinct), *Myotis insularum, Pteropus allenorum*†, *Pteropus coxi*†, *Pteropus samoensis, Pteropus tonganus*

Introduced

Canis familiaris, Felis catus, Mus musculus, Rattus exulans, Rattus norvegicus, Rattus rattus, Sciurus carolinensis, Sus scrofa

Marine

Balaenoptera edeni, Feresa attenuata, Globicephala macrorhynchus, Grampus griseus, Kogia breviceps, Kogia sima, Lagenodelphis hosei, Megaptera novaeangliae, Mesoplodon ginkgodens, Orcinus orca, Peponocephala electra, Physeter macrocephalus, Pseudorca crassidens, Stenella coeruleoalba, Stenella longirostris, Steno bredanensis, Tursiops truncatus, Ziphius cavirostris

Solomon Islands

Terrestrial

Anthops ornatus, Aselliscus tricuspidatus, Dobsonia inermis, Emballonura dianae, Emballonura raffrayana, Hipposideros calcaratus, Hipposideros cervinus, Hipposideros demissus, Hipposideros diadema, Hipposideros dinops, Macroglossus minimus, Melomys bougainville, Nesonycteris fardoulisi, Nesonycteris maccoyi, Nesonycteris mengermani, Nesonycteris woodfordi, Miniopterus australis, Miniopterus blepotis, Miniopterus macrocneme, Miniopterus tristis, Mops solomonis, Mosia nigrescens, Myotis moluccarum, Nyctimene bougainville, Nyctimene major, Nyctimene malaitensis, Pipistrellus angulatus, Pteralopex atrata, Pteralopex flanneryi, Pteralopex pulchra, Pteralopex taki, Pteropus admiralitatum, Pteropus cognatus, Pteropus mahaganus, Pteropus nitendiensis, Pteropus rayneri, Pteropus rennelli, Pteropus solomonis, Pteropus tonganus, Pteropus tuberculatus, Pteropus woodfordi, Rousettus amplexicaudatus, Saccolaimus saccolaimus, Solomys ponceleti, Solomys salamonis, Solomys salebrosus, Solomys sapientis, Uromys imperator, Uromys porculus, Uromys rex, Uromys vika

Introduced

Canis familiaris, Felis catus, Mus musculus, Phalanger breviceps, Rattus exulans, Rattus norvegicus, Rattus praetor, Rattus rattus, Sus Scrofa

Marine

Balaenoptera edeni, Balaenoptera musculus, Balaenoptera omurai, Dugong dugon, Feresa attenuata, Globicephala macrorhynchus, Grampus griseus, Kogia breviceps, Kogia sima, Lagenodelphis hosei, Megaptera novaeangliae, Mesoplodon densirostris, Orcinus orca, Peponocephala electra, Physeter macrocephalus, Pseudorca crassidens, Stenella attenuata, Stenella coeruleoalba, Stenella longirostris, Steno bredanensis, Tursiops aduncus, Tursiops truncatus, Ziphius cavirostris

Tokelau

Terrestrial

No known extant indigenous terrestrial mammals.

Introduced

Canis familiaris, Felis catus, Rattus exulans, Sus scrofa

Marine

Feresa attenuata, Kogia breviceps, Kogia sima, Megaptera novaeangliae, Mesoplodon densirostris, Mesoplodon ginkgodens, Pseudorca crassidens

Tonga

Terrestrial

Emballonura semicaudata (locally extinct), *Pteropus tonganus*

Introduced

Canis familiaris, Capra aegagrus hircus, Felis catus, Mus musculus, Rattus exulans, Rattus norvegicus, Rattus rattus, Sus scrofa

Marine

Balaenoptera edeni, Feresa attenuata, Globicephala macrorhynchus, Grampus griseus, Kogia breviceps, Kogia sima, Megaptera novaeangliae, Mesoplodon densirostris, Mesoplodon ginkgodens, Orcinus orca,

Peponocephala electra, Physeter macrocephalus, Pseudorca crassidens, Stenella attenuata, Stenella coeruleoalba, Stenella longirostris, Steno bredanensis, Tursiops truncatus, Ziphius cavirostris

Tuvalu

Terrestrial

No known extant indigenous terrestrial mammals.

Introduced

Felis catus, Mus musculus, Rattus exulans, Rattus norvegicus, Rattus rattus

Marine

Balaenoptera edeni, Balaenoptera musculus, Globicephala macrorhynchus, Kogia breviceps, Kogia sima, Megaptera novaeangliae, Mesoplodon ginkgodens, Physeter macrocephalus, Stenella attenuata, Stenella longirostris

Vanuatu

Terrestrial

Aselliscus tricuspidatus, Hipposideros cervinus, Miniopterus australis, Miniopterus macrocneme, Miniopterus tristis, Mops bregullae, Myotis moluccarum, Notopteris macdonaldi, Pteropus anetianus, Pteropus fundatus, Pteropus tonganus

Introduced

Canis familiaris, Capra hircus, Felis catus, Mus musculus, Rattus exulans, Rattus norvegicus, Rattus rattus, Sus scrofa

Marine

Balaenoptera edeni, Balaenoptera musculus, Dugong dugon, Globicephala macrorhynchus, Kogia breviceps, Kogia sima, Megaptera novaeangliae, Mesoplodon densirostris, Mesoplodon ginkgodens, Orcinus orca, Peponocephala electra, Physeter macrocephalus, Pseudorca crassidens, Stenella attenuata, Stenella longirostris, Tursiops truncatus, Ziphius cavirostris

Wallis and Futuna

Terrestrial

Pteropus tonganus

Introduced

Felis catus, Rattus exulans, Rattus norvegicus, Rattus rattus, Sus scrofa

Marine

Balaenoptera edeni, Feresa attenuata, Globicephala macrorhynchus, Kogia breviceps, Kogia sima, Megaptera novaeangliae, Mesoplodon densirostris, Mesoplodon ginkgodens, Orcinus orca, Peponocephala electra, Pseudorca crassidens, Stenella coeruleoalba, Stenella longirostris, Steno bredanensis, Tursiops truncatus, Ziphius cavirostris

References

Adler GH, Counsell E, Seamon JO, Bentivenga SP (2018) Exotic rats consume sporocarps of arbuscular mycorrhizal fungi in American Samoa. *Mammalia* **82**, 197–200. doi:10.1515/mammalia-2016-0135

Ali JR, Heaney L (2021) Wallace's line, Wallacea, and associated divides and areas: history of a tortuous tangle of ideas and labels. *Biological Reviews of the Cambridge Philosophical Society* **96**, 922–942. doi:10.1111/brv.12683

Allen H (1890) Description of a new species of *Pteropus. Proceedings of the American Philosophical Society* **28**, 70–72.

Allen G (1933) Two new bats from Australia. *Journal of Mammalogy* **14**, 149–151.

Allen G (1936) Two new races of Indian bats. *Records of the Indian Museum* **38**, 343–346.

Allen J (1996) The pre-Austronesian settlement of island Melanesia: implications for Lapita archaeology. *Transactions of the American Philosophical Society* **86**, 11–27.

Allen J (2000) From beach to beach: the development of maritime cultures. In *East of Wallace's Line: Studies of Past and Present Maritime Cultures of the Indo-Pacific Region*. (Eds S O'Connor and P Veth) pp. 139–176. Balkema, Rotterdam.

Allen RB (2021) *Rusa timorensis*. In *The Handbook of New Zealand Mammals*. 3rd edn. (Eds CM King and DM Forsyth) pp. 447–527. CSIRO Publishing, Melbourne.

Allen G, Barbour T (1909) A new marsupial from the Netherlands New Guinea. *Proceedings of the New England Zoölogical Club* **4**, 43–46.

Allen J, Gosden C, Jones R, White JP (1988) Pleistocene dates for the human occupation of New Ireland, northern Melanesia. *Nature* **331**, 707–709.

Allison A, Eldredge L (1999) Polynesia and Micronesia. In *Hotspots. Earth's Biologically Richest and Most Endangered Terrestrial Ecoregions*. (Eds RA Mittermier, N Myers, CG Mittermeier and P Robles Gil) pp. 390–401. Cemex and Conservation International, Mexico City.

Almeida FC, Giannini NP, Desalle R, Simmons NB (2009) The phylogenetic relationships of cynopterine fruit bats (Chiroptera: Pteropodidae: Cynopterinae). *Molecular Phylogenetics and Evolution* **53**, 772–783.

Almeida FC, Giannini NP, Simmons NB, Helgen KM (2014) Each flying fox on its own branch: a phylogenetic tree for *Pteropus* and related genera (Chiroptera: Pteropodidae). *Molecular Phylogenetics and Evolution* **77**, 83–95. doi:10.1016/j.ympev.2014.03.009

Almeida FC, Simmons NB, Giannini NP (2020) A species-level phylogeny of Old World fruit bats with a new higher-level classification of the Family Pteropodidae. *American Museum Novitates* **3950**.

Alston ER (1874) On a new species of *Pteropus* from Samoa. *Proceedings of the Linnean Society of London* **1874**, 96 (Plate XIV).

Alston ER (1877) On the rodents and marsupials collected by Rev. G. Brown in Duke-of-York Island, New Britain, and New Ireland. *Proceedings of the Zoological Society of London* **1877**, 123–127 (Plates XVIII and XIX).

Amante C, Eakins B (2009) ETOPO1 1 Arc-Minute Global Relief Model: Procedures, Data Sources and Analysis. NOAA Technical Memorandum NESDIS NGDC-24. 19.

Andersen K (1905a) On *Hipposideros diadema* and its closest allies. *Annals and Magazine of Natural History, London* (7)**16**, 497–508. doi:10.1080/03745480509443077

Andersen K (1905b) Bats of the Rhinolophus arcuatus Group, with descriptions of five new forms. *Annals and Magazine of Natural History, London* (7)**16**, 281–292.

Andersen K (1905c) On some bats of the genus Rhinolophus, with remarks on their mutual affinities, and descriptions of twenty-six new forms. *Proceedings of the Zoological Society, London* **2**, 75–145.

Andersen K (1906) On some new or little-known bats of the genus *Rhinolophus* in the collection of the Museo Civico. *Annali del Museo Civico di Storia Naturale di Genova* (3)**2**, 173–195.

Andersen K (1907) Chiropteran notes. *Annali del Museo Civico di Storia Naturale di Genova* (3)**3**, 5–45.

Andersen K (1908) Twenty new forms of *Pteropus. Annals and Magazine of Natural History, London* (8)**2**, 361–370.

Andersen K (1909a) On the fruit bats of the genus *Dobsonia. Annals and Magazine of Natural History, London* (8)**4**, 528–536.

Andersen K (1909b) Two new bats from the Solomon Islands. *Annals and Magazine of Natural History, London* (8)**3**, 266–270.

Andersen K (1909c) A new species of *Pteropus* from the Loyalty Islands. *Annals and Magazine of Natural History, London* (8)**3**, 233.

Andersen K (1910) Ten new fruit bats of the genera *Nyctimene*, *Cynopterus* and *Eonycterus. Annals and Magazine of Natural History, London* (8)**6**, 621–625.

Andersen K (1911) Six new fruit bats of the genera *Macroglossus* and *Syconycteris*. *Annals and Magazine of Natural History, London* (8)**7**, 641–643.

Andersen K (1912a) *Catalogue of the Chiroptera in the collection of the British Museum. 1. Megachiroptera.* British Museum (Natural History), London.

Andersen K (1912b) A new Nyctimene from New Guinea. *Annals and Magazine of Natural History, London* (8)**9**, 95–96.

Andersen K (1913) A new *Pteropus* from the Polynesian subregion. *Annals and Magazine of Natural History, London* (8)**11**, 469–472.

Andersen K (1918) Diagnoses of new bats of the families Rhinolophidae and Megadermatidae. *Annals and Magazine of Natural History, London* (9)**2**, 374–384.

Anderson A (2008) The rat and the octopus: initial human colonization and the prehistoric introduction of domestic animals to Remote Oceania. *Biological Invasions* **11**, 1503–1519. doi:10.1191/0959683604hl760ft

Anderson T, Berry A, Amos J, Cook J (1988) Spool-and-line tracking of the New Guinea spiny bandicoot, *Echymipera kalubu* (Marsupialia: Peramelidae). *Journal of Mammalogy* **69**, 114–120.

Anthony N (2001) Small mammals: an annotated checklist. In *New Britain biological survey.* (Ed. G Allez) pp. 34–45. University of Wisconsin New Britain Biological Survey, Middleton.

Aplin K (2015) Family Acrobatidae (feather-tailed gliders and feather-tailed possum). In *Handbook of the Mammals of the World. Vol. 5. Monotremes and Marsupials.* (Eds DE Wilson and RA Mittermeirer) pp. 574–591. Lynx Edicions, Barcelona.

Aplin K, Armstrong K (2016) *Nyctimene albiventer*. The IUCN Red List of Threatened Species 2016: e.T14962A22007665. doi:10.2305/IUCN.UK.2016-2. RLTS.T14962A22007665.en>

Aplin K, Opiang M (2011) The mammal fauna of the Nakanai Mountains, East New Britain Province, Papua New Guinea. In *Rapid Biological Assessments of the Nakanai Mountains and the Upper Strickland Basin: Surveying the Biodiversity of Papua New Guinea's Sublime Karst Environments.* RAP Bulletin of Biological Assessment 60. (Eds S Richards and B Gamui) pp. 85–103. Conservation International, Arlington, VA.

Aplin K, Pasveer J (2006) Mammals and other vertebrates from late Quaternary archaeological sites on Pulau Kobroor, Aru Islands, Eastern Indonesia. In *The Archaeology of the Aru Islands, Eastern Indonesia.* (Eds S O'Connor, M Spriggs and P Veth) pp. 41–62. ANU Press, Canberra.

Aplin K, Helgen K, Lunde D (2010) A review of *Peroryctes broadbenti*, the giant bandicoot of Papua New Guinea. *American Museum Novitates* **3696**.

Aplin KP, Suzuki H, Chinen AA, Chesser RT, ten Have J, Donnellan SC, Austin J, Frost A, Gonzalez JP, Herbreteau V, *et al.* (2011) Multiple geographic origins of commensalism and complex dispersal history of Black Rats. *PloS One* **6**, e26357. doi:10.1371/journal.pone.0026357

Aplin K, Novera J, Armstrong K (2015) Mammals of Manus and Mussau Islands. In *A Rapid Biological Survey of Papua New Guinea's Manus and Mussau Islands.* (Ed. N Whitmore) pp. 50–68. Wildlife Conservation Society Papua New Guinea Program, Goroka.

Armstrong K (2019) *Emballonura beccarii.* The IUCN Red List of Threatened Species 2019: e.T7672A22135577. doi:10.2305/IUCN.UK.2019-3.RLTS.T7672A22135577.en

Armstrong K, Aplin K (2011) Bats of the Muller Range, Papua New Guinea. In *Rapid Biological Assessments of the Nakanai Mountains and the Upper Strickland Basin: Surveying the Biodiversity of Papua New Guinea's Sublime Karst Environments.* RAP Bulletin of Biological Assessment 60. (Eds S Richards and B Gamui) pp. 85–103. Conservation International, Arlington, VA.

Armstrong K, Aplin K (2017) *Emballonura furax*. The IUCN Red List of Threatened Species 2017: e.T7667A22135664. doi:10.2305/IUCN.UK.2017-2. RLTS.T7667A22135664.en

Armstrong KN, Novera J, Aplin KP (2015) Acoustic survey of the echolocating bats of Manus and Mussau Islands. In *A Rapid Biological Survey of Papua New Guinea's Manus and Mussau Islands.* (Ed. N Whitmore) pp. 69–85. Wildlife Conservation Society Papua New Guinea Program, Goroka.

Armstrong K, Wiantoro S, Aplin K (2019a) *Miniopterus macrocneme*. The IUCN Red List of Threatened Species 2019: e.T136579A21998194. doi:10.2305/IUCN.UK.2019-3.RLTS.T136579A21998194.en

Armstrong KD, Wiantoro S, Aplin K (2019b) *Miniopterus medius*. The IUCN Red List of Threatened Species 2019: e.T13567A22105101. doi:10.2305/IUCN.UK.2019-3. RLTS.T13567A22105101.en

Ash J (1988) Stunted cloud-forest in Taveuni, Fiji. *Pacific Science* **41**, 191–199.

Athens J (2018) Archaeology of the Eastern Caroline Islands, Micronesia. In *The Oxford Handbook of Prehistoric Oceania*. (Eds E Cochrane and T Hunt) pp. 26–47. Oxford University Press, New York.

Aziz SA, Clements GR, Peng LY, Campos-Arceiz A, McConkey KR, Forget P-M, Gan HM (2017) Elucidating the diet of the island flying fox (*Pteropus hypomelanus*) in Peninsular Malaysia through Illumina Next-Generation Sequencing. *PeerJ* **5**, e3176. doi:10.7717/peerj.3176/supp-9

Baker RH (1948) Comments on conservation problems in Micronesia. In *Conservation in Micronesia.*

(Ed. HJ Coolidge) pp. 53–55. National Research Council, Washington, DC.

Baker JR, Baker Z (1936) The seasons in a tropical rain-forest (New Hebrides). Part 3. Fruit-bats (Pteropidae). *Zoological Journal of the Linnean Society* **40**, 123–141. doi:10.1111/j.1096-3642.1936.tb01681.x

Baker J, Bird T (1936) The seasons in a tropical rain-forest (New Hebrides). Part 4. Insectivorous bats (Vespertilionidae and Rhinolophidae). *Zoological Journal of the Linnean Society* **40**, 143–161.

Balouet C (1991) The fossil vertebrate record of New Caledonia. In *Vertebrate Palaeontology of Australasia.* (Eds P Vickers, J Monaghan, R Bajrd and T Rich) pp. 1383–1409. Monash University Publications, Melbourne.

Banack SA (1998) Diet selection and resource use by flying foxes (genus *Pteropus*). *Ecology* **79**, 1949–1967.

Banack SA, Grant GS (2009) Reproduction and behaviour of the Samoan flying fox, *Pteropus samoensis* (Chiroptera: Pteropodidae). *Mammalia* **67**, 419–437.

Barbehenn KR (1974) Recent invasions of Micronesia by small mammals. *Micronesica* **10**, 41–50.

Barun A, Hanson CC, Campbell KJ, Simberloff D (2011) A review of small Indian mongoose management and eradications on islands. In *Island Invasives: Eradication and Management.* (Eds CR Veitch, M Clout and DR Towns) pp. 17–25. IUCN, Gland, Switzerland.

Batcheler C (1983) The possum and rata-kamahi dieback in New Zealand: a review. *Pacific Science* **37**, 415–426.

Bates PJJ, Rossiter SJ, Suyanto J, Kingston T (2007) A new species of *Hipposideros* (Chiroptera: Hipposideridae) from Sulawesi. *Acta Chiropterologica* **9**, 13–26.

Bell E, Bell B, Merton D (2016) The legacy of Big South Cape: rat irruption to rat eradication. *New Zealand Journal of Ecology* **40**, 212–218. doi:10.20417/nzjecol.40.24

Bengsen AJ, West P, Krull CR (2017) Feral pigs in Australia and New Zealand: range, trend, management, and impacts of an invasive species. In *Ecology, Conservation and Management of Wild Pigs and Peccaries.* (Ed. M Melletti) pp. 325–338. Cambridge University Press, Cambridge, UK.

Bergman S (1961) *My Father is a Cannibal.* Robert Hale, London.

Bergmans W (1975) A new species of *Dobsonia* Palmer, 1898 (Mammalia, Megachiroptera) from Waigeo, with notes on other members of the genus. *Beaufortia* **23**, 1–13.

Bergmans W (2001) Notes on distribution and taxonomy of Australian bats. I. Pteropodinae and Nyctimeninae (Mammalia, Megachiroptera, Pteropodidae). *Beaufortia* **51**, 119–152.

Bergmans W, Sarbini S (1985) Fruit-bats of the genus *Dobsonia* Palmer, 1898 from the islands of Biak, Owi, Numfoor and Yapen, Irian Jaya (Mammalia: Megachiroptera). *Beaufortia* **34**, 181–189.

Berkenhout J (1769) *Outlines of the Natural History of Great Britain and Ireland. Containing a Systematic Arrangement and Concise Description of all the Animals, Vegetables, and Fossils Which Have Hitherto Been Discovered in These Kingdoms.* Elmsly, London.

Best S (1984) *Lakeba: The Prehistory of a Fijian Island.* University of Auckland, Auckland.

Bettink KA (2016) Shedding light on rakali: genetic and morphological differentiation in the Australo-Papuan golden-bellied water rat (*Hydromys chrysogaster*), with notes on the Barrow Island population. PhD thesis. The University of Western Australia, Australia.

Blainville MH-M (1822) Mémoire sur les caractères distinctifs des espèces de Cerfs. *Journal de Physique, de Chimie, d'Histoire Naturelle et des Arts* **94**, 254–283.

Blyth E (1844) Notes of various Mammalia, with description of many new species. *Journal of the Asiatic Society of Bengal* **13**, 463–494.

Bonaccorso F (1998) *Bats of Papua New Guinea.* Conservation International, Washington, DC.

Bonaccorso FJ, Winkelmann JR (2005) Home range, territoriality, and flight time budgets in the black-bellied fruit bat, *Melonycteris melanops* (Pteropodidae). *Journal of Mammalogy* **86**, 931–936.

Bonaccorso FJ, Winkelmann JR, Dumont ER (2002) Home range of *Dobsonia minor* (Pteropodidae): a solitary, foliage-roosting fruit bat in Papua New Guinea. *Biotropica* **34**, 127–135.

Bonaccorso F, Leary T, Hamilton S (2008) *Emballonura dianae.* The IUCN Red List of Threatened Species 2008: e.T7673A12842653. doi:10.2305/IUCN.UK.2008.RLTS.T7673A12842653.en

Bonnaud E, Medina FM, Vidal E, Nogales M, Tershy B, Zavaleta E, Donlan CJ, Keitt B, Le Corre M, Horwath SV (2010) The diet of feral cats on islands: a review and a call for more studies. *Biological Invasions* **13**, 581–603. doi:10.1007/s10530-010-9851-3.

Bouillard N (2018) *Bats of New Caledonia*, <https://www.bigbatyear.com/bats-new-caledonia/>

Bowen-Jones E, Abrutat D, Markham B, Bowe S (1997) Flying foxes on Choiseul (Solomon Islands) – the need for conservation action. *Oryx* **31**, 209–217.

Brass LJ (1956) *1956 Diary. Section 1, 1956–57 Expedition Diary of L. J. Brass.* The American Museum of Natural History, New York.

Brass LJ (1959) Results of the Archbold Expeditions. No. 79. Summary of the fifth Archbold expedition to New Guinea (1956–1957). *Bulletin of the American Museum of Natural History* **118**, 1–84.

Breed AC, Field HE, Smith CS, Edmonston J, Meers J (2010) Bats without borders: long-distance movements and implications for disease risk management. *EcoHealth* **7**, 204–212. doi:10.1007/s10393-010-0332-z

Brescia F (2008) *Pteropus vetulus*. The IUCN Red List of Threatened Species 2008: e.T18767A8595636. doi:10.2305/IUCN.UK.2008.RLTS.T18767A8595636.en

Brockie RE (2010) Distribution and abundance of the hedgehog (*Erinaceus europaeus*) L. in New Zealand, 1869–1973. *New Zealand Journal of Zoology* **2**, 445–462. doi:10.1038/1801372a0

Brooke AP, Tschapka M (2002) Threats from overhunting to the flying fox, *Pteropus tonganus*, (Chiroptera: Pteropodidae) on Niue Island, South Pacific Ocean. *Biological Conservation* **103**, 343–348.

Brown VA, Brooke A, Fordyce JA, Mccracken GF (2011) Genetic analysis of populations of the threatened bat *Pteropus mariannus*. *Conservation Genetics* **12**, 933–941. doi:10.1046/j.1523-1739.2000.99082.x

Bruner P, Pratt H (1979) Notes on the status and natural history of Micronesian bats. *Elepaio* **40**, 1–4.

Bryant LM, Donnellan SC, Hurwood DA, Fuller SJ (2011) Phylogenetic relationships and divergence date estimates among Australo-Papuan mosaic-tailed rats from the *Uromys* division (Rodentia: Muridae). *Zoologica Scripta* **40**, 433–447. doi:10.1111/j.1463-6409.2011.00482.x

Buchanan G, Butchart S, Dutson G, Pilgrim J, Steininger M, Bishop K, Mayaux P (2008) Using remote sensing to inform conservation status assessment: estimates of recent deforestation rates on New Britain and the impacts upon endemic birds. *Biological Conservation* **141**, 55–66.

Buden D (1996a) Reptiles, birds, and mammals of Ant Atoll, Eastern Caroline Islands. *Micronesica* **29**, 21–36.

Buden D (1996b) Reptiles, birds, and mammals of Pakin Atoll, Eastern Caroline Islands. *Micronesica* **29**, 37–48.

Buden D, Helgen KM, Wiles G (2013) Taxonomy, distribution, and natural history of flying foxes (Chiroptera, Pteropodidae) in the Mortlock Islands and Chuuk State, Caroline Islands. *ZooKeys* **345**, 97–135. doi:10.3897/zookeys.345.5840.figure7

Burbidge A, Woinarski J (2016) *Macropus eugenii*. The IUCN Red List of Threatened Species 2016: e.T41512A21953803. doi:10.2305/IUCN.UK.2016-2.RLTS.T41512A21953803.en

Burgin C (2017) Family Muridae (true mice and rats, gerbils and relatives). In *Handbook of the Mammals of the World. Vol. 7 Rodents II*. (Eds D Wilson, T Lacher and R Mittermeier) pp. 536–884. Lynx Edicions in association with Conservation International and IUCN, Barcelona.

Burgin C (2019) Family Pteropodidae. In *Handbook of the Mammals of the World. Vol.9 Bats* (Eds D Wilson and R Mittermeier) pp. 16–162. Lynx Edicions in association with Conservation International and IUCN, Barcelona.

Burley DV (1998) Tongan archaeology and the Tongan past, 2850–150 B.P. *Journal of World Prehistory* **12**, 337–392.

Buys J, Spandler C, Holm RJ, Richards SW (2014) Remnants of ancient Australia in Vanuatu: implications for crustal evolution in island arcs and tectonic development of the southwest Pacific. *Geology* **42**, 939–942. doi:10.1130/g36155.1

Cabrera A (1920) Dos nuevos murciélagos frugivoros. *Boletín de la Real Sociedad Española de Historia Natural* **20**, 106–108.

Caughley G (1963) Dispersal rates of several ungulates introduced into New Zealand. *Nature* **200**, 280–281.

Chambers M, Esrom D (1988) *The Flying Foxes of Vanuatu, with Notes on their Cultural and Social Importance*. Ministry of Lands, Port Vila, Vanuatu.

Chapman NG, Chapman DI (1980) The distribution of fallow deer: a worldwide review. *Mammal Review* **10**, 61–138. doi:10.1111/j.1365-2907.1980.tb00234.x

Chasen F (1940) A handlist of Malaysian mammals. *Bulletin of the Raffles Museum* **15**, 1–209.

Christie A, Andrews J (1965) Introduced ungulates in New Zealand (b) Virginia deer. *Tuatara* **13**, 1–7.

Christie A, Andrews J (1966) Introduced ungulates in New Zealand (d) Fallow Deer. *Tuatara* **14**, 82–88.

Churchill S (2008) *Australian Bats*. Allen & Unwin, Crows Nest.

Chynoweth MW, Litton CM, Lepczyk CA, Hess SC, Cordell S (2013) Biology and impacts of Pacific island invasive species. 9. *Capra hircus*, the feral goat (Mammalia: Bovidae). *Pacific Science* **67**, 141–156. doi:10.2984/67.2.1

Cibois A, Thibault J-C, Watling D (2019) Birds and bats of Rotuma, Fiji. *Notornis* **66**, 139–149.

Clapperton BK (2001) Advances in New Zealand mammalogy 1990–2000: feral ferret. *Journal of the Royal Society of New Zealand* **31**, 185–203. doi:10.1080/03014223.1995.9518054

Clark G (2005) A 3000 year culture sequence from Palau, Western Micronesia. *Asian Perspective* **44**, 349–380.

Clark PU, Dyke AS, Shakun JD, Carlson AE, Clark J, Wohlfarth B, Mitrovica JX, Hostetler SW, McCabe AM (2009) The Last Glacial Maximum. *Science* **325**, 710–714. doi:10.1126/science.1172873

Clarke C, Dzieciolowski R (1991) Feral pigs in the northern South Island, New Zealand: I. Origin, distribution, and density. *Journal of the Royal Society of New Zealand* **21**, 237–247.

Cochrane E (2006) Archaeology in the Yasawa Islands, Western Fiji: a Report on Fieldwork from 1978 to 2003. *Domodomo* **18**, 1–67.

Cohn L (1910) Die Papuanischen Perameles – Arten. *Zoologischer Anzeiger* **35**, 718–727.

Cohn L (1914) Ein neue Varietät von *Phalanger maculatus*. *Zoologischer Anzeiger* **44**, 507–516.

Cole J, Graham IJ, Gibson IL (1990) Magmatic evolution of Late Cenozoic volcanic rocks of the Lau Ridge, Fiji. *Contributions to Mineralogy and Petrology* **104**, 540–554.

Colgan D, Flannery T (1992) Biochemical systematic studies in the genus *Petaurus* (Marsupialia, Petauridae). *Australian Journal of Zoology* **40**, 245–256.

Colgan D, Soheili S (2008) Evolutionary lineages in *Emballonura* and *Mosia* bats (Mammalia: Microchiroptera) from the southwestern Pacific. *Pacific Science* **62**, 219–232.

Collett R (1897) On a collection of mammals from north and north-west Australia. *Proceedings of the Zoological Society of London* **1897**, 317–336.

Cooper S, Day PR, Reardon TB, Shulz M (2001) Assessment of species boundaries in Australian *Myotis* (Chiroptera: Vespertilionidae) using mitochondrial DNA. *Journal of Mammalogy* **82**, 328–338.

Coultas W (1931) Kusaie collections. In *Unpublished Journal and Letters of William F. Coultas in the Whitney South Sea Expedition Journals*. American Museum of Natural History, New York.

Cousins JA, Compton SG (2005) The Tongan flying fox *Pteropus tonganus*: status, public attitudes and conservation in the Cook Islands. *Oryx* **39**, 196–203. doi:10.1017/S003060530500044X

Cowan PE, Glen AS (2021) *Trichosurus vulpecula*. In *The Handbook of New Zealand Mammals*. 3rd edn. (Eds CM King and DM Forsyth) pp. 43–77. CSIRO Publishing, Melbourne.

Cowan P, Moeed A (1987) Invertebrates in the diet of brushtail possums, *Trichosurus vulpecula*, in lowland podocarp/broadleaf forest, Orongorongo Valley, Wellington, New Zealand. *New Zealand Journal of Zoology* **14**, 163–177.

Cox PA (1983) Observations on the natural history of Samoan bats. *Mammalia* **47**, 519–523.

Cox PA (1984) Flying fox nearly extinct in Samoa. *Bats* **1984**, 1–2.

Cox MP (2008) The genetic environment of Melanesia: clines, clusters and contact. In *Population Genetics Research Progress*. (Ed. VT Koven) pp. 45–83. Nova Science Publishers Inc., New York.

Cox P, Elmqvist T, Pierson E, Rainey W (1992) Flying foxes as pollinators and seed dispersers in Pacific Island ecosystems. In *Pacific Island Flying Foxes: Proceedings of an International Conservation Conference*. 1–2 February 1990, Honolulu. (Eds DE Wilson and GL Graham) pp. 18–23. U. S. Department of the Interior, Fish and Wildlife Service, Washington, DC.

Craig P, Trail P, Morrell TE (1994) The decline of fruit bats in American Samoa due to hurricanes and overhunting. *Biological Conservation* **69**, 261–266.

Cremona T, Baker AM, Cooper S, Montague-Drake R, Stobo-Wilson AM, Carthew SM (2020) Integrative taxonomic investigation of *Petaurus breviceps* (Marsupialia: Petauridae) reveals three distinct species. *Zoological Journal of the Linnean Society* **191**, 503–527. doi:10.1093/zoolinnean/zlaa060/5870721

Cronin S, Neall VE (2001) Holocene volcanic geology, volcanic hazard, and risk on Taveuni, Fiji. *New Zealand Journal of Geology and Geophysics* **44**, 417–437.

Cruz J, Thomson C, Parkes JP, Gruner I, Forsyth DM (2016) Long-term impacts of an introduced ungulate in native grasslands: Himalayan tahr (*Hemitragus jemlahicus*) in New Zealand's Southern Alps. *Biological Invasions* **19**, 339–349. doi:10.1007/s10530-016-1283-2

Cuthbert RJ, Denny MJH (2014) Aspects of the ecology of the kalubu bandicoot (*Echymipera kalubu*) and observations on Raffray's bandicoot (*Peroryctes raffrayanus*), Eastern Highlands Province, Papua New Guinea. *Australian Mammalogy* **36**, 21–28. doi:10.2307/1377911

Cuvier F (1837) Rapport sur un mémoire de M. Jourdan, de Lyon, convernant quelques mammiféres nouveaux. *Comptes Rendus Hebdomadaires des Séances de l'Académie des Sciences* **6**, 2–6.

Czenze ZJ, Tucker JL, Clare EL, Littlefair JE, Hemprich Bennett D, Oliveira HFM, Brigham RM, Hickey AJR, Parsons S (2018) Spatiotemporal and demographic variation in the diet of New Zealand lesser short-tailed bats (*Mystacina tuberculata*). *Ecology and Evolution* **8**, 7599–7610. doi:10.1111/mec.13770

Dahruddin H, Farida WR, Rohman AS (2014) Plants species as feed sources and nesting site of cuscus (Family Phalangeridae) in Northern Biak Nature Reserve, Papua. *Biodiversitas Journal of Biological Diversity* **6**, 253–258.

Daniel MJ (1990) Order Chiroptera. In *The Handbook of New Zealand Mammals*. (Ed CM King) pp. 114–137. Oxford University Press, Auckland.

Davidson M, Fraser K (1991) Official hunting patterns and trends in the proportions of sika (*Cervus nippon*) and red deer (*C. elaphus*) in the Kaweka Range, New Zealand, 1958–1988. *New Zealand Journal of Ecology* **15**, 31–40.

Davies H (2005) The geology of Bougainville. In *Bougainville Before the Conflict*. (Eds A Regan and H Griffin) pp. 20–30. Pandanus Books, Canberra.

Davies TE, Ruzicka F, Lavery TH, Walters CL, Pettorelli N (2016) Ultrasonic monitoring to assess the impacts of forest conversion on Solomon Island bats. *Remote Sensing in Ecology and Conservation* **2**, 107–118.

De Vis C (1884) On apparently new species of *Halmaturus*. *Proceedings of the Royal Society of Queensland* **1**, 107–112.

De Vis C (1895) A review of the fossil jaws of the Macropodidae in the Queensland Museum. *Proceedings of the Linnean Society of New South Wales* **10**, 75–133.

De Vis C (1905) Bats. *Annals of the Queensland Museum* **6**, 36–40.

Dekrout A (2009) Monitoring New Zealand long-tailed bats (*Chalinolobus tuberculatus*) in urban habitats: ecology, physiology and genetics. The University of Auckland, Auckland.

Delhey K (2017) Gloger's rule. *Current Biology* **27**, R689–R691. doi:10.1016/j.cub.2017.04.031

Demos TC, Webala PW, Lutz HL, Kerbis Peterhans JC, Goodman SM, Cortés Delgado N, Bartonjo M, Patterson BD (2019) Multilocus phylogeny of a cryptic radiation of Afrotropical long-fingered bats (Chiroptera, Miniopteridae). *Zoologica Scripta* **49**, 1–13. doi:10.1093/molbev/msu279

Denham T (2018) The 'Austronesian' dispersal in island Southeast Asia. In *The Oxford Handbook of Prehistoric Oceania*. (Eds E Cochrane and T Hunt) pp. 26–47. Oxford University Press, New York.

Desmarest AG (1804) *Nouveau dictionaire d'histoire naturelle, appliquée aux arts, à l'Agriculture, a' l'économie rurale et domestique, à l'médecine, etc.* Deterville, Paris.

Desmarest AG (1817) *Nouveau dictionaire d'histoire naturelle, appliquée aux arts, à l'Agriculture, a' l'économie rurale et domestique, à l'médecine, etc.* Deterville, Paris.

Desmarest AG (1818) *Nouveau dictionaire d'histoire naturelle, appliquée aux arts, à l'Agriculture, a' l'économie rurale et domestique, à l'médecine, etc.*,Deterville, Paris.

Desmarest AG (1820) *Mammalogie ou description des espèces de mammifères. Première partie, contenant les ordres des bimanes, des quadrumanes et des carnassiers. Encyclopédie Méthodique.* Veuve Agasse, Paris.

Desmarest AG (1822) *Mammalogie ou description des espèces de mammifères. Seconde partie, contenant les ordres des rongeurs, des édentés, des pachydermes, des ruminans et des cétacés. Encyclopédie Méthodique.* Veuve Agasse, Paris.

Diamond JM (1983) Survival of bird populations stranded on land-bridge islands. *National Geographic Society Research Reports* **15**, 127–141.

Dobson G (1871) Description of four new species of Malayan bats, from the collection of Dr Stoliczka. *Journal of the Asiatic Society of Bengal* **40**, 260–266.

Dobson G (1873) On the Pteropodidae of India and its islands, with descriptions new and little-known species. *Journal of the Asiatic Society of Bengal* **42**, 194–205.

Dobson G (1875) A monograph of the Genus *Taphozous*, Geoff. *Proceedings of the Linnean Society of London* **1875**, 546–556.

Dobson G (1877) On a collection of Chiroptera from Duke-of-York Island and the adjacent parts of New Ireland and New Britain. *Proceedings of the Zoological Society of London* **1877**, 114–123 (Plate XVII).

Dobson G (1878) *Catalogue of the Chiroptera in the Collection of the British Museum, London.* British Museum, London.

Dobson GE (1879) Notes on recent additions to the collection of Chiroptera in the Muséum d'Histoire Naturelle at Paris, with descriptions of new and rare species. *Proceedings of the Zoological Society of London* **1878**, 873–880.

Doherty TS, Dickman CR, Glen AS, Newsome TM, Nimmo DG, Ritchie EG, Vanak AT, Wirsing AJ (2017) The global impacts of domestic dogs on threatened vertebrates. *Biological Conservation* **210**, 56–59. doi:10.1016/j.biocon.2017.04.007

DON (Depart. of the Navy) (2019) Integrated Natural Resources Management Plan for Joint Region Marianas. Prepared for Joint Region Marianas and NAVFAC Marianas, Guam. Cardno, Honolulu.

Donne T (1924) *The Game Animals of New Zealand.* John Murray, London.

Du Bois A (1882) Remarques sur l'*Acanthoglossus bruijnii. Bulletin de la Société Zoologique de France* **6**, 266–270.

Du Bois A (1884) Description d'un Échidné et d'un Perroquet inédits de la Nouvelle-Guinée. *Bulletin du Musée Royal d'Histoire Naturelle de Belgique* **3**, 109.

Duckworth J, Timmins R, Choudhury A, Chutipong W, Willcox D, Mudappa D, Rahman H, Widmann P, Wilting A, Xu W (2016) *Paradoxurus hermaphroditus*. The IUCN Red List of Threatened Species 2016: e.T41693A45217835. doi:10.2305/IUCN.UK.2016-1.RLTS.T41693A45217835.en

Duffy D, Capece P (2012) Biology and impacts of Pacific Island invasive species. 7. The domestic cat (*Felis catus*). *Pacific Science* **66**, 173–212.

Duncan RA, Clague DA (1985) Pacific Plate motion recorded by linear volcanic chains. In *The Ocean Basins and Margins, 7A*. (Eds AEM Nairn, FG Stehli and S Uyeda) pp. 89–121. Plenum Publishing, New York.

Dwyer P (1962) Studies on the two New Zealand bats. *Zoology Publications from Victoria University of Wellington* **28**, 1–28.

Egloff B (1975) Archaeological investigations in the coastal Madang area and on Eloaua Islands of the St. Matthias Group. *Records of the Papua New Guinea Public Museum and Art Gallery* **5**, 15–31.

Ekstrom J, Jones J, Willis J, Isherwood I (2000) *The Humid Forests of New-Caledonia: Biological Research and Conservation Recommendations for the Vertebrate Fauna of Grande Terre.* CSB Conservation Publications, Cambridge, UK.

Eldridge MDB, Browning TL, Close RL (2001) Provenance of a New Zealand brush-tailed rock-wallaby (*Petrogale penicillata*) population determined by mitochondrial DNA sequence analysis. *Molecular Ecology* **10**, 2561–2567.

Elliott TF, Vernes K (2021) Notes on the diets of four rodent species from Goodenough Island. *Australian Mammalogy* **43**, 256–259. doi:10.1071/AM20022

Elliott TF, Travouillon KJ, Warburton NM, Danks MA, Vernes K (2022) New Guinean bandicoots: new insights into diet, dentition and digestive tract morphology and a dietary review of all extant non-Australian Peramelemorphia. *Australian Mammalogy* **44**, 266–279.

Emmons L, Kinbag F (2001) Survey of the mammals of southern New Ireland. In *Southern New Ireland, Papua New Guinea: A Biodiversity Assessment*. (Eds BA Beehler and LE Alonso) pp. 67–69. Conservation International, Center for Applied Biodiversity Science, Deptartment of Conservation Biology, Washington, DC.

Engbring J (1989) *Observations of Fruit Bats in Samoa with Emphasis on the Status of the Samoan Fruit Bats (*Pteropus samoensis*)*. U.S. Fish and Wildlife Service, Honolulu.

Erxleben J (1777a) *Anfangsgründe der Naturlehre and Systema regni animalis*. Johann Hans Dieterich, Göttingen.

Erxleben J (1777b) *Systema regni animalis per classes, ordines, genera, species, varietates: cvm synonymia et historia animalivm: Classis I. Mammalia*. Impensis Weygandianis, Lipsiae.

Esselstyn JA, Wiles GJ, Amar A (2004) Habitat use of the Pacific Sheath-Tailed Bat (*Emballonura semicaudata*) on Aguiguan, Mariana Islands. *Acta Chiropterologica* **6**, 303–308. doi:10.3161/001.006.0208

Fabre P-H, Fitriana YS, Semiadi G, Pagès M, Aplin K, Supriatna N, Helgen KM (2018) New record of *Melomys burtoni* (Mammalia, Rodentia, Murinae) from Halmahera (North Moluccas, Indonesia): a review of Moluccan *Melomys*. *Mammalia* **82**, 218–247. doi:10.1093/nar/21.22.5264

Falanruw M (1988) On the status, reproductive biology and management of fruit bats of Yap. *Micronesica* **21**, 39–51.

Falvey D, Pritchard T (1982) Preliminary paleomagnetic results from northern Papua New Guinea: evidence for large microplate rotations. *Transactions of the 3rd Circum-Pacific Energy and Mineral Resources Conference*. 22–28 August, Honolulu (Ed S Watson) pp. 593–600. American Association of Petroleum Geologists, Tulsa.

Farr TG, Rosen PA, Caro E, Crippen R, Duren R, Hensley S, Kobrick M, Paller M, Rodriguez E, Roth L, *et al.* (2007) The Shuttle Radar Topography Mission. *Reviews of Geophysics* **45**, 1485.

Feiler A (1978a) Bemerkungen über Phalanger der 'orientalis–Gruppe' nache Tate (1945). *Zoologische Abhandlungen* **34**, 385–395.

Feiler A (1978b) Über artliche abgrenzung und innerartliche aus formung bei *Phalanger maculatus* (Mammalia: Marsupialia). *Zoologische Abhandlungen* **35**, 1–30.

Felten H (1961) Eine neue Unterart von *Pteropus neohibernicus*. *Senckenbergiana Biologica* **42**, 417–418.

Felten H (1964a) Flughunde der Gattung Pteropus von Neukaledonien und den Loyalty–Inseln. *Senckenbergiana Biologica* **45**, 671–683.

Felten H (1964b) Zur Taxonomie indo-australischer Fledermäuse der Gattung Tadarida (Mammalia, Chiroptera). *Senckenbergiana Biologica* **45**, 1–13.

Felten H, Kock D (1972) Weitere flughunde der Gattung *Pteropus* von den Neuen Hebriden, sowie den Banks- und Torres- Inseln, Pazifischer Ozean. *Senckenbergiana Biologica* **53**, 179–188.

Findley JS (1972) Phenetic relationships among bats of the genus *Myotis*. *Systematic Zoology* **21**, 31–52.

Fischer J (1829) *Synopsis Mammalium*. Sumtibus J. C. Cottae, Stuttgardtiae.

Fisher D, Tasker E (1997) Natural history of the New Georgia monkey-faced bat *Pteralopex* sp. nov. from the Solomon Islands. *Pacific Conservation Biology* **3**, 134–142.

Fisher R, Uili M, Iese C, Evaimalo M, Tipamaa F, Bonin M, Atherton J, Tye A (2011) Rapid response and eradication of an invading mongoose in Samoa. *Oryx* **45**, 15–16.

Flannery TF (1987) An historic record of the New Zealand Greater Short-tailed Bat *Mystacina robusta* (Michrochiroptera: Mystacinidae) from the South Island, New Zealand. *Australian Mammalogy* **10**, 45–46.

Flannery TF (1990a) *Echymipera davidi*, a new species of Perameliformes (Marsupialia) from Kiriwina Island, Papua New Guinea, with notes on the systematics of the genus *Echymipera*. In *Bandicoots and Bilbies*. (Eds J Seebeck, R Wallis, P Brown and C Kemper) pp. 29–35. Surrey Beatty and Sons, Sydney.

Flannery TF (1990b) Flying foxes in Melanesia: populations at risk. *Bats* **7**, 5–7.

Flannery TF (1991a) Emporer, king and little pig: three rats of Guadacanal. *Australian Natural History* **23**, 634–641.

Flannery TF (1991b) A new species of *Pteralopex* (Chiroptera: Pteropidae) from montane Guadacanal, Solomon Islands. *Records of the Australian Museum* **43**, 123–130.

Flannery TF (1992) Taxonomic revision of the *Thylogale brunii* complex (Macropodidae: Marsupialia) in Melanesia, with description of a new species. *Australian Mammalogy* **15**, 7–23.

Flannery TF (1993) Revision of the genus *Melonycteris* (Pteropodidae: Mammalia). *Records of the Australian Museum* **45**, 59–80.

Flannery TF (1994a) *Possums of the World: A Monograph of the Phalangeroidea.* GEO Productions, Sydney.

Flannery TF (1994b) Systematic revision of *Emballonura furax* Thomas, 1911 and *E. dianae* Hill, 1956 (Chiroptera: Emballonuridae), with description of new species and subspecies. *Mammalia* **58**, 601–612.

Flannery TF (1995a) *The Mammals of New Guinea.* Cornell University Press, Ithaca, NY.

Flannery TF (1995b) *Mammals of the South-west Pacific and Moluccan Islands.* Reed Books, Chatswood.

Flannery TF (1996) Mammalian zoogeography of New Guinea and the southwestern Pacific. In *The Origin and Evolution of Pacific Island Biotas, New Guinea to Eastern Polynesia: Patterns and Processes.* (Eds A Keast and S Miller) pp. 399–406. SPB Academic Publishing, Amsterdam.

Flannery TF, Calaby J (1987) Notes on the species of *Spilocuscus* (Marsupialia: Phalangeridae) from northern New Guinea and the Admiralty and St. Matthias island groups. In *Possums and Opossums: Studies in Evolution Vol. 2.* (Ed. M Archer) pp. 547–558. Surrey Beatty and Sons, Chipping Norton.

Flannery TF, White JP (1991) Animal translocation. *National Geographic Research & Exploration* **7**, 96–113.

Flannery TF, Wickler S (1990) Quaternary murids (Rodentia: Muridae) from Buka Island, Papua New Guinea, with descriptions of two new species. *Australian Mammalogy* **13**, 127–139.

Flannery TF, Kirch PV, Specht J, Spriggs M (1988) Holocene mammal faunas from archaeological sites in island Melanesia. *Archaeology in Oceania* **23**, 89–94.

Flannery TF, Colgan D, Trimble J (1994) A new species of *Melomys* from Manus Island, Papua New Guinea. *Proceedings of the Linnean Society of New South Wales* **114**, 39–43.

Fleming C (1962) New Zealand biogeography: a paleontologist's approach. *Tuatara* **10**, 53–108.

Forster JR (1844) Descritiones Animalium quae in Itinere ad Maris Australis Terras per Annos 1772, 1773 et 1774 Suscepto Collegit Observavit et Delineavit. Ex Officina Academica, Berolini.

Förster B (1913) Neue Marsupialia vom Huongolf–Inland (Deutsch–Neuguinea). *Zoologischer Anzeiger* **42**, 177–178.

Förster B, Rothschild W (1907) Description of a new Tree-kangaroo. *Novitates Zoologicae* **14**, 506.

Förster B, Rothschild W (1914) Ein neues Känguruh aus Neuguinea. *Novitates Zoologicae* **XXI**, 261–262.

Forsyth DM (2021) *Rupicapra rupicapra.* In *The Handbook of New Zealand Mammals.* 3rd edn. (Eds CM King and DM Forsyth) pp. 393–445. CSIRO Publishing, Melbourne.

Forsyth DM, Hickling GJ (1998) Increasing Himalayan tahr and decreasing chamois densities in the eastern Southern Alps, New Zealand: evidence for interspecific competition. *Oecologia* **113**, 377–382. doi:10.1007/s004420050389

Forsyth DM, Tustin KG (2021) *Hemitragus jemlahicus.* In *The Handbook of New Zealand Mammals.* 3rd edn. (Eds CM King and DM Forsyth) pp. 393–445. CSIRO Publishing, Melbourne.

Francis G (1988) Stratigraphy of Manus Island, western New Ireland basin, Papua New Guinea. In *Geology and Offshore Resources of Pacific Islands Arcs – New Ireland and Manus Region, Papua New Guinea.* Earth Science Series 9. (Eds N Marlow, S Dadisman and N Exon) pp. 31–40. Circum-Pacific Council for Energy and Mineral Resources, Texas.

Francis CM, Habersetzer J (1998) Interspecific and intraspecific variation in echolocation call frequency and morphology of horseshoe bats, *Rhinolophus* and *Hipposideros.* In *Bat Biology and Conservation.* (Eds TH Kunz and P Racey) pp. 169–179. Smithsonian Institution Press, Washington DC.

Frechkop S (1932) Notes sur les Mammiferes 9. Un grand rat de l'ile Waigeu (Nouvelle Guinée). *Bulletin du Musée Royal d'Histoire Naturelle de Belgique* **24**, 1–11.

Frechkop S (1948) Notes sur les Mammifères XXXI. Sur une collection provenant de l'île de Bougainville. *Bulletin du Musée Royal d'Histoire Naturelle de Belgique* **24**, 1–12.

Fredericksen C, Spriggs M, Ambrose W (1993) Pamwak reockshelter: a Pleistocene site on Manus Island, Papua New Guinea. In *Sahul in Review, Pleistocene Archaeology in Australia, New Guinea and Island Melanesia, Occasional Papers in Prehistory, 24.* (Eds M Smith, M Sprigg and, B Frankhauser) Department of Prehistory, Research School of Pacific Studies, Australian National University, Canberra.

Fritts T, Rodda G (1998) The role of introduced species in the degradation of island ecosystems. *Annual Review of Ecology and Systematics* **29**, 113–140.

Gaimard J (1824) Descriptions des trois nouvelles especes de mammiferes de la famile marsupiaux, reportées par M.M. Quoy et Gaimard. *Bulletin des Sciences Naturelles et de Géologie* **1**, 271.

Gallina S, Lopez Arevalo H (2016) *Odocoileus virginianus.* The IUCN Red List of Threatened Species 2016: e.T42394A22162580. doi:10.2305/IUCN.UK.2016-2.RLTS.T42394A22162580.en

Garvey PM, Byrom AE (2021) *Mustela furo.* In *The Handbook of New Zealand Mammals.* 3rd edn. (Eds CM King and DM Forsyth) pp. 285–341. CSIRO Publishing, Melbourne.

Geoffroy É (1803) *Catalogue des Mammiféres du Musée National d'Histoire Naturelle, Paris.* Imprimerie Nationale, Paris.

Geoffroy É (1804) Note sur un nouveau genre de Mammiféres, de l'orde des rongeurs, sous le non d'hydromys. *Bulletin de la Société Philomathique de Paris* **3**, 353–354.

Geoffroy É (1810) Description des Roussettes et des Cephalotes, deux nouveaux genres de la famille des Chauve-souris. *Annales du Musée D'Histoire Naturelle, Paris* **15**, 86–108.

Geoffroy É (1813) Sur un Genre de Chauve-souris, sous le nom De Rhinolophes. *Annales du Musée D'Histoire Naturelle, Paris* **20**, 254–266.

Gilchrist J, Jennings A, Veron G, Cavallini P (2009) Family Herpestidae (mongooses). In *Handbook of the Mammals of the World. Vol. 1. Carnivores.* (Eds D Wilson and R Mittermeier) pp. 262–328. Lynx Edicions, Barcelona.

Gill BJ (2002) Records of bats (Mammalia: Chiroptera) from late Holocene dune-sands at Te Werahi Beach, Northland, New Zealand. *Records of the Auckland Museum* **39**, 45–47.

Gilliard ET, Lecroy M (1967) Results of the 1958–1959 Gilliard New Britain Expedition. 4. Annotated list of birds of the Whiteman Mountains, New Britain. *Bulletin of the American Museum of Natural History* **135**, 1–48.

Gillingham N (1996) The behaviour and ecology of long-tailed bats (*Chalinolobus tuberculatus* Gray) in the central North Island. Master of Science thesis. Massey University, New Zealand.

Giribet G, Baker CM (2019) Further discussion on the Eocene drowning of New Caledonia: discordances from the point of view of zoology. *Journal of Biogeography* **46**, 1912–1918. doi:10.1111/jbi.13635

Glass P, Taisacan EM (1988) Marianas fruit bat surveys and research. In *Five-year Progress Report, Fiscal Year 1982–87.* pp. 1–22. Division of Fish and Wildlife, Commonwealth of the Northern Mariana Islands, Saipan.

Global Forest Watch (2020) World Resources Institute, <https://www.globalforestwatch.org>.

Global Witness (2018) *Paradise Lost: How China can Help the Solomon Islands Protect its Forests.* Global Witness, London.

Glover I (1986) Archaeology in Eastern Timor, 1966–67. Dept of Prehistory, Research School of Pacific Studies, Australian National University, Canberra.

Gmelin J (1789) In *Systema Naturae, Per Regna Tria Naturae: Secundum Classes, Ordines, Genera, Species Cum Characteribus, Differentiis, Synonymis, Locis.* Ed. 13. Typis Ioannis Thomae, Vindobonae [Vienna].

Gordon G, Lawrie B (1978) The Rufescent Bandicoot, *Echymipera rufescens* (Peters & Doria). *Wildlife Research* **5**, 41–45.

Gorman ML (1975) The diet of feral *Herpestes auropunctatus* (Carnivora: Viverridae) in the Fijian Islands. *Journal of Zoology* **175**, 273–278.

Gould J (1842a) Descriptions of four new species of kangaroos. *Proceedings of the Zoological Society of London* **9**(1841), 80–83.

Gould J (1842b) Descriptions of two new species of kangaroos. *Proceedings of the Zoological Society of London* **10**(1842), 57–58.

Gould J (1850) On a new species of Mammalia and birds from Australia. *Proceedings of the Zoological Society of London* **1850**, 109–112.

Gould J (1853) *The Mammals of Australia, Pt 5.* John Gould, London.

Gould J (1854) *The Mammals of Australia, Pt 6.* John Gould, London.

Gould J (1856) *The Mammals of Australia, Pt 8.* John Gould, London.

Gould J (1863) *The Mammals of Australia, Pt 13.* John Gould, London.

Goulding W, Moss PT, McAlpine CA (2016) Cascading effects of cyclones on the biodiversity of Southwest Pacific islands. *Biological Conservation* **193**, 143–152. doi:10.1016/j.biocon.2015.11.022

Grandcolas P, Murienne J, Robillard T, Desutter-Grandcolas L, Jourdan H, Guilbert E, Deharveng L (2008) New Caledonia: a very old Darwinian island? *Philosophical Transactions of the Royal Society of London. Series B, Biological Sciences* **363**, 3309–3317. doi:10.1098/rstb.2008.0122

Gray JE (1827) *Synopsis of the Species of the Class Mammalia, as Arranged With Reference to Their Organization, by Cuvier, and Other Naturalists, with Specific Characters, Synonyma, &c. &c.* G. B. Whittaker, London.

Gray JE (1834) Characters of a new species of bat (*Rhinolophus* Geoffr.) from New Holland. *Proceedings of the Zoological Society of London* **1834**, 52–53.

Gray JE (1837) Description of some new or little known Mammàlia, principally in the British Museum collection. *Magazine of Natural History* **n.s. 1**, 577–587.

Gray JE (1843a) Descriptions of some new genera and species of Mammalia in the British Museum. *Annals & Magazine of Natural History* **11**, 117–119.

Gray JE (1843b) *List of the Specimens of Mammalia in the Collection of the British Museum.* The Trustees, London.

Gray JE (1843c) Fauna of New Zealand. In *Travels in New Zealand; With Contributions to the Geography, Geology, Botany and Natural History of That Country.* (Ed. E Dieffenbach) pp. 296ff. John Murray, London.

Gray JE (1858) List of species of mammalia sent from the Aru Islands by Mr A.R. Wallace to the British Museum. *Proceedings of the Zoological Society of London* **26**, 106–113.

Gray JE (1859) Notice of *Notopteris*, a new genus of Pteropine bat from the Feejee Islands. *Proceedings of the Zoological Society of London* **1859**, 36–38.

Gray JE (1863) Description of some new species of Mammalia. *Proceedings of the Zoological Society of London* **1862**, 261–263.

Gray JE (1866a) Notes on some mammalia from Port Albany (Cape York Peninsula), North Australia, with descriptions of some new species. *Proceedings of the Zoological Society of London* **34**, 219–221.

Gray JE (1866b) A revision of the genera of Rhinolophidae, or horseshoe bats. *Proceedings of the Zoological Society of London* **1866**, 81–83.

Gray JE (1867) Notes on the variegated or yellow-tailed rats of Australia. *Proceedings of the Zoological Society of London* **1867**, 597–600.

Gray JE (1870) *Catalogue of Monkeys, Lemurs, and Fruit-eating Bats in the* collection *of the British Museum.* British Museum, London.

Gray JE (1873) Notes on the rats; with the description of some new species from Panama and the Aru Islands. *Annals & Magazine of Natural History* (4)**12**, 416–419.

Green R, Anson D (1991) The Reber–Rakival Lapita site on watom. Implications of the 1985 excavations at SAC and DSI localities. In *Report of the Lapita Homeland Project*. Occasional Papers in Prehistory No. 20. (Eds J Allen and C Gosden) pp. 170–181. Department of Prehistory, Research School of Pacific Studies, Australian National University, Canberra.

Greig K, Rawlence NJ (2021) The contribution of kurī (Polynesian dog) to the ecological impacts of the human settlement of Aotearoa New Zealand. *Frontiers in Ecology and Evolution* **9**, 757988. doi:10.3389/fevo.2021.757988

Greig K, Gosling A, Collins CJ, Boocock J, McDonald K, Addison DJ, Allen MS, David B, Gibbs M, Higham CFW, *et al.* (2018) Complex history of dog (*Canis familiaris*) origins and translocations in the Pacific revealed by ancient mitogenomes. *Scientific Reports* **8**, 9130. doi:10.1038/s41598-018-27363-8

Griffiths T, Koopman KF, Starett A (1991) The systematic relationships of *Emballonura nigriscens* to other species of *Emballonura* and *Coleura* (Chiroptera: Emballonuridae). *American Museum Novitates* **2996**.

Griffiths R, Cranwell S, Derand D, Ghestemme T, Will D, Zito J, Hall T, Pott M, Coulston G (2019) Multi island, multi invasive species eradication in French Polynesia demonstrates economies of scale. In *Island invasives: scaling up to meet the challenge.* Occasional Paper SSC no. 62. (Eds CR Veitch, MN Clout, AR Martin, JC Russell and CJ West) pp. 611–617. IUCN, Gland, Switzerland.

Grossek GL, Woolley PA, Menzies JI (2010) The diet of some New Guinean dasyurid marsupials. *Australian Mammalogy* **32**, 145–155. doi:10.1071/AM10002

Groves CP (1982) The systematics of tree-kangaroos (*Dendrolagus*: Marsupialia, Macropodidae). *Australian Mammalogy* **5**, 157–186.

Groves CP (2005) Order Diprotodontia. In *Mammal Species of the World: A Taxonomic and Geographic Reference. Vol. 1.* (Eds D Wilson, D Reeder) pp. 43–70. Johns Hopkins University Press, Baltimore, MD.

Groves CP, Flannery TF (1989) Revision of the Genus *Dorcopsis* (Macropodidae: Marsupialia). In *Kangaroos, Wallabies and Rat-kangaroos.* (Eds G Grigg, P Jarman and I Hume) pp. 117–128. Surrey Beatty & Sons, Chipping Norton.

Groves CP, Flannery TF (1994) A revision of the genus *Uromys* Peters, 1867 (Muridae: Mammalia) with descriptions of two new species. *Records of the Australian Museum* **46**, 145–169. doi:10.3853/j.0067-1975.46.1994.12

Grubb PJ (1971) Interpretation of the 'Massenerhebung'effect on tropical mountains. *Nature* **229**, 44–45.

Günther A (1883) Description of a new species of *Perameles* from New Britain. *Annals and Magazine of Natural History, London* **11**, 247.

Hackman B (1973) The Solomon Islands fractured arc. In *The Western Pacific: Island Arcs; Marginal Seas; Geochemistry*. (Ed. P Coleman) pp. 179–191. University of Western Australia Press, Perth.

Hall R (2002) Cenozoic geological and plate tectonic evolution of SE Asia and the SW Pacific: computer-based reconstructions, model and animations. *Journal of Asian Earth Sciences* **20**, 353–431.

Hand SJ, Grant-Mackie JA (2011) Late-Holocene bats of Mé Auré Cave, New Caledonia: evidence of human consumption and a new species record from the recent past. *The Holocene* **22**, 79–90. doi:10.22459/TA31.12.2009.10

Hand SJ, Lee DE, Worthy TH, Archer M, Worthy JP, Tennyson A, Salisbury SW, Scofield RP, Mildenhall DC, Kennedy EM, Lindqvist JK (2015) Miocene fossils reveal ancient roots for New Zealand's endemic *Mystacina* (Chiroptera) and its rainforest habitat. *PLoS One* **10**, e0128871. doi:10.1371/journal.pone.0128871'domain=pdf

Hand SJ, Beck RMD, Archer M, Simmons NB, Gunnell GF, Scofield RP, Tennyson AJD, De Pietri VL, Salisbury SW, Worthy TH (2017) A new, large-bodied omnivorous bat (Noctilionoidea: Mystacinidae) reveals lost morphological and ecological diversity since the Miocene in New Zealand. *Scientific Reports* **8**, 1–11. doi:10.1038/s41598-017-18403-w

Handasyde K (2008) Striped possum, *Dactylopsila trivirgata*. In *The Mammals of Australia*. (Eds S Van Dyck and R Strahan) pp. 224–225. Reed New Holland, Sydney.

Harper GA, Bunbury N (2015) Invasive rats on tropical islands: their population biology and impacts on native species. *Global Ecology and Conservation* **3**, 607–627. doi:10.1016/j.gecco.2015.02.010

Harris R (2015) *Cervus nippon*. The IUCN Red List of Threatened Species 2015: e.T41788A22155877. doi:10.2305/IUCN.UK.2015-2.RLTS.T41788A22155877.en

Hawkins S (2015) *Human Behavioural Ecology, Anthropogenic Impact and Subsistence Change at the Teouma Lapita Site, Central Vanuatu, 3000–2500 BP*. Australian National University, Canberra.

Hawkins S, Worthy TH, Bedford S, Spriggs M, Clark G, Irwin G, Best S, Kirch P (2016) Ancient tortoise hunting in the southwest Pacific. *Scientific Reports* **6**, 20152399. doi:10.1016/0305-4403(84)90013-X

Hayes FE, Engbring J (2020) Historic and recent status of the Kosrae flying fox (*Pteropus ualanus*) (Chiroptera: Pteropidae) on Kosrae, Micronesia. *Journal of Asia-Pacific Biodiversity* **13**, 141–150. doi:10.1016/j.japb.2020.02.008

Hayes FE, Wiles G (2020) *Pteropus ualanus*. The IUCN Red List of Threatened Species 2020: e.T136531A21979719. doi:10.2305/IUCN.UK.2020-3.RLTS.T136531A21979719.en

Healey C (1995) Traps and trapping in the Aru Islands. *Cakalele: Maluku Research Journal* **6**, 51–65.

Heideman P, Utzurrum R (2003) Seasonality and synchrony of reproduction in three species of nectarivorous Philippines bats. *BMC Ecology* **3**, 11. doi:10.1186/1472-6785-3-11

Heinsohn T (2002) Possum extinctions at the marsupial frontier: the status of the northern common cuscus *Phalanger orientalis* on Santa Ana Island, Makira Province, Solomon Islands. *Australian Mammalogy* **24**, 247–248.

Heinsohn TE (2005) Wallaby extinctions at the macropodid frontier: the changing status of the northern pademelon *Thylogale browni* (Marsupialia: Macropodidae) in New Ireland Province, Papua New Guinea. *Australian Mammalogy* **27**, 175–183.

Heinsohn TE (2010a) Animal translocation: long-term human influences on the vertebrate zoogeography of Australasia (natural dispersal versus ethnophoresy). *Australian Zoologist* **32**, 351–376.

Heinsohn TE (2010b) Marsupials as introduced species: long-term anthropogenic expansion of the marsupial frontier and its implications for zoogeographic interpretation. In *Altered Ecologies: Fire, Climate and Human Influence on Terrestrial Landscapes. Terra Australis 32.* (Eds S Haberle, J Stevenson and M Prebble) pp. 133–176. ANU Press, Canberra.

Helgen K (2005) Systematics of the Pacific monkey-faced bats (Chiroptera: Pteropodidae), with a new species of *Pteralopex* and a new Fijian genus. *Systematics and Biodiversity* **3**, 433–453.

Helgen KM (2007) A reassessment of taxonomic diversity and geographic patterning in the Melanesian mammal fauna. PhD thesis. The University of Adelaide, Australia.

Helgen KM, Flannery TF (2004) Notes on the phalangerid marsupial genus *Spilocuscus*, with description of a new species from Papua. *Journal of Mammalogy* **85**, 825–833.

Helgen K, Leary T, Doria G, Amori G (2008) Catalogue of Melanesian rodents in the Museum of Genova (Mammalia: Rodentia). *Annali del Museo Civico di Storia Naturale Giacomo Doria* **99**, 653–686.

Helgen KM, Helgen LE, Wilson DE (2009) Pacific flying foxes (Mammalia: Chiroptera): two new species of *Pteropus* from Samoa, probably extinct. *American Museum Novitates* **3646**, 1–37. doi:10.1206/614.1

Heller K (1897) *Zwei neue Beutelthiere aus Deutsch New Guinea, nebst einer Aufzahlung der bekannten papuanischen Säugethiere*. Abhandungen Zoologishe Museum, Dresden.

Heude P (1896) *Memoires concernant l'Histoire Naturelle de l'Empire Chinosis*. Impr. De la Mission catholique, Shanghai, China.

Hide RL, Pernetta JC, Senabe T (1984) Exploitation of wild animals. In *The Research Report of the Simbu Land Use Project. Vol. 4. South Simbu: Studies in Demography, Nutrition and Subsistence* (Ed. RL Hide) pp. 291–380. Institute of Applied Social and Economic Research, Port Moresby.

Hill JE (1956) The mammals of Rennell Island. In *The Natural History of Rennell Island, British Solomon Islands. Vol. 1.* (Ed. T. Wolff) pp. 73–84. Danish Science Press Ltd, Copenhagen.

Hill JE (1962) A little-known fruit-bat from Rennell Island. In *The Natural History of Rennell Island, British Solomon Islands. Vol. 4.* (Ed. T Wolff) pp. 7–9. Danish Science Press, Ltd, Copenhagen.

Hill JE (1963) A revision of the genus *Hipposideros. Bulletin of the British Museum Natural History (Zoology)* **11**, 1–129, <http://biostor.org/reference/1291>

Hill JE (1965) Asiatic bats of the genera *Kerivoula* and *Phoniscus* (Vespertilionidae), with a note on *Kerivoula aerosa* Tomes. *Mammalia* **29**, 524–556.

Hill JE (1968) Notes on mammals from the islands of Rennell and Bellona. In *The Natural History of Rennell Island, British Solomon Islands. Vol. 5.* (Ed. T Wolff) pp. 53–60. Danish Science Press, Ltd, Copenhagen.

Hill JE (1971) Bats from the Solomon Islands. *Journal of Natural History* **5**, 573–581. doi:10.5962/bhl.part.26839

Hill JE (1983) Bats (Mammalia: Chiroptera) from Indo-Australia. *Bulletin of the British Museum (Natural History) Zoology* **45**, 103–208.

Hill JE, Beckon W (1978) A new species of *Pteralopex* Thomas, 1888 (Chiroptera: Pteropodidae) from the Fiji Islands. *Bulletin of the British Museum (Natural History) Zoology* **34**, 65–82.

Hill JE, Daniel M (1985) Systematics of the New Zealand short-tailed bat *Mystacina* Gray, 1843 (Chiroptera: Mystacinidae). *Bulletin of the British Museum (Natural History) Zoology* **48**, 279–300.

Hill K, Hall R (2003) Mesozoic-Cenozoic evolution of Australia's New Guinea margin in a west Pacific context. *Geological Society of Australia Special Publication* **22**, 259–283. doi:10.1130/0-8137-2372-8.265

Hill JE, Rozendaal F (1989) Records of bats (Microchiroptera) from Wallacea. *Zoologische Mededelingen* **63**, 97–122.

Hill EJ, Baldwin SL, Lister GS (1992) Unroofing of active metamorphic core complexes in the D'Entrecasteaux Islands, Papua New Guinea. *Geology* **20**, 907–910.

Hinton M (1925) Two new leaf-nosed bats from Buru. *Annals & Magazine of Natural History* **16**, 255–256.

Hodgson B (1836) Synoptical description of sundry new animals, enumerated in the Catalogue of Nepalese Mammals. *Journal of the Asiatic Society of Bengal* **5**, 231–238.

Hodgson B (1845) On the rats, mice and shrews of the central region of Nepal. *Annals & Magazine of Natural History* **15**, 266–270.

Holm RJ, Rosenbaum G, Richards SW (2016) Post 8 Ma reconstruction of Papua New Guinea and Solomon Islands: microplate tectonics in a convergent plate boundary setting. *Earth-Science Reviews* **156**, 66–81. doi:10.1016/j.earscirev.2016.03.005

Hood CS, Smith J (1989) Sperm storage in a tropical nectar-feeding bat, *Macroglossus minimus* (Pteropodidae). *Journal of Mammalogy* **70**, 404–406.

Horsfield T (1823) *Zoological Researches in Java and Neighbouring Islands.* Printed for Kingsbury, Parbury, & Allen, London.

Hume I, Jazwinski E, Flannery TF (1993) Morphology and function of the digestive tract in New Guinea possums. *Australian Journal of Zoology* **41**, 85–100.

Hunt TL, Kirch PV (1988) An archaeological survey of the Manu'a Islands, American Samoa. *The Journal of the Polynesian Society (N. Z.)* **97**, 153–183.

Husson A (1958) Notes on *Protemnodon agilis papuanus* (Peters & Doria) from New Guinea (Mammalia: Marsupialia). *Nova Guinea* **9**, 245–252.

Hutterer R, Molur S, Heaney L (2016) *Suncus murinus.* The IUCN Red List of Threatened Species 2016: e.T41440A22287830. doi:10.2305/IUCN.UK.2016-2.RLTS.T41440A22287830.en

Hutton F (1871) On the bats of New Zealand. *Transactions and Proceedings of the New Zealand Institute, Wellington* **4**, 184–186.

Ibáñez J, Juste J (2019) Family Miniopteridae (long-fingered bats). In *Handbook of the Mammals of the World. Vol. 9. Bats.* (Eds DE Wilson and RA Mittermeier) pp. 674–709. Lynx Edicions, Barcelona.

Ingleby S, Colgan D (2003) Electrophoretic studies of the systematic and biogeographic relationships of the Fijian bat genera *Pteropus*, *Pteralopex*, *Chaerephon* and *Notopteris*. *Australian Mammalogy* **25**, 13–29.

Ingleby S, Flannery TF (1991) 'A survey of the mammals of Fiji'. Report to the Government of Fiji. Australian Museum in collaboration with the National Trust of Fiji and the Ministry of Primary Industries, Suva, Fiji.

Iredale T, Troughton ELG (1934) A check-list of the mammals recorded from Australia. *Australian Museum Memoir* **6**, 1–122. doi:10.3853/j.0067-1967.6.1934.516

Irwin N (2017) A new tube-nosed fruit bat from New Guinea, *Nyctimene wrightae* sp. nov., a re-diagnosis of *N. certans* and *N. cyclotis* (Pteropodidae: Chiroptera), and a review of their conservation status. *Records of the Australian Museum* **69**, 73–100. doi:10.3853/j.2201-4349.69.2017.1654

IUCN (2021) The IUCN Red List of Threatened Species. Version 2021-2, <https://www.iucnredlist.org>

Jackson W (1962) Area of study. In *Pacific Island Rat Ecology*. (Ed. T Storer) pp. 14–20. B. P. Bishop Museum, Honolulu, Hawaii.

Jackson S, Groves CP (2015) *Taxonomy of Australian Mammals.* CSIRO Publishing, Clayton South.

Jacquinot H, Pucheran J (1853) Mammifères et oiseaux. In *Zoologie: III. Voyage au pole sud et dansl'Océanie sur les corvettes l'Astrolabe et la Zélée; overnm par ordre du roi pendant les années 1837–1838–1839–1840, sous le commandement de M. J.Dumont-d'Urville, capitaine de vaisseau; publié par ordre du overnment, sous la direction supérieure de M. Jacquinot, capitaine de vaisseau, commandant de la Zélée.* (Eds J Hombron and H Jacquinot) pp. 1–166. Gide and Baudry, Paris.

Jentink F (1902) Revision of *Macroglossus* and *Syconycteris*, and description of a new genus and species, *Odontonycteris meijeri*. *Notes from the Leyden Museum* **23**, 131–142.

Jentink F (1907) Mammals collected by the members of the Humboldt Bay and the Merauke River Expeditions. *Nova Guinea* **5**, 361–374.

Jentink F (1908) Mammals – collected by the Dutch New Guinea Expedition 1907. *Nova Guinea* **9**, 1–13.

Johnson D (1959) Four new mammals from the Northern Territory of Australia. *Proceedings of the Biological Society of Washington* **72**, 183–187.

Jones C (2021) *Erinaceus europaeus occidentalis*. In *The Handbook of New Zealand Mammals*. 3rd edn. (Eds CM King and DM Forsyth) pp. 79–93. CSIRO Publishing, Melbourne.

Jones G, Webb PI, Sedgeley JA, O'Donnell CFJ (2003) Mysterious *Mystacina*: how the New Zealand short-tailed bat (*Mystacina tuberculata*) locates insect prey. *The Journal of Experimental Biology* **206**, 4209–4216. doi:10.1242/jeb.00678

Jones C, Moss K, Sanders M (2005) Diet of hedgehogs (*Erinaceus europaeus*) in the upper Waitaki Basin, New Zealand: implications for conservation. *New Zealand Journal of Ecology* **29**, 29–35.

Jouan C (1863) Additions á la fauna de la Nouvelle Calédonie. *Mémoires de la Société Impériale des Sciences Naturelles de Cherbourg* **1864**, 301–311.

Jourdan M (1837) Mémoire sur quelques mammiféres nouveaux. *Comptes Rendus Hebdomadaires des Séances de l'Académie des Sciences* **5**, 521–524.

Kamp PJJ, Green PF, White SH (1989) Fission track analysis reveals character of collisional tectonics in New Zealand. *Tectonics* **8**, 169–195.

Karels TJ, Dobson FS, Trevino HS, Skibiel AL (2008) The biogeography of avian extinctions on oceanic islands. *Journal of Biogeography* **35**, 1106–1111. doi:10.1214/aoms/1177732676

Katovai E, Edwards W, Laurance WF (2015) Dynamics of logging in Solomon Islands: the need for restoration and conservation alternatives. *Tropical Conservation Science* **8**, 718–731.

Kealy S, Donnellan SC, Mitchell KJ, Herrera M, Aplin K, O'Connor S, Louys J (2019) Phylogenetic relationships of the cuscuses (Diprotodontia: Phalangeridae) of island Southeast Asia and Melanesia based on the mitochondrial ND2 gene. *Australian Mammalogy* **42**, 266–276. doi:10.1071/AM18050

Kelaart E (1850) Description of new species and varieties of mammals found in Ceylon. *Journal of the Ceylon Branch of the Royal Asiatic Society* **2**, 207–215.

Kemp L (2010) Establishment, behaviour and ecology of the SA mainland tammar wallaby (*Macropus eugenii eugenii*) following an experimental reintroduction. PhD thesis. The University of Adelaide, Australia.

Kerr R (1792) *The Animal Kingdom, or Zoological System, of the Celebrated Sir Charles Linnæus. Containing a Complete Systematic Description, Arrangement, and Nomenclature, of all the Known Species and Varieties of the Mammalia, or Animals Which Give Suck to their Young Class I, Mammalia.* Printed for A Strahan, T Cadell, London, and W Creech, Edinburgh.

Ketebengang H, Gupta A (2011) *State of Palau's Birds 2010: A Conservation Guide for Communities and Policymakers.*Palau Conservation Society, Koror, Republic of Palau.

King C (2017a) Pandora's box down-under: origins and numbers of mustelids transported to New Zealand for biological control of rabbits. *Biological Invasions* **19**, 1811–1823. doi:10.1007/s10530-017-1392-6

King CM (2017b) Liberation and spread of stoats (*Mustela erminea*) and weasels (*M. nivalis*) in New Zealand, 1883–1920. *New Zealand Journal of Ecology* **41**, 163–177.

King CM, Forsyth D (2021) *The Handbook of New Zealand Mammals.* 3rd edn. CSIRO Publishing, Melbourne.

King CM, Lloyd BD (2021) *Mystacina robusta.* In *The Handbook of New Zealand Mammals.* 3rd edn. (Eds CM King and DM Forsyth) pp. 95–130. CSIRO Publishing, Melbourne.

King CM, Murphy EC (2021) *Mustela nivalis vulgaris.* In *The Handbook of New Zealand Mammals.* 3rd edn. (Eds CM King and DM Forsyth) pp. 285–341. CSIRO Publishing, Melbourne.

King CM, Veale AJ (2021) *Mustela erminea.* In *The Handbook of New Zealand Mammals.* 3rd edn. (Eds CM King and DM Forsyth) pp. 285–341. CSIRO Publishing, Melbourne.

Kirch P (1987) Lapita and Oceanic cultural origins: excavations in the Mussau Islands, Bismarck Archipelago, 1985. *Journal of Field Archaeology* **14**, 163–180.

Kirch P (1988) The Talepakemalai Lapita site and Oceanic prehistory. *National Geographic Research & Exploration* **4**, 328–342.

Kirch P (2000) *On the Road of the Winds: An Archaeological History of the Pacific Islands Before European Contact.* University of California Press, Berkeley, CA.

Kirch P, Yen D (1982) *Tikopia, the Prehistory and Ecology of a Polynesian Outlier.* Bernice P. Bishop Museum Bulletin 238. Bishop Museum Press, Honolulu.

Kirch P, Flenley J, Steadman D, Lamont F, Dawson S (1992) Ancient environment degradation. Prehistoric human impacts on an island system: Mangaia, Central Polynesia. *National Geographic Research & Exploration* **8**, 166–179.

Kirsch R, Tupinier Y, Beuneux G, Rainho A (2002a) 'Contributions à l'inventaire chiroptérologique de la Nouvelle-Calédonie: Chiroptera Pacifca, Missions 2000 & 2001. Rapport final et recommandations'. SFEPM, Groupe Chiroptères outre-Mer.

Kirsch R, Tupinier Y, Rainho A (2002b) 'Chiroptera Pacifica: Missions Nouvelle-Calédonie 2000–2001: Rapport d'Activité'. Société Française pour l'Etude et la Protection des Mammifères, Paris.

Kitchener D, Caputi N, Jones B (1986) Revision of Australo-Papuan *Pipistrellus* and of *Falsistrellus* (Microchiroptera: Vespertilionidae). *Records of the Western Australian Museum* **12**, 435–495.

Kitchener D, How R, Cooper N, Suyanto A (1992) *Hipposideros diadema* (Chiroptera: Hipposideridae) in the Lesser Sunda Islands, Indonesia: taxonomy and geographic morphological variation. *Records of the Western Australian Museum* **16**, 1–60.

Kitchener D, Hisheh S, Schmitt L, Suyanto A (1994) Shrews (Soricidae: *Crocidura*) from the Lesser Sunda

Islands and southeast Maluku, eastern Indonesia. *Australian Mammalogy* **17**, 7–17.

Kitchener DJ, Cooper N, Maryanto I (1995) The *Myotis adversus* (Chiroptera: Vespertilionidae) species complex in Eastern Indonesia, Australia, Papua New Guinea and the Solomon Islands. *Records of the Western Australian Museum* **17**, 191–212.

Kittlitz F (1836) Observations zoologiques faites pendant l'expédition de la corvette Le Séniavine. In *Voyage Autour de Monde, Execute par Ordre de sa Majeste l'Empereur Nicolas I, sur la Corvette Le Seniaville, Dans los Annees 1826, 1827, 1828, et 1829, vol. 3.* (Ed. F Lutke) pp. 237–330. Didot, Paris.

Kloss C (1916) Notes on the *hypomelanus* fruit-bats of the straits of Malacca, with the description of a new race *Pteropus hypomelanus fretensis. Journal of the Federated Malay States Museum* **6**, 245–248.

Kobayashi K (2000) Horozontally-moving subducted slab may generate enigmatic features of the Palau and Yap Trench-Arcs. *Proceedings of the Japan Academy, Series B* **76**. doi:10.2183/pjab.76.133

Kock D (1981) *Philetor brachypterus* auf New-Britannien und den Philippinen (Mammalia: Chiroptera: Vespertilionidae). *Senckenbergiana Biologica* **61**, 313–319.

Kompanje E, Moeliker CW (2001) Some fruit bats from remote Moluccan and West-Papuan Islands, with the description of a new subspecies of *Macroglossus minimus* (Megachiroptera: Pteropodidae). *DEINSEA* **8**, 143–167.

Koopman KF (1979) Zoogeography of mammals from islands off the northeastern coast of New Guinea. *American Museum Novitates* **2690**, 1–17.

Koopman KF (1982) Results of the Archbold Expeditions. No. 109. Bats From Eastern Papua and the East Papuan Islands. *American Museum Novitates* **2747**, 1–34.

Koopman KF (1984) Taxonomic and distributional notes on tropical Australian bats. *American Museum Novitates* **2778**, 1–48.

Koopman KF (1993) Order Chiroptera. In *Mammal Species of the World: A Taxonomic and Geographic Reference.* 2nd edn. (Eds DE Wilson and DM Reeder) pp. 137–241. Smithsonian Institution Press, Washington, DC.

Koopman KF (1997) The subspecies of *Emballonura semicaudata* (Chiroptera: Emballonuridae). *Journal of Mammalogy* **78**, 358–360.

Koopman KF, Steadman DW (1995) Extinction and biogeography of bats on 'Eua, Kingdom of Tonga. *American Museum Novitates* **3125**.

Koppers AAP, Russell JA, Jackson MG, Konter J, Staudigel H, Hart SR (2008) Samoa reinstated as a primary hotspot trail. *Geology* **36**, 435–438. doi:10.1130/2008109

Krefft G (1864) *Catalogue of the Mammalia in the Collection of the Australian Museum.* Australian Museum Catalogue No. 2.Trustees of the Australian Museum, Sydney.

Krefft G (1867) Notes on the mammals and birds of Cape York, with description of two new rodents *Hapalotis. Proceedings of the Zoological Society of London* **1867**, 316–319.

Kress WJ (1985) Bat pollination of an old world *Heliconia. Biotropica* **17**, 302–308.

Kroenke LW with a contribution by Peter Rodda (1984) *Cenozoic Tectonic Development of the Southwest Pacific.* CCOP/SOPAC, Suva, Fiji.

Kuroda N (1938) *A list of Japanese mammals.* Self-published, Tokyo.

Lacépède B (1799) *Tableau des divisions, sous-divisions, ordres et genres des Mammifères* . Chez Plassan, Paris.

Lamaris J (2021) *Life History of the Admiralty Cuscus* Spilocuscus kraemeri *and Customary Management on Manus Island, Papua New Guinea.* University of Queensland, Gatton.

Lamaris J, Whitmore N (2018) Forest connectivity is important for sustaining Admiralty cuscus (*Spilocuscus kraemeri*) in traditional terrestrial no-take areas on Manus Island, Papua New Guinea. *Pacific Conservation Biology* **24**, 55–62. doi:10.13057/biodiv/d140207

Larson G, Cucchi T, Fujita M, Matisoo-Smith E, Robins J, Anderson A, Rolett B, Spriggs M, Dolman G, Kim TH, *et al.* (2007) Phylogeny and ancient DNA of Sus provides insights into neolithic expansion in Island Southeast Asia and Oceania. *Proceedings of the National Academy of Sciences* **104**, 4834–4839. doi:10.1073/pnas.0607753104

Latham ADM, Nugent G (2021) *Cervus canadensis.* In *The Handbook of New Zealand Mammals.* 3rd edn. (Eds CM King and DM Forsyth) pp. 447–527. CSIRO Publishing, Melbourne.

Latham ADM, Warburton B (2021) *Macropus eugenii, M. r. rufogriseus, M. parma, M. dorsalis, Petrogale penicillata, Wallabia bicolor.* In *The Handbook of New Zealand Mammals.* 3rd edn. (Eds CM King and DM Forsyth) pp. 1–26. CSIRO Publishing, Melbourne.

Latham ADM, Latham MC, Warburton B (2018) Current and predicted future distributions of wallabies in mainland New Zealand. *New Zealand Journal of Zoology* **46**, 31–47. doi:10.1016/j.ecoinf.2008.08.004

Laurie EMO (1952) Mammals collected by Mr Shaw Mayer in New Guinea 1932–1949. *Bulletin of the British Museum (Natural History) Zoology* **1**, 269–318.

Laurie EMO, Hill JE (1954) *List of Land Mammals of New Guinea and Celebes, and Adjacent Islands, 1758–1952.* British Museum (Natural History) Trustees, London.

Laurie EMO, Hill JE (1957) Renaming of *Miniopterus australis minor* Laurie & Hill. *Journal of Mammalogy* **38**, 128. doi:10.1093/jmammal/38.1.128

Lavery TH (2013) A reassessment of the type locality of the giant rat *Solomys salamonis* (Rodentia: Muridae) from the Solomon Islands. *Australian Mammalogy* **36**, 99–102. doi:10.1071/AM13019

Lavery TH, Fasi J (2017) Buying through your teeth: traditional currency and conservation of flying foxes *Pteropus* spp. in Solomon Islands. *Oryx* **53**, 505–512. doi:10.1017/S0030605317001004

Lavery TH, Judge H (2017) A new species of giant rat (Muridae, *Uromys*) from Vangunu, Solomon Islands. *Journal of Mammalogy* **98**, 1518–1530. doi:10.1093/jmammal/gyx116

Lavery T, Pikacha P, Sirikolo M (2013a) A record of reproduction in *Anthops ornatus* (Chiroptera: Hipposideridae), Ysabel Island, Solomon Islands. *Memoirs of the Queensland Museum* **56**, 84–85.

Lavery TH, Fisher DO, Flannery TF, Leung LK-P (2013b) Higher extinction rates of dasyurids on Australo-Papuan continental shelf islands and the zoogeography of New Guinea mammals. *Journal of Biogeography* **40**, 747–758.

Lavery TH, Leung LK-P, Seddon JM (2014) Molecular phylogeny of hipposiderid bats (Chiroptera: Hipposideridae) from Cape York Peninsula, Australia and northern Melanesia. *Zoologica Scripta* **43**, 429–442.

Lavery TH, Olds AD, Seddon JM, Leung LK-P (2016) The mammals of northern Melanesia: speciation, ecology, and biogeography. *Mammal Review* **46**, 60–76. doi:10.1111/mam.12057

Lavery TH, Posala C, Tasker E, Fisher DO (2020a) Ecological generalism and resilience of tropical island mammals to logging: a 23 year test. *Global Change Biology* **26**, 3285–3293. doi:10.1111/gcb.15038

Lavery TH, Alabai M, Esau T, Fuiberi S, Furina Di, Johnson RN, Kekeubata E, MacLaren D, Waneagea J, Flannery TF (2020b) Integrating traditional knowledge, science and conservation in the search for undescribed mammals on Malaita, Solomon Islands. *Pacific Conservation Biology* **26**, 404–411. doi:10.1071/PC19037

Lavery TH, Alabai M, Holland P, Qaqara C, Vatohi N (2020c) Feral cat abundance, density and activity in tropical island rainforests. *Wildlife Research* **47**, 660–668. doi:10.1071/WR19205

Lavery TH, Leary TN, Shaw C, Tahi M, Posala C, Pierce R (2020d) Ecology and conservation of bats in Temotu Province, Solomon Islands and Torba Province, Vanuatu. *Pacific Conservation Biology* **27**, 27–38. doi:10.1071/PC20035

Lavery TH, DeCicco LH, Richmond JQ, Tigulu IG, Andersen MJ, Boseto D, Moyle RG (2021) New faunal records from a World Heritage Site in danger: Rennell Island, Solomon Islands. *Pacific Science* **75**, 407–420.

Law BS, Chidel M (2004) Roosting and foraging ecology of the golden-tipped bat (*Kerivoula papuensis*) on the south coast of New South Wales. *Wildlife Research* **31**, 73–82. doi:10.1071/WR03001

Lawrence B (1945) Three new *Pteropus* from New Caledonia and the Solomons. *Proceedings of the New England Zoölogical Club* **23**, 59–69.

Leary T (2003) Solomon Islands. In *A Directory of Wetlands in Oceania*. (Ed. DA Scott) pp. 1–370. IWRB, Slimbridge, UK and AWB, Kuala Lumpur, Malaysia.

Leary T, Pennay M (2011) Echolocation calls of eight microchiroptera from Papua New Guinea. In *The Biology and Conservation of Australasian Bats*. (Eds B Law, P Eby, D Lunney and L Lumsden) pp. 106–127. Royal Zoological Society of New South Wales, Mosman.

Leavesley M (2005) Prehistoric hunting strategies in New Ireland, Papua New Guinea: the evidence of the cuscus. *Asian Perspective* **44**, 207–218.

Leavesley M, Chappell J (2004) Buang Marabek: additional early radiocarbon evidence of the colonization of the Bismarck Archipelago, Papua New Guinea. *Antiquity* Project Gallery 78(301), <http://www.antiquity.ac.uk/projgall/leavesley301>.

Lees A (1990) *A Representative Protected Forests System for the Solomon Islands*. Maruia Society, Nelson.

Lemke TO (1986) Distribution and status of the sheath-tailed bat (*Emballonura semicaudata*) in the Mariana Islands. *Journal of Mammalogy* **67**, 743–746.

Lesson R, Garnot P (1826) Description des Mammiféres. In *Voyage Autour de Monde, Exécuté par Ordre du Roi, Sur la Corvette de Sa Majesté, la Coquille, Pendant les Années 1822, 1823, 1824, et 1825* (Ed. L-I Duperrey) pp. 137–176. Arthus Bertrand, Paris.

Lever C (1985) *Naturalized Mammals of the World*. Longman Group Limited, London.

Lidicker JWZ, Ziegler AC (1968) Report on a collection of mammals from eastern New Guinea, including species keys for fourteen genera. *Proceedings of the California Academy of Sciences* **36**, 33–71.

Liem D (1977) 'A preliminary report on the ecosystem assessment and the possible re-establishment of agile wallaby (*Macropus agilis*) on Goodenough Island, Milne Bay Province'. Department of Natural Resources, Wildlife Division, Konedobu, Papua New Guinea.

Liem D (1983) Survey and management of wildlife resources along the Purari River. In *The Purari – Tropical Environment of a High Rainfall River Basin. Monographiae Biologicae. Vol 51*. (Ed. T Petr) pp. 269–282. Springer Netherlands, Dordrecht.

Lilley I (1986) *Prehistoric Exchange in the Vitiaz Strait, Papua New Guinea*. Australian National University, Canberra.

Lim V-C, Ramli R, Bhassu S, Wilson J-J (2017) A checklist of the bats of Peninsular Malaysia and progress towards a DNA barcode reference library. *PLoS One* **12**, e0179555. doi:10.1371/journal.pone.0179555.s011

Lindley I (1988) Early Cainozoic stratigraphy and structure of the Gazelle Peninsula, east New Britain: an example of extensional tectonics in the New Britain arc–trench complex. *Australian Journal of Earth Sciences* **35**, 231–244.

Linnaeus C (1758) *Sytema Naturæ per Regna Tria Naturæ, Secundumclasses, Ordines, Genera, Species, cum Characteribus, Differentiis, Synonymis, Locis.* Editio decima. Laurentius Salvius, Sweden.

Linnaeus C (1766) *Sytema Naturæ per Regna Tria Naturæ, Secundumclasses, Ordines, Genera, Species, cum Characteribus, Differentiis, Synonymis, Locis.* Editio duodecima. Laurentius Salvius, Sweden.

Liske-Clark J (2015) Wildlife Action Plan for the Commonwealth of the Northern Mariana Islands, 2015–2025. CNMI DLNR-Division of Fish and Wildlife, Saipan.

Long J (2003) Introduced mammals of the world: their history, distribution and influence. CABI Publishing, Wallingford, UK and Cambridge, MA.

Lönneberg E (1913) Results of Dr E. Mjöberg's Swedish Scientific Expedition to Australia, 1910–1913. I. Mammals. *Kunglika Svenska Vetenskapsakademiens Handlingar* **52**, 1–10.

Lydekker R (1896) *A Geographical History of Mammals.* Cambridge University Press, Cambridge, UK.

Macdonald L (2002) Going bats: an archaelogical investigation of flying fox usage in prehistoric Vanuatu. PhD Thesis. University of Auckland, New Zealand.

MacGillivray J (1860) Zoological notes from Aneiteum, New Hebrides. *The Zoologist* **18**, 7133–7142.

Mack AL, Wright DD (2005) The frugivore community and the fruiting plant flora in a New Guinea rainforest: identifying keystone frugivores. In *Tropical Fruits and Frugivores.* (Eds JL Dew and JP Boubli) pp. 183–203. Springer, Dordrecht.

MacKinnon J, Ong P, Gonzales J (2015) *Rusa marianna.* The IUCN Red List of Threatened Species 2015: e.T4274A22168586. doi:10.2305/IUCN.UK.2015-2.RLTS.T4274A22168586.en

Macqueen P, Seddon JM, Austin JJ, Hamilton S, Goldizen AW (2010) Phylogenetics of the pademelons (Macropodidae: *Thylogale*) and historical biogeography of the Australo-Papuan region. *Molecular Phylogenetics and Evolution* **57**, 1134–1148. doi:10.1016/j.ympev.2010.08.010

Maeda K (1982) *Studies on the Classification of Miniopterus in Eurasia, Australia and Melanesia. Honyurui kagaku (Mammalian Science) Suppl. 1.* Mammal Research Association, Kyoto.

Malekian M, Cooper SJB, Norman JA, Christidis L, Carthew SM (2010) Molecular systematics and evolutionary origins of the genus *Petaurus* (Marsupialia: Petauridae) in Australia and New Guinea. *Molecular Phylogenetics and Evolution* **54**, 122–135. doi:10.1016/j.ympev.2009.07.026

Manne LL, Brooks TM, Pimm SL (1999) Relative risk of extinction of passerine birds on continents and islands. *Nature* **399**, 258–261. doi:10.1038/380115a0

Marshall B, Allen J (1991) Excavations at Panakiwuk Cave, New Ireland. In *Report of the Lapita Homeland Project.* Occasional Papers in Prehistory No. 20. (Eds J Allen and C Gosden) pp. 55–91. Department of Prehistory, Research School of Pacific Studies, Australian National University, Canberra.

Martin HB (1981) *The Polynesian Journal of Captain Henry Byam Martin, R.N. in Command of H.M.S. Grampus-50 Guns at Hawaii and on Station in Tahiti and the Society Islands, August 1846 to August 1847.* Australian National University Press, Canberra.

Martin R (2005) *Tree-kangaroos of Australia and New Guinea.* CSIRO, Melbourne.

Matisoo-Smith E (2007) Animal translocations, genetic variation and the human settlement of the Pacific. In *Genes, Language and Culture History in the Southwest Pacific.* (Ed. JS Friedlaender) pp. 157–170. Oxford University Press, New York.

Matisoo-Smith E, Robins JH (2004) Origins and dispersals of Pacific peoples: evidence from mtDNA phylogenies of the Pacific rat. *Proceedings of the National Academy of Sciences of the United States of America* **101**, 9167–9172.

Matschie P (1899) Die Fledermäuse des Berliner Museums fur Naturkunde.*1. Lieferung.* In *Die Megachiroptera des Berliner Museums für Naturkunde.* G. Reimer, Berlin.

Matschie P (1901) Die Säugetiere der von W. Kükenthal auf Halmahera, Batjan und Nord-Celebes gemachten Ausbeute. *Abhandlungen herausgegeben von der Senckenbergischen Naturforschenden Gesellschaft* 55(2), 247–296.

Matschie P (1915) Eine Beiträge zur Kenntis der Garrung *Pseudocheirus. Sitzungsberichte der Gesellschaft Naturforschender Freunde zu Berlin* **13**, 83–95.

Matschie P (1916) Die verbreitung der beutelthiere auf Neuguinea. *Mitteilungen aud den Zoologischen Museum Berlin* **8**, 259–308.

Maurizot P, Campbell H (2020) Paleobiogeography of New Caledonia. *Geological Society, London, Memoirs* **51**, 189–213.

Maynes G (2008) Parma Wallaby *Macropus parma.* In *The Mammals of Australia.* (Eds S Van Dyck and R Strahan) pp. 341–343. Reed New Holland, Sydney.

Mayr E, Diamond J (2001) *The Birds of Northern Melanesia: Speciation, Ecology & Biogeography.* Oxford University Press, New York.

McConkey KR, Drake DR (2006) Flying foxes cease to function as seed dispersers long before they become rare. *Ecology* **87**, 271–276.

McCracken GF, Wilkinson GS (2000) Bat mating systems. In *Reproductive Biology of Bats*. (Eds EG Crichton and PH Krutzsch) pp. 321–362. Academic Press, London.

McDougall I (2010) Age of volcanism and its migration in the Samoa Islands. *Geological Magazine* **147**, 705–717. doi:10.1130/GSAB-55-1279

McGregor M, Bishop R (2011) *A Technical Assessment of the Current Agricultural Conditions of Angaur Island Palau: With Recommendations for the Sustainable Use of the Island's Natural Resources*. Government of Palau, Palau.

McKean J (1972) *Notes on some collections of bats (Order Chiroptera) from Papua New Guinea and Bougainville Island. CSIRO Division of Wildlife Research Technical Paper No. 26*. CSIRO, Melbourne.

McKean J, Calaby JH (1968) A new genus and species of bat from New Guinea. *Mammalia* **32**, 372–378.

Mearns E (1905) Descriptions of new mammals from the Philippine Islands. *Proceedings of the United States National Museum* **28**, 69–72.

Medina FM, Bonnaud E, Vidal E, Tershy BR, Zavaleta ES, Josh Donlan C, Keitt BS, Corre M, Horwath SV, Nogales M (2011) A global review of the impacts of invasive cats on island endangered vertebrates. *Global Change Biology* **17**, 3503–3510. doi:10.1111/j.1365-2486.2011.02464.x

Medway L, Marshall AG (1975) Terrestrial vertebrates of the New Hebrides: origin and distribution. *Philosophical Transactions of the Royal Society of London. Series B, Biological Sciences* **272**, 423–465.

Meffre S, Crawford A (2001) Collisional tectonics in the New Hebrides arc (Vanuatu). *The Island Arc* **10**, 33–50.

Meier G, Varnham K (2004) Rat eradication as part of a green turtle (*Chelonia mydas*) conservation programme in Indonesia. *Marine Turtle Newsletter* **106**, 11–12.

Meijer A, Reagan M (1981) Petrology and geochemistry of the island of Sarigan in the Mariana Arc; Calcalcaline volcanism in an oceanic setting. *Contributions to Mineralogy and Petrology* **77**, 337–354.

Menzies J (1971) The lobe-lipped bat (*Chalinolobus nigrogriseus* Gould) in New Guinea. *Records of the Papua New Guinea Museum* **1**, 6–8.

Menzies JI (1996) A systematic revision of *Melomys* (Rodentia: Muridae) of New Guinea. *Australian Journal of Zoology* **44**, 367–426.

Menzies J (2011) *A Handbook of New Guinea's Marsupials and Monotremes*. Masalai Press, Oakland, CA.

Menzies JI, Dennis E (1979) *Handbook of New Guinea Rodents*. Wau Ecology Institute, Wau, Papua New Guinea.

Menzies J, Parnetta J (1986) A taxonomic revision of cuscuses allied to *Phalanger orientalis* (Marsupialia: Phalangeridae). *Journal of Zoology* **1**, 551–618.

Merchant J (2008) Agile wallaby *Macropus agilis*. In *The Mammals of Australia*. (Eds S Van Dyck and R Strahan) pp. 323–324. Reed New Holland, Chatswood.

Meredith RW, Mendoza MA, Roberts KK, Westerman M, Springer MS (2010) A phylogeny and timescale for the evolution of Pseudocheiridae (Marsupialia: Diprotodontia) in Australia and New Guinea. *Journal of Mammalian Evolution* **17**, 75–99. doi:10.1007/s10914-010-9129-7

Meyen F (1833) Beiträge zur Zoologie gesamelt auf einer Reise um die Erde. *Abhandlungen Nova Acta Akademie Caes. Leopoldino-carolinischen. Halle* **16**, 549–610.

Meyer A (1876) On the habitat of *Uromys aruensis* (Gray) and its allies. *Annals & Magazine of Natural History* **17**, 145.

Meyer A (1899) Säugethiere vom Celébes- und Philippinen-Archipel. II. Celébes-Sammlungen der Herren Sarasin. *Abhandlungen und Berichte des königlichen zoologischen und anthropologisch-ethnographischen Museums zu Dresden*. **7**(7), 1–55.

Mickleburgh S, Hutson A, Racey P (1992) *Old World Fruit Bats: An Action Plan for their Conservation*. IUCN, Gland, Switzerland.

Miklouho-Maclay N (1884) Notes on the zoology of the Maclay-Coast in New Guinea 1. On a new sub-genus of Peramelidae *Brachymelis*. *Proceedings of the Linnean Society of New South Wales* **9**, 713–720.

Miklouho-Maclay N (1885a) Notes on the zoology of the Maclay–Coast of New Guinea II. On a new species of *Macropus*. *Proceedings of the Linnean Society of New South Wales* **10**, 141–144.

Miklouho-Maclay N (1885b) On two new species of *Macropus* from the south coast of New Guinea. *Proceedings of the Linnean Society of New South Wales* **9**, 890–895.

Mildenstein T, Tanshi I, Racey P (2016) Exploitation of bats for bushmeat and medicine. In *Bats in the Anthropocene: Conservation of Bats in a Changing World*. (Eds C Voigt and T Kingston) pp. 325–375. Springer Open, London.

Miller G (1900) Mammals collected by Dr W.L. Abbott on islands in the North China Sea. *Proceedings of the Washington Academy of Sciences* **2**, 203–246.

Miller G (1902) Two new tropical old world bats. *Proceedings of the Biological Society of Washington* **15**, 245–246.

Miller G (1903) Seventy new Malayan mammals. *Smithsonian Miscellaneous Collections* **90**, 1–73.

Miller G (1906) The mammals of Engano Island, West Sumatra. *Proceedings of the United States National Museum* **30**, 819–825.

Miller G (1910) Descriptions of two new genera and sixteen new species of mammals from the Philippine Islands. *Proceedings of the United States National Museum* **38**, 391–410.

Miller G (1911) A new bat from the Caroline Islands. *Proceedings of the Biological Society of Washington* **35**, 93–104.

Milne D (2002) 'Key to the bat calls of the top end of the Northern Territory'. Technical Report No. 71. Parks and Wildlife Commission of the Northern Territory, Darwin.

Milne-Edwards A (1877) Note sur quelques Mammiféres nouveaux provenant de la Nouvelle-Guinée. *Comptes Rendus Hebdomadaires des Séances de l'Académie des Sciences* **85**, 1079–1081.

Milne-Edwards A (1878) Description d'une Nouvelle Espéce de Perameles–Provenant de la Nouvelle-Guinée. *Annales des Sciences Naturelles, Paris* (6)**7**, 1–2.

Milne-Edwards A (1880) On a new species of *Dasyurus* from New Guinea. *Annals and Magazine of Natural History, London* **6**, 171–172.

Milsom J (1985) New Guinea and the western Melanesian arcs. In *The Ocean Basins and Margins.* (Eds A Nairn, F Stehli and S Uyeda) pp. 551–606. Plenum Press, New York.

Mitchell A, Warden A (1971) Geological evolution of the New Hebrides island arc. *Journal of the Geological Society* **127**, 501–529.

Mittermeier R, Robles Gil P, Hoffman M, Pilgrim J, Brooks T, Mittermeier C, Lamoreux J, da Fonseca G (2004) *Hotspots Revisited.* CEMEX, Mexico City.

Molnar R (1982) A longirostrine crocodilian from Murua (Woodlark), Solomon Sea. *Memoirs of the Queensland Museum* **20**, 675–685.

Moore L (2008) Giant white-tailed rat *Uromys caudimaculatus*. In *The Mammals of Australia.* (Eds S Van Dyck and R Strahan) pp. 675–677. Reed New Holland, Chatswood.

Moore L, Leung LK-P (2008) Cape York rat *Rattus leucopus*. In *The Mammals of Australia.* (Eds S Van Dyck and R Strahan) pp. 688–689. Reed New Holland, Chatswood.

Morat P, Jaffré T, Tronchet F, Munzinger J, Pillon Y, Veillon J, Chalopin M (2012) Le référentiel taxonomique Florical et les caractéristiques de la flore vasculaire indigène de la Nouvelle-Calédonie. *Adansonia sér* **3**, 177–219. doi:10.5252/a2012n2a1

Morley CG, McLenachan PA, Lockhart P (2007) Evidence for the presence of a second species of mongoose in the Fiji Islands. *Pacific Conservation Biology* **13**, 29–34.

Mortimer N (2004) New Zealand's geological foundations. *Gondwana Research* **7**, 261–272. doi:10.1016/S1342937X(05)70324-5

Mudappa D (2002) Observations of small carnivores in the Kalakad-Mundanthurai Tiger Reserve, Western Ghats, India. *Small Carnivore Conservation : the Newsletter and Journal of the IUCN/SSC Mustelid, Viverrid & Procynid Specialist Group* **27**, 4–5.

Müeller-Dombois D, Fosberg F (2013) *Vegetation of the Tropical Pacific Islands.* Springer, New York.

Müller P (1776) *Von Linnés Vollständiges Natursystem, Nach der Zwölften Lateinschen Ausgab und nach Anleitung des Holländischen Houttynischen Werks, mit einer Ausführlichen Erklärung Ausfertiget. Supplement.* Raspe, Nuremberg.

Müller S (1840) Bijdragen tot de kennis van Nieuw-Guinea. In *Verhandlingen over De Natuurlijke Geschiedenis der Nederlandische Overzeesche Bezittingen, door de Leden der Natuurkundige Commissie in Indie en Andere Schrijvers.* (Ed. CJ Temminck) p. 20. In commissie bij. S. en J. Luchtmans en C.C. van der Hoek, Leiden.

Murphy S (2002) Observations of the critically endangered bare-rumped shetath-tail bat *Saccolaimus saccolaimus* Temminck (Chiroptera: Emballonuridae) on Cape York Peninsula, Queensland. *Australian Mammalogy* **23**, 185–187.

Murphy EC, Nathan HW (2021) *Mus musculus*. In *The Handbook of New Zealand Mammals.* 3rd edn. (Eds CM King and DM Forsyth) pp. 161–240. CSIRO Publishing, Melbourne.

Murphy E, Maddigan F, Edwards B, Clapperton K (2008) Diet of stoats at Okarito Kiwi Sanctuary, South Westland, New Zealand. *New Zealand Journal of Ecology* **32**, 41–45.

Murray P (1978) A Pleistocene spiny anteater from Tasmania (Monotremata: Tachyglossidae, *Zaglossus*). *Papers and Proceedings of the Royal Society of Tasmania* **112**, 39–68.

Musser G, Carleton M (2005) Superfamily Muroidea. In *Mammal Species of the World: A Taxonomic and Geographic Reference.* (Eds DE Wilson and DM Reeder) pp. 894–1531. Johns Hopkins University, Baltimore, MD.

Neall VE, Trewick SA (2008) The age and origin of the Pacific islands: a geological overview. *Philosophical Transactions of the Royal Society of London. Series B, Biological Sciences* **363**, 3293–3308. doi:10.1186/BF03351733

Neaves LE, Danks M, Lott MJ, Dennison S, Frankham GJ, King A, Eldridge MDB, Johnson RN, Divljan A (2018) Unmasking the complexity of species identification in Australasian flying-foxes. *PLoS One* **13**, e0194908. doi:10.1371/journal.pone.0194908.s007

Nedachi M, Maeda T, Shigeyoshi R, Inoue A, Shimada K, Higashi M, Habano A, Azuma T, Hatta A (2001) Geological environments of Yap islands, Micronesia. *Kagoshima University Research Center for the Pacific Islands Occasional Papers* **34**, 69–76.

Nelson J, Hamilton-Smith E (1982) Some observations on *Notopteris macdonaldi* (Chiroptera: Pteropodidae). *Australian Mammalogy* **5**, 247–252.

Nicholl M (1904) Observations on natural history made during the voyage round the world of the R.Y.S. Valhalla, 1902–1903. *The Zoologist* **4**, 401–420.

Norbury G, Duckworth JA (2021) *Oryctolagus cuniculus*. In *The Handbook of New Zealand Mammals*. 3rd edn. (Eds CM King and DM Forsyth) pp. 131–159. CSIRO Publishing, Melbourne.

Norbury G, Flux J (2021) *Lepus europaeus occidentalis*. In *The Handbook of New Zealand Mammals*. 3rd edn. (Eds CM King and DM Forsyth) pp. 131–159. CSIRO Publishing, Melbourne.

Norbury G, McGlinchy A (1996) The impact of rabbit control on predator sightings in the semi-arid high country of the South Island, New Zealand. *Wildlife Research* **23**, 93–97.

Norris A (1999) *Phalanger lullulae*. *Mammalian Species* **620**, 1–4. doi:10.2307/3504488/2600455

Norris C, Musser G (2001) Systematic revision within the *Phalanger orientalis* complex (Diprotodontia, Phalangeridae): a third species of lowland gray cuscus from New Guinea and Australia. *American Museum Novitates* **3356**, 1–20. doi:10.1206/0003-0082(2001)356<0001:SRWTPO>2.0.CO;2

Novick A (1955) Laryngeal muscles of the bat and production of ultrasonic sounds. *The American Journal of Physiology* **183**, 64.

Nugent G (2021) *Rusa unicolor*. In *The Handbook of New Zealand Mammals*. 3rd edn. (Eds CM King and DM Forsyth) pp. 447–527. CSIRO Publishing, Melbourne.

Nugent G, Asher GW (2021) *Dama dama*. In *The Handbook of New Zealand Mammals*. 3rd edn. (Eds CM King and DM Forsyth) pp. 447–527. CSIRO Publishing, Melbourne.

Nugent G, Challies CN (1988) Diet and food preferences of white-tailed deer in north-eastern Stewart Island. *New Zealand Journal of Ecology* **11**, 61–71.

Nugent G, Forsyth DM (2021) *Cervus elaphus*. In *The Handbook of New Zealand Mammals*. 3rd edn. (Eds CM King and DM Forsyth) pp. 447–527. CSIRO Publishing, Melbourne.

Nugent G, Speedy C (2021) *Cervus nippon*. In *The Handbook of New Zealand Mammals*. 3rd edn. (Eds CM King and DM Forsyth) pp. 447–527. CSIRO Publishing, Melbourne.

O'Connor S, Hiscock P (2018) The peopling of Sahul and Near Oceania. In *The Oxford Handbook of Prehistoric Oceania*. (Eds E Cochrane and T Hunt) pp. 26–47. Oxford University Press, New York.

O'Donnell CFJ (2001) Home range and use of space by *Chalinolobus tuberculatus*, a temperate rainforest bat from New Zealand. *Journal of Zoology* **253**, 253–264. doi:10.1017/S095283690100022X

O'Donnell CFJ (2002) Timing of breeding, productivity and survival of long-tailed bats *Chalinolobus tuberculatus* (Chiroptera: Vespertilionidae) in cold-temperate rainforest in New Zealand. *Journal of Zoology* **257**, 311–323.

O'Donnell CFJ, Borkin KM (2021) *Chalinolobus tuberculatus*. In *The Handbook of New Zealand Mammals*. 3rd edn. (Eds CM King and DM Forsyth) pp. 95–130. CSIRO Publishing, Melbourne.

O'Donnell CFJ, Weston KA, Monks JM (2017) Impacts of introduced mammalian predators on New Zealand's alpine fauna. *New Zealand Journal of Ecology* **41**, 1–22.

Obidzinski K, Kusters K, Gnych S (2015) Taking the bitter with the sweet: sugarcane's return as a driver of tropical deforestation. *Conservation Letters* **8**, 449–455. doi:10.1111/conl.12172

Oedin M, Brescia F, Boissenin M, Vidal E, Cassan J-J, Hurlin J-C, Millon A (2019) Monitoring hunted species of cultural significance: estimates of trends, population sizes and harvesting rates of flying-fox (*Pteropus* sp.) in New Caledonia. *PLoS One* **14**, e0224466. doi:10.1371/journal.pone.0224466.s004

Oedin M, Brescia F, Millon A, Murphy BP, Palmas P, Woinarski JCZ, Vidal E (2021) Cats *Felis catus* as a threat to bats worldwide: a review of the evidence. *Mammal Review* **51**, 323–327. doi:10.1111/mam.12240

Oken L (1816) *Lehrbuch der Naturgeschchichte Th. 3. Zoologie*. Schmidt, Jena.

Olsen PD (1982) Reproductive biology and development of the water rat, *Hydromys chrysogaster*, in captivity. *Wildlife Research* **9**, 39–53. doi:10.1071/WR9820039

Opiang M (2009) Home ranges, movement, and den use in long-beaked echidnas, *Zaglossus bartoni*, from Papua New Guinea. *Journal of Mammalogy* **90**, 340–346.

Owen R (1839) Outlines of a classification of the Marsupialia. *Proceedings of the Zoological Society of London* **7**, 3–19.

Oyler-McCance S, Valdex E, O'Shea T, Fike J (2013) Genetic characterization of the Pacific sheath-tailed bat (*Emballonura semicaudata rotensis*) using mitochondrial DNA sequence data. *Journal of Mammalogy* **94**, 1030–1036.

Paijmans K (Ed.) (1976) *New Guinea Vegetation*. Australian National University Press, Canberra.

Pallas P (1766) *Miscellania Zoologica, Quibus Novæ Imprimis Atque Obscuræ Animalum Species Describuntur et Observationibus Iconibusque Illustrantur.* van Cleef, Den Haag.

Pallas P (1777) Das Zwitterstinktier Viverra hermaphrodita. In *Die Säugethiere in Abbildungen nach der Natur, mit Beschreibungen. Vol. 3*. (Ed. J Schreber) p. 426. Wolfgang Walther, Erlangen.

Pallas P (1778) *Novae Species Quadrupedum e Glirium Ordine*. Wolfgangi Waltheri, Erlangen.

Palmas P, Jourdan H, Rigault F, Debar L, De Meringo H, Bourguet E, Mathivet M, Lee M, Adjouhgniope R, Papillon Y, Bonnaud E, Vidal E (2017) Feral cats threaten the outstanding endemic fauna of the New Caledonia biodiversity hotspot. *Biological Conservation* **214**, 250–259. doi:10.1016/j.biocon.2017.08.003

Palmeirim JM, Champion A, Naikatini A, Niukula J, Tuiwawa M, Fisher M, Yabaki-Gounder M, Thorsteinsdóttir S, Qalovaki S, Dunn T (2007) Distribution, status and conservation of the bats of the Fiji Islands. *Oryx* **41**, 509–519. doi:10.1017/S0030605307004036

Parkes JP (2021a) *Bos taurus*. In *The Handbook of New Zealand Mammals*. 3rd edn. (Eds CM King and DM Forsyth) pp. 393–445. CSIRO Publishing, Melbourne.

Parkes JP (2021b) *Ovis aries*. In *The Handbook of New Zealand Mammals*. 3rd edn. (Eds CM King and DM Forsyth) pp. 393–445. CSIRO Publishing, Melbourne.

Parkes J, Murphy E (2003) Management of introduced mammals in New Zealand. *New Zealand Journal of Zoology* **30**, 335–359. doi:10.1890/0012-9615(2001)071[0587:IBMINZ]2.0.CO;2

Parnaby H (2002a) A taxonomic review of the genus *Pteralopex* (Chiroptera: Pteropodidae), the monkey-faced bats of the south-western Pacific. *Australian Mammalogy* **23**, 145–162.

Parnaby H (2002b) A new species of long-eared bat (*Nyctophilus*: Vespertilionidae) from New Caledonia. *Australian Mammalogy* **23**, 115–124.

Parnaby H, Ingleby S, Diviljan A (2017) Type specimens of non-fossil mammals in the Australian Museum, Sydney. *Records of the Australian Museum* **69**, 277–420.

Parnaby H, Divljan A, Hoye G (2020) A review of the taxonomic status of the New Caledonia Wattled Bat *Chalinolobus neocaledonicus* Revilliod, 1914 (Chiroptera: Vespertilionidae) and *Chalinolobus gouldii* venatoris Thomas, 1908 from northern Australia. *Zootaxa* **4778**, 135–158.

Parsons S, Toth C (2021) *Mystacina tuberculata*. In *The Handbook of New Zealand Mammals*. 3rd edn. (Eds CM King and DM Forsyth) pp. 95–130. CSIRO Publishing, Melbourne.

Parsons JG, Cairns A, Johnson CN, Robson SKA, Shilton LA, Westcott DA (2006) Dietary variation in spectacled flying foxes (*Pteropus conspicillatus*) of the Australian Wet Tropics. *Australian Journal of Zoology* **54**, 417–428. doi:10.2307/2997325

Pattabiraman N, Morgan-Richards M, Powlesland R, Trewick SA (2022) Unrestricted gene flow between two subspecies of translocated brushtail possums (*Trichosurus vulpecula*) in Aotearoa New Zealand. *Biological Invasions* **24**, 247–260. doi:10.1007/s10530-021-02635-z

Pattiselanno F, Iriansul, Barnes PA, Arobaya AYS (2021) Using local ecological knowledge to locate the western long-beaked echidna *Zaglossus bruijnii* on the Vogelkop Peninsula, West Papua, Indonesia. *Oryx* doi:10.1017/S0030605321000351

Pattiselanno F, Arobaya AYS (2012) A rapid survey of small mammals from the Northern Tamrau Nature Reserve, Papua. *Tiger Paper* **39**, 20–24.

Pavey CR, Burwell CJ (1997) The diet of the diadem leaf-nosed bat *Hipposideros diadema*: confirmation of a morphologically-based prediction of carnivory. *Journal of Zoology* **243**, 295–303.

Pavey C, Young R (2008) Eastern horseshoe bat, *Rhinolophus megaphyllus*. In *The Mammals of Australia*. 3rd edn. (Eds S Van Dyck and R Strahan) pp. 451–453. Reed New Holland, Sydney.

Pavlides C, Gosden C (1994) 35,000-year-old sites in the rainforests of West New Britain, Papua New Guinea. *Antiquity* **68**, 604–610. doi:10.1007/BF00889209

Peale T (1848) Mammalogy and ornithology. In *United States Exploring Expedition, During the Years 1838, 1839, 1840, 1841, 1842, Under the Command of Charles Wilkes, U.S.N.* C. Sherman, Philadelphia.

Pelletier B (2006) Geology of the New Caledonia region and its implications for the study of the New Caledonian biodiversity. In *Compendium of Marine Species of New Caledonia. Documents Scientifiques et Techniques 117*. 2nd edn. (Eds C Pyri and B Richer de Forges) pp. 19–32. IRD, Nouméa.

Pelseneer P (1904) La 'Ligne de Weber,' limite zoologique de l'Asie et de l'Australie. *Bulletin de l'Académie Royale des Sciences de Belgique Classe des Sciences* **1904**, 1001–1022.

Pennay M, Lavery TH (2017) 'Identification guide to bat echolocation calls of Solomon Islands & Bougainville'. Unpublished report, <https://www.ausbats.org.au/bat-calls-of-the-solomon-islands.html>

Perez G (1968) Notes on Palau fruit bats. *Journal of Mammalogy* **49**, 758.

Perez G (1972) Observations of Guam bats. *Micronesica* **8**, 141–149.

Perez G (1973) Notes on the ecology and life history of Pteropodidae on Guam. *Periodicum Biologorum* **75**, 163–168.

Peters W (1861) Über die von Hrn. F. Jagor bisher auf Malacca, Borneo, Java und den Philippinen gesammelten Säugethiere aus den Ordnung dar Halbaffen, Pelzflatterer und Fledertiere. *Monatsberichte der Königlichen Preussischen Akademie der Wissenschaften zu Berlin* **1861**, 706–712.

Peters W (1862) Mittheilung über einer neuen Flederhunde, *Pteropus scapulatus*, aus Neuholland. *Monatsberichte der Königlichen Preussischen Akademie der Wissenschaften zu Berlin* **1862**, 574–577.

Peters W (1867) Über die Flederhunde, *Pteropi*, und insbesondere über die Arten der Gattung *Pteropus*. *Monatsberichte der Königlichen Preussischen Akademie der Wissenschaften zu Berlin* **1867**, 319–333.

Peters W (1869) Ueber neue oder weniger bekannte Flederthiere, insbesondere des Pariser Museums. *Monatsberichte der Königlichen Preussischen Akademie der Wissenschaften zu Berlin* **1869**, 391–406.

Peters W (1871) Über die gattungen und arten der hufeisennasen, *Rhinolophi*. *Monatsberichte der Königlichen Preussischen Akademie der Wissenschaften zu Berlin* **1871**, 301–338.

Peters W (1874) Diagnosi di tre nuovi mammiferi della Nuova Guinea ed Isole Key. *Annali del Museo Civico di Storia Naturale di Genova* **6**, 303.

Peters W (1875) Diagnosi di una nuova species di marsupial del genre *Chaetocercus*. *Annali del Museo Civico di Storia Naturale di Genova* **7**, 420.

Peters W (1876) Mittheilung uber die Pelzrobbe von den Inselen St Paul und Amsterdam und uber die von S.M.S. Gazelle mitgebrachten Flederthiere. *Monatsberichte der Königlichen Preussischen Akademie der Wissenschaften zu Berlin* **18**, 315–319.

Peters W (1880) Mittheilung über eine neue Art von Flederthieren, *Vesperugo angulatus*, von der Papua-Insel Duke of York. *Sitzungsberichte der Gesellschaft Naturforschender Freunde zu Berlin* **1880**, 122.

Peters W (1881) Über die Chiropterengattung *Mormopterus* und die dahin gehörigen Arten. *Monatsberichte der Königlichen Preussische Akademie des Wissenschaften zu Berlin* **1881**, 482–485.

Peters W (1883) Über die von Herren Dr Finsch von den Carolinin Inseln ein-gasandten flederhunde. *Sitzungsberichte der Gesellschaft Naturforschender Freunde zu Berlin* **1883**, 1.

Peters W (1906) *Museum Zoologicum Berolinzuse. Chiroptera*. Berlin.

Peters W, Doria G (1875) Diagnosi di alcune nuove specie di marsupiali appartenenti fauna papuana. *Annali del Museo Civico di Storia Naturale di Genova* **7**, 541.

Peters W, Doria G (1876a) Descrizione di una nuova specie di *Tachyglossus* proveniente dalla Nuova Guinea settentrionale. *Annali del Museo Civico di Storia Naturale di Genova* **9**, 183–187.

Peters W, Doria G (1876b) Diagnosi di tre nuove specie de mammiferi della Nuova Guine e di Salawatti. *Annali del Museo Civico di Storia Naturale di Genova* **8**, 335–336.

Peters W, Doria G (1881) Enumerazione dei Mammiferi raccolti da O. Beccari, L.M. D'Albertis, ed A.A. Bruijn, nella Nuova Guinea propriamente detta. *Annali del Museo civico di storia naturale di Genova* **16**, 664–707 (Plates V–XVIII).

Petersen R (1981) Systematic variation in the *tristis* group of the bent-wing bats of the genus *Miniopterus* (Chiroptera: Vespertilionidae). *Canadian Journal of Zoology* **59**, 828–843.

Petersen RL (1991) Systematic variation in the Megachiropteran Tube-nosed Bats *Nyctimene cyclotis* and *N. certans*. *Bulletin of the American Museum of Natural History* **206**, 26–41.

Peterson GD, Jr (1956a) *Suncus murinus*, a recent Introduction to Guam. *Journal of Mammalogy* **37**, 278–279. doi:10.2307/1376699

Peterson RF (1956b) *5th Archbold Expedition to New Guinea: Field Catalog April 11 to December 17, 1956*. Field Book Collection. Unpublished, American Museum of Natural History, New York.

Petit S, Scanlon AT, Naikatini A, Pukala T, Schumann R (2022) A novel bat pollination system involving obligate flower corolla removal has implications for global *Dillenia* conservation. *PLoS One* **17**, e0262985. doi:10.1371/journal.pone.0262985

Petterson M, Neal C, Mahoney J, Kroenke L, Saunders A, Babbs T, Duncan R, Tolia D, McGrail B (1997) Structure and deformation of north and central Malaita, Solomon Islands: tectonic implications for the Ontong Java Plateau-Solomon arc collision, and for the fate of oceanic plateaus. *Tectonophysics* **283**, 1–33.

Phillips C (1966) A new species of bat of the genus *Melonycteris* from the Solomon Islands. *Journal of Mammalogy* **47**, 23–27.

Phillips C (1967) A new subspecies of horseshoe bat (*Hipposideros diadema*) from the Solomon Islands. *Proceedings of the Biological Society of Washington* **80**, 35–40.

Phillips C (1968) Systematics of Megachiropteran bats in the Solomon Islands. *University of Kansas Publications. Museum of Natural History* **16**, 779–837.

Pierce R, Anterea N, Coulston G, Gardiner C, Shilton L, Taabu K, Wragg G (2011) Atoll restoration in the Phoenix Islands, Kiribati: survey results in November–December 2009. Critical Ecosystem Partnership Fund (CEPF) and Conservation International Pacific Islands Program (CI-Pacific), Apia, Samoa.

Pierson E, Rainey W (1992) The biology of flying foxes of the genus *Pteropus*: a review. In *Pacific Island Flying Foxes: Proceedings of an International Conference*. 1–2 February 1990, Honolulu. (Eds DE Wilson and GL Graham) pp. 1–17. U.S. Department of the Interior Fish and Wildlife Service, Washington, DC.

Pierson ED, Elmqvist T, Rainey WE, Cox PA (1996) Effects of tropical cyclonic storms on flying fox populations on the South Pacific Islands of Samoa. *Conservation Biology* **10**, 438–451.

Pigram C, Davies P (1987) Terranes and the accretion history of the New Guinea orogen. *BMR Journal of Australian Geology and Geophysics* **10**, 193–212.

Pikacha P (2005) *The Conservation Status and Ethnobiology of Small Mammals and Amphibians in Solomon Islands: A Case Study of Sene Village, Sou.* University of the South Pacific, School of Social and Economic Development, Suva.

Pikacha P (2008) 'Distribution, habitat preference, and conservation status of the endemic giant rats *Solomys ponceleti* and *S. salebrosus* on Choiseul Island, Solomon Islands'. BP Conservation 2005 Final Report, Project No. 700305. BP Conservation Leadership Programme. Conservation International, Honiara.

Pine RH, Mack AL, Timm RM (2017) Marsupials and rodents of the Admiralty Islands, Papua New Guinea. *Occasional Papers Museum of Texas Tech University* **352**, 1–32.

Pohle H (1952) Über die Fledertiere von Bougainville. *Zeitschrift für Säugetierkunde* **17**, 127–137.

Poirier FE, Smith EO (1974) The crab-eating macaques (*Macaca fascicularis*) of Angaur Island, Palau, Micronesia. *Folia Primatologica* **22**, 258–282. doi:10.1159/000317609

Porolak G (2008) Home range of the Huon tree kangaroo, *Dendrolagus matschiei*, in cloud forest on the Huon peninsula, Papua New Guinea. Masters thesis. James Cook University, Australia.

Porolak G, Dabek L, Krockenberger AK (2014) Spatial requirements of free-ranging huon tree kangaroos, *Dendrolagus matschiei* (Macropodidae), in upper montane forest. *PLoS One* **9**, e91870. doi:10.1371/journal.pone.0091870.t002

Pracy L (1974) *Introduction and Liberation of the Opossum (*Trichosurus vulpecula*) in New Zealand.* NZ Forest Service, Wellington.

Prié V (2011) Focus on bats. In *The Natural History of Santo.* (Eds P Bouchet, H Le Guyader and O Pascal) pp. 316–322. Muséum National d'Histoire Naturelle; IRD, Marseille; Pro-Natura International, Paris.

Pryde MA, O'Donnell CFJ, Barker RJ (2005) Factors influencing survival and long-term population viability of New Zealand long-tailed bats (*Chalinolobus tuberculatus*): implications for conservation. *Biological Conservation* **126**, 175–185. doi:10.1016/j.biocon.2005.05.006

Pulvers J, Colgan D (2007) Molecular phylogeography of the fruit bat genus *Melonycteris* in northern Melanesia. *Journal of Biogeography* **34**, 713–723.

Quoy J, Gaimard J (1824) *Voyage Autor du Monde, Enterpris par Ordre Duroi, Execute sur la Corvette i'Urania et al Physicienne, Pendant les Années 1817–1820.* Chez Pillet aîné, Paris.

Quoy J, Gaimard J (1830) *Voyage de la Corvette l'Astrolabe: Exécuté par Ordre du roi, Pendant les Années 1826-1827-1828-1829, Zoologie.* J. Tastu, Paris.

Raffles TS (1821) Descriptive catalogue of a zoological collection, made on account of the Honourable East India Company, in the island of Sumatra and its vicinity, under the direction of Sir Thomas Stamford Raffles, Lieutenant-Governor of Fort Marlborough. *Transactions of the Linnean Society of London* **13**, 246–247.

Rainey W (1990) The flying fox trade: becoming a rare commodity. *Bats* **8**, 6–9.

Rainey WE, Pierson ED (1992) Distribution of Pacific island flying foxes. In *Pacific Island Flying Foxes: Proceedings of an International Conservation Conference.* 1–2 February 1990, Honolulu. (Eds DE Wilson and GL Graham) pp. 111–121. Department of the Interior, Fish and Wildlife Service, Washington, DC.

Ramme W (1914) *Dactylopsila hindenburgi,* ein neuer Streifenbeutler aus Kaiser-Wilhelms-Land. *Sitzungsberichte der Gesellschaft Naturforschender Freunde zu Berlin* **1914**, 413–419.

Ramsay EP (1876) Description of a new species of kangaroo, from New Guinea. *Proceedings of the Linnean Society of New South Wales* (1)**1**, 162–164.

Ramsay EP (1877a) Description of a new species of *Perameles*, from New Ireland. *Proceedings of the Linnean Society of New South Wales* (1)**1**, 310.

Ramsay EP (1877b) Description of a new species of *Halmatrus*, from New Ireland. *Proceedings of the Linnean Society of New South Wales* (1)**1**, 307–309.

Ramsay EP (1877c) Notes on a collection of birds from Port Moresby, with descriptions of some new species. *Proceedings of the Linnean Society of New South Wales* (1)**1**, 386–395.

Ramsay EP (1877d) Descriptions of two supposed new species of *Mus* and of a Pteropine bat, of a new genus, from Duke of York Island. *Proceedings of the Linnean Society of New South Wales* (1)**2**, 15–19.

Ramsay EP (1877e) Zoology of the 'Chevert'. Mammals Pt 1. *Proceedings of the Linnean Society of New South Wales* (1)**2**, 7–15.

Ramsay EP (1882) On a new species of *Mus* from the island of Ugi, Solomon Islands. *Proceedings of the Linnean Society of New South Wales* (1)7, 43–44.

Ramsay EP (1883) Contributions to the zoology of New Guinea, Part 7. *Proceedings of the Linnean Society of New South Wales* (1)**8**, 15–29.

Ramsay EP (1887) Description of two new species of marsupials (*Perameles* and *Antechinus*), and of a new species of *Mus* (*M. burtoni*), from the neighbourhood of Derby, N.W.A. *Proceedings of the Linnean Society of New South Wales* (2)**2**, 551–553.

Ramsay EP (1891) On a new species of pteropine bat from the New Britain group. *Records of the Australian Museum* **1**, 105–106.

Ramsey DSL, Forsyth DM (2018) 'Estimates of Himalayan Tahr (*Hemitragus jemlahicus*) abundance in New Zealand: results from aerial surveys'. Unpublished client report for the New Zealand Department of Conservation. Arthur Rylah Institute for Environmental Research, Department of Environment, Land, Water and Planning, Heidelberg.

Rand A, Brass LJ (1940) Results of the Archbold Expeditions. No. 29. Summary of the 1936–1937 New Guinea Expedition. *Bulletin of the American Museum of Natural History* **77**, 341–380.

Reagan M, Meijer A (1984) Geology and geochemistry of early arc-volcanic rocks from Guam. *Geological Society of America Bulletin* **95**, 701–713.

Reardon T (2017) *Mormopterus beccarii* (errata version published in 2019). The IUCN Red List of Threatened Species 2017: e.T13880A154818942. doi:10.2305/IUCN.UK.2017-2.RLTS.T13880A154818942.en)

Reardon TB, Mckenzie NL, Cooper SJB, Appleton B, Carthew S, Adams M (2014) A molecular and morphological investigation of species boundaries and phylogenetic relationships in Australian free-tailed bats *Mormopterus* (Chiroptera: Molossidae). *Australian Journal of Zoology* **62**, 109. doi:10.1071/ZO13082

Reinhold L, Law B, Ford G, Pennay M (2001) Key to the bat calls of south-east Queensland and north-east New South Wales. Department of Natural Resources and Mines, Brisbane.

Revilliod P (1914) Les mammiféres de la Nouvelle-Calédonie et des îsles Loyalty. In *Nova Caledonia. A. Zoologie* Vol. 4. (Eds F Sarasin and J Roux) pp. 340–365. C.W. Kreidels Verlag, Weisenbad.

Robin C, Monzier M, Crawford AJ, Eggins SM (1993) *The Geology, Volcanology, Petrology-Geochemistry, and Techtonic Evolution of the New Hebrides Island Arc, Vanuatu. Excursion Guide: Ancient Volcanism and Modern Analogues.* IAVCEI, Canberra.

Robinet O, Craig J, Chardonnet L (1998) Impact of rat species in Ouvea and Lifou (Loyalty Islands) and their consequences for conserving the endangered Ouvea Parakeet. *Biological Conservation* **86**, 223–232.

Robson S, Inkster TE, Krockenberger A (2012) *Bats of the YUS conservation area Papua New Guinea.* School of Marine & Tropical Biology James Cook University, Townsville.

Roe D (1992) *Investigations into the Prehistory of the Central Solomons: Some Old and Some New Data from Northwest Guadalcanal.* (Ed. J-Ce Galipaud) pp. 91–102. Orstom, Nouméa.

Rose J, Koppers AAP (2019) Simplifying age progressions within the Cook-Austral islands using ARGUS-VI high-resolution 40Ar/39Ar incremental heating ages. *Geochemistry Geophysics Geosystems* **20**, 4756–4778. doi:10.1029/2019GC008302

Rothschild W (1892) Description of two new mammals from New Guinea. *Proceedings of the Zoological Society of London* **1892**, 545–546.

Rothschild W (1922) In Thomas, O. and Rothschild W. On a new subspecies of *Zaglossus*, with remarks on other species of the genus. *Annals and Magazine of Natural History, London* **10**, 129–131.

Rothschild W, Rothschild N (1898) Descriptions of three new kangaroos, and notes on the skull of *Dendrolagus bennettianus* De Vis. *Novitates Zoologicae* **5**, 511–513.

Rouys S, Theuerkauf J (2003) Factors determining the distribution of introduced mammals in nature reserves of the southern province, New Caledonia. *Wildlife Research* **30**, 187–191. doi:10.1071/WR01116

Rowe KC, Aplin KP, Baverstock PR, Moritz C (2011) Recent and rapid speciation with limited morphological disparity in the genus *Rattus. Systematic Biology* **60**, 188–203. doi:10.1093/sysbio/syq092

Rümmler H (1935) Neue Muriden aus Neuguinea. *Zeitschrift für Säugetierkunde* **10**, 105–118.

Russell AL, Brown VA, Utzurrum RCB, Brooke AP, Wolf LA, Mccracken GF (2016) Comparative Phylogeography of *Pteropus samoensis* and *P. tonganus* (Pteropodidae: Chiroptera) in the South Pacific. *Acta Chiropterologica* **18**, 325–335. doi:10.3161/15081109 ACC2016.18.2.002.s1

Sanborn C (1930) Two new fruit bats collected by the Whitney south sea expedition. *American Museum Novitates* **435**.

Sanborn C (1931) Bats from Polynesia, Melanesia and Malaysia. *Field Museum of Natural History Zoological Series* **18**, 7–29.

Sanborn C, Beecher W (1947) Bats from the Solomon Islands. *Journal of Mammalogy* **28**, 387–391.

Sanborn CC, Nicholson AJ (1950) Bats from New Caledonia, the Solomon Islands, and New Hebrides. *Fieldiana* **31**, 313–338.

Sánchez-Bayo F, Wyckhuys KAG (2019) Worldwide decline of the entomofauna: a review of its drivers. *Biological Conservation* **232**, 8–27. doi:10.1016/j.biocon.2019.01.020

Sand C (1998) Archaeological report on localities WKO013A and WKO013B at the site of Lapita (Koné, New Caledonia). *The Journal of the Polynesian Society (N. Z.)* **107**, 7–33.

Scanlon A (2009) The long-tailed flying-fox (*Notopteris macdonaldi*): Viti Levu, Fiji. *The Australasian Bat Society Newsletter* **32**, 17–19.

Scanlon AT (2019) *Mirimiri acrodonta*. The IUCN Red List of Threatened Species 2019: e.T18655A22071017.

doi:10.2305/IUCN.UK.2019-3.RLTS.T18655A22071017.en

Scanlon A, Petit S, Bottroff G (2013a) The conservation status of bats in Fiji. *Oryx* **48**, 451–459. doi:10.3161/150811011X624785

Scanlon AT, Petit S, Sternberg LS (2013b) Insectivory in Fijian flying foxes (Pteropodidae). *Australian Journal of Zoology* **61**, 342–349. doi:10.1071/ZO13047

Scanlon AT, Petit S, Tuiwawa M, Naikatini A (2014) High similarity between a bat-serviced plant assemblage and that used by humans. *Biological Conservation* **174**, 111–119. doi:10.1016/j.biocon.2014.03.023

Schlegel H (1866) Observations Zoologiques. II + III. Mammiféres. *Nederlandsch Tijdschrift voor de Dierkunde* **3**, 353–357.

Schlegel H (1880) On *Dasyurus albopunctatus*. *Notes from the Leyden Museum* **2**, 51–53.

Schlitter D, Williams S, Hill J (1983) Taxonomic review of Temminck's Trident Bat, *Aselliscus tricuspidatus* Temminck, (1834) (Mammalia: Hipposideridae). *Annals of the Carnegie Museum* **52**, 337–358.

Schreber J (1778) *Die Saügethiere in Abbildungen nach der Natur, mit Beschreibungen.* Wolfgang Walther, Erlangen.

Schulz M (2000) Diet and foraging behavior of the Golden-Tipped Bat, *Kerivoula papuensis*: a spider specialist? *Journal of Mammalogy* **81**, 948–957. doi:10.1644/1545-1542(2000)081<0948:Dafbot>2.0.Co;2.

Schulz M, Richards G, Coles R, Spencer H (2008) Flute-nosed Bat *Murina florium*. In *The Mammals of Australia*. (Eds S Van Dyck and R Strahan) pp. 514–515. Reed New Holland, Sydney.

Schwartz E (1934) On a Wallaby and a *Phalanger* brought by Mr. Wilfred Frost from the Islands West of New Guinea. With Notes on the Evolution of Coat, Colour, and Pattern in the Genus *Phalanger*. *Proceedings of the Zoological Society of London* **104**, 87–92.

Schwarz E (1910) Ein anscheinend neuer Fleckenkuskus von den Admiralitäts-Inseln. *Gesellschaft Naturforschender Freunde Berliner Sitzungberichte* **1910**, 406–408.

Seabra Ad (1898) Noticia sobre uma nova especie do genero 'Cynonycteris' e annatação das especies d'este genero que existem nas colleçoes do Museu Nacional de Lisboa. *Journal de Sciencias, Mathematicas Physicas et Naturales Academia Sciencas, Lisboa* (2)**5**, 157–162.

Secretariat of the Pacific Regional Environment Programme (SPREP) (2016) 'State of conservation in Oceania:regional report'. SPREP, Apia, Samoa.

Sedgeley JA (2003) Roost site selection and roosting behaviour in lesser short-tailed bats (*Mystacina tuberculata*) in comparison with long-tailed bats (*Chalinolobus tuberculatus*) in Nothofagusforest, Fiordland. *New Zealand Journal of Zoology* **30**, 227–241. doi:10.1111/j.1469-7998.1999.tb00991.x

Shamel H (1940) The insectivorous bats collected by H. C. Raven in Celebes. *Journal of Mammalogy* **21**, 352–354.

Shevill DI, Johnson C (2008) Diet and breeding of the rufous spiny bandicoot *Echymipera rufescens australis*, Iron Range, Cape York Peninsula. *Australian Mammalogy* **29**, 169–175.

Shiels AB, Pitt WC, Sugihara RT, Witmer GW (2014) Biology and impacts of Pacific island invasive species. 11. *Rattus rattus*, the Black Rat (Rodentia: Muridae). *Pacific Science* **68**, 145–184. doi:10.2984/68.2.1

Simmons N (2005) Order Chiroptera. In *Mammal Species of the World: A Taxonomic and Geographic Reference*. (Eds D Wilson and D Reeder) pp. 312–529. Johns Hopkins University Press, Baltimore, MD.

Sinaga U, Maryanto I (2008) *Scotophilus collinus*. The IUCN Red List of Threatened Species 2008: e.T136612A4318302. doi:10.2305/IUCN.UK.2008.RLTS.T136612A4318302.en

Smith C (1826) *The Seventh Order of the Mammalia. The Ruminantia.* In *The Animal Kingdom Arranged in Conformity with its Organization, by the Baron Cuvier, Member of the Institute of France, &c. &c. &c. with Additional Descriptions of all the Species Hitherto Named, and of Many not Before Noticed. Vol. 4.* (Eds E Griffith, C Smith and E Pidgeon) pp. 1–498. G. B. Whittaker, London.

Smith I (2000) Terrestrial fauna from excavations at the Kainapirina (SAC) locality, Watom Island, Papua New Guinea. *New Zealand Journal of Archaeology* **20**, 137–147.

Smith J, Hill J (1981) A new species and subspecies of bat of the *Hipposideros bicolor*-group from Papua New Guinea, and the systematic status of *Hipposideros calcaratus* and *Hipposideros cupidus* (Mammalia: Chiroptera: Hipposideridae). *Contributions in Science* **331**, 1–19.

Smith J, Hood CS (1981) Preliminary notes on bats from the Bismarck Archipelago (Mammalia: Chiroptera). *Science in New Guinea* **8**, 81–121.

Smith J, Hood C (1983) A new species of tube-nosed fruit-bat (*Nyctimene*) from the Bismarck Archipelago, Papua New Guinea. *Occasional Papers. The Museum, Texas Tech University* **81**, 1–14.

Smith A, Xie Y (2008) *A Guide to the Mammals of China*. Princeton University Press, Princeton, NJ.

Sody H (1936) Seventeen new generic, specific and subspecific names for Dutch East Indian mammals. *Natuurkundig tijdschrift voor Nederlandsch-Indië* **96**, 42–55.

Spennemann D (1997) Distribution of rat species (*Rattus* spp.) on the atolls of the Marshall Islands: past and present dispersal. *Atoll Research Bulletin* **446**, 1–18.

Spriggs M (1991) Nissan, the island in the middle. Summary report on excavation at the north end of the Solomons and the south end of the Bismarcks. In *Report of the*

Lapita Homeland Project, Occasional Papers in Prehistory No. 20. (Eds J Allen and C Gosden) pp. 222–243. Department of Prehistory, Research School of Pacific Studies, Australian National University, Canberra.

Srinivasulu C, Srinivasulu B (2006) First record of *Hipposideros ater* Templeton, 1848 from Andhra Pradesh, India, with a description of a new subspecies. *Zoos' Print Journal* **21**, 2241–2244.

Steadman DW (1991) Extinct and extirpated birds from Aitutaki and Atiu, Cook Islands. *Pacific Science* **45**, 325–347.

Steadman DW (1995) Prehistoric extinction of Pacific island birds: Biodiversity meets zooarchaeology. *Science* **267**, 1123–1131.

Steadman DW, Kirch PV (2016) Biogeography and prehistoric exploitation of birds in the Mussau Islands, Bismarck Archipelago, Papua New Guinea. *Emu – Austral Ornithology* **98**, 13–22. doi:10.1017/S0003598X00075104

Stein G (1932) Eiige neue Beuteltiere aus Neuguinea. *Zeitschrift für Säugetierkunde* **7**, 254–257.

Stein G (1933) Weitere Mitteilungen zur Systematik Papuanischer Sauger. *Zeitschrift für Säugetierkunde* **8**, 87–95.

Stewart GH, Burrows LE (1989) The impact of white-tailed deer *Odocoileus virginianus* on regeneration in the coastal forests of Stewart Island, New Zealand. *Biological Conservation* **49**, 275–293.

Stewart W, Sandy M (1988) Geology of New Ireland and Djaul Islands, northeastern Papua New Guinea. In *Geology and Offshore Resources of Pacific Islands Arcs – New Ireland and Manus Region, Papua New Guinea.* Circum-Pacific Council for Energy and Mineral Resources Earth Science Series 9. (Eds N Marlow, S Dadisman and N Exon) pp. 31–40.

Stewart GH, Wardle J, Burrows L (1987) Forest understorey changes after reduction in deer numbers, Northern Fiordland, New Zealand. *New Zealand Journal of Ecology* **10**, 35–42.

Stinson D, Glass P, Taisacan E (1992) Declines and trade in fruit bats on Saipan, Tinian, Aguijan and Rota. *Pacific Island Flying Foxes: Proceedings of an International Conservation Conference*. 1–2 February 1990, Honolulu. (Eds DE Wilson and GL Graham). pp. 18–23. Department of the Interior, Fish and Wildlife Service, Washington, DC.

Summerhayes G, Szabó K, Fairbairn A, Horrocks M, McPherson S, Crowther A (2019) Early Lapita subsistence: the evidence from Kamgot, Anir Islands, New Ireland Province, Papua New Guinea. In *Debating Lapita: Distribution, Chronology, Society and Subsistence*. (Eds S Bedford and M Spriggs) pp. 379–402. ANU Press, Canberra.

Sutherland R, Dickens G, Blum P, Agnini C, Alegret L, Asatryan G, Bhattacharya J, Bordenave A, Chang L, Collot J, *et al.* (2020) Continental scale geo-graphic change across Zealandia during Paleogene subductioninitiation. *Geology* **48**, 419–424. doi:10.1130/G47008.1

Tarburton MK (2002) Demise of the Polynesian sheath-tailed bat *Emballonura semicaudata* in Samoa. *Micronesica* **34**, 105–108.

Tarragon L (1847) Description du male adulte du Colobe guereza et du *Phascogale virginiae*. *Revue de Zoologie* **1847**, 177–178.

Tate GHH (1934a) An Apparently New Fruit Bat of the *Pteropus hypomelanus* Group from Gower Island, Solomon Islands. *American Museum Noviates* **718**.

Tate GHH (1934b) Bats from the Pacific Islands, including a new fruit bat from Guam. *American Museum Novitates* **713**.

Tate GHH (1941a) Results of the Archbold expeditions. No. 36. Remarks on some old world leaf-nosed bats. *American Museum Novitates* **1140**.

Tate GHH (1941b) A review of the genus *Hipposideros* with special reference to Indo-Australian species. *Bulletin of the American Museum of Natural History* **78**, 353–393.

Tate GHH (1942a) Results of the Archbold Expeditions. No. 46. A new genus and species of fruit bats, allied to *Nyctimene*. *American Museum Novitates* **1204**, 1–2.

Tate GHH (1942b) An apparently undescribed member of the *Pteropus pselaphon* group from New Caledonia. *American Museum Novitates* **1213**.

Tate GHH (1945a) Results of the Archbold Expeditions. No. 52. The marsupial genus *Phalanger*. *American Museum Novitates* **1283**.

Tate GHH (1945b) Results of the Archbold Expeditions. No. 55. Notes on the squirrel-like and mouse-like possums (Marsupialia). *American Museum Novitates* **1305**.

Tate GHH (1947) On the anatomy and classification of the Dasyuridae (Marsupialia). *Bulletin of the American Museum of Natural History* **88**, 1–64.

Tate GHH (1948) Results of the Archbold Expeditions. No. 60. Studies in the Peramelidae (Marsupialia). *Bulletin of the American Museum of Natural History* **92**, 313–346.

Tate GHH (1951) Results of the Archbold Expeditions. No. 65. The rodents of Australia and New Guinea. *Bulletin of the American Museum of Natural History* **97**, 189–430.

Tate GHH, Archbold R (1935a) Results of the Archbold Expeditions. No. 5. Seven apparently new forms of Phalangeridae from the New Guinea region. *American Museum Novitates* **810**.

Tate GHH, Archbold R (1935b) Results of the Archbold Expeditions. No. 3. Twelve apparently new forms of Muridae (other than *Rattus*) from the Indo-Australian region. *American Museum Novitates* No. **803**.

Tate GHH, Archbold R (1936) Results of the Archbold Expeditions. No. 8. Four apparently new polyprotodont marsupials from New Guinea. *American Museum Novitates* **823**.

Taylor E (1934) *Philippine Land Mammals*. Monograph. Bureau of Science, Manilla.

Taylor RH (2010) What limits kiore (*Rattus exulans*) distribution in New Zealand? *New Zealand Journal of Zoology* **2**, 473–477. doi:10.1080/03036758.1973.1042 1869

Taylor A, Cooper D (1999) Microsatellites identify introduced New Zealand tammar wallabies (*Macropus eugenii*) as an 'extinct' taxon. *Animal Conservation* **2**, 41–49.

Taylor JM, Calaby JH, Van Deusen HM (1982) A revision of the genus *Rattus* (Rodentia, Muridae) in the New Guinean region. *Bulletin of the American Museum of Natural History* **173**, 178–336.

Taylor F, Frohlich C, Lecolle J, Strecker M (1987) Analysis of partially emerged corals and reef terraces in the central Vanuatu arc: comparison of contemporary coseismic and nonseismic with Quaternary vertical movements. *Journal of Geophysical Research* **92**, 4905–4933.

Taylor JM, Calaby JH, Smith SC (1990) Reproduction in New Guinean *Rattus* and comparison with Australian *Rattus*. *Australian Journal of Zoology* **38**, 587–602.

Temminck CJ (1824–1841) *Monographies de Mammalogie, ou Description de Quelques Genres de Mammifères, Dont les Espèces ont été Observées dans les Différens Musées de l'Europe*. Vol. 1 1824–1827; Vol. 2 1835–1841. G. Dufour et E. d'Ocagne, Paris.

Temminck CJ (1838a) Over die geslatschen *Taphozous, Emballonura, Urocryptus* en *Diclidurus*. *Tijdschrift voor Natuurlijke Geschiedenis en Physiologie*. **5**, 3–34.

Temminck CJ (1838b) In *Fauna Japonica, Coup d'Oeil Faune Iles Sonde Emp*. (Eds P Siebold, C Temminck and H Schlegel) p. xxii. A. Arnz & Co, Amsterdam.

Temminck CJ (1844) Aflevering 2. In *Fauna Japonica. Aperçu Général et Spécifique sur les Mammifères qui Habitent le Japon et les îles qui en Dépendent*. (Eds P von Siebold, C Temminck and H Schlegel) p. 51. A. Arnz & Co, Amsterdam.

Temminck CJ (1853) *Esquisses Zoologiques Sur la Côte de Guiné*. E. J. Brill, Leiden.

Templeton W (1848) In Blyth, E., Report of Curator, Zoological Department. *Journal of the Asiatic Society of Bengal* **17**, 247–255.

Tennent WJ (2009) A cat among the pigeons! Known specimens and supposed distribution of the extinct Solomons Crested Pigeon *Microgoura meeki* Rothschild, 1904. *Bulletin of the British Ornithologists' Club* **129**, 241–253.

Thibault J-C, Bretagnolle V (1999) Breeding seabirds of Gambier Islands, Eastern Polynesia: numbers and changes during the 20th Century. *Emu* **99**, 100–107. doi:10.1071/MU99013

Thomas O (1883) Description of two new species of *Pteropus* from the Caroline Islands. *Proceedings of the Zoological Society of London* **1882**, 755–757.

Thomas O (1887a) On the specimens of *Phascogale* in the Museo Civicio, Genoa, with notes on the allied species of the genus. *Annali del Museo Civico di Storia Naturale di Genova* (2)**4**, 502–510.

Thomas O (1887b) On the bats collected by Mr. C. M. Woodford in the Solomon Islands. *Proceedings of the Zoological Society of London* **1887**, 320–328 (Plates XXV and XXVI).

Thomas O (1887c) Diagnoses of two new fruit-eating bats from the Solomon Islands. *Annals and Magazine of Natural History, London* (5)**19**, 147.

Thomas O (1888a) *Catalogue of the Marsupialia and Monotremata in the Collection of the British Museum (Natural History)*. British Museum, London.

Thomas O (1888b) Description of a new genus and species of rat from New Guinea. *Proceedings of the Zoological Society of London* **1888**, 237–240.

Thomas O (1888c) Diagnoses of six new Mammals from the Solomon Islands. *Annals and Magazine of Natural History* (6)**1**, 155–158.

Thomas O (1888d) The mammals of the Solomon Islands, based on the collections made by Mr. CM Woodford during his second expedition to the Archipelago. *Proceedings of the Zoological Society of London* **56**, 470–484.

Thomas O (1888e) Description of a new bat of the genus *Nyctophilus*. *Annals and Magazine of Natural History, London* (6)**2**, 226.

Thomas O (1894) Diagnosis of a new *Pteropus* from the Admiralty Islands. *Annals and Magazine of Natural History, London* (6)**13**, 293.

Thomas O (1895) On some mammals collected by Mr. Albert Meek on Fergusson Island, D'Entrecasteaux Group. *Novitates Zoologicae* **2**, 163–166.

Thomas O (1896) On mammals collected by Mr. Albert Meek on Woodlark Island, and on Kiriwina in the Trobriand Group. *Novitates Zoologicae* **3**, 526–529.

Thomas O (1897) On the mammals collected in British New Guinea by Dr. Lamberto Loria. *Annali del Museo Civico di Storia Naturale di Genova* (2)**18**, 606–622.

Thomas O (1898a) Descriptions of two new cuscuses (*Phalanger*) obtained by Dr Loria in British New Guinea. *Annali del Museo Civico di Storia Naturale di Genova* (2)**19**, 5–8.

Thomas O (1898b) Descriptions of two new cuscuses (*Phalanger*). *Novitates Zoologicae* **5**, 433–434.

Thomas O (1900) Description of a new fruit-bat from New Guinea. *Annals and Magazine of Natural History, London* (7)**5**, 216–217.

Thomas O (1902a) Description of a new *Uromys* from the Solomon Islands. *Annals and Magazine of Natural History, London* (7)**9**, 446–447.

Thomas O (1902b) On a new genus of vespertilionine bat from New Guinea. *Annals and Magazine of Natural History, London* (7)**9**, 220–222.

Thomas O (1904a) On some mammals from British New Guinea presented to the National Museum by Mr. C. A. W. Monckton, with descriptions of other species from the same region. *Annals and Magazine of Natural History, London* (7)**14**, 397–403.

Thomas O (1904b) New bats and rodents from West Africa, the Malay Peninsula, and Papuasia. *Annals and Magazine of Natural History, London* (7)**14**, 196–202.

Thomas O (1904c) New species of *Pteropus*, *Mus* and *Pogonomys* from the Australian region. *Novitates Zoologicae* **11**, 597–600.

Thomas O (1904d) In Schneider, G. Ergebnisse zoologischer Forschungsreisen in Sumatra. *Zoologischer Anzeiger* **27**, 722–724.

Thomas O (1905) On some Australasian mammals. *Annals and Magazine of Natural History, London* (7)**16**, 422–428.

Thomas O (1906) New mammals from the Australian region. *Annals and Magazine of Natural History, London* (7)**17**, 324–332.

Thomas O (1907) A new *Acanthoglossus* from the island of Salawatti. *Annals and Magazine of Natural History, London* (7)**20**, 498–499.

Thomas O (1908a) New bats and rodents in the British Museum collection. *Annals and Magazine of Natural History, London* (8)**2**, 370–375.

Thomas O (1908b) The species of the genus *Dactylopsila*. *Annals and Magazine of Natural History, London* (8)**1**, 122–124.

Thomas O (1909a) Two new mammals from N. Australia. *Annals and Magazine of Natural History, London* (8)**3**, 197–198.

Thomas O (1909b) Some mammals from the N.E. Kimberley, northern Australia. *Annals and Magazine of Natural History, London* (8)**3**, 149–152.

Thomas O (1910a) Three new Asiatic mammals. *Annals and Magazine of Natural History* (8)**5**, 534–536.

Thomas O (1910b) New mammals from the East Indian Archipelago. *Annals and Magazine of Natural History, London* (8)**5**, 383–387.

Thomas O (1911a) Three new mammals from Dutch New Guinea. *Annals and Magazine of Natural History, London* (8)**7**, 384–387.

Thomas O (1911b) Two new eastern bats. *Annals and Magazine of Natural History, London* (8)**8**, 378–380.

Thomas O (1913a) On new mammals obtained by the Utakwa Expedition to Dutch New Guinea. *Annals and Magazine of Natural History,* London (8)**12**, 205–212.

Thomas O (1913b) Two new Australian mammals. *Annals and Magazine of Natural History, London* (8)**11**, 79–80.

Thomas O (1913c) On some mammals collected from Bali and Mysol, collected by Mr E. Streseman during the 'Freiburger Molukken-Expedition'. *Annals and Magazine of Natural History, London* (8)**11**, 504–507.

Thomas O (1913d) Some new species of *Uromys*. *Annals and Magazine of Natural History, London* (8)**12**, 212–217.

Thomas O (1914a) New Asiatic and Australasian bats and a new bandicoot. *Annals and Magazine of Natural History, London* (8)**13**, 439–444.

Thomas O (1914b) On mammals from Manus Island Admiralty Group, and Ruk Island, Bismarck Archipelago. *Annals and Magazine of Natural History, London* (8)**13**, 434–439.

Thomas O (1915a) On some pteropine bats from Vulcan and Dampier Islands off the N.E. coast of New Guinea. *Annals and Magazine of Natural History, London* (8)**15**, 387–389.

Thomas O (1915b) Notes on *Emballonura*, with descriptions of new species. *Annals and Magazine of Natural History, London* (8)**15**, 137–140.

Thomas O (1915c) Two new species of *Leuconoe*. *Annals and Magazine of Natural History, London* (8)**15**, 170–172.

Thomas O (1915d) Notes on the genus *Nyctophilus*. *Annals and Magazine of Natural History, London* (8)**15**, 493–499.

Thomas O (1920) New small mammals from New Guinea. *Annals and Magazine of Natural History* (9)**6**, 533–537.

Thomas O (1921) Notes on Australasian rats, with a selection of lectotypes of Australasian Muridae. *Annals and Magazine of Natural History, London* (9)**8**, 425–433.

Thomas O (1922a) New mammals from New Guinea and neighbouring islands. *Annals and Magazine of Natural History, London* (9)**9**, 261–265.

Thomas O (1922b) On mammals collected by C. Keysser in the Saruwaged and Rawlinson Moutains region of N.E. New Guinea. *Annals and Magazine of Natural History, London* (9)**9**, 669–676.

Thomas O (1922c) Two new subspecies of *Phalanger orientalis*. *Annals and Magazine of Natural History, London* (9)**9**, 680–681.

Thomas O (1923a) The Godman Exploration Fund. List of mammals from north Queensland collected by Mr T. Sherrin. *Annals and Magazine of Natural History, London* (9)**11**, 171–178.

Thomas O (1923b) A new *Uromys* from the Kei Islands. *Treubia* **3**, 422.

Thomas O (1923c) On some mammals from Simalur Island, West of Sumatra, collected by Mr E. Jacobson. *Annals and Magazine of Natural History* (9)**12**, 591–593.

Thomas O (1923d) On some small mammals, chiefly bats, from the east Indian Archipelago. *Annals and Magazine of Natural History, London* (9)**11**, 250–255.

Thomas O (1925) On some mammals from the New Hebrides. *Annals and Magazine of Natural History, London* (9)**16**, 240–241.

Thomas O, Dollman G (1909) Mammals from Inkerman, north Queensland, presented to the National Museum by Sir William Ingram and the Hon. John Forrest. *Proceedings of the Zoological Society of London* **1908**, 788–794.

Thomas O, Doria G (1886) Nuove specie di chirotteri. *Phyllorina papua. Annali del Museo Civico di Storia Naturale di Genova* **4**, 204.

Thomas O, Wroughton RC (1909a) Diagnoses of new mammals collected by Mr H. C. Robinson in the islands of the Straits of Malacca. *Annals and Magazine of Natural History* (8)**4**, 534–536.

Thomas O, Wroughton R (1909b) On a collection of mammals from Western Java presented to the National Museum by Mr. W. E. Balston. *Proceedings of the Zoological Society of London* **1909**, 371–393.

Tidemann C (1986) Morphological variation in Australian and island populations of Gould's wattled bat, *Chalinolobus gouldii* (Gray) (Chiroptera: Vespertilionidae). *Australian Journal of Zoology* **34**, 503–514.

Timm RM, Weijola V, Aplin KP, Donnellan SC, Flannery TF, Thomson V, Pine RH (2016) A new species of *Rattus* (Rodentia: Muridae) from Manus Island, Papua New Guinea. *Journal of Mammalogy* **97**, 861–878. doi:10.1093/jmammal/gyw034

Timmins R, Kawanishi K, Giman B, Lynam A, Chan B, Steinmetz R, Sagar Baral H, Samba Kumar N (2015) *Rusa unicolor*. The IUCN Red List of Threatened Species 2015: e.T41790A85628124. doi:10.2305/IUCN.UK.2015-2.RLTS.T41790A22156247.en)

Tipene P (1980) A stranger among us. *New Zealand Journal of Agriculture* **141**, 71–73.

Tiraa A (1992) The moa kirikiri of the Cook Islands. In *Pacific Island Flying Foxes: Proceedings of an International Conservation Conference*. 1–2 February 1990, Honolulu. (Eds DE Wilson and GL Graham) pp. 18–23. Department of the Interior, Fish and Wildlife Service, Washington, DC.

Tomes R (1858) A monograph of the genus *Miniopterus*. *Proceedings of the Zoological Society of London* **1858**, 115–128.

Tomes R (1859) Notice of five species of bats in the collection of L.L. Dillwin, Esq., M.P.; collected in Labuan by Mr James Motley. *Proceedings of the Zoological Society of London* **1858**, 536–540.

Toth CA, Dennis TE, Pattemore DE, Parsons S (2015) Females as mobile resources: communal roosts promote the adoption of lek breeding in a temperate bat. *Behavioral Ecology* **26**, 1156–1163. doi:10.1016/0003-3472(92)90059-I

Trouessart EL (1908) *Notopteris macdonaldi di neocaledonica nov. subsp.* Chiroptére nouveau pour la fauna de la Nouvelle Calédonie. *Bulletin Musée d'Histoire Naturelle, Paris* **14**, 257–259.

Troughton ELG (1927) Fixation of the habitat, and extended description of *Pteropus tuberculatus*, Peters. *Records of the Australian Museum* **15**, 355–359. doi:10.3853/j.0067-1975.15.1927.820

Troughton ELG (1929a) A new genus and species of bat (Kerivoulinae) from the Solomons, with a review of the genera of the sub-family. *Records of the Australian Museum* **17**, 85–99.

Troughton ELG (1929b) A new fruit bat (*Pteropus rayneri* Group) from the Solomons. *Records of the Australian Museum* **17**, 193–198. doi:10.3853/j.0067-1975.17.1929.761

Troughton ELG (1930) A new species and subspecies of fruit bats (*Pteropus*) from the Santa Cruz Group. *Records of the Australian Museum* **18**, 1–4.

Troughton ELG (1931) Three new bats of the genera *Pteropus*, *Nyctimene* and *Chaerephon* from Melanesia. *Proceedings of the Linnean Society of New South Wales* **56**, 204–209.

Troughton ELG (1935a) Five new rats of the genera *Hydromys* and *Melomys* from Northern Australia. *Records of the Australian Museum* **19**, 251–258.

Troughton ELG (1935b) A new genus and species of Giant Rat from the Solomons. *Records of the Australian Museum* **19**, 259–262. doi:10.3853/j.0067-1975.19.1935.702

Troughton ELG (1936a) The mammalian fauna of Bougainville Island, Solomons Group. *Records of the Australian Museum* **19**, 341–353.

Troughton ELG (1936b) A redescription of *Solomys* ('*Mus*') *salamonis* Ramsay. *Proceedings of the Linnean Society of New South Wales* **61**, 128–130.

Troughton ELG (1937a) Six new bats (Microchiroptera) from the Australasian region. *Australian Zoologist* **8**, 274–281.

Troughton ELG (1937b) Descriptions of some New Guinea mammals. *Records of the Australian Museum* **20**, 117–127. doi:10.3853/j.0067-1975.20.1937.254

Troughton ELG (1945) Diagnosis of new mammals from the South-West Pacific. *Records of the Australian Museum* **21**, 373–375.

Troughton ELG (1946) Diagnoses of new rats from the New Guinea area. *Records of the Australian Museum* **21**, 406–410.

Troughton ELG, Le Souef A (1929) New forms of mosaic-tailed rats (*Melomys* and *Uromys*) from Hinchinbrook Island, Queensland. *Australian Zoologist* **6**, 96–99.

Trusdell F (2009) Marianas, geology. In *Encyclopedia of Islands.* (Eds R Gillespie and D Clague) pp. 598–603. University of California Press, Berkeley and Los Angeles, CA.

Tsang S (2019a) *Pteropus hypomelanus.* In *Handbook of the Mammals of the World – Volume 9 Bats.* (Eds DE Wilson and RA Mittermeier) p. 143. Lynx Edicions in association with Conservation International and IUCN, Barcelona.

Tsang S (2019b) *Pteropus samoensis.* In *Handbook of the Mammals of the World – Volume 9 Bats.* (Eds DE Wilson and RA Mittermeier) p. 150. Lynx Edicions in association with Conservation International and IUCN, Barcelona.

Tsang SM, Wiantoro S, Veluz MJ, Sugita N, Nguyen YL, Simmons NB, Lohman DJ (2019) Dispersal out of Wallacea spurs diversification of *Pteropus* flying foxes, the world's largest bats (Mammalia: Chiroptera). *Journal of Biogeography* **47**, 527–537. doi:10.1038/srep12213

Tustin K, Challies C (1978) The effects of hunting on the numbers and group sizes of Himalayan thar (*Hemitragus jemlahicus*) in Carneys Creek, Rangitata catchment. *New Zealand Journal of Ecology* **1**, 153–157.

Ulmer F (1940) Zoological results of the Denison–Crockett South Sea Expedition for the academy of natural sciences of Philadelphia, 1937–1938. Part 6 – A new race of the New Guinea short-headed flying phalanger. *Notulae Naturae* **52**, 1–3.

U.S. Fish and Wildlife Service (USFWS) (2009) Draft revised recovery plan for the Marianna Fruit Bat or fanihi (*Pteropus mariannus mariannus*). U.S. Fish and Wildlife Service, Portland, OR.

Valdez EW, Wiles GJ, O'Shea TJ (2011) Diets of the sympatric Pacific Sheath-Tailed Bat (*Emballonura semicaudata rotensis*) and Mariana Swiftlet (*Aerodramus bartscht*) on Aguiguan, Mariana Islands. *Pacific Science* **65**, 301–309. doi:10.2984/65.3.301

Van Deusen HM (1957) Results of the Archbold Expeditions. No. 76 – A new species of wallaby (Genus *Dorcopsis*) from Goodenough Island, Papua. *American Museum Novitates* **1826**, 1–25.

Van Deusen HM (1961) New Guinea record of the tube-nosed insectivorous bat, *Murina. Journal of Mammalogy* **42**, 1531–1533.

Van Deusen HM (1969) Results of the 1958–1959 Gilliard New Britain Expedition 5. A new species of *Pteropus* (Mammalia, Pteropodidae) from New Britain, Bismarck Archipelago. *American Museum Novitates* **2371**, 1–16.

Van Deusen HM, George GG (1969) Results of the Archbold Expeditions. No. 90. Notes on the echidnas (Mammalia, Tachyglossidae) of New Guinea. *American Museum Novitates* **2383**, 1–23.

Van Deusen HM, Koopman KF (1971) Results of the Archbold Expeditions. No. 95. The genus *Chalinolobus* (Chiroptera, Vespertilionidae). Taxonomic review of *Chalinolobus picatus, C. nigrogriseus*, and *C. rogersi. American Museum Novitates* **2468**, 1–30.

Van Dyck S (2002) Morphology-based revision of *Murexia* and *Antechinus* (Marsupialia: Dasyuridae). *Memoirs of the Queensland Museum* **48**, 239–330.

Van Dyck S, Strahan R (2008) *The Mammals of Australia.* Reed New Holland, Chatswood.

Van Strien N (1996) Mammals of the Aru Islands. In *The Aru Archipelago: Plants, Animals, People and Conservation.* (Ed. H Nooteboom) pp. 87–106. Nederlandse Commissie Voor Internationale Natuurbescherming, Amsterdam.

Vander Velde N (2009) Marshall Islands. In *Encyclopedia of Islands.* (Eds R Gillespie and D Clague) pp. 610–612. University of California Press, Berkeley and Los Angeles, CA.

Vernes K, Lebel T (2011) Truffle consumption by New Guinea forest wallabies. *Fungal Ecology* **4**, 270–276. doi:10.1016/j.funeco.2011.02.001

Veron G, Patou M-L, Simberloff D, McLenachan PA, Morley CG (2009) The Indian brown mongoose, yet another invader in Fiji. *Biological Invasions* **12**, 1947–1951. doi:10.1111/j.1463-6409.2006.00261.x

Veron G, Patou M-L, Tóth M, Goonatilake M, Jennings AP (2015) How many species of *Paradoxurus* civets are there? New insights from India and Sri Lanka. *Journal of Zoological Systematics and Evolutionary Research* **53**, 161–174. doi:10.1111/jzs.12085

Vestjens W, Hall LS (1977) Stomach contents of forty-two species of bats from the Australasian region. *Wildlife Research* **4**, 25–35.

Von Rosenberg C (1867) *Reis Naar de Zuidoostereilander Gedaan in 1865 op last der Regering van Nederlandsch–Indie.* Martinus Nijhoff, s-Gravenhage.

Waldien DL, Scanlon A (2021) *Emballonura semicaudata.* The IUCN Red List of Threatened Species 2021: e.T7669A22135085. doi:10.2305/IUCN.UK.2021-1.RLTS.T7669A22135085.en

Waldien DL, Scanlon A, Thompson BL, Sherwin RE, Naikatini A, Tikoca S (2019) *Chaerephon bregullae* (errata version published in 2021). The IUCN Red List of Threatened Species 2019: e.T4309A209550994. doi:10.2305/IUCN.UK.2019-3.RLTS. T4309A209550994.en

Waldien DL, Wiles G, Buden D, Obispo B (2020) *Pteropus molossinus*. The IUCN Red List of Threatened Species 2020: e.T18741A22084572. doi:10.2305/ IUCN.UK.2020-2.RLTS.T18741A22084572.en

Wallace D, Johnson R, Chappell B, Arculus R, Perfit M, Crick I (1983) Cainozoic volcanism of the Tabar, Lihir, Tanga, and Feni Islands, Papua New Guinea; geology, whole-rock analyses, and rock-forming mineral compositions. Australian Geological Survey Organisation, Canberra.

Walter R (1990) *The Southern Cook Islands in Eastern Polynesian Prehistory*. PhD thesis. University of Auckland, New Zealand.

Waterhouse G (1838a) Observations on certain modifications observed in the dentition of the flying opossums (of the genus *Petaurus* of authors). *Proceedings of the Zoological Society of London* **1838**, 149–156.

Waterhouse G (1838b) On two new species of Mammalia, from the Society's collection, belonging to the genera *Gerbillus* and *Herpestes*. *Proceedings of the Zoological Society of London* **VI**, 55–56.

Waterhouse G (1845) Descriptions of bats collected in the Philippine Islands and presented to the society by H. Cuming, Esq. *Proceedings of the Zoological Society of London* **1845**, 3–10.

Waterhouse G (1846) *A Natural History of the Mammalia. Vol. I. Containing the order Marsupiata, or Pouched Animals*. London, Hippolyte Bailliere.

Watling D (2011) *Conservation of the Endangered Fiji Flying Fox* Mirimiri acrodonta *on Taveuni Island, Fiji*. Biodiversity Conservation Lessons Learned Technical Series 1.Conservation International, Apia, Samoa.

Weber M (1902) *Der Indo-Australische Archipel und die Geschichte seiner Tierwelt*. Verlag von Gustav Fischer, Jena.

Wehr NH, Hess SC, Litton CM (2018) Biology and impacts of Pacific Islands invasive species. 14. *Sus scrofa*, the feral pig (Artiodactyla: Suidae). *Pacific Science* **72**, 177–198. doi:10.2984/72.2.1

Weisler M (2001) *On the Margins of Sustainability: Prehistoric Settlement of Utrok Atoll. Northern Marshall Islands*. BAR International Series 967, Oxford.

Weisler M, Bollt R, Findlater A (2006) A new eastern limit of the Pacific flying fox, *Pteropus tonganus* (Chiroptera: Pteropodidae), in prehistoric Polynesia: a case of possible human transport and extirpation. *Pacific Science* **60**, 403–411.

Welch JN, Fordyce JA, Simberloff DS (2016) Indirect impacts of invaders: a case study of the Pacific sheath-tailed bat (*Emballonura semicaudata rotensis*). *Biological Conservation* **201**, 146–151. doi:10.1016/j. biocon.2016.07.004

Westerman M, Kear BP, Aplin K, Meredith RW, Emerling C, Springer MS (2012) Phylogenetic relationships of living and recently extinct bandicoots based on nuclear and mitochondrial DNA sequences. *Molecular Phylogenetics and Evolution* **62**, 97–108. doi:10.1016/j. ympev.2011.09.009

Wheatley A, Anderson P, Callebaut J, Reupena L (Eds) (2020) 'State of environment and conservation in the Pacific Islands: 2020 regional report Secretariat of the Pacific Regional Environment Programme'. Apia, Samoa.

White C, Brimm D, Clunie F (1988) A study of peregrines in the Fiji Islands, South Pacific Ocean. In *Peregrine Falcon Populations: Their Management and Recovery*. (Eds T Cade, J Enderson, C Thelander and C White) pp. 275–288. The Peregrine Fund, Boise, ID.

White J, Flannery TF, O'Brien R, Hanock R, Pavlish P (1991) The Balof shelters, New Ireland. In *Report on the Lapita Project*. (Eds J Allen and C Gosden) pp. 46–58. Australian National University, Canberra.

White JP, Clark G, Bedford S (2000) Distribution, present and past, of *Rattus praetor* in the Pacific and its implications. *Pacific Science* **54**, 105–117.

Whitehouse IE, Pearce AJ (1992) Shaping the mountains of New Zealand. In *Landforms of New Zealand*. 2nd edn. (Eds JM Soons and MJ Selby) pp. 144–160. Longman Paul, Auckland.

Wickler S (2001) *The Prehistory of Buka: A Stepping Stone Island in the Northern Solomons*. Department of Archaeology and Natural History and Centre for Archaeological Research, Australian National University, Canberra.

Wiles G (1987a) Current research and future management of Marianas fruit bats (Chiroptera; Pteropodidae) on Guam. *Australian Mammalogy* **10**, 93–95.

Wiles GJ (1987b) The status of fruit bats on Guam. *Pacific Science* **41**, 148–157.

Wiles GJ (1992a) *Pteropus insularis* Carolines fruit bat, Chuuk (Truk) fruit bat. In *Old World Fruit Bats: An Action Plan for their Conservation*. (Eds S Mickelburgh, A Hutson and P Racey) p. 105. IUCN, Gland, Switzerland.

Wiles GJ (1992b) *Pteropus mariannus mariannus* Marianas fruit bat. In *Old World Fruit Bats: An Action Plan for their Conservation*. (Eds S Mickelburgh, A Hutson and P Racey) pp. 108–111. IUCN, Gland, Switzerland.

Wiles GJ (1992c) *Pteropus molossinus*, Pohnpei fruit bat. In *Old World Fruit Bats: An Action Plan for their Conservation*. (Eds S Mickelburgh, A Hutson and P Racey) p. 117. IUCN, Gland, Switzerland.

Wiles GJ (1992d) *Pteropus mariannus ualanus*, Ualan fruit bat. In *Old World Fruit Bats: An Action Plan for their Conservation*. (Eds S Mickelburgh, A Hutson and P Racey) p. 112. IUCN, Gland, Switzerland.

Wiles GJ (1992e) Recent trends in the fruit bat trade on Guam. In *Pacific Island Flying Foxes: Proceedings of an International Conservation Conference*. 1–2 February 1990, Honolulu. (Eds DE Wilson and GL Graham) pp. 53–60. Biological Report 90 (23). US Fish and Wildlife Service, Washington, DC.

Wiles G (2005) Decline of a population of wild seeded breadfruit (*Artocarpus mariannensis*) on Guam, Mariana Islands. *Pacific Science* **59**, 509–522.

Wiles G (2020) *Pteropus pelagicus*. The IUCN Red List of Threatened Species 2020: e.T85043053A22081930. doi:10.2305/IUCN.UK.2020-2.RLTS. T85043053A22081930.en

Wiles GJ, Brooke AP (2009) Conservation threats to bats in the tropical Pacific islands and insular Southeast Asia. In *Island Bats: Evolution, Ecology, and Conservation*. (Eds TH Fleming and PA Racey) pp. 405–459. University of Chicago Press, Chicago, IL.

Wiles GJ, Conry P (1990) Terrestrial vertebrates of the Ngerukewid Islands Wildlife Preserve, Palau Islands. *Micronesia* **23**, 41–66.

Wiles G, Glass P (1990) Interisland movements of fruit bats (*Pteropus mariannus*) in the Mariana Islands. *Atoll Research Bulletin* **343**, 1–6.

Wiles GJ, Johnson NC (2004) Population size and natural history of Mariana Fruit Bats (Chiroptera: Pteropodidae) on Sarigan, Mariana Islands. *Pacific Science* **58**, 585–596.

Wiles GJ, Payne N (1986) The trade in fruit-bats, *Pteropus* spp. on Guam and other Pacific Islands. *Biological Conservation* **38**, 143–161.

Wiles G, Waldien DL (2020) *Pteropus pelewensis*. The IUCN Red List of Threatened Species 2020: e.T118093652A118356424. doi:10.2305/IUCN. UK.2020-2.RLTS.T118093652A118356424.en

Wiles G, Lemke T, Payne N (1989) Population estimates of fruit bats (*Pteropus mariannus*) in the Mariana Islands. *Conservation Biology* **3**, 66–76.

Wiles G, Amerson A, Beck R (1990) The mammals of Tinian, Mariana Islands. *Micronesica* **23**, 167–180.

Wiles GJ, Engbring J, Falanruw M (1991) Population status and natural history of *Pteropus mariannus* on Ulithi Atoll, Caroline Islands. *Pacific Science* **45**, 76–84.

Wiles GJ, Engbring J, Otobed D (1997) Abundance, biology, and human exploitation of bats in the Palau Islands. *Journal of Zoology* **241**, 203–227.

Wiles G, Buden D, Worthington D (1999) History of introduction, population status, and management of Philippine deer (*Cervus mariannus*) on Micronesian islands. *Mammalia* **63**, 193–215.

Wiles GJ, O'Shea TJ, Worthington DJ, Esselstyn JA, Valdez EW (2011) Status and Natural History of *Emballonura semicaudata* Rotensison Aguiguan, Mariana Islands. *Acta Chiropterologica* **13**, 299–309. doi:10.3161/150811011X624785

Williams C (1999) Faunal composition of Pamwak site, Manus Island, PNG. In *Le Pacific de 5000 a` 2000 avant le present* (Eds J-C Galipaud and I Lilley) pp. 241–249. Institut de Recherche pour le Développement, Paris.

Wilmshurst JM, Ruscoe WA (2021) *Rattus exulans*. In *The Handbook of New Zealand Mammals*. 3rd edn. (Eds CM King and DM Forsyth) pp. 161–240. CSIRO Publishing, Melbourne.

Wilmshurst JM, Hunt TL, Lipo CP, Anderson AJ (2011) High-precision radiocarbon dating shows recent andrapid initial human colonization of East Polynesia. *Proceedings of the National Academy of Sciences of the United States of America* **108**, 1815–1820. doi:10.1073/ pnas.1015876108/-/DCSupplemental/st01.xlsx

Wilson D, Engbring J (1992) The flying foxes *Pteropus samoensis* and *Pteropus tonganus*: status in Fiji and Samoa. In *Pacific Island Flying Foxes: Proceedings of an International Conservation Conference*. 1–2 February 1990, Honolulu. (Eds DE Wilson and GL Graham) pp. 18–23. Department of the Interior, Fish and Wildlife Service, Washington, DC.

Wilson D, Reeder D (2005) *Mammal Species of the World: a Taxonomic and Geographic Reference*. Johns Hopkins University Press, Baltimore, MD.

Winkelmann JR, Bonaccorso FJ (2003) Home range and territoriality in the least blossom bat, *Macroglossus minimus*, in Papua New Guinea. *Journal of Mammalogy* **84**, 561–570.

Winkelmann JR, Bonaccorso FJ, Strickler TL (2000) Home range of the southern blossom bat, *Syconycteris australis*, in Papua New Guinea. *Journal of Mammalogy* **81**, 408–414.

Winter J, Leung LK-P (2008) Southern common cuscus *Phalanger mimicus*. In *The Mammals of Australia*. (Eds S Van Dyck and R Strahan) pp. 268–270. Reed New Holland, Chatswood.

Wodzicki K, Flux J (1967) Guide to introduced wallabies in New Zealand. *Tuatara* **15**, 48–58.

Wollstein A, Lao O, Becker C, Brauer S, Trent RJ, Nürnberg P, Stoneking M, Kayser M (2010) Demographic History of Oceania Inferred from Genome-wide Data. *Current Biology* **20**, 1983–1992. doi:10.1016/j.cub.2010.10.040

Wood BL (2012) Geology of the Cook Islands. *New Zealand Journal of Geology and Geophysics* **10**, 1429–1445. doi:10.1029/JZ071i014p03379

Wood JR, Alcover JA, Blackburn TM, Bover P, Duncan RP, Hume JP, Louys J, Meijer H, Rando JC, Wilmhurst JM (2017) Island extinctions: processes, patterns, and potential for ecosystem restoration. *Environmental Conservation* **44**, 348–358. doi:10.1017/S037689291700039X

Woolley PA (1989) Nest location by spool-and-line tracking of dasyurid marsupials in New Guinea. *Journal of Zoology* **218**, 689–700.

Woolley PA (1994) The dasyurid marsupials of New Guinea: use of museum specimens to assess seasonality of breeding. *Science in New Guinea* **20**, 49–55.

Woolley PA (2001) Observations on the reproductive biology of *Myoictis wallacei*, *Neophascogale lorentzi*, *Dasyurus albopunctatus* and *Dasyurus spartacus*, dasyurid marsupials endemic to New Guinea. *Australian Mammalogy* **23**, 63–66.

Woolley PA (2004) Reproductive biology of some dasyurid marsupials of New Guinea. In *Predators with Pouches: The Biology of Carnivorous Marsupials*. (Eds M Jones, M Dickman and M Archer) pp. 169–182. CSIRO Publishing, Melbourne.

Woolley PA (2005) Revision of the three-striped dasyures, genus *Myoictis* (Marsupialia: Dasyuridae), of New Guinea, with description of a new species. *Records of the Australian Museum* **57**, 321–340.

Woolley PA (2008) Red-cheeked Dunnart *Sminthopsis virginiae*. In *The Mammals of Australia*. (Eds S Van Dyck and R Strahan) pp. 158–159. Reed New Holland, Chatswood.

Woolley PA (2019) The type specimen of *Murexia longicaudata* (Dasyuromorphia: Dasyuridae): an enigma. *Australian Mammalogy* **42**, 359–360. doi:10.1644/06-MAMM-A-310R.1

Woolley PA, Allison A (1996) Reproduction in *Pipistrellus papuanus* (Chiroptera: Vespertilionidae) in Wau, Papua New Guinea. *Science in New Guinea* **21**, 131–138.

Woolley P, Krajewski C, Westerman M (2020) The endemic New Guinean genus *Murexia* (Dasyuromorphia: Dasyuridae). How many species? An analysis based on morphological, distributional and molecular data. *Australian Journal of Zoology* **67**, 134–144. doi:10.1071/ZO20013

Worthy TH, Clark G (2009) Bird, mammal and reptile remains. In *The Early History of Fiji, Terra Australis 31*. (Eds G Clark and A Anderson) pp. 231–258. ANU Press, Canberra.

Worthy TH, Scofield P (2004) Skeletal and dental variation within and between *Mystacina* species in southern New Zealand. *New Zealand Journal of Zoology* **31**, 351–361. doi:10.1080/03014223.1996.951 8071

Worthy TH, Daniel M, Hill JE (1996) An analysis of skeletal size variation in *Mystacina robusta* Dwyer, 1962 (Chiroptera: Mystacinidae). *New Zealand Journal of Zoology* **23**, 99–110.

Worthy TH, Tennyson AJD, Archer M, Musser AM, Hand SJ, Jones C, Douglas BJ, McNamara JA, Beck RMD (2006) Miocene mammal reveals a Mesozoic ghost lineage on insular New Zealand, southwest Pacific. *Proceedings of the National Academy of Sciences of the United States of America* **103**, 19419–19423.

Yamashima M (1932) New species of bat from the Japanese South Pacific mandated islands. *Transactions of the Natural History Society of Formosa* **22**, 240–241.

Yamashima M (1943) On a new subspecies of *Emballonura semicaudata* (Chiroptera) found in the Marianas, Micronesia. *Bulletin of the Biogeographical Society of Japan* **13**, 99–100.

Yan C, Kroenke L (1993) A plate tectonic reconstruction of the SW Pacific 0–100 Ma. In *Proceedings of the Ocean Drilling Program, Scientific Results 130*. (Eds T Berger, L Kroenke and L Mayer) pp. 697–709.

Yockney IJ, Hickling GJ (2000) Distribution and diet of chamois (*Rupicapra rupicapra*) in Westland forests, South Island, New Zealand. *New Zealand Journal of Ecology* **24**, 31–38.

Ziegler A (1982) An ecological check-list of New Guinea recent mammals. In *Biogeography and Ecology of New Guinea. Vol. 2*. (Ed. J Gressitt) pp. 863–894. Junk, The Hague.

Zimmermann E (1780) *Geographische Geschichte des Menschen, und der Allgemein Verbreiteten Vierfüßigen Thiere*. In der Weygandschen Buchhandlung, Leipzig.

Index

Acrobatidae 56
acrodonta, Mirimiri 133–5
admiralitatum, Pteropus 160–1
aello, Nyctimene 142–3
agilis, Notamacropus 61–2
agnella, Kerivoula 250
albertisii, Pseudochirops 84
albiventer, Nyctimene 143–4
albopunctatus, Dasyurus 42–3
allenorum, Pteropus 161
amplexicaudatus, Rousettus 201–3
anceps, Pteralopex 153–4
anderseni, Dobsonia 120–1
anetianus, Pteropus 161–3
angulatus, Pipistrellus 259–61
Anthops 216
Anthops ornatus 216–17
arcium, Melomys 91–2
aries, Ovis 267
Artiodactyla 266
aruensis, Pteropus 163
Aselliscus 217
Aselliscus tricuspidatus 217–18
ater, Hipposideros 218–20
atrata, Dorcopsis 59–60
atrata, Pteralopex 154–6
auropunctatus, Herpestes 276
australis, Miniopterus 229–31
australis, Syconycteris 203–4

beauforti, Dobsonia 121–2
beccarii, Emballonura 205–6
beccarii, Ozimops 240–1
bicolor, Wallabia 286
blepotis, Miniopterus 231–2
boeadii, Uromys 110
Bos 266
Bos taurus 266
bougainville, Melomys 92–4
bougainville, Nyctimene 144–5
brachypterus, Philetor 257–8
bregullae, Mops 236–7
breviceps, Phalanger 65–6
browni, Thylogale 62–3
bruijnii, Pogonomelomys 99
bruijnii, Zaglossus 40–1
brunii, Thylogale 64
bubalis, Bubalus 266
Bubalus 266
Bubalus bubalis 266
burtoni, Melomys 94

calcaratus, Hipposideros 220–1
canadensis, Cervus 269–70
canescens, Pseudochirulus 84–5
Canis 280
Canis familiaris 280–1
capistratus, Pteropus 163–5
Capra 266
Capra hircus 266–7
Carnivora 276
catus, Felis 279–80
caudimaculatus, Uromys 111–12
Cercopithecidae 291
certans, Nyctimene 146
Cervidae 268
cervinus, Hipposideros 221–2
Cervus 268
Cervus canadensis 269–70
Cervus elaphus 268–9
Cervus nippon 270
Chalinolobus 246
Chalinolobus neocaledonicus 246–7
Chalinolobus nigrogriseus 248
Chalinolobus tuberculatus 249–50
Chiroptera 119
Chiruromys 86
Chiruromys forbesi 86–8
Chiruromys sp. 88
chrysogaster, Hydromys 88–90
clara, Echymipera 49–50
cognatus, Pteropus 165–6
collinus, Pipistrellus 261–2
collinus, Scotophilus 263–4
conspicillatus, Pteropus 166–7
coxi, Pteropus 167–8
Crocidura 297
Crocidura maxi 297
cuniculus, Oryctolagus 289–90

Dactylopsila 78
Dactylopsila tatei 78–9
Dactylopsila trivirgata 80–1
Dama 272
Dama dama 272–3
dama, Dama 272–3
Dasyuridae 42
Dasyuromorphia 42
Dasyurus 42
Dasyurus albopunctatus 42–3
davidi, Echymipera 50–1
demissus, Hipposideros 222–3
Dendrolagus 57, 282
Dendrolagus inustus 57–8
Dendrolagus matschiei 282
detentus, Rattus 101–2
diadema, Hipposideros 224–5
dianae, Emballonura 206–7

dinops, Hipposideros 225–7
Diprotodontia 56, 282
Distoechurus 56
Distoechurus pennatus 56–7
Dobsonia 120
Dobsonia anderseni 120–1
Dobsonia beauforti 121–2
Dobsonia emersa 122–3
Dobsonia inermis 123–4
Dobsonia magna 124–5
Dobsonia minor 125–7
Dobsonia moluccensis 127–8
Dobsonia pannietensis 128–9
Dobsonia praedatrix 129–30
Dorcopsis 59
Dorcopsis atrata 59–60
Dorcopsis muelleri 60–1
dorsalis, Notamacropus 282–3

Echymipera 49
Echymipera clara 49–50
Echymipera davidi 50–1
Echymipera kalubu 51–3
Echymipera rufescens 53–4
elaphus, Cervus 268–9
Emballonura 205
Emballonura beccarii 205–6
Emballonura dianae 206–7
Emballonura furax 207–8
Emballonura raffrayana 208–9
Emballonura semicaudata 209–11
Emballonura serii 211–12
Emballonuridae 205
emersa, Dobsonia 122–3
emmae, Uromys 112
ennisae, Pteropus 168–9
Erinaceidae 288
Erinaceomorpha 288
Erinaceus 288
Erinaceus europaeus 288
erminea, Mustela 277
eugenii, Notamacropus 284
europaeus, Erinaceus 288
europaeus, Lepus 289
euryotis, Rhinolophus 244–5
exulans, Rattus 292–4

familiaris, Canis 280–1
fardoulisi, Nesonycteris 135–6
fascicularis, Macaca 291
Felis 279
Felis catus 279–80
fergussoniensis, Pogonomelomys 99–100
flanneryi, Pteralopex 156–7
florium, Murina 251–2
forbesi, Chiruromys 86–8
fundatus, Pteropus 169–70
furax, Emballonura 207–8
furo, Mustela 277–8
fuscus, Herpestes 276–7

gilliardorum, Pteropus 170–1
gymnotis, Phalanger 67

Hemitragus 267
Hemitragus jemlahicus 267
hermaphroditus, Paradoxurus 279
Herpestes 276
Herpestes auropunctatus 276
Herpestes fuscus 276–7
Hipposideridae 216
Hipposideros 218
Hipposideros ater 218–20
Hipposideros calcaratus 220–1
Hipposideros cervinus 221–2
Hipposideros demissus 222–3
Hipposideros diadema 224–5
Hipposideros dinops 225–7
Hipposideros maggietaylorae 227–8
Hipposideros papua 228
hircus, Capra 266–7
Hydromys 88
Hydromys chrysogaster 88–90
Hydromys neobritannicus 90
hypomelanus, Pteropus 171–3

imperator, Uromys 112–13
inermis, Dobsonia 123–4
insularum, Myotis 253
intercastellanus, Phalanger 68
inustus, Dendrolagus 57–8

jemlahicus, Hemitragus 267
jobensis, Mops 237–9
jobiensis, Rattus 102–3

kalubu, Echymipera 51–3
Kerivoula 250
Kerivoula agnella 250
Kerivoula myrella 250–1
kraemeri, Spilocuscus 72–4

Lagomorpha 289
Leporidae 289
Lepus 289
Lepus europaeus 289
leucopus, Rattus 103
longicaudata, Murexia 43–5
lullulae, Phalanger 68–70

Macaca 291
Macaca fascicularis 291
maccoyi, Nesonycteris 136–7
macdonaldii, Notopteris 140–1
macrocneme, Miniopterus 232–3
Macroglossus 130
Macroglossus minimus 130–2
Macropodidae 57, 282

macrotis, Pteropus 173
macrourus, Pogonomys cf. 100–1
maculatus, Spilocuscus 74–6
maggietaylorae, Hipposideros 227–8
magna, Dobsonia 124–5
mahaganus, Pteropus 173–5
major, Nyctimene 146–8
malaitensis, Nyctimene 150–1
Mammelomys 90
Mammelomys rattoides 90–1
marianna, Rusa 270–1
mariannus, Pteropus 175–6
masalai, Nyctimene 149–50
matambuai, Melomys 95
matschiei, Dendrolagus 282
maxi, Crocidura 297
medius, Miniopterus 233–4
megaphyllus, Rhinolophus 245–6
melanops, Melonycteris 132–3
melas, Myoictis 46–7
Melomys 91
Melomys arcium 91–2
Melomys bougainville 92–4
Melomys burtoni 94
Melomys matambuai 95
Melomys rufescens 96
Melonycteris 132
Melonycteris melanops 132–3
mengermani, Nesonycteris 137–8
microtis, Nyctophilus 256
mimicus, Phalanger 70–1
minimus, Macroglossus 130–2
Miniopteridae 228–9
Miniopterus 229
Miniopterus australis 229–31
Miniopterus blepotis 231–2
Miniopterus macrocneme 232–3
Miniopterus medius 233–4
Miniopterus robustior 234–5
Miniopterus tristis 235–6
minor, Dobsonia 125–7
Mirimiri 133
Mirimiri acrodonta 133–5
Molossidae 236
molossinus, Pteropus 177–8
moluccarum, Myotis 253–4
moluccensis, Dobsonia 127–8
moncktoni, Paramelomys 97
Mops 236
Mops bregullae 236–7
Mops jobensis 237–9
Mops solomonis 239–40
mordax, Rattus 104
Mosia 212
Mosia nigrescens 212–14
muelleri, Dorcopsis 60–1
murex, Murexia 45–6
Murexia 43
Murexia longicaudata 43–5
Murexia murex 45–6
Muridae 86, 292
Murina 251
Murina florium 251–2
murinus, Suncus 297
Mus 292
Mus musculus 292
musculus, Mus 292
Mustela 277
Mustela erminea 277
Mustela furo 277–8
Mustela nivalis 278–9
Myoictis 46
Myoictis melas 46–7
Myoictis wallacei 47–8
Myotis 253
Myotis insularum 253
Myotis moluccarum 253–4
Myotis stalkeri 255–6
myrella, Kerivoula 250–1
Mystacina 241
Mystacina robusta 241–2
Mystacina tuberculata 242–3
Mystacinidae 241

naso, Paramelomys 97–8
nebulosus, Nyctophilus 256–7
neobritannicus, Hydromys 90
neobritannicus, Uromys 113–14
neocaledonicus, Chalinolobus 246–7
neocaledonicus, Notopteris 141–2
neohibernicus, Pteropus 178–9
Nesonycteris 135
Nesonycteris fardoulisi 135–6
Nesonycteris maccoyi 136–7
Nesonycteris *mengermani* 137–8
Nesonycteris woodfordi 138–9
nigrescens, Mosia 212–14
nigrogriseus, Chalinolobus 248
nippon, Cervus 270
nitendiensis, Pteropus 179–80
nitidus, Rattus 294
nivalis, Mustela 278–9
norvegicus, Rattus 294–5
Notamacropus 61, 282
Notamacropus agilis 61–2
Notamacropus dorsalis 282–3
Notamacropus eugenii 284
Nota*macropus parma* 284–5
Notamacropus rufogriseus 285
Notopteris 140
Notopteris macdonaldii 140–1
Notopteris neocaledonicus 141–2
Nyctimene 142
Nyctimene aello 142–3
Nyctimene albiventer 143–4
Nyctimene bougainville 144–5
Nyctimene certans 146
Nyctimene major 146–8

Nyctimene malaitensis 150–1
Nyctimene masalai 149–50
Nyctimene vizcaccia 151–2
Nyctophilus 256
Nyctophilus microtis 256
Nyctophilus nebulosus 256–7

Odocoileus 273
Odocoileus virginianus 273–4
orientalis, Phalanger 71–2
ornatus, Anthops 216–17
ornatus, Pteropus 180–2
Oryctolagus 289
Oryctolagus cuniculus 289–90
Ovis 267
Ovis aries 267
Ozimops 240
Ozimops beccarii 240–1

pannietensis, Dobsonia 128–9
papua, Hipposideros 228
papuanus, Petaurus 81–3
papuanus, Pipistrellus 262–3
papuensis, Phoniscus 258–9
papuensis, Spilocuscus 76–7
Paradoxurus 279
Paradoxurus hermaphroditus 279
Paramelomys 97
Paramelomys moncktoni 97
Paramelomys naso 97–8
Paramelomys platyops 98
Paranyctimene 152
Paranyctimene raptor 152–3
parma, Notamacropus 284–5
pelagicus, Pteropus 182–3
pelewensis, Pteropus 183–5
penicillata, Petrogale 285–6
pennatus, Distoechurus 56–7
Peramelemorphia 49
Peroryctes 54
Peroryctes raffrayana 54–5
Peroryctidae 49
Petauridae 78
Petaurus 81
Petaurus papuanus 81–3
Petrogale 285
Petrogale penicillata 285–6
Phalanger 65
Phalanger breviceps 65–6
Phalanger gymnotis 67
Phalanger intercastellanus 68
Phalanger lullulae 68–70
Phalanger mimicus 70–1
Phalanger orientalis 71–2
Phalangeridae 64, 286
Philetor 257
Philetor brachypterus 257–8
Phoniscus 258
Phoniscus papuensis 258–9
pilosus, Pteropus 186
Pipistrellus 259
Pipistrellus angulatus 259–61
Pipistrellus collinus 261–2
Pipistrellus papuanus 262–3
Pipistrellus wattsi 263
platyops, Paramelomys 98
Pogonomelomys 99
Pogonomelomys bruijnii 99
Pogonomelomys fergussoniensis 99–100
Pogonomys cf. *macrourus* 100–1
pohlei, Pteropus 186–7
ponceleti, Solomys 105–6
porculus, Uromys 114–16
praedatrix, Dobsonia 129–30
praetor, Rattus 104–5
Primates 291
Pseudocheiridae 83
Pseudochirops 84
Pseudochirops albertisii 84
Pseudochirulus 84
Pseudochirulus canescens 84–5
Pteralopex 153
Pteralopex anceps 153–4
Pteralopex atrata 154–6
Pteralopex flanneryi 156–7
Pteralopex pulchra 157–8
Pteralopex taki 158–9
Pteropodidae 119–20
Pteropus 160
Pteropus admiralitatum 160–1
Pteropus allenorum 161
Pteropus anetianus 161–3
Pteropus aruensis 163
Pteropus capistratus 163–5
Pteropus cognatus 165–6
Pteropus conspicillatus 166–7
Pteropus coxi 167–8
Pteropus ennisae 168–9
Pteropus fundatus 169–70
Pteropus gilliardorum 170–1
Pteropus hypomelanus 171–3
Pteropus macrotis 173
Pteropus mahaganus 173–5
Pteropus mariannus 175–6
Pteropus molossinus 177–8
Pteropus neohibernicus 178–9
Pteropus nitendiensis 179–80
Pteropus ornatus 180–2
Pteropus pelagicus 182–3
Pteropus pelewensis 183–5
Pteropus pilosus 186
Pteropus pohlei 186–7
Pteropus rayneri 187–9
Pteropus rennelli 189–90
Pteropus samoensis 190–2
Pteropus solomonis 192–3
Pteropus tokudae 193–4
Pteropus tonganus 194–7

Pteropus tuberculatus 197–8
Pteropus ualanus 198–9
Pteropus vetula 199–200
Pteropus woodfordi 200–1
pulchra, Pteralopex 157–8

raffrayana, Emballonura 208–9
raffrayana, Peroryctes 54–5
raptor, Paranyctimene 152–3
rattoides, Mammelomys 90–1
Rattus 101, 292
Rattus detentus 101–2
Rattus exulans 292–4
Rattus jobiensis 102–3
Rattus leucopus 103
Rattus mordax 104
Rattus nitidus 294
Rattus norvegicus 294–5
Rattus praetor 104–5
Rattus rattus 295
Rattus tanezumi 295–6
rattus, Rattus 295
rayneri, Pteropus 187–9
rennelli, Pteropus 189–90
rex, Uromys 116–17
Rhinolophidae 244
Rhinolophus 244
Rhinolophus euryotis 244–5
Rhinolophus megaphyllus 245–6
robusta, Mystacina 241–2
robustior, Miniopterus 234–5
Rodentia 86, 292
Rousettus 201
Rousettus amplexicaudatus 201–3
rufescens, Echymipera 53–4
rufescens, Melomys 96
rufogriseus, Notamacropus 285
Rupicapra 268
Rupicapra rupicapra 268
rupicapra, Rupicapra 268
Rusa 270
Rusa marianna 270–1
Rusa timorensis 271–2
Rusa unicolor 272

Saccolaimus 214
Saccolaimus saccolaimus 214–15
saccolaimus, Saccolaimus 214–15
salamonis, Solomys 106–7
salebrosus, Solomys 107–9
samoensis, Pteropus 190–2
sapientis, Solomys 109–10
Scotophilus 263
Scotophilus collinus 263–4
scrofa, Sus 274–5
semicaudata, Emballonura 209–11
serii, Emballonura 211–12
Sminthopsis 48
Sminthopsis virginiae 48
solomonis, Mops 239–40
solomonis, Pteropus 192–3
Solomys 105
Solomys ponceleti 105–6
Solomys salamonis 106–7
Solomys salebrosus 107–9
Solomys sapientis 109–10
Soricidae 297
Soricomorpha 297
Spilocuscus 72
Spilocuscus kraemeri 72–4
Spilocuscus maculatus 74–6
Spilocuscus papuensis 76–7
Spilocuscus wilsoni 77–8
stalkeri, Myotis 255–6
Suidae 274
Suncus 297
Suncus murinus 297
Sus 274
Sus scrofa 274–5
Syconycteris 203
Syconycteris australis 203–4

Tachyglossidae 40
taki, Pteralopex 158–9
tanezumi, Rattus 295–6
tatei, Dactylopsila 78–9
taurus, Bos 266
Thylogale 62
Thylogale browni 62–3
Thylogale brunii 64
timorensis, Rusa 271–2
tokudae, Pteropus 193–4
tonganus, Pteropus 194–7
Trichosurus 286
Trichosurus vulpecula 286–7
tricuspidatus, Aselliscus 217–18
tristis, Miniopterus 235–6
trivirgata, Dactylopsila 80–1
tuberculata, Mystacina 242–3
tuberculatus, Chalinolobus 249–50
tuberculatus, Pteropus 197–8

ualanus, Pteropus 198–9
unicolor, Rusa 272
Uromys 110
Uromys boeadii 110
Uromys caudimaculatus 111–12
Uromys emmae 112
Uromys imperator 112–13
Uromys neobritannicus 113–14
Uromys porculus 114–16
Uromys rex 116–17
Uromys vika 117–18

Vespertilionidae 246
vetula, Pteropus 199–200
vika, Uromys 117–18
virginiae, Sminthopsis 48

virginianus, Odocoileus 273–4
Viveriidae 276
vizcaccia, Nyctimene 151–2
vulpecula, Trichosurus 286–7

Wallabia 286
Wallabia bicolor 286
wallacei, Myoictis 47–8
wattsi, Pipistrellus 263
wilsoni, Spilocuscus 77–8
woodfordi, Nesonycteris 138–9
woodfordi, Pteropus 200–1

Zaglossus 40
Zaglossus bruijnii 40–1